W9-ATC-548

BUSINESS **TELECOMMUNICATIONS**

BUSINESS TELECOMMUNICATIONS

SECOND EDITION

STANFORD H. ROWE, II
DOW CORNING CORPORATION

MACMILLAN PUBLISHING COMPANY
New York

COLLIER MACMILLAN CANADA
Toronto

MAXWELL MACMILLAN INTERNATIONAL
New York Oxford Singapore Sydney

Editor: Charles Stewart
Production Supervisor: Dora Rizzuto
Production Manager: Richard Fischer
Cover Designer: Jane Edelstein
Photo Researcher: Chris Migdol

This book was set in Palatino and Helvetica Type by Graphic Typesetting Service, printed and bound by Halliday Lithographers.
The cover was printed by Lehigh Press, Inc.

Macmillan Publishing Company
866 Third Avenue, New York, New York 10022

Collier Macmillan Canada, Inc.
1200 Eglinton Avenue East, Suite 200
Don Mills, Ontario, M3C 3N1

Library of Congress Cataloging-in-Publication Data

Rowe, Stanford H.
Business telecommunications / Stanford H. Rowe II.
p. cm.
Includes bibliographical references.
ISBN 0-02-404104-1
1. Telecommunication. 2. Business—Communication systems.
I. Title.
HE7631.R69 1990 90-32981
651.7—dc20 CIP

Printing: 4 5 6 7 8 Year: 3 4 5 6 7 8 9 0

To Pam

PREFACE

Business Telecommunications, second edition, presents a complete introduction to the fast-paced world of telecommunications. Designed for a first course in this field, it covers all facets of telecommunications as it is used in business, including both data and voice communications—applications, technical details, and managerial aspects.

Audience

Business Telecommunications is aimed at the individual who has no background in telecommunications other than that which is obtained in the course of daily living. Some knowledge of data processing is assumed, but no more than a student would gain by working with a personal computer or by taking an introductory data processing course.

Students at all levels from community colleges through graduate programs at universities will be able to learn about telecommunications from this book. In addition, it will be helpful to people in industry who need to understand more about telecommunications concepts and terminology. Its comprehensive coverage will make it a useful reference tool.

Highlights

It has been said that telecommunications is one of those subjects about where you need to know everything before you can learn anything. *Business Telecommunications* reverses the cycle by presenting the material in a logical, building-block fashion. Words and terms are defined and explained when they are first used. Examples are used as a foundation and then expanded with more detail. By the end of the book, the student will have an excellent understanding of the subject.

Business Telecommunications takes an outside-in approach to its subject. Telecommunications applications, familiar to everyone, are discussed first as a way of easing the student into the material and explaining *what* the subject of telecommunications includes. The examples, references, and case studies all come from real-life business settings and illustrate the uses of telecommunications in business. The student will learn *why* business feels that telecommunications is vitally important as well as how the regulatory environment affects the telecommunications industry.

With that background, the text leads the reader into the technical details of telecommunications. The technology is explained in an easy-to-understand, yet thorough manner. Current and emerging technologies are covered as well as traditional material. The student will gain an in-depth understanding of *how* telecommunications works.

Equipped with an understanding of the applications and technical details of telecommunications, the student is then introduced to the management of telecommunications. This material broadly covers all facets of telecommunications department management. The student will learn *why* it is necessary to manage telecommunications. The functions of the telecommunications department are examined in detail, and alternative ways of organizing the department are shown. One chapter is dedicated to the subject of network design, and another covers network operations. The book concludes with a quick look at several emerging applications of telecommunications that the student will be dealing with in the next few years.

Changes in the Second Edition

The second edition of *Business Telecommunications* contains additional material to make the book even more useful and relevant than the first edition.

- ***Networks*** A new chapter has been added that is dedicated to telecommunications networks. Network material, which was included with the material on circuits in the first edition, has been updated, and the section on *Local Area Networks (LANs)* has been rewritten and greatly expanded.
- ***ISO-OSI Model*** The ISO-OSI reference model is introduced earlier in the book to provide a framework into which the student can fit the technical material. After the technical material is presented, the ISO-OSI model is studied in greater detail.
- ***Personal Computers*** The material about the uses of personal computers in telecommunications networks has been expanded. Terminal emulation is covered in more detail.
- ***Data Compression*** Data compression, as a way of providing more throughput on a telecommunications line, has been emphasized.
- ***International Telecommunications*** A new appendix, containing reprints of recent articles about the status of telecommunications in other countries, has been added.
- ***General Update*** The book has been updated throughout to provide the most current information about telecommunications applications, techniques, and products.

Pedagogical Features

Business Telecommunications contains many pedagogical features designed to assist both the student and the instructor.

- A set of objectives at the beginning of each chapter, outlining what the student should learn;

- A running case study at the end of most chapters, which illustrates how the concepts and techniques have been applied by a real company;
- An extensive word list at the end of each chapter, serving as a check list of important terms, concepts, and ideas;
- Review questions for each chapter, which give the student an opportunity to test his or her knowledge of the material;
- Problems and Projects at the end of each chapter designed to get the student "thinking" and "doing." The problems are challenging questions, which will lead the student beyond the text. In many cases, real-world situations are presented for his or her consideration. The projects often require the student to talk to telecommunications professionals and users;
- A comprehensive glossary of terms;
- A separate list of all of the acronyms used in the book;
- A list of references to standard telecommunications books as well as current articles on topical subjects.

Compared to the first edition, the second edition of *Business Telecommunications* contains more questions at the end of every chapter, additional problems and projects, and more illustrations and photographs to aid the student's understanding.

For the instructor there is a comprehensive instructor's guide that includes:

- Suggestions for several ways to organize the course, depending on the desired emphasis and focus;
- Supplemental textual material that elaborates on some of the topics covered in the text;
- Transparency masters of the art in the book and the chapter outlines;
- Answers to the review questions in the text;
- Suggested solutions to the problems in the text;
- Hints for the presentation of material in the classroom;
- Test bank questions for examinations.

Organization

Business Telecommunications is divided into three parts. Part One deals with telecommunications applications and the environment.

- Chapter 1 introduces the subject matter and leads the student to realize that he or she may know more about telecommunications than he or she realizes, by virtue of daily living experiences.
- Chapter 2 examines several telecommunications applications in detail. This material serves as a reference set of applications for the student throughout the rest of the text.

- Chapter 3 discusses the environment in which telecommunications exists within a company, covering users' and managers' expectations of telecommunications.
- Chapter 4 discusses the external environment. Deregulation and divestiture are covered, and the nature of the telecommunications industry is explained. The ISO-OSI model is introduced to provide the student a framework for the technical material of Part Two.

Based on the foundation established in Part One, Part Two delves into the technical details of telecommunications.

- Chapter 5 explains voice telecommunications, with particular emphasis on the business setting.
- Chapter 6 describes various types of communications terminals. This material, familiar to many students, serves the purpose of setting a common level of knowledge for the rest of Part Two.
- Chapter 7 discusses how data is coded for computing or telecommunications. Voice digitization is discussed, and the student begins to see how digitized voice and coded data can be transmitted together in an integrated network.
- Chapter 8 explains how data is transmitted and how the terminal is interfaced to the communications circuit. The workings of modems are examined in detail.
- Chapter 9 describes communications circuits. Various media are studied, and the attributes of each are listed. Error detection and correction are explained. Network topologies for both wide area and local area networks are described.
- Chapter 10 explains data link control protocols, the "rules of the road" for communications circuits. Both wide area and local area network protocols are emphasized. Public and value-added networks are discussed. LAN software and network management issues are described.
- Chapter 11 presents telecommunications networks. The ways that circuits can be combined to form networks are shown. Several ways that networks can be classified are described. Wide area and local area networks are emphasized.
- Chapter 12 discusses the way in which a circuit is connected to a computer. The functions of the front-end processor and communications software are covered.
- Chapter 13 discusses network architectures. The chapter puts all of the technical material into perspective. The ISO-OSI reference model, which was introduced in Chapter 4, is studied in more detail. IBM's SNA and Digital Equipment Corporation's DNA are examined. The need for network architectures and their advantages and disadvantages are explained.

Part Three deals with telecommunications management in the broadest sense.

- Chapter 14 focuses on the need for management, the functions of the telecommunications department, and management's line responsibilities.
- Chapter 15 describes the process of network design and implementation. The process is broken into phases, each of which is discussed in detail.
- Chapter 16 describes the management and operation of the network. Day-to-day operational procedures, as well as the functions of problem management, performance management, configuration control, and change management, are explained. The critical role of the communications technical support staff is also covered.
- Chapter 17 provides a glimpse of the future. Several applications of telecommunications that will be upon us in the near future are described. These are the types of applications that the students of today will be supporting and managing within just a few years.

Acknowledgments

Accurate and up-to-date coverage of a fast-changing field such as telecommunications requires the input of many people. I want to thank the following individuals for their excellent suggestions and assistance during the preparation of the text: Bruce Estes and John Burkhart of AT&T; Howard Super of IBM; Jim Bradley of The Dow Chemical Company; Louie Lubahn, independent consultant; Robby Bosko of Dow Corning Europe; Amado Reboucas of Dow Corning Brazil; and Mark Doepel, Theresa Srebinski, Ken Morris, Frank Aymer, Noel Vandewoestijne, Edward Steinhoff, and J. Kermit Campbell of Dow Corning Corporation.

I want to express my appreciation to my employer, Dow Corning Corporation, for permission to use the case study material and a number of photographs and for their support in many other ways. Thanks also go to Dow Corning Japan, where I am currently assigned, and in particular to my manager, Richard MacKenzie, and my secretary, Rumiko Tsukada.

Warm thanks go also to the people who reviewed the first edition of *Business Telecommunications* and made many helpful suggestions that improved the content of the second edition: Christine Bullen, Sloan School of Management, Massachusetts Institute of Technology; Curt H. Hartog, Washington University; Jim Koerlin, School of Telecommunications Management, Golden Gate University; Patricia Mayer Milligan, Baylor University; and Roger A. Zaremba, Barney School of Business and Public Administration, The Univeristy of Hartford. Special thanks go to Marjorie M. Leeson (CIS Consultant) who first encouraged me to write this book and who offered significant help along the way.

The team at Macmillan deserves special mention. Charles Stewart provided assistance and useful advice throughout the project. Barbara C. Newman provided excellent administrative support and research for appendices B, C, and D. Others employed by Macmillan who made substantial contributions for which I am grateful are Stewart Kenter and Dora Rizzuto.

Finally, I want to express my thanks to my wife, Pam. Pam not only gave me complete support during this project but also typed, helped prepare the manuscript, and made many suggestions that improved the content of the book. It is because of her love and ongoing encouragement that the work was possible.

Stanford H. Rowe, II

CONTENTS

PART ONE

Introduction to the Business Telecommunications Environment **1**

CHAPTER 9 Communications Circuits 293

PART ONE

Introduction to the Business Telecommunications Environment

The purpose of Part One is to introduce you to telecommunications concepts, to give you background material about uses of telecommunications in business, and to explain the environment in which business telecommunications operates.

Chapter 1 introduces the subject, explains some basic vocabulary, and gives some simple examples of ways in which telecommunications is used. A brief history of telecommunications and an introduction to telecommunications regulation also are included.

Chapter 2 delves into telecommunications applications in greater detail, giving categories of applications and examples of each type. The intention of this chapter is to show the varying ways that telecommunications is used in business.

Chapter 3 deals with the internal influences on telecommunications within a company and the telecommunications department. Users and managers both have sets of expectations about what telecommunications can do for them. The importance of understanding and striving to meet these expectations is stressed.

Chapter 4 examines the external forces that shape the way telecommunications is used within business. The history of telecommunications regulation and the movement to a largely deregulated industry are described. An overview of today's telecommunications industry and major companies is given. The rapidly changing technology and its impact, as well as the emergence of telecommunications standards, are discussed. Finally, the International Standards Organization's model for Open Systems Interconnection is introduced. This model will give you a framework on which to build your telecommunications knowledge and will serve you well as you study the technical material in Part Two.

After studying Part One, you will be familiar with basic telecommunications terms and concepts, understand the many ways that telecommunications can be used in business, and be aware of the forces inside and outside the company that shape the way telecommunications technology and techniques can be applied.

CHAPTER

1

Telecommunications
Basics of a Fast-Paced Industry

OBJECTIVES

After you complete your study of this chapter, you should be able to

- give a definition of telecommunications;
- describe the basic elements of a telecommunications system and give examples of all of them;
- explain why telecommunications is an integral part of the contemporary business environment;
- describe why it is important to study and understand something about telecommunications;
- describe several examples of the way telecommunications is used in business;
- give a brief summary of telecommunications history;
- understand why it is likely to be difficult to keep your telecommunications knowledge up-to-date.

INTRODUCTION

As the world moves into the information age, the ability to share and communicate knowledge and information becomes more important. Every passing year brings new discoveries: more to be learned about the sky above, the depths of the oceans, human and animal life, incredible inventions. Although information known only to one person can be very useful to that person in decision making, knowledge and information are most useful when they are shared with others. Sharing means communicating, and communicating can occur in many ways. A raised eyebrow, a shrug, a quizzical expression, a posture, a stance—all are very effective ways of *body language* communication in certain situations. Communication can occur in other forms, too. Music, art, and dance come immediately to mind as ways of expressing feeling and emotion that sometimes stir and inspire the soul.

Another example is *written* communication. Books, newspapers, magazines, and graffiti scrawled on a wall all exist to convey messages to readers. In some cases, the intent is to convey information that the reader wants to receive; in other cases, the writer wants to express his or her thoughts, ideas, or feelings on a topic of importance to him.

Public radio and television are two forms of *broadcast* communication that cannot be ignored. Every day, most people in civilized countries are bombarded by broadcasts designed to entertain, sell, or provide information. Although one can ignore the specific broadcasts, it is virtually impossible to deny the strong impact this type of communication has on our daily lives.

The amount of information that is available to us is accelerating rapidly. More books are being published and magazine articles written than ever before, to say nothing of the increasing numbers of research reports and all of the information generated by computers. Learning to cope with this vast array of information is a challenge that must be addressed continually. Communication technology will ensure that the information is available and accessible when needed, but sifting and selecting the right information to meet our needs is our responsibility. Computers can be of significant help in the information-screening process.

DEFINITION OF COMMUNICATION

Communication is defined as "a process that allows information to pass between a sender and one or more receivers" or "the transfer of meaningful information from one location to a second location." *Webster's New World Dictionary* adds, "the art of expressing ideas especially in speech and writing" and "the science of transmitting information, especially in symbols." Each of these definitions has important elements. Communication is a process that is ongoing, and it obviously can occur between a sender and one or more receivers. The word *meaningful* in the second definition is significant. It is clear that communication is not effective if the information is not meaningful. One could also argue that if the information is not meaningful, no communication has occurred at all. There is a strong analogy here to the traditional question posed to beginning physics students: If a tree falls deep in the forest and no one is within earshot, does the tree make a sound as it falls? The answer is that there is no sound because sound also requires a source and a receiver. Without the receiver, there is no sound.

The last definition of communication relates most closely to the focus of this book. You will be studying the science of communication using electrical or electromagnetic techniques. The information being communicated will be in coded form so that it is compatible with the transmitting and receiving technologies. For our purposes, the practical application of the communication science in the business environment is essential to the definition and study of communication.

DEFINITION OF TELECOMMUNICATIONS

Webster also tells us that the prefix *tele* means far off, distant, or remote. Practically speaking, the word *telecommunications* means communication by electrical or electromagnetic means, usually (but not necessarily) over a distance. Not long ago, telecommunications meant communication by wire. Although this is still accurate in many situations today, it is not complete, for telecommunications can also occur using optical fiber or radio waves.

BASIC ELEMENTS OF A TELECOMMUNICATIONS SYSTEM

A telecommunications system contains three basic elements: the *source, medium,* and *sink.* These are illustrated in Figure 1–1. More common terms would be *transmitter, medium,* and *receiver.* Examples of each element in the system are abundant: voicebox, air, ear; telephone, telephone lines, telephone; or terminal, data circuit, computer. There are many possible combinations of each of the basic elements.

We will immediately begin referring to the medium as the *communication line.* You can think of it as the telephone wire that comes into your house to connect your telephone to the telephone company office. The line can be implemented in many physical forms, such as copper wire or microwave radio, but visualizing it as a copper wire is convenient and not inaccurate.

communication networks

Communication lines are connected together in many ways to build *communication networks.* It is easy to think of the telephone network as a series of lines connecting telephones to telephone company offices. As you undoubtedly realize, there are also many lines that connect the telephone company offices with each other. Together, all of these lines make up the *telephone network.* In a similar way, businesses frequently build their own networks, which connect all of their locations together. Lines and networks are precisely defined and more thoroughly discussed in Chapters 9 and 11.

communication rules

To ensure that a communication will be successful and effective, rules must guide its progress. Although these rules are not technically a basic element of telecommunications, they are absolutely necessary to prevent chaos. For example, there are unwritten rules that guide our use of the telephone. Most of us learned them when we were very young. When receiving a telephone call, the answering party traditionally initiates the conversation by saying "hello." The caller and answerer

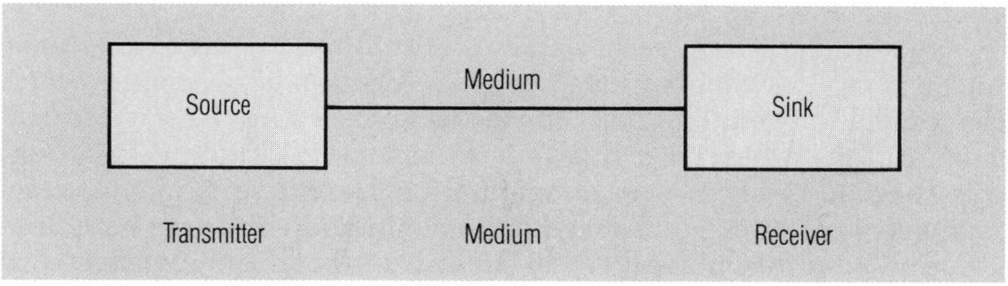

Source — Medium — Sink

Transmitter — Medium — Receiver

Figure 1–1
Basic elements of a telecommunications system.

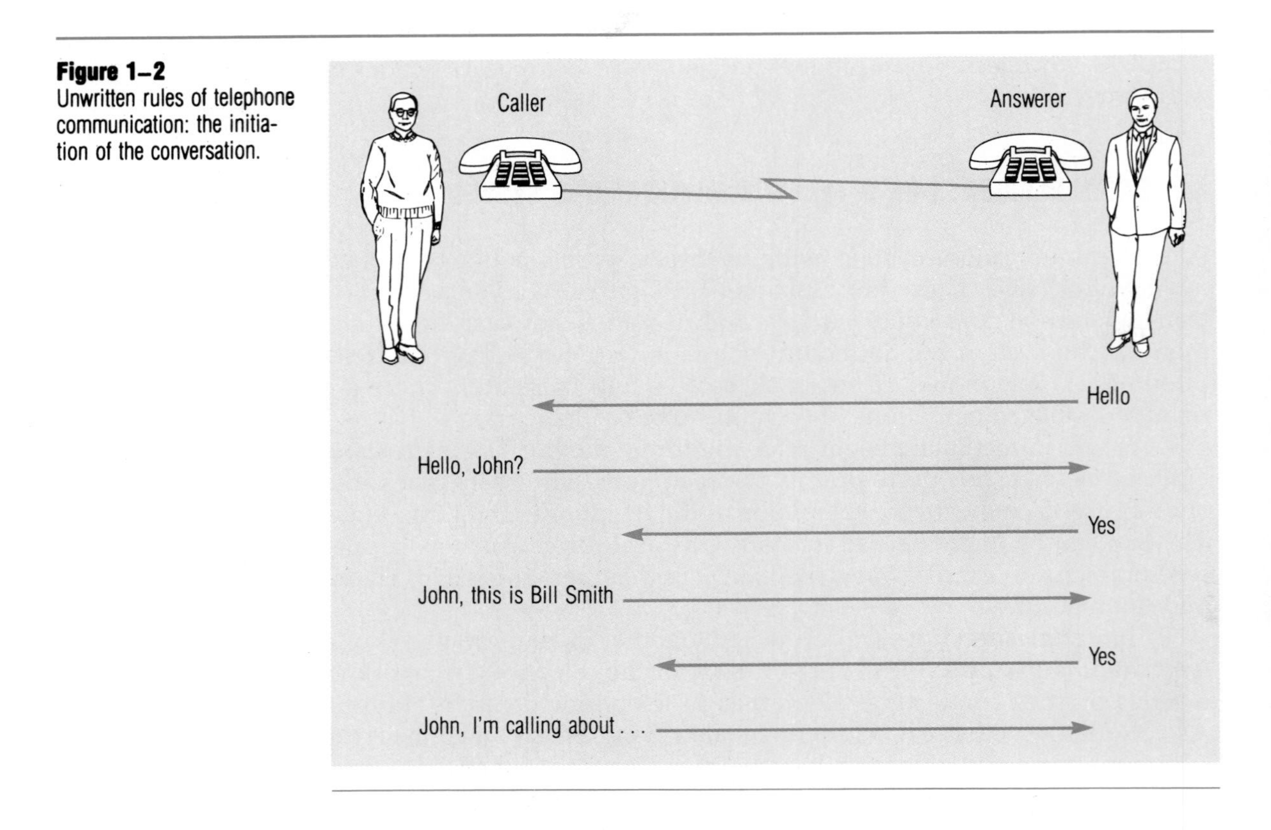

Figure 1–2
Unwritten rules of telephone communication: the initiation of the conversation.

then go through a brief dialogue to identify each other before they launch into the purpose of the call. This is illustrated in Figure 1–2. The process is often shortcut by combining some of the exchanges or when one of the parties recognizes the voice of the other. Some examples of this process are shown in Figure 1–3.

In some aspects of a telephone communication, well-defined rules don't exist. For example, what happens if the conversation is cut off due to some fault in the telephone equipment? Who calls whom to reinitiate the conversation? When you have been cut off, have you received a busy signal when you redialed the call because the person you were talking to also tried to redial? Eventually one person waits, and the problem is solved.

In a similar way, when two pieces of equipment such as a terminal and a computer communicate, rules are needed to determine which device will transmit first, how the terminal and computer will be identified to each other, what happens if the communication gets cut off, and so forth. Unlike the voice communication example, however, when equipment is communicating automatically, all of the rules must be defined precisely, and they must cover all situations—usual and unusual—that can occur. The rules of communication are called *protocol*.

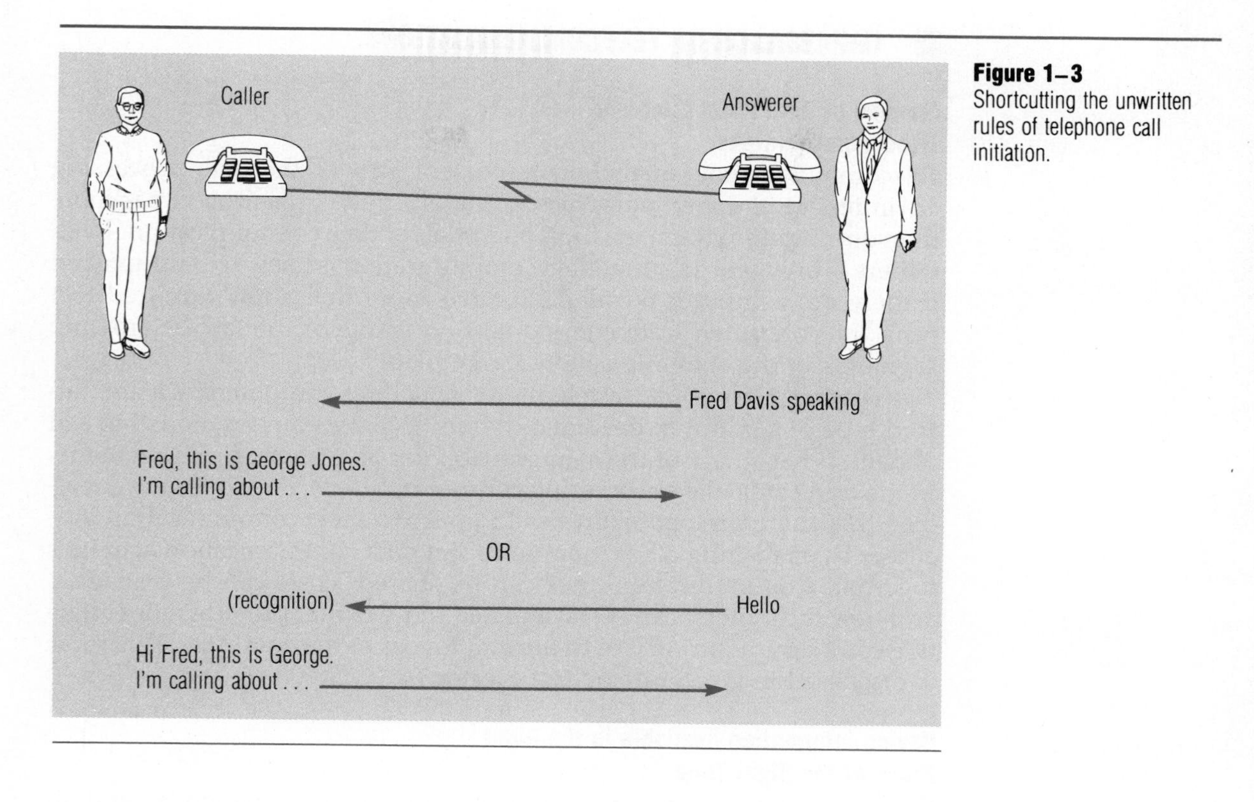

Figure 1–3
Shortcutting the unwritten rules of telephone call initiation.

■ SCOPE

For the purposes of this book, telecommunications includes the transmission and reception of information using electrical or electromagnetic means from a transmitter to a receiver over a medium. Various types of transmitters, receivers, and media, and alternative sets of rules that can be used for those communications, will be introduced and described in some detail. The primary focus of the book is the communication of information in commerce or business taken in its broadest sense. Although most of the examples will come from business and industry, the points being made will in most cases apply equally well to government, educational, medical, or other not-for-profit institutions. The types of communication that will be discussed involve the transmission and reception of voice, data, text, graphics or images, or combinations of these forms. You will see that when they are reduced to their most basic electrical form, all of these communications look alike. The book does not specifically describe or analyze commercial radio or television or uses of telecommunications in the home. However, most of the principles and techniques discussed here apply equally well in those settings, too.

IMPORTANCE OF TELECOMMUNICATIONS

Melding of Data Processing and Telecommunications

Telecommunications is also an important part of the data processing environment in many companies. Certainly, if the organization has more than one location, there is a high potential for data communication. Even if there is only one location, data communication is likely to be employed if there are computer terminals located more that a few hundred feet from the computer. Data communication is one of the fastest growing segments of the communication marketplace.

Furthermore, modern telephone switching equipment is a specialized type of computer, designed to handle voice conversations instead of data. It has many of the same capabilities and physical requirements as the traditional data processing computer. Many companies have realized that there are opportunities to improve their communication efficiency by managing all of their voice *and* data communication activities together. Communication lines can be shared, costs can be managed, and new technology can be assimilated in ways that provide more effective communication service to the employees of the company. This topic is discussed in greater detail in Chapter 14.

Having Information Available in the Right Place at the Right Time

Businesses in the information age generate more information faster than ever before. At the same time, most companies are realizing that information is an increasingly valuable asset that must be managed with the same care and attention as the company's finances, buildings, machines, and people. Having the right information in the right place at the right time can mean the difference between profitability and unprofitability of the business—and ultimately determine its success or failure. Having information available to the right people isn't a new requirement of business success. What is new is the ability to deliver information near or far at speeds that were unthinkable a few years ago. For years, companies have had telephones for voice communication, and reports and analyses were hand-generated and mailed to arrive days later. Now, with the marriage of communication and computer techniques, you can make the detail report or summary analysis available anywhere virtually instantaneously. People in widely different locations can look at the data and discuss it soon after it becomes available.

Capturing Basic Data About Business Operations As They Occur

basic business transactions

Another reason that telecommunications is becoming increasingly important is that, combined with the computer, it is being used more and more as an input mechanism for capturing data about the basic operations of the business as soon as they occur. Online computer applications, in which a terminal is connected directly to a computer with virtually instantaneous communication, are being used by businesses to enter customer orders, record customer payments, give notification of product shipments, and track inventory. These operations are the *basic business transactions* that are fundamental units of business operations. Once data about the business transaction is captured and stored in a computer, it is available, via telecommunications, to others.

A classic example is the airline reservation system, in which the traveler's airline representative or travel agent makes a flight reservation. Information about this reservation is recorded in a computer database and has the effect of reducing the number of seats that are available for the particular flight. This reduction of seat inventory is a basic business transaction of the airline. As soon as the database is updated, other airline reservation agents or travel agents in other locations around the country or around the world can check to see if seats are still available on the flight for their customers.

In another case, a customer who wishes to order a product from a company calls the company and talks to an order processing clerk. Through her computer terminal, the clerk can check to see if the required product is available or, if not, when it will be. If the product is available and the customer places the order, shipping instructions can be processed in the warehouse nearest the customer to ensure the most rapid delivery. Later, the reports detailing this and all other sales transactions can be generated and transmitted to analysts who can spot trends or detect inventory problems quickly. These reports often generate follow-up questions that analysts and management confer about over the telephone or through terminals.

As businesses implement online systems, they are finding that they are increasingly dependent on the computing and telecommunications technologies on which online systems are based. In the transportation, insurance, and finance industries, the dependence on telecommunications and computer systems has become critical. If a massive failure occurred, business could not be conducted, and financial failure would follow. As a result, some companies have gone to great lengths to ensure that they have complete backup facilities in place so that major computer "downtimes" cannot occur. Most companies find that whether or not the computer and telecommunications systems are vital to their operation, it is prudent to make contingency plans for how business operations would be conducted if a prolonged outage did occur.

Allowing Geographic Dispersion of Facilities and People

Telecommunications allows people in diverse locations to work together as if they were in close proximity. Branch banking clerks, car rental agents, and insurance agents can all share common information and have most of the same capabilities they would have if they were located in the home office. At the same time, companies in industries that are not *required* to have operations in widely separated geographic regions have a relatively inexpensive opportunity and the flexibility to do so, thanks to telecommunications. While corporate headquarters remain in a major metropolitan area, such as New York City, other facilities, such as sales and marketing offices, can be located close to the customer, and manufacturing plants can be located close to sources of raw materials or natural resources used in the manufacturing process. With telecommunications, the company can still operate as a single, coordinated entity and, for most purposes, ignore the geographic dispersion.

In summary, telecommunications is becoming an integral part of the way companies conduct business because its capabilities provide efficient, effective ways of conducting business. In some cases, the combination of telecommunications and computing allows business to be performed in ways that are impossible using manual techniques. Business can be conducted faster and more accurately, and decisions can be made with more timely information than was previously possible. It is important to point out, however, that making decisions faster and with better information does not necessarily lead to making "better" decisions, although it should contribute to their overall quality.

REASONS FOR STUDYING TELECOMMUNICATIONS

There are several important reasons for learning about telecommunications, its terminology, and its applications.

Wide Use at Home

Telecommunications techniques and products are becoming more widely used every day. Nearly everyone has or will have some contact with them. At home, using the telephone is perfectly natural, but recently we have been faced with some new decisions about the type of telephone service we want to have. We have been given the option of purchasing our own telephones rather than renting them from the telephone company. We have been asked to select which company we want to carry our long distance telephone calls.

Some knowledge of telecommunications is desirable to help us make intelligent decisions about the communications services we want to purchase. Many people have cable television at home, and this is another form of telecommunications. Subscribers must select from a wide variety of available programming, some of which is included in the basic fee paid each month and some of which is available only at extra cost. In some sections of the country, it is possible to do banking using the telephone or a personal computer in the home. In the future, we will be faced with ever-increasing alternatives and choices, and a knowledge of telecommunications principles and terminology will be even more useful.

Direct Use on the Job

In businesses of all types, telecommunications is becoming an integral part of the way work is done. Whether it be the automated teller machine at a bank, the supermarket checkout scanner that reads bar codes on food products, the wand and computer terminal that the librarian uses to check books out of the public library, or the terminal that the clerk at the Internal Revenue Service uses to enter the income tax information from a tax form into the computer, telecommunications often is involved. More and more people are using computer terminals connected to computers via telecommunications lines.

terminals per capita

A 1980 study by the Bureau of Labor Statistics showed that there was 1 terminal for every 23 people in the United States work force. The same study estimated that by 1989, the ratio would be 1 terminal for every 2 workers. More recently, market research analysts have revised the estimate to a 1:1 ratio. That is, it is now estimated that on average, there was a computer terminal for *every* person in the work force in 1989. In some jobs, more than one computer terminal is used. Scientists, for example, may have a terminal in their laboratory, another in their office, and a third at home so that they can dial into their laboratory computer in the evening to check on the progress of a long-running experiment.

Indirect Use on the Job

Even if your job doesn't deal directly with telecommunications, you will most likely work with people who do. Knowledge of the subject and its vocabulary will help you communicate with "telecommunications workers." In business, it could help you request new information or services and understand problems. In some cases, people find that because of a job change or promotion, they are suddenly thrust into a new job directly working with telecommunications terminals or equipment. They are thankful for any telecommunications background they have obtained in the past.

A Possible Career

Another reason for learning about telecommunications is to help you assess whether you might want to consider making telecommunications your career. A partial list of telecommunications jobs is shown in Figure 1–4. As the use of telecommunications grows, so does the need for knowledgeable people to design, install, repair, maintain, and operate systems. We can all relate to the job of telephone operator, computer terminal operator, or repairperson because we have seen these people at work.

Figure 1–4
A partial list of telecommunications career opportunities.

Telephone operator
Terminal operator
Network control operator
Telecommunications administrative support specialist
Communication repair technician
Network analyst
Network designer
Telecommunications programmer
Telecommunications technical support specialist

related careers

Not so obvious however, are the thousands of people who work "behind the scenes" to sell, design, and install telecommunications systems and keep them operating reliably and properly. Network analysts and designers are people who are knowledgeable about the types of telecommunications hardware and services that are available. They identify the communication requirements for a company and then design a telecommunications solution that will provide the required capabilities. Communication programmers write special computer software that enables computers to communicate with one another or with terminals. Network operators monitor the day-to-day operation of a communication network, solve problems when they occur, capture statistics about the network's performance, and assist the users of the network to take advantage of its capabilities.

All of these jobs, and others, are described in detail in Part Three of the text. Suffice it to say now that good telecommunications people—who creatively apply technology and techniques to help solve a company's communication problems or to take advantage of new business opportunities—are always in demand. They are especially valuable if they can communicate well with others in the company and have a solid business background in addition to their specialized telecommunications knowledge. Good telecommunications people can earn a very good income because the value of their skills and the services they provide are well recognized by the companies and other organizations they serve.

Even if you don't work directly in the telecommunications field, you will undoubtedly work with people who are telecommunications specialists. Studying telecommunications will help you develop knowledge and vocabulary and make communication with specialists easier. This is significant because many telecommunications people today do not know the language of business. If you, as a nontelecommunications person, can understand some of the telecommunications terminology, it will help make your discussions more productive. Whether you are selecting a terminal for your use at work, discussing the level of service you need in order to do your job properly, or selecting a new telephone to use at home, you will be in a better position to question the alternatives, discuss options, and make a more informed, intelligent decision.

NEW TERMINOLOGY

Like any other subject, telecommunications has its own vocabulary to master. When you first learned about computers, you were probably overwhelmed and mystified by such words as *disk*, *CPU*, and *file*. After some study, however, these words became a comfortable part of your vocabulary.

It is the same with telecommunications. There are many new words to learn, but your computer background will stand you in good stead and provide a foundation on which to build. Unfortunately, many people assume that since they know the language of computers, they also know and understand the language of telecommunications. This is usually not the case. There are even some words, such as *dataset*, that mean one thing in the computer sense and something entirely different in a telecommunications context. Like the computer vocabulary you learned, there are also quite a few acronyms to be mastered. Some of these are directly related to specific vendor's products, but many are used more generally. A list of commonly used telecommunications acronyms can be found in Appendix F.

There are three words used rather generically throughout the book that need to be explained up front. *Business* generically includes the business, education, government, and medical fields. A sentence that begins, "Telecommunications is becoming more widely used in," could end with business, education, government, or medicine and be equally correct.

Company is used generically to mean organization, bureau, hospital, college, or farm. A company is a specific organizational unit within the business field, just as a hospital is a specific organization within the medical field.

Virtually anyplace the pronoun *he* is used, *she* could be substituted. Telecommunications is truly an equal opportunity career field. Whether someone is a terminal operator, network designer, or repair technician, either a man or woman can do the work.

COMMON EXAMPLES OF TELECOMMUNICATIONS

You have been exposed to telecommunications in many forms—probably more than you realize. Let's look at some common activities that rely on telecommunications technology and techniques.

Telephone Call

Figure 1–5 is a simplified diagram of the components involved when you make a telephone call. The line that looks like a small lightning bolt connecting the telephone to the telephone company is the common

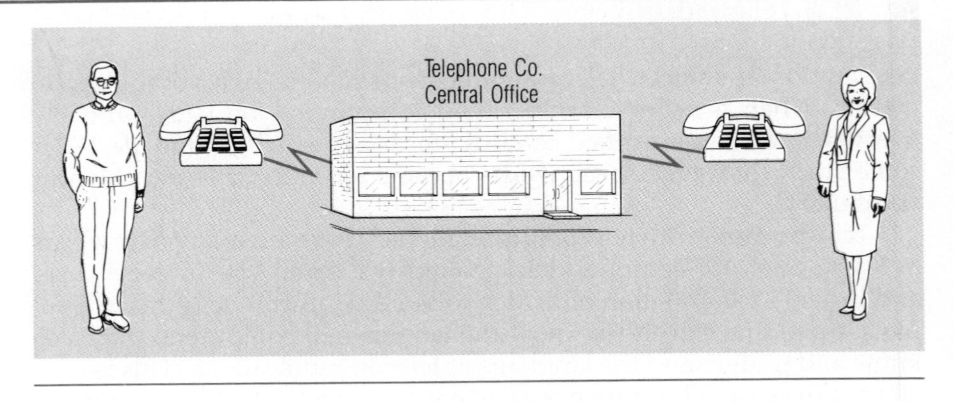

Figure 1–5
Simplified diagram of the components of a standard telephone call.

symbol used to indicate a telecommunications connection. This connection can take many forms, as you'll see in Chapter 9. As an experienced telephone user, you probably have a number of perceptions about what is involved when a telephone call is made. Probably most of your perceptions are correct. You know that when you touch the buttons or dial a rotary telephone, a signal is transmitted to the telephone company that causes a connection to be made to the party to whom you wish to talk (assuming that you dialed the number correctly). Normally, you will soon hear a buzzing sound that tells you that the other telephone is ringing. With luck, the person you want to talk to will answer the telephone, and the conversation will begin.

You probably have the perception that no people (telephone operators) are involved at the telephone company in making the connection or otherwise completing the call, and you are correct. Today's telephone switching equipment in the telephone company central office is completely automatic, and the vast majority of calls are completed without human intervention. You probably also feel that the telephone system works fine most of the time, except maybe when the weather is very bad or on Christmas or Mother's Day. Again you are correct, for the vast majority of calls are completed accurately and with high quality. Occasionally, the telephone system gets overloaded, but such occurrences are rare.

"The telephone company" in Figure 1–5 is probably a bit more nebulous. You know that you have seen the telephone company building in your town, telephone poles with wires strung on them, repair trucks, telephone stores, and certainly the monthly bill, but you are probably unclear about how all of these pieces fit together to provide the telephone service you have come to expect. Suffice it to say that in addition to the parts of the telephone company and its operations that you know about, there are at least as many parts that you probably have never heard about or been exposed to. The details are described in Chapter 5.

To put this example in terms defined earlier, the calling party in the example is the *source;* the connection to the telephone company, the telephone company itself, and the connection to the called party are the *medium;* and the called party is the *sink*. This is an example of person-to-person communication. The caller and receiver are the direct users of the telecommunications system, which in this case is the telephone system. They operate the terminal (telephone), and that operation is so routine and familiar that they don't think twice about it. Most people learned how to operate this telecommunications system when they were very young.

Airline Reservation System

Figure 1–6 illustrates a slightly more complicated use of telecommunications: making an airline reservation, in which you and the airline reservation agent use telecommunications. There are several variations on how the airline reservation can be made; this example illustrates the main points. The first part of this figure is the same as for Figure 1–5: you make a telephone call through the telephone company to a person—the airline reservation agent. The reservation agent operates a *video display terminal (VDT),* a terminal with a screen that looks like a small television set and a keyboard. The VDT is connected via a telecommunications line to the airline's central computer. You describe the reservation you would like to make, and the agent requests information, such as flight schedules, alternative connections, and ticket prices from the computer using the terminal. The information displayed on the terminal is in a highly coded form, which the reservation agent interprets and relays to you on the telephone. Two dialogues are taking place: one between you and the reservation agent and one between the reservation agent and the computer.

Ultimately, you decide on the flight, and the reservation agent enters your personal information, along with the flight number and the time and date of the flight, through the terminal into the computer's memory. The data is copied from there onto disk storage. A ticket may be printed and mailed to you, or you may pick the ticket up at the airport when you arrive for the flight.

The telecommunications line between the reservation agent's terminal and the computer is used so much that it is usually advantageous for the airline to install special communication wiring if the terminal and computer are in the same building. If the terminal and computer are in different locations, the airline may lease the telecommunications line full-time as a *private* or *leased line* from the telephone company. The airline company pays the telephone company a flat monthly fee for its exclusive use. By contrast, the connection between your telephone and

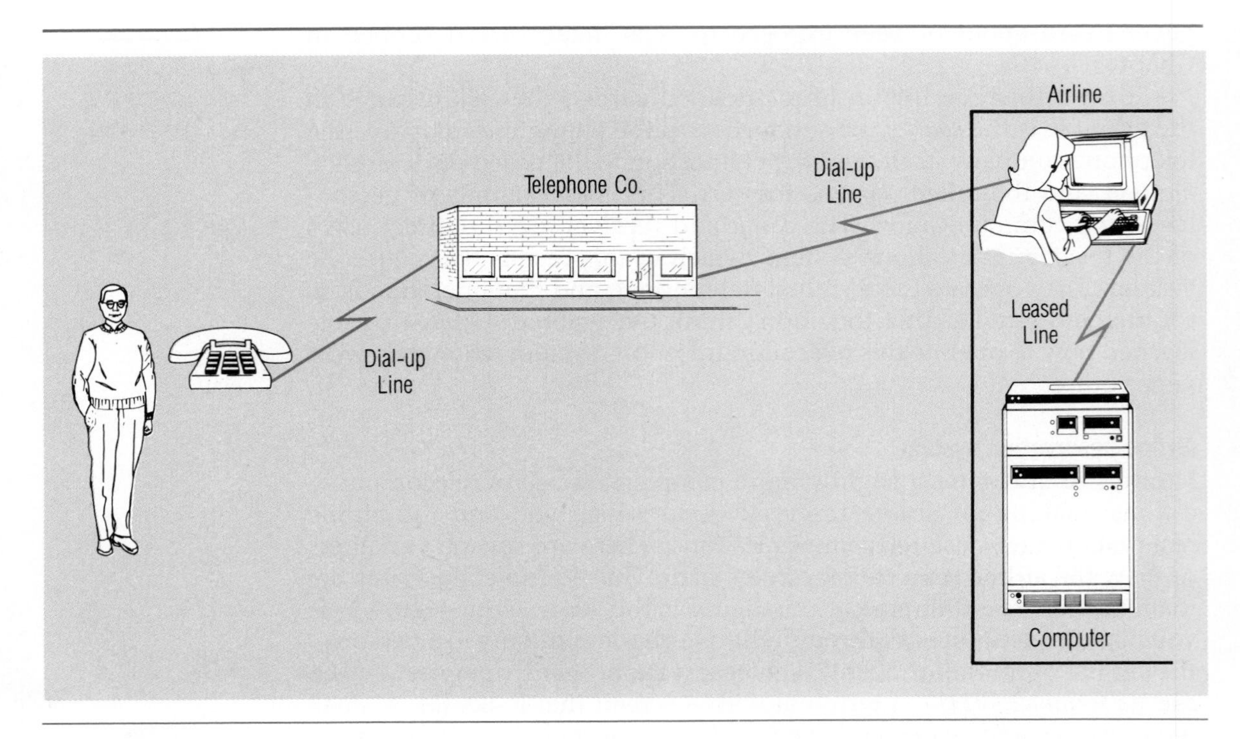

Figure 1–6
The telecommunications connections between a traveler and an airline reservation computer.

the airline is called a *dial-up* or *switched line*. You pay a basic charge to make telephone calls, plus a usage charge based on the number, distance, and duration of the calls you make. Dial-up lines are discussed in more detail in Chapter 5.

Banking with an Automatic Teller Machine (ATM)

Figure 1–7 illustrates the modern way of banking: interacting directly with the bank's computer through a computer terminal called an *automatic teller machine (ATM)*. The ATM may be located at the bank, shopping center, airport, grocery store—in fact, just about anywhere. Telecommunications allows the ATM to be connected to the bank's computer many miles away.

With ATMs, consumers with little knowledge about telecommunications or computers operate computer terminals. The ATM contains a small television-like screen similar to the airline agent's VDT. On the screen are instructions or questions, such as "Insert your card," "Please enter your password," and "Would you like to make a deposit, a withdrawal, or pay a bill?" The machine "walks users through" the process of performing banking transactions. The ATM interacts with customers and with the computer, getting the information about the transaction the customer wants to perform, then relaying the information to the

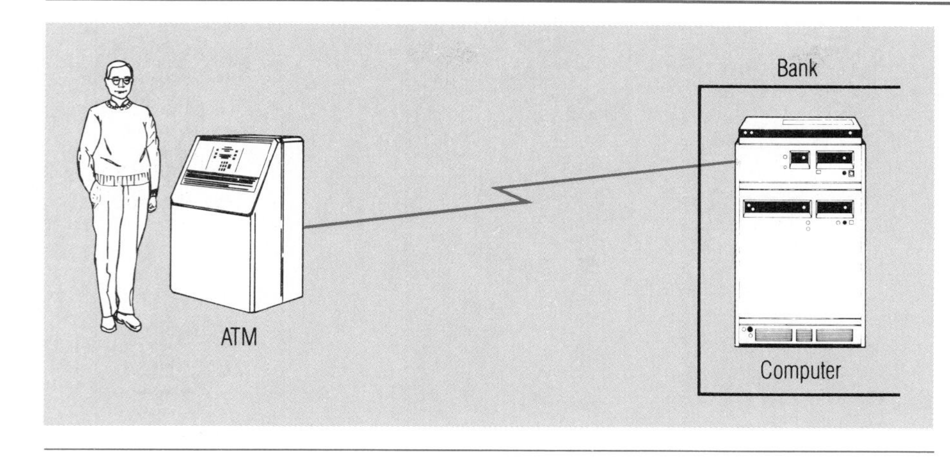

Figure 1–7
An ATM connected to a computer by a telecommunications line.

computer. For withdrawals, the ATM verifies that the customer has enough money in the account by requesting the computer to check the balance.

ATMs may contain a small computer that can be programmed to provide the user with instructions. A programmable ATM is an example of an *intelligent terminal*. The intelligent ATM handles all interactions with the user and only communicates with a central computer to check account balances or to relay the results of the ATM's processing. Alternately, the ATM may receive all of its capability from a computer to which it is connected via a telecommunications line. In that case, the ATM is considered to be a *dumb terminal,* and the instructions to the user come from the computer and all input from the keyboard of the ATM is passed to the computer for processing.

Initially, many people are somewhat afraid to use ATMs. Some people never become comfortable using an ATM and prefer the human contact that a live teller provides. The bank prefers to have customers use ATMs because it reduces the number of human tellers, which in turn reduces the bank's costs. ATMs also improve the bank's service by allowing customers to bank 24 hours a day, 7 days a week.

Automatic Remote Water Meter Reading

Figure 1–8 illustrates another example of the use of telecommunications: water meters with telecommunications capability in residences and commercial buildings. In one city, a computer at the water company office dials the telephone line connected to the water meter in the residence. An encoder on the meter is activated by an interrogation signal. The encoder senses the meter reading and sends back meter data to start the billing process.

The cost advantage to the water company is having no meter readers walk to all of the houses and buildings in the city to read the water meters every month. Although the initial cost to install these automatic

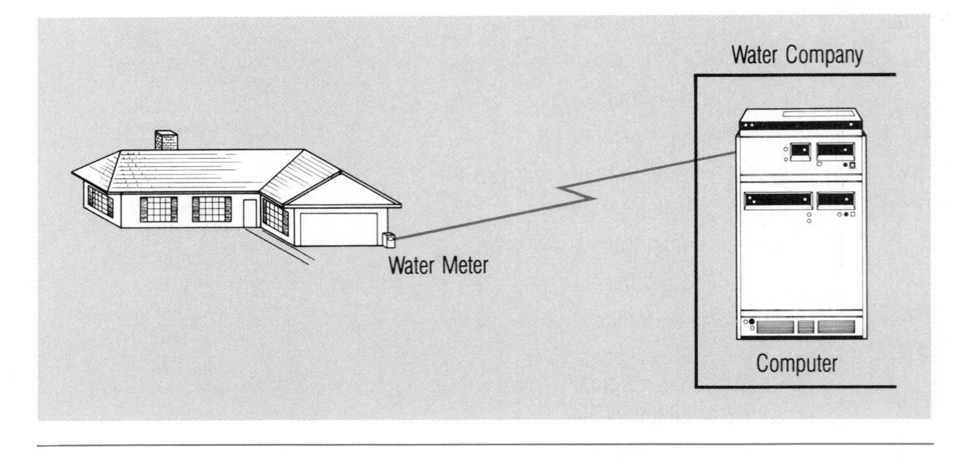

Figure 1–8
Remote reading of a water meter using telecommunications.

water meters is high, the payback comes rather quickly when fewer people are needed to read meters. In addition, the automatically generated meter readings are more accurate than those read manually.

In some cases, there are practical alternatives to the telecommunications technologies that are described and used as examples in this book. For example, a water company may be able to substantially increase the accuracy of meter readings by equipping its human meter readers with electronic data recorders. Whereas the meter reader still walks to all of the locations on the route, the time it takes to record the meter reading and the data entry cost are significantly reduced. The key to finding the best solution for any business problem is to first define the problem and then explore several alternative solutions before selecting one to implement. This process is discussed more fully in Chapter 15.

HISTORY OF TELECOMMUNICATIONS

What do you think of first when you think of telecommunications history? Alexander Graham Bell? Samuel F. B. Morse? Telegraphers sitting with telegraph keys to send messages across the continent? The old-style, dial-up telephone your grandmother used to have? All of these images are certainly a part of telecommunications history. The timetable in Figure 1–9 lists many of the significant events of recent telecommunications history (since communication was first performed using wire as the medium).

Let's go back further in time. Long distance or *tele* communications were used in ancient times, when cavemen used fires to communicate simple messages over long distances. Similarly, drums, smoke signals, and the printing press were used by cultures long before electricity or

1837	Samuel F. B. Morse invents the telegraph.
1838	Telegraph demonstrated to government. Government declines to use.
1845	Morse forms a company with private money to exploit the telegraph.
1851	50 telegraph companies in operation.
1856	Western Union Telegraph Company formed.
1866	Western Union Telegraph Company was the largest communications company in the U.S.
1876	Patent issued to Alexander Graham Bell for telephone.
1876	Bell offers to sell telephone patents to Western Union for $100,000. Western Union declines.
1877	Bell Telephone Company formed.
1878	First telephone exchange with operator installed. Western Union Telegraph Company sets up its own phone company, sued by Bell for patent infringement, gets out of phone business, and sells network to Bell.
1885	American Telephone and Telegraph Company (AT&T) formed to build and operate long lines interconnecting regional telephone companies.
1893–94	Original Bell patents expire; independent telephone companies enter market.
1911	Bell Telephone franchise companies reorganize into larger organizations known as the Bell Associated Companies. Beginning of the Bell System.
1913	Invention of the vacuum tube.
1941	First marriage of computer and communication technology.
1943	Development of submersible amplifier/repeaters.
1947	Invention of the transistor.
1956	First trans-Atlantic telephone cable installed.
1957	First satellite launched.
1968	Carterfone decision.
1971	Computer Inquiry I.
1981	Computer Inquiry II.
1982	Modified Final Judgment.
1984	Divestiture.

(Note: These events are discussed in Chapter 4.)

Figure 1–9
Significant events in telecommunications history.

magnetism was known. Humans have always had a need to communicate over long distances and have found ways to do it without sophisticated tools.

Invention of the Telegraph

The history of modern telecommunications as we think of it really begins with Samuel F. B. Morse, who invented the telegraph and first demonstrated it on September 2, 1837. In 1845, Morse formed a company with private money to exploit the telegraph, and his idea caught on. By 1851, there were 50 telegraph companies in the United States, and the invention was beginning to be used by railroads, newspapers, and the

A re-creation of Alexander Graham Bell's original laboratory. (Courtesy of AT&T Bell Laboratories)

government! In 1856, the Western Union Telegraph Company was formed. By 1866, it was the largest communication company in the United States and had absorbed all of the other telegraph companies.

Invention of the Telephone

On March 7, 1876, a patent entitled Improvements in Telegraphy was issued to Alexander Graham Bell. This patent, which did not mention the word *telephone*, discussed only a method for the electrical transmission of "vocal or other sounds." At the time Bell's patent was issued, he did not have a working model of the telephone—only plans and drawings—but within a week of the patent's being issued, he and his assistant, Thomas Watson, got the telephone to work in their laboratory. In July of 1877, one of Bell's financial backers, Gardinar Hubbard, created the Bell Telephone Company. By the fall of that year, there were approximately 600 telephone subscribers.

first telephone company

This early telephone "system" was a little different than we are used to today. Telephone switching equipment, which could connect any telephone to any other telephone, had not yet been invented. Telephone

The first successful telephone call in 1876. The first complete sentence to be transmitted electrically was, "Mr. Watson, come here, I want to see you!" (Courtesy of AT&T Bell Laboratories)

subscribers in 1877 had one pair of wires coming into their home or business for each telephone they wanted to connect to! Since there were 600 subscribers, it is possible, though not likely, that one individual could have had 600 pairs of wires coming into the home and then had to find the right wires and connect them to the telephone before making the call! There were also no ringers or bells on the telephone. If the other party didn't just happen to pick up the telephone when the caller wanted to talk, there was no conversation!

The technology advanced quickly. By January of 1878, the first telephone exchange with an operator was in place. With that innovation, it was possible to have just one pair of wires connecting a telephone to the central switchboard. The operator, with a series of plugs and jacks, could make the connection from one telephone to any other telephone that was connected to that switchboard.

Later in 1878, Western Union Telegraph Company set up its own telephone company and took advantage of its network of telegraph wires, which were beginning to blanket the country. Bell sued Western Union for patent infringement. After studying the situation, the Western Union attorney became convinced that Bell would win the suit. Western Union settled out of court, got out of the telephone business, and sold its network to the Bell Telephone Company.

In 1885, American Telephone and Telegraph Company (AT&T) was formed to build and operate long distance telephone lines. These lines interconnected the regional telephone companies that had established franchises with Bell to provide telephone service in various parts of the country.

competition begins

In 1893 and 1894, the original Bell patents for the telephone expired. Many independent telephone companies entered the market and started providing telephone service. For the first time there was substantial competition in the telephone business as these independent telephone companies competed with the regional telephone companies that were franchised by Bell.

Bell System

In 1911, the regional companies with Bell franchises were reorganized into larger organizations, which became known as the Bell associated companies. This marked the beginning of the Bell System, as the collection of companies, headed by AT&T, came to be known.

computers, communications, cables, and satellites

Since the structure of the telecommunications industry and its basic capabilities were established in the early 1900s, technological advances have allowed new capability, new services, and reduced cost almost continually. The marriage of computers and communications in the early 1940s was a major milestone that had synergistic effects on both technologies as they developed. The development of the submersible amplifier/repeater made undersea telephone cables possible and greatly expanded long distance calling capabilities. The launch of the first satellite in 1957 opened up a totally new type of communication capability.

transistors and integrated circuits

The invention of the transistor in 1947 has undoubtedly had the most profound impact of all of the technological advancements to date. The transistor—and subsequently integrated circuits made up of thousands of transistors—made it possible to develop miniaturized devices that operated with lower power requirements than their predecessors. Without this technology, communication and computer capabilities would be nothing like we have today. We would long ago have reached the limits of the physical size, required power, and reliability of devices that could be built with vacuum tubes.

Regulatory actions have been another major area of activity in the communications industry during this century. The regulatory posture was first oriented to protecting the fledgling telecommunications industry and allowing a nationwide compatible network to get on its feet. Recently, the movement has been toward deregulation in order to encourage competition, innovation, and a lowering of costs. Since we will be studying the events and impact of regulation extensively in Chapter 4, they are not included in the present discussion.

Telecommunications and the Computer

It is interesting to note that the connection between telecommunications and computers first occurred in 1941, before computers had even emerged from development laboratories. In that year, a message recorded in telegraph code on punched paper tape was converted to a code used to represent the message data on punched cards to be read into a computer. Although it was certainly not a sophisticated process, this process demonstrated a way that these two technologies could work together.

THE CHALLENGE OF STAYING CURRENT

One of the things that make telecommunications such an interesting field is that change is occurring rapidly. The rapid technological change is very similar to the type of change that has occurred in the computer field. The basic building blocks, the electronic chips and circuits, are very similar. However, in the telecommunications field, we also have experienced the advent of satellites and fiberoptic technology, both of which present new alternatives for connecting pieces of telecommunications equipment to one another.

The other dimension of change impacting telecommunications is the legislative and regulatory process. The movement of the industry from a monopolistic to a competitive structure has had a profound impact on the companies within the industry.

Both the rapid technological and regulatory changes are making it possible, practical, and economical to use telecommunications techniques in innovative ways. The good news is that this change keeps work interesting and exciting for people in the field. The bad news is that it is very difficult to understand all the new developments and to be current with what's happening.

Serious telecommunications students and professionals must constantly educate and reeducate themselves about products, technology, and trends that are appropriate and relevant for their companies. Fortunately, there is no limit to the number of opportunities to do so. Seminars are taught frequently in major cities, numerous telecommunications trade magazines exist, and specialized books on a wide variety of telecommunications topics are available.

This book distinguishes between the basic telecommunications principles and concepts that are relatively constant and stable and current products or implementations that change rapidly. Understand, however, that if you are going to stay current in the telecommunications field, you must be prepared to continue your study.

SUMMARY

The focus of this book is telecommunications in a business environment. Telecommunications is important to study because it is used widely and is an important part of our lives at work and in the home. The technologies on which telecommunications is based are changing rapidly.

Advances in microelectronics and transmission media, such as fiber-optics and satellite communication, exemplify this change. Telecommunications is changing in another dimension—the regulatory one. Recently, the movement has been toward deregulation. Today the United States has a mixed environment, with part of the telecommunications industry being regulated and part unregulated. Like any discipline, telecommunications has its own terminology to be learned. We have begun building our telecommunications vocabulary in this chapter and will continue to build on it throughout the book.

Review Questions

1. Define telecommunications. What are the advantages of using a telecommunications system?
2. Distinguish between communications and telecommunications.
3. Why is it important to know something about telecommunications?
4. Why was the invention of the transistor important to the progress of telecommunications technology?
5. Explain the terms:
   source
   medium
   sink
   private line
   switched line
   VDT
   ATM
6. Describe the human protocol of making a telephone call.
7. Discuss reasons why telecommunications is becoming increasingly important to business.
8. In what ways can telecommunications education be obtained?
9. List some of the careers that are available in the telecommunications industry. Which ones do you think might be of interest to you? Why?
10. Briefly describe the history of the telephone industry from its inception through the early 1900s. 1984
11. What company passed up the opportunity to own the basic patents on the telephone?
12. In what situations is it necessary to stay current with developments in telecommunications? Why is it difficult to stay current?

Problems and Projects

1. Telecommunications is one way of making information available at the right place at the right time. Identify other ways that information can be transported rapidly without using telecommunications.
2. List the basic business transactions that are fundamental to the operation of a retail store. Which ones are candidates for telecommunications?
3. The chapter described reading water meters automatically via telecommunications lines. Can you think of any ways in which the readings could be tampered with so that the automatic reading is incorrect?
4. Survey your school (or, if you are working, your company) and find as many examples as you can of telecommunications in use.
5. Visit a bank in your community and find out how the ATMs work. Are they intelligent, or do they completely rely on a computer? How often do they fail? Has the bank reduced its costs as a result of installing ATMs? Are the ATMs connected into a network with other banks? How many of the ATMs are located at locations other than banks?
6. Visit a travel agency in your community and find out about the type of terminal and data communication equipment used there for checking airline schedules and making reservations. Are they tied into a telecommunications network? Which one? Has the use of the terminal reduced the agency's telephone usage? How do agents conduct business when the terminal fails and is inoperable?
7. Describe how airline reservations would be made without the use of computers and telecommunications systems.

Vocabulary

communications
telecommunications
source
medium
sink
protocol
basic business transaction
video display terminal (VDT)
private line
leased line
dial-up line
switched line
automatic teller machine (ATM)
intelligent terminal
dumb terminal
Bell System

References

Blyth, W. John, and Mary M. Blyth. *Telecommunications: Concepts, Development, and Management.* Indianapolis, IN: The Bobbs-Merrill Company, Inc., 1985.

Brooks, John. *Telephone: The First Hundred Years.* New York: Harper & Row Publishers, 1975.

Fitzgerald, Jerry. *Business Data Communications,* 2nd ed. New York: John Wiley & Sons, 1988.

Friend, George E., John L. Fike, H. Charles Baker, and John C. Bellamy. *Understanding Data Communications.* Dallas: Texas Instruments, 1984.

Kleinfield, Sonny. *The Biggest Company on Earth: A Profile of AT&T.* New York: Holt, Rinehart & Winston, 1981.

———. "Meters Are Dialed Up in Florida," *Communications News,* June 1989, p. 13.

Naisbitt, John. *Megatrends: Ten New Directions Transforming Our Lives.* New York: Warner Books, Inc., 1982.

Reynolds, George W. *Introduction to Business Telecommunications.* Columbus, OH: Charles E. Merrill Publishing Company, 1984.

Stamper, David A. *Business Data Communications,* 2nd ed. Menlo Park, CA: The Benjamin/Cummings Publishing Company, 1989.

CHAPTER

Telecommunications Applications

How the Enterprise Uses Telecommunications

OBJECTIVES

After you complete your study of this chapter, you should be able to

- categorize telecommunications applications in several ways;
- describe in nontechnical terms how a telephone call is completed;
- explain how data applications evolved from early message switching applications;
- compare and contrast electronic mail to older administrative message systems;
- describe several telecommunications applications, such as an airline reservation system and a supermarket checkout system;
- discuss the special factors to consider when telecommunications applications are planned;
- explain the importance of planning for telecommunications system outages.

INTRODUCTION

This chapter looks at some practical uses (applications) of telecommunications in the business environment. The purposes of this chapter are

1. to give you a better feeling for the importance of telecommunications to business;
2. to give you some specific knowledge of telecommunications applications for later reference as you learn the technical details later in the book;
3. to introduce some additional words and concepts and further build your telecommunications vocabulary;
4. to lay a practical foundation for the technical material in later chapters.

CATEGORIES OF APPLICATIONS

Human-Machine Interaction

Telecommunications applications can be categorized in many ways, as shown in Figure 2–1. One way is according to how people and machines interact with each other in the process of communicating. A "person-to-person" communication or application occurs when one person com-

Figure 2–1
Several ways to categorize telecommunications applications.

Human–Machine Interaction
- Person-to-Person
- Person-to-Machine
- Machine-to-Person
- Machine-to-Machine

Type of Information
- Voice
- Structured Data
- Unstructured Data
- Image

Timeliness
- Online Realtime
- Store-and-Forward
- Batch

municates directly with another. Machines may be used in the middle of the conversation, but their presence is transparent to the people. This type of communication is typified by the standard telephone conversation. The machine in this example is the telephone company's central office computer. Its role is transparent to the two people conversing.

person-to-person

person-to-machine
machine-to-person

Person-to-machine and machine-to-person applications usually go hand in hand. They are typified by the user of a data terminal who carries on a dialogue with a computer. First the person, through the terminal, sends a message or command to the computer. Then the computer sends a response to the user. This interchange normally continues until the user gets the result. Examples of this application are the airline reservation agent checking seat inventory on the computer, an ATM terminal user making a deposit or withdrawal, and a warehouse employee determining the location of certain merchandise in the warehouse.

machine-to-machine

A third type of communication, the machine-to-machine interaction, occurs when one machine automatically communicates with another, without human intervention. It is typified by the automated instrumentation found in a modern chemical or manufacturing plant. Intelligent, microprocessor-based instruments gather data about manufacturing processes and relay the data via a telecommunications line to a control computer. Here the data is analyzed, and any appropriate action is taken. No person is directly involved in the communication between the instrument and the control computer, although results of data collected over time or from several instruments may be analyzed and displayed for operator interpretation. Companies such as Monsanto and Ford have thousands of these communicating instruments installed in their plants around the world.

Type of Information

Another way to categorize telecommunications applications is by the type of information carried. The telephone call is the simplest example of a "voice" application. "Data" applications are often grouped according to whether the data is "structured" into records and fields in typical data processing fashion or "unstructured" text such as is found in word processing or text manipulation. Applications that transmit "images" can be divided into those that transmit static images, such as facsimile and freeze-frame television, or dynamic images, such as normal, full-motion television.

Timeliness

A third way to categorize telecommunications applications is by the timeliness of the transmission and reception. *Online, realtime* applications, such as traditional computer timesharing or inquiry-response, require that the data or information be delivered virtually instantaneously. Airline reservation systems certainly fit in this category.

online, realtime

Store-and-forward applications are those where input is transmitted, usually to a computer, where it is stored, and then later delivered to the recipient. Many electronic mail applications (Western Union's Easylink and Tymnet's OnTyme) work this way.

store and forward

Batch applications are usually thought of as computer applications that don't require telecommunications. In some cases this is true, but in many batch applications, the data is collected via a telecommunications system before it is processed by the batch computer programs. One method of collecting the data is to have it entered through VDTs or other terminals, one record at a time. Another method is to have the data collected into a batch and then transmitted to the computer as a unit. This method is called *remote batch* or *remote job entry (RJE)*. The name comes from the fact that frequently the data is prefaced by control cards that instruct the computer to execute a certain program or job as soon as the data is received. In this type of application, the results of the computer processing are frequently returned to the terminal that submitted the data. When the terminal is located off-site from the computer, it is most probably connected to the computer by a communication line. Hence, even batch applications may use telecommunications.

batch

With virtually any method of categorization, there is overlap. Many applications have elements from several of the categories. For example, the airline reservation system discussed in Chapter 1 has an element of person-to-person communication between the person wanting to travel and the airline reservation agent. There is also an element of person-to-machine application between the reservation agent and the computer. Both of those elements occur in realtime. Since the data terminal is connected to the computer, it is *online*. If the reservation agent instructs the computer to print a ticket for the traveler at the airport on the day of departure, the computer stores the data until the day of departure, when it is forwarded to the airport and printed on a ticket form.

TELEPHONE COMMUNICATIONS

Let us now look again at the telephone application briefly described in Chapter 1. This time we will look at what goes on behind the scenes when the call is placed.

Figure 2–2 shows the equipment and the facilities involved when a call is made. The connection between the telephone in a residence or business and the local telephone company's central office is called the *local loop*. The building that houses the telephone equipment for a specific geographical area is called the *central office*. When a caller lifts the telephone handset from the cradle, an electrical signal is sent to the central office signaling it that the person is placing a call. The central office sends the dial tone back on the local loop to the telephone. Assuming things are working normally, the person hears the tone by the time the telephone handset reaches his or her ear.

When the caller presses the buttons to dial a number (or the dial is turned on a rotary-dial telephone), the number is stored in the telephone switching equipment at the central office. The switching equipment usually is a computer but in older offices may still be an electromechanical device. In either case it is called the *central office switch*. The first three digits of the telephone number determine whether the call is long distance or local. If it is local, the switch determines whether it can complete the call itself or needs to forward it to another nearby central office that handles that telephone number. If the switch can complete the call (because the number being called is also handled through the same central office), the switch sends an electrical signal to the receiving telephone, causing the bell to ring. At the same time, a ringing signal

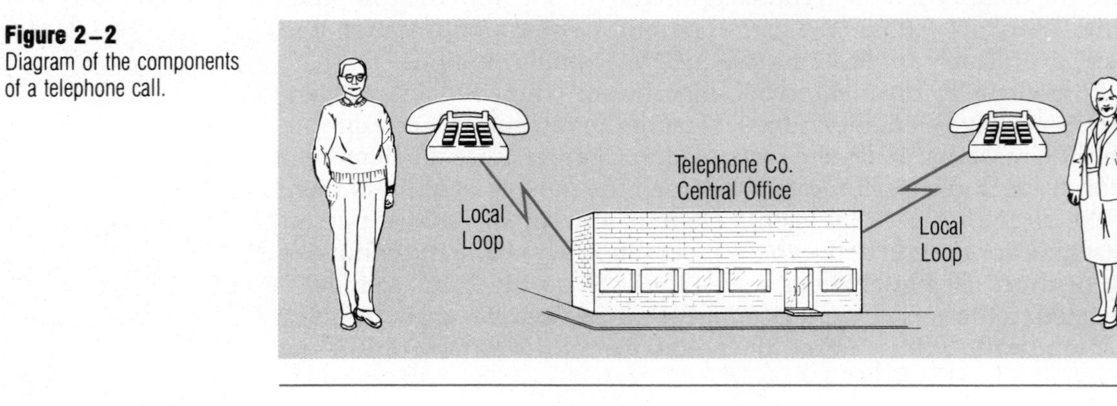

Figure 2–2
Diagram of the components of a telephone call.

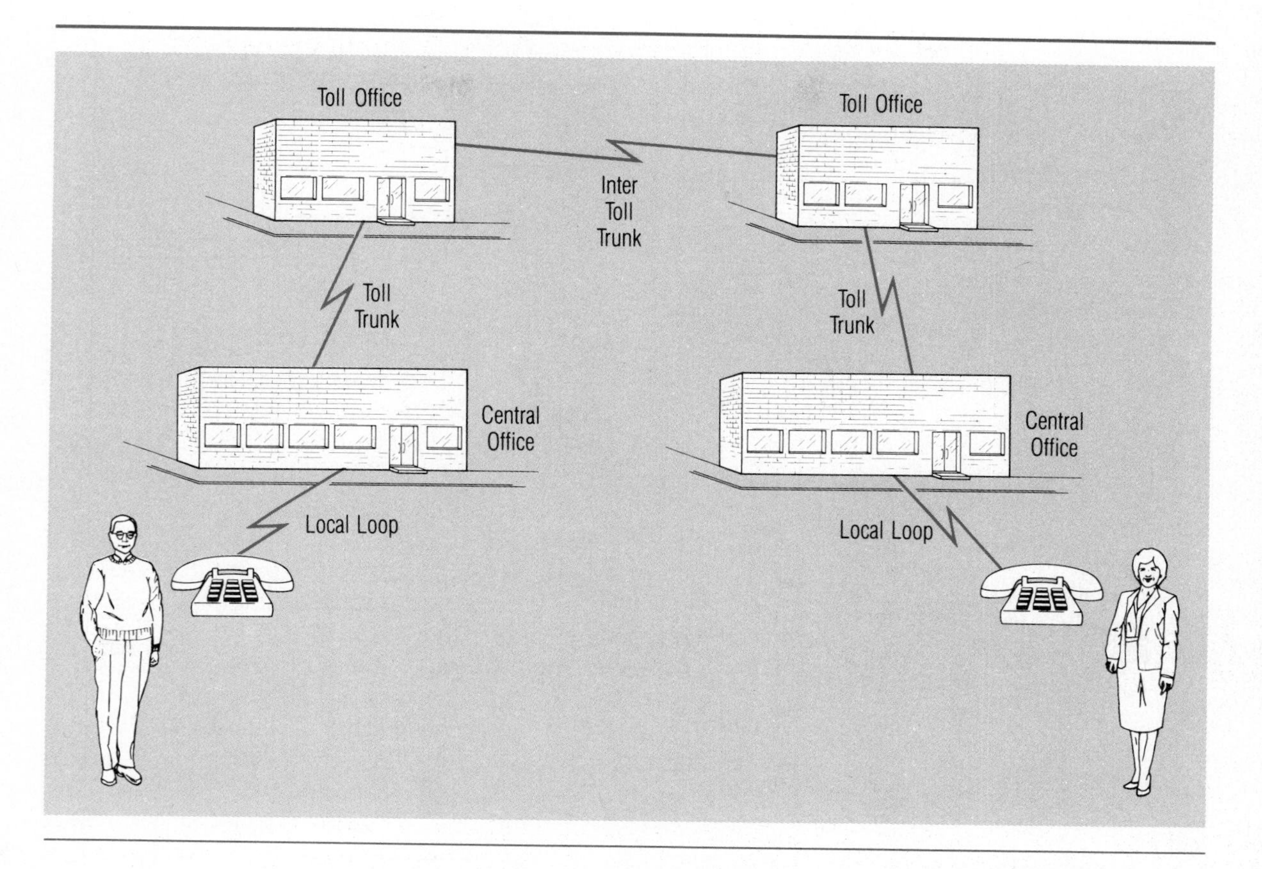

Figure 2–3
Diagram of a telephone call involving toll trunks and toll offices.

is sent back to the caller to indicate that the phone being dialed is ringing. The ringing signal that the caller hears is not, however, directly connected to the ring at the called telephone, nor is the timing the same. That is why occasionally when you make a call, the telephone sometimes seems to be answered before you hear ringing start.

If the caller's central office determines from the first three digits of the number that this is a long distance call, it passes the call on a communication line called a *toll trunk* to another telephone company central office called the *toll office,* as shown in Figure 2–3. From the toll office the call may be passed directly to the central office that handles the receiving telephone or to another toll office. Each toll office examines the telephone number to determine whether it will pass the call to a central office or to the next toll office on the route to the ultimate destination. Ultimately, a toll office passes the call to the central office that is connected to the receiving telephone, and the call is complete. It is impressive to realize that all of these "decisions" take place in a matter of seconds.

Figure 2–4 shows how the toll offices are interconnected into a network, with each office connected to several others. This provides

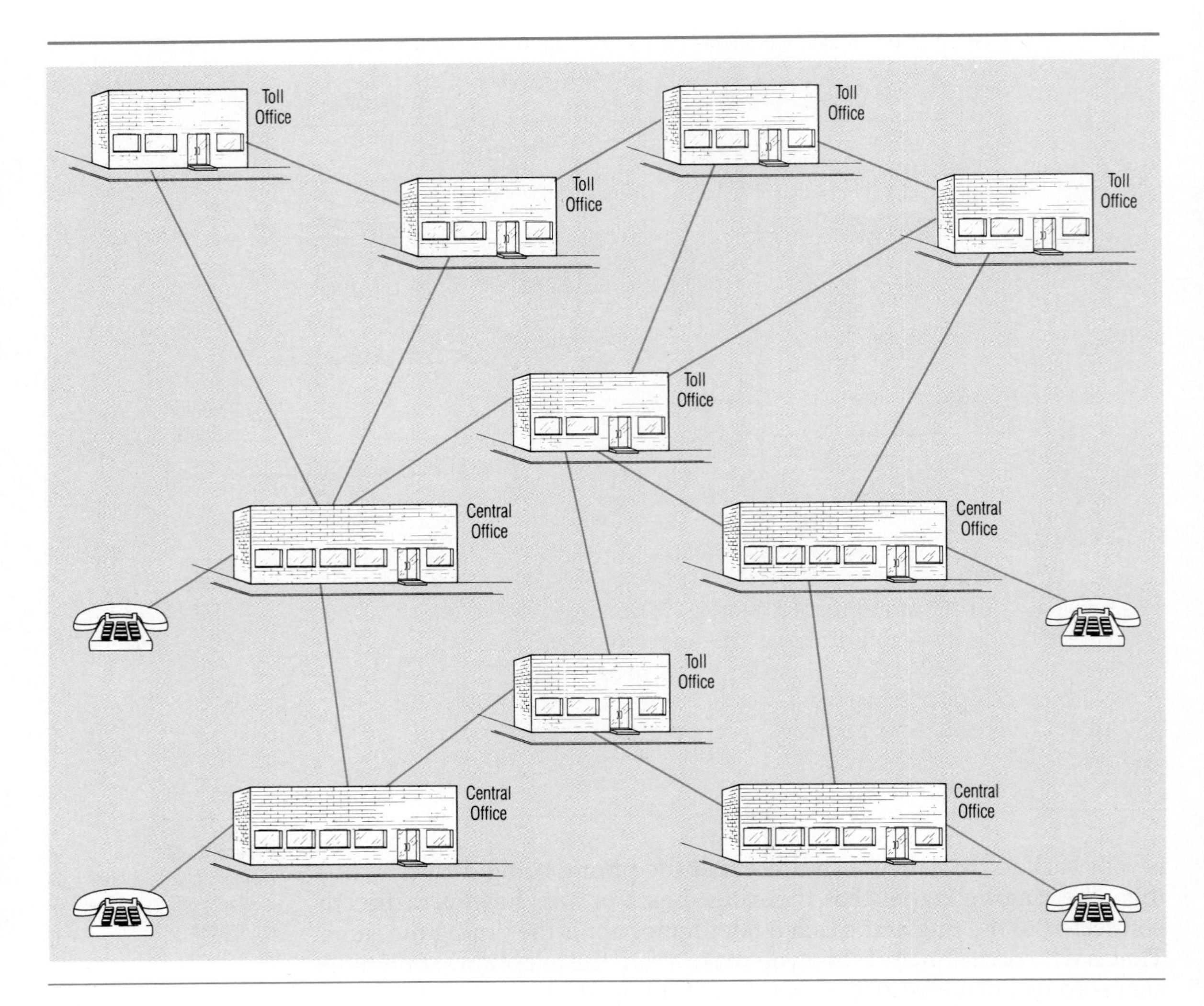

Figure 2–4
A network of toll offices.

multiple paths over which a telephone call could be completed. When a long distance telephone call is placed, it may or may not be routed over the most direct path to the destination. If two calls are placed to the same destination simultaneously, each might be routed over a different path. This redundancy is what makes the telephone system in the United States so reliable and able to withstand the failure of central offices, switches, and lines.

Although the telephone system is generally viewed as a person-to-person application of telecommunications, the switches in most central offices are specialized computers designed for processing and switching voice communications. In the last few years, network and switch designers have begun to use techniques that are more suited for carrying data.

However, most of the telephone systems in the United States today are of the older style, designed for voice communication.

ADVANCED TELEPHONE FEATURES

So far we've explored what is known as *plain old telephone service (POTS)*. In recent years, many advanced features have become available to consumers as the central office switches have become more sophisticated. Now, in many parts of the country, it is possible to add a third party into a call (conference call), to indicate an incoming call while you are talking to someone else (call waiting), and to put one party on hold while you talk to a second person.

In the business world, many more sophisticated capabilities frequently are available. Often businesses find that they have so many telephone calls to a particular city or area of the country that they rent special lines from the telephone company for their exclusive use. Alternately, they may acquire special services that give access to a particular area of the country at a reduced charge. When these special services are employed, it is desirable to have the telephone system in the business route the telephone call because it can be programmed to minimize the cost. Many telephone systems provide a function called *least cost routing* to meet this need. In a medium to large company, the savings that can be generated by properly routing the calls on the least expensive line can be substantial. Chapter 5 examines these and many other capabilities.

DATA APPLICATION EVOLUTION

Before we take a more detailed look at typical data applications that use telecommunications, let's look at how these applications have evolved over the years. The earliest type of telecommunications equipment (other than smoke signals and drums) that did not use voice was the telegraph. Telegraphs sent a message from one person, group, or location to another. Business use of the telegraph evolved quickly. Many companies required rapid communication between their staff at widely separated locations. Originally, companies employed telegraph operators to send the messages using Morse code. In the last 50 years, *teletypewriters* have come into widespread use. Remembering the definition of *tele* as distant, we can think of the teletypewriter as being a device on which a person types a message; it prints on a similar machine at a distant location.

teletypewriter

Of course, teletypewriters are connected to each other via telecommunications lines, and this application of telecommunications has become known as *administrative message switching*. It was one of the first appli-

An older style teletypewriter. The paper tape punching mechanism is at the center of the machine inside the glass door. The paper tape reader is on the flat surface at the left side of the machine below the two black buttons. (Courtesy of The Dow Chemical Company)

cations of telecommunications and is still the most prevalent application for international communication in many companies.

Administrative Message Switching

Figure 2–5 shows the way a company might have set up its telecommunications network for administrative message switching thirty-five years ago. The company headquarters are in Chicago, and branch offices are in Detroit, New York, and Washington. A telecommunications line connects each of the outlying locations with company headquarters. This type of connection is called a *point-to-point line*. The line runs from one point to another. In the Chicago office there would be three teletypewriter machines—one connected to each of the three lines. If a person in the Chicago office wanted to send a message to Detroit, the message would have been typed on the teletypewriter connected to the Detroit line, a message to Washington would have been typed on the teletypewriter on the Washington line, and so on. Conversely, when a person in Detroit sent a message to company headquarters, the message

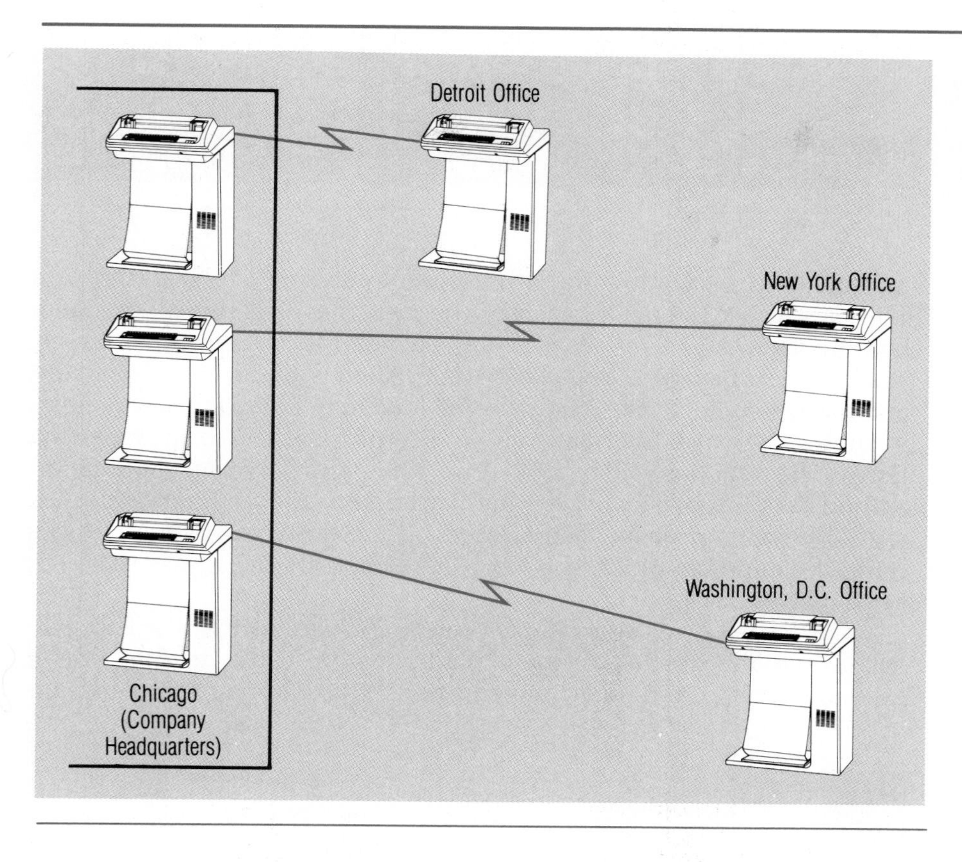

Figure 2–5
A simple network with point-to-point lines connecting offices in Detroit, New York, and Washington, D.C., with company headquarters in Chicago.

would have printed in the Chicago office on the teletypewriter connected to the Detroit line.

If someone in Detroit wanted to send a message to someone in Washington, the message would have been typed in Detroit, received in Chicago, and printed. The printed copy of the message would then have been taken to the machine connected to the Washington line, retyped, and then sent to Washington, where it would have printed. Time consuming? Yes. Labor intensive? Yes, yet at the time, this type of system was faster and more effective than any message communication system that had existed before.

The first improvement to such a system was to install an additional piece of hardware on the teletypewriters that could punch and read paper tape. Figure 2–6 shows a typical early punched paper tape. Each column of holes across the tape represents one letter, number, or special character. The advantage of the paper tape was that incoming messages from the outlying offices could be simultaneously printed on the teletypewriter and punched into the paper tape. If the message was to be forwarded to another location, the paper tape could be torn off one

Figure 2–6
Five-level, Baudot coded paper tape from a teletypewriter.

machine and read into another machine to be sent to the destination. This eliminated the need to retype the message. This type of operation became known as a *torn tape message* system. In large companies, the *message center* became a large noisy room filled with dozens of machines and many people. Racks were often needed to hold the tapes containing incoming messages that had not yet been forwarded to their destinations. The whole system took on a store-and-forward connotation. Although outmoded by today's standards, torn tape administrative message switching systems vastly improved the efficiency of the message center by eliminating the need to retype all of the messages that were to be relayed.

Getting back to our example, Figure 2–7 shows that the company has grown and opened a new office in Boston. If the network was to expand as it had in the past, an additional point-to-point telecommun-

An early message switching center in a large industrial company. Short sections of punched paper tape can be seen in the rack in the foreground. Storage bins of punched tape are on the back wall. (Courtesy of The Dow Chemical Company)

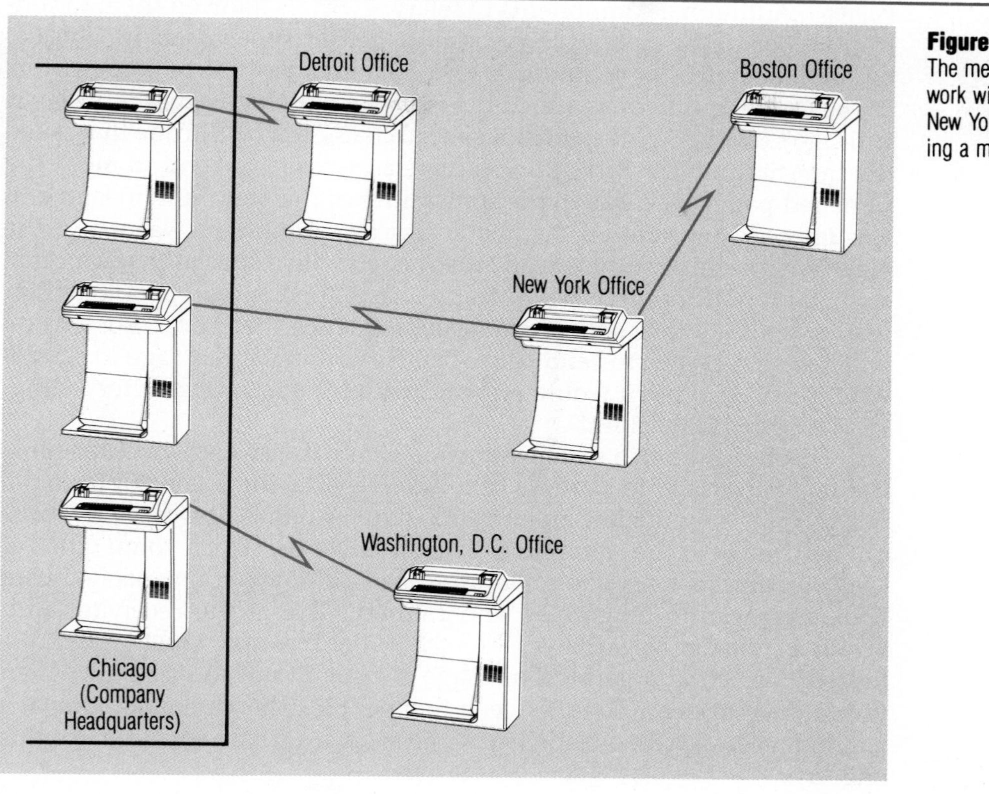

Figure 2–7
The message switching network with the offices in New York and Boston sharing a multipoint line.

ications line would be added between Chicago and Boston. However, since Boston is close to New York, it is reasonable to consider running a new line from New York to Boston and having the two locations share the line between New York and Chicago. Indeed, this was the economical thing to do, particularly if the line between Chicago and New York was not always busy and could handle the additional messages to and from the Boston office. This new type of line, which is shared between several terminals, is called a *multi-point line.*

Complications arise, however, when multipoint lines are used. On a point-to-point line using teletypewriters, when a message is typed at one end, it is printed at the other end as it is being typed. For example, when a message is typed on a machine in Washington, it is simultaneously printed and/or punched on the machine in Chicago. But what happens when a message is typed in Chicago destined for the Boston office? Is the message also printed in New York? For some messages this might be okay, but clearly for many messages it is desirable that they are only received and printed at the intended destination. To meet the need, teletypewriters had to have additional sophistication.

Polling Additional components in teletypewriters allowed them to control the use of the line. In our example, the teletypewriters in Chicago became the *control* or *master station* on each line, and the teletypewriters in the outlying offices assumed the role of *subordinate,* or *slave, stations.* Chicago's teletypewriter sent out special characters on the line that asked the question, "New York, do you have a message to send to me?" If a punched paper tape was in the transmitter of the New York machine, it was immediately sent on the line to Chicago. If no tape was ready, the New York machine automatically responded with a special character that said, "No, nothing right now." In that case, the Chicago teletypewriter would send another special character sequence that said, "Boston, do you have a message to send to me?" If Boston had a message to send it would send it; if not it would respond with the special character saying, "No, nothing right now."

This technique, in which a control terminal asks each slave terminal if it has a message to send, is called *polling.* Having a control terminal poll terminals on the line ensures that two terminals do not try to transmit a message at the same time. This occurrence, which some types of telecommunications systems allow, is called a *collision.* It invariably causes both messages to be garbled and unintelligible at the receiving end. When a collision or garbling is detected, the message must be retransmitted. Detecting a garbled message is usually not too difficult when textual messages are sent between two people. The person looking at a garbled message can usually figure out what a garbled message like this one says:

> I'm cxming tw Lrs Anceles on Moy 17sh. Arriving Nzrtqwest, flijt 762. P-eabe pick me up ut thx azrport.

Obviously, the problem is more difficult when the message contains numeric data. One hopes that the digits in the date and flight number in the above message were received correctly, but asking the sender to retransmit the message would be prudent.

The primary disadvantage of a system that uses polling is the additional number of characters that have to be transmitted (the polling characters) on the line. Another disadvantage is that the slave terminals must wait until they are polled before they send data. If there are a large number of terminals on the line, the delay can be long.

Addressing A complementary line control technique that evolved at the same time as polling is called *addressing.* When the control terminal has a message to send to one of the subordinate terminals, it first sends special control characters, called *addressing characters,* on the line. The addressing characters are recognized only by the terminal that has the specified address. To that terminal the addressing characters say, in effect, "Slave terminal, I have a message for you. Are you ready to receive it?" The addressing characters also tell all other terminals on the line that the following message is not for them and that they should not print it. If the slave terminal is ready, it responds with a character that says, "Ready to receive," and the control terminal immediately sends the message. After the message is sent, the control terminal either addresses another terminal (if it has more messages to send) or resumes polling the slave terminals to solicit messages from them.

Addressing has the added benefit of preventing the transmission of messages to a terminal that is not turned on, is out of paper, or is otherwise not ready. If a message is sent to a terminal that is not ready to receive, the message usually is lost. Assuming there is a way of detecting when messages have been lost and that the originals have been saved by the sending terminal, they would have to be retransmitted later. If the slave terminal is not ready when it is addressed, it simply does not respond, and the control terminal knows not to send the message.

The control station *polls* the *subordinate* stations, asking them if they have messages to *send*.
The control station *addresses* a *subordinate* station, asking if it is ready to *receive* a message.

The complementary techniques of polling and addressing are a simple form of line *protocol,* rules under which the line operates. Most telecommunications systems today use some sort of protocol to control the lines; various protocols are examined in detail in Chapter 10.

protocol

Public Messaging Services

As teletypewriter usage expanded, competing public networks were developed to provide teletypewriter message service between subscribers. Internationally, the de facto standard became the *telex* network, whereas in the United States, AT&T created the *teletypewriter exchange (TWX)* network. (TWX was later sold to Western Union Corp.) Operating much like the public telephone system, these networks carry messages from attached teletypewriters. A subscriber pays a monthly fee to be connected to the network and either rents or purchases a teletypewriter. After that, the cost of the service is based on the number of messages, length of transmission, and distance over which messages are sent.

The telex and TWX networks use different standards for transmission. Until the 1960s, they were totally incompatible. During that decade, the networks were attached to computers that perform the translations between the different codes and speeds used by the two networks. Now the two operate as one.

Today the telex network is operated by government agencies in most other countries and by Western Union in the United States. There are approximately one million telex subscribers worldwide—about half of these in the United States. The telex network operates at the relatively

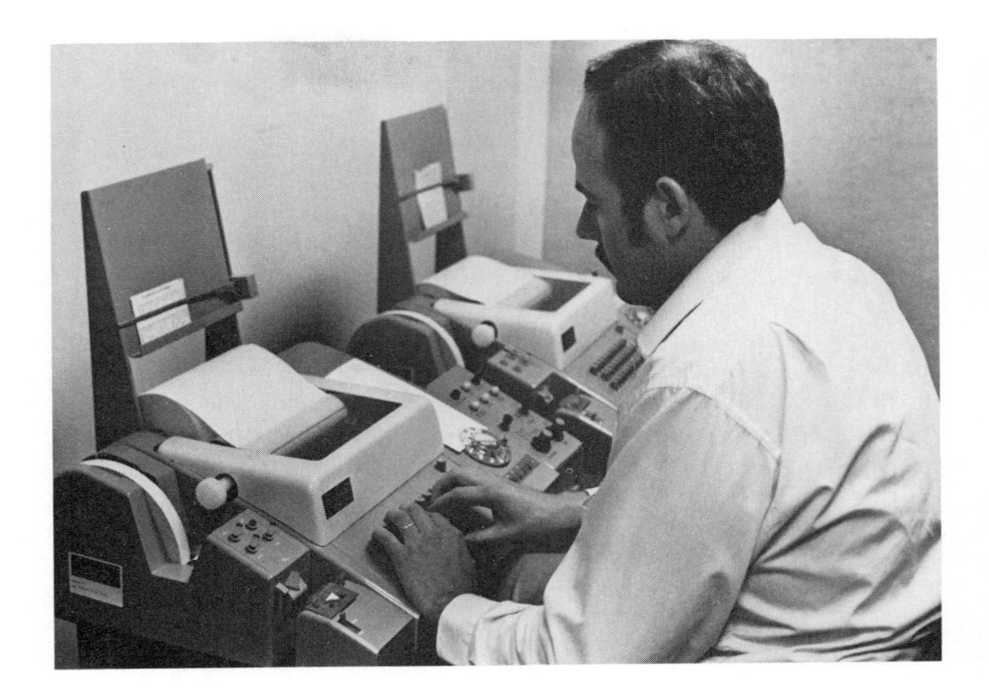

These telex machines are very much like teletypewriters in appearance and function but are connected to the public telex network.

low speed of approximately 66 words per minute, has a somewhat limited character set, and has a relatively high error rate. Nonetheless, it is still the most prevalent way of sending messages throughout the world.

The replacement technology for telex/TWX is called *teletex*. Teletex allows transmissions between word processors at speeds 48 times greater than telex. Furthermore, because of the coding used, the character set is much greater, and the transmissions are essentially error free. Teletex messages are typed and edited on word processors in standard business letter format.

teletex

Teletex was primarily developed in Europe and is used much more widely there than in the rest of the world. It is well defined by standards and thus allows machines from different manufacturers to communicate with each other. Because of the faster transmission speed, the cost of sending a teletex message is less than sending the same message by telex. Teletex remains relatively unpopular because the cost of the terminals is higher than for telex, and there are hundreds of thousands of telex terminals already installed. Teletex certainly has the potential of replacing telex if more organizations start to take advantage of its capabilities.

The Evolution to Computers

Soon after computers started to be used in business, hardware was developed that allowed telecommunications lines to be connected to the

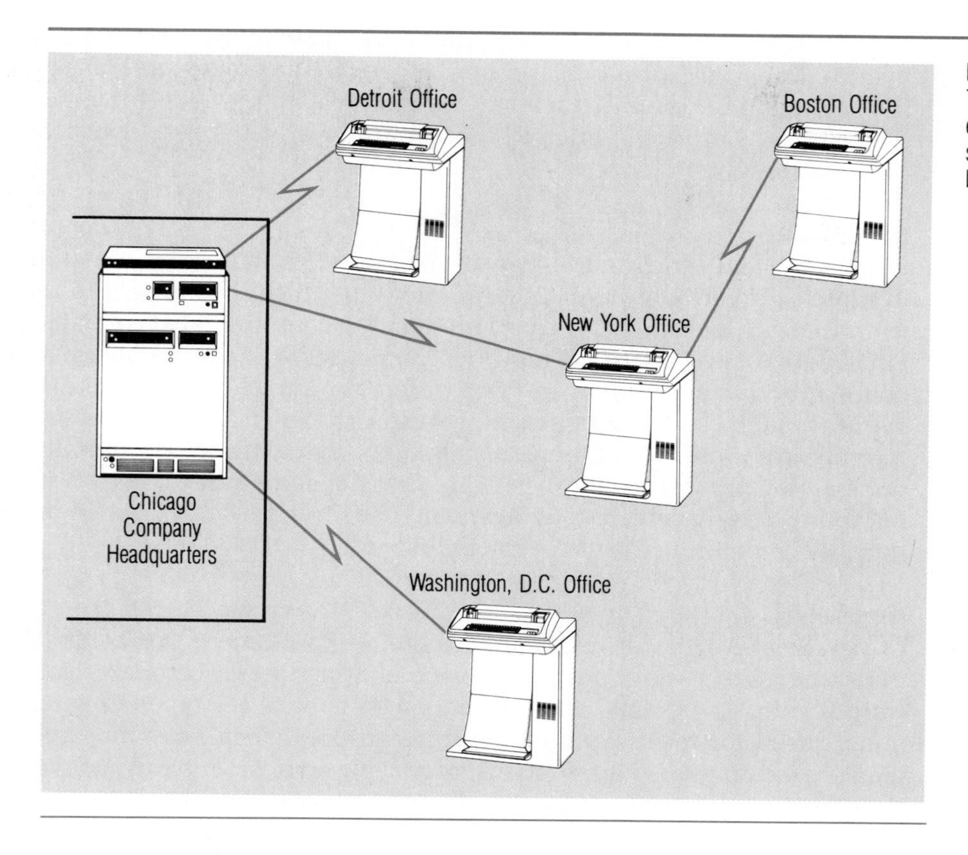

Figure 2–8
The control terminals on each line of the message switching network replaced by a computer.

computer. With special programming, data could be read from the line or sent out on the line. It became obvious that administrative message switching was an ideal application for the computer. It could perform all of the functions of the control terminal and, because of its speed, could handle many lines simultaneously. All that was needed was to connect the telecommunications lines to the computer, as illustrated in Figure 2–8.

With proper programming, the computer could read a message from one line and send it out on another line without human intervention. In addition, as computer programming grew in sophistication, computer controlled message switching systems provided added features, such as adding the time and date to each message, collecting statistics about the number of messages sent and received, logging of all messages, and storing messages for later retrieval or retransmission.

Inquiry-Response

Soon there came the realization that a message could be sent to the computer, itself (or, more specifically, to a program running in the computer). A very simple example is a coded message to the computer asking it to send back the current time. A more useful example is a coded message asking for the computer program to look for certain data in a computer file, format it, and send it back to the requestor. This was the beginning of computerized inquiry systems. Figure 2–9 shows two examples of simple inquiries to the computer and the responses the computer might generate.

File Updating

From simple inquiries and responses, it is easy to imagine the evolution to more elaborate computer programming in which data is sent to the computer, where it is checked and ultimately a computer file is updated. Finally, combining a series of inquiries, responses, and file updates gives us the foundation for many modern online computer-based processing systems, such as airline reservation, customer order entry, and inventory applications. Applications of this type, marrying telecommunications and computer technology, were first developed in the late 1950s but did not really come into widespread use until the late 1960s. Today, they are common in almost every industry.

Timesharing

In parallel with the development of online applications for business transaction processing, such as those mentioned here, computer software (control programs) was developed to control teletypewriter terminals used for relatively unstructured activities, such as writing programs, executing existing programs, or playing games. This was handled by a process commonly called *timesharing* and first came into its own in colleges and universities for educational purposes, where there were large computers that could handle multiple applications for multiple uses. One student might run a program for an economics class while another student wrote a program for a FORTRAN class, and a third student wrote a program to solve a mathematics problem.

Transaction Processing Systems

By way of contrast, *transaction processing systems* are generally characterized as those in which the users use prewritten programs to perform business transactions, generally of a somewhat repetitive nature. Typical examples are the airline reservation process, an order entry application, or an online inventory application.

similarities between processing systems

In point of fact, timesharing and transaction processing systems have more similarities than differences. Both share the time of a computer among many users. Furthermore, it is entirely possible that a university system designed primarily for student use might also be used by administrative staff to update student grade records or tuition payments—that is, for transaction processing as well. Similarly, the computer in a business that primarily handles the company's accounting, inventory, and other business transactions might also be used by the scientists in the research department or programmers in the data

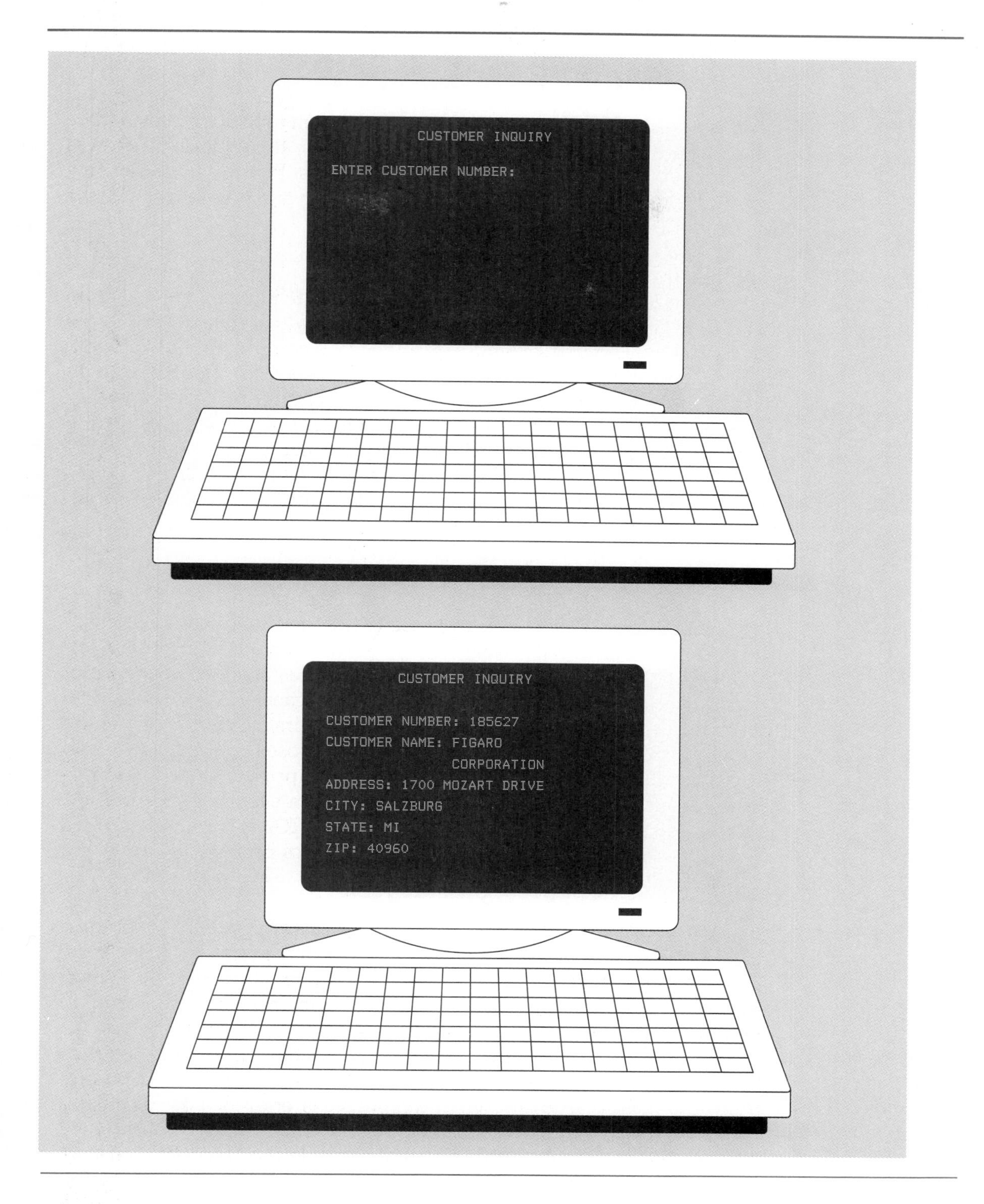

Figure 2–9
Two examples of simple inquiries to a computer system and the response from the computer.

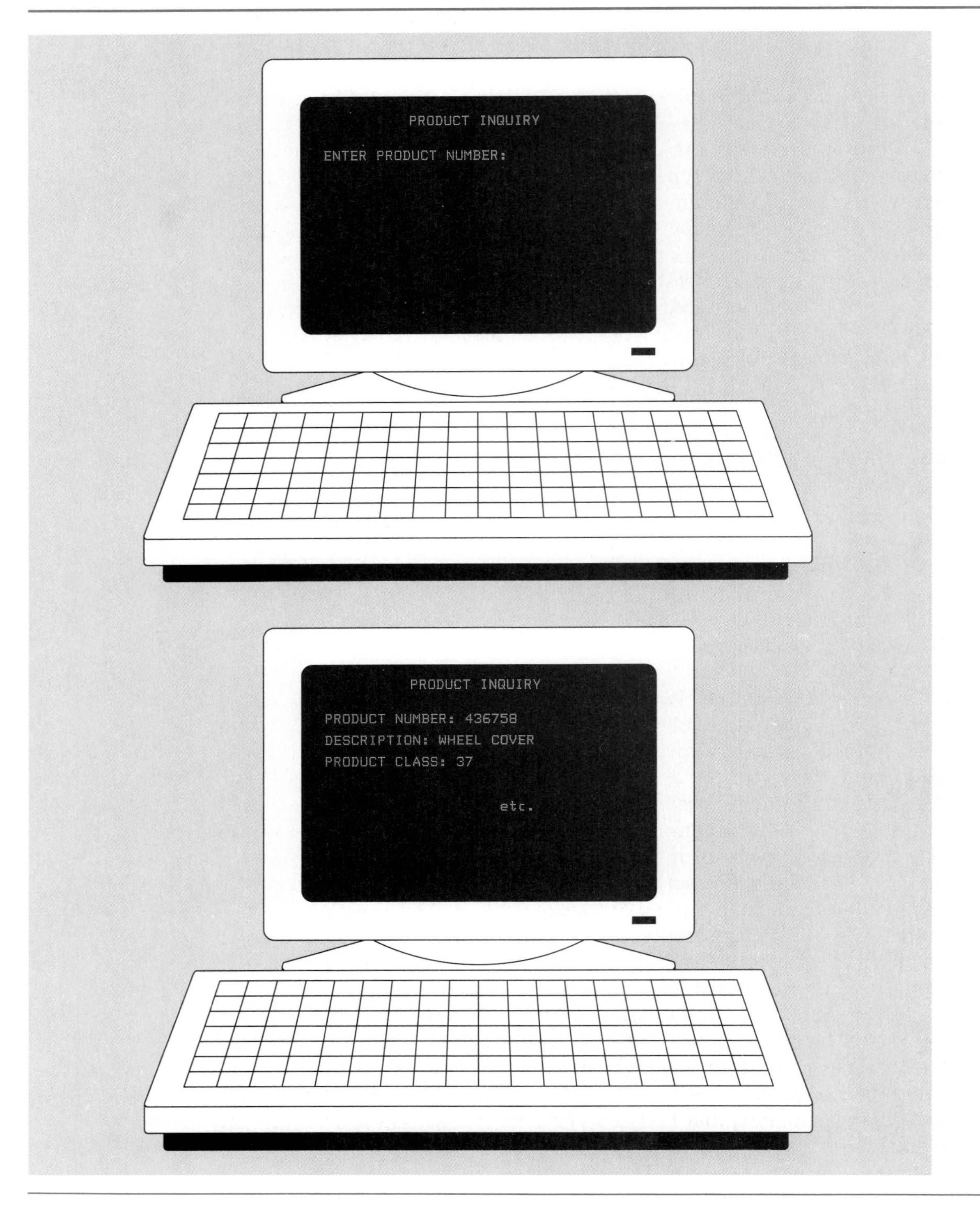

Figure 2–9
(Continued)

processing department for data analysis or programming—typical time-sharing tasks.

From the standpoint of the computer and telecommunications requirements, the two types of systems have many similar elements and a number of different requirements as well. The workload on a timesharing system is relatively less structured, less defined, and less predictable than the workload on a commercial business transaction processing system. Generally, the workload in transaction processing systems is somehow related to the business volume of the company. As airline customers buy more tickets, more reservation transactions are generated. As bank customers use ATMs more extensively, more transactions are generated.

TYPICAL APPLICATIONS OF DATA COMMUNICATIONS

Now that you've seen the differences between processing methods, you're ready to explore several applications that make extensive use of telecommunications. The applications are grouped according to the primary type of information processed.

Structured Data Applications

Airline Reservation System With your new knowledge of the way in which data applications have evolved, look again at the airline reservation and ATM applications that were presented in Chapter 1. Figure 2–10 shows an expanded view of the airline reservation application. Multiple travelers can call multiple reservation agents, each using a terminal connected to the central computer. Between the callers and reservation agents is a new piece of hardware, an *automatic call distribution (ACD) unit*. The ACD is an adjunct to the telephone system in the airline office. Its purpose is to route the next incoming call to the next available reservation agent. If all reservation agents are busy, the ACD unit can play a voice recording to the caller stating that the next available reservation agent will serve him or her as quickly as possible.

ACD units work well in any situation where multiple callers must be routed to multiple employees who handle the calls sequentially. Examples include other types of reservation systems, such as rental car companies; customer service lines at utility or manufacturing companies; and companies that have a large number of customers calling to place orders for merchandise, such as catalog sales or toll-free order centers.

The data portion of this application shows multiple terminals connected to the central computer of the airline. Also, the airline's computer is connected via a telecommunications line to the computers of other airlines. In this way, reservation agents from United Airlines can make

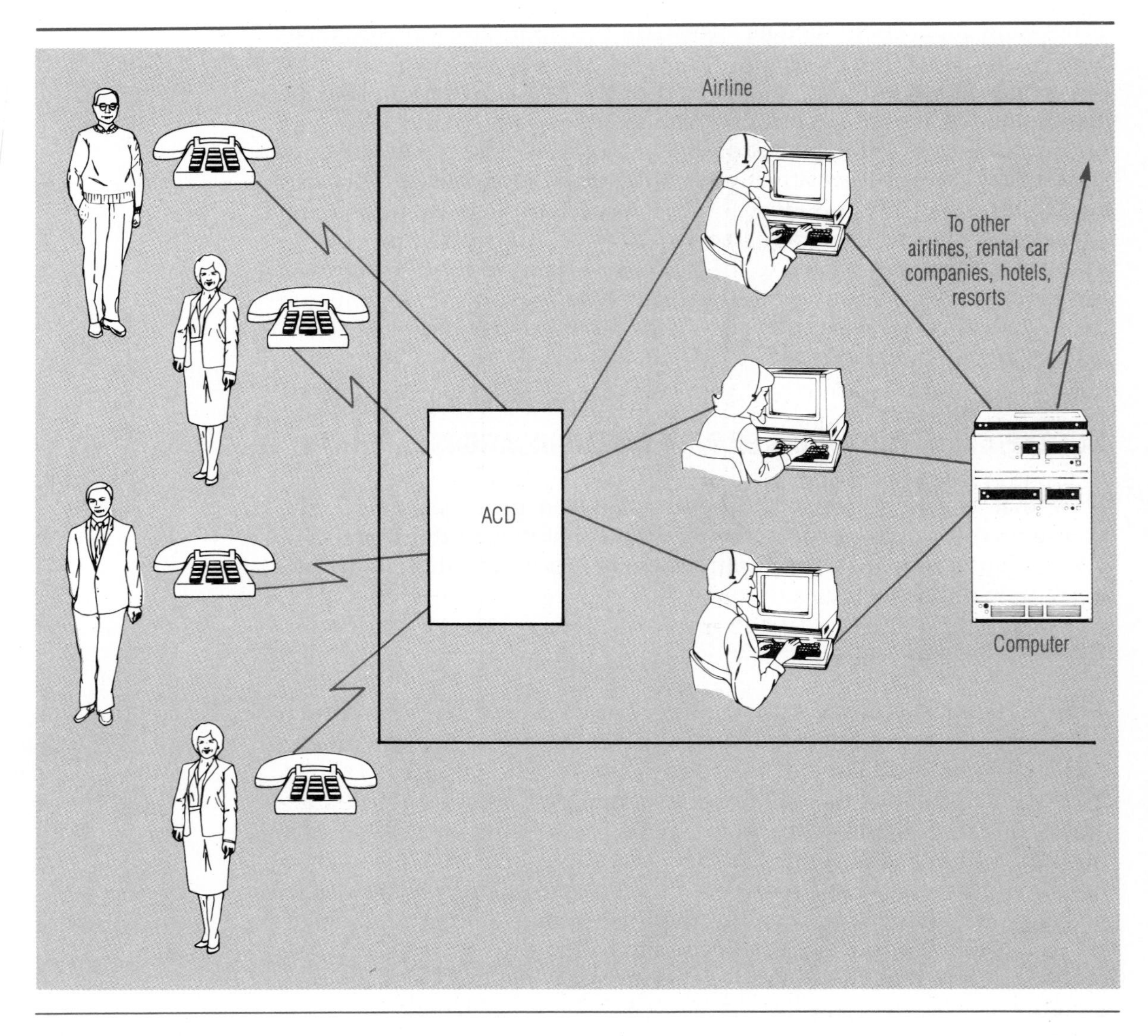

Figure 2–10
The telecommunications of an airline reservation system.

reservations on the flights of American Airlines or any other airline whose computer is connected into the network. All of the airlines in the world, except the very smallest, have interconnected their telecommunications networks and computers so that travelers can call one airline and make all of the reservations for a trip, even though several airlines may provide the transportation. Furthermore, the computers of the rental car companies and many hotels and resorts also are connected to the same network so that all of the reservations for an entire trip can be made with one telephone call.

American Airlines' Sabre system and United Airlines' Apollo reservation system have been significantly enhanced with additional computer programming. Both airlines have actively sold the right to access the systems to travel agencies and large corporations whose employees travel extensively. The services these systems offer are extensive and go far beyond making airline reservations. Hotel accommodations and theater tickets can be ordered in addition to transportation. The revenue generated by the reservations systems has become a substantial part of the total income of the airlines, sometimes exceeding the amount of revenue generated by the airplane flights themselves.

An airline reservation system is a high-volume transaction processing system that must provide fast response time to the reservation agents. Reservation systems are critical to airline operations, and the airlines spend a great deal of money to ensure that their computer system will always be available by having backup computers ready to take over instantly if a primary computer fails.

The operational statistics of airline reservation systems are amazing. In 1984, TWA had almost 12,000 communication terminals connected to its network worldwide. The typical daily volume was close to 7 million transactions. The transaction rate during peak times was normally approximately 170 transactions per second, although higher rates were encountered. The response time goal of the system was to ensure that 90 percent of the transactions were responded to in 3 seconds or less, with an average response time of 1.5 seconds.

In 1985, Delta Airlines had over 4,000 voice communication lines and 18 reservation centers throughout the United States. Delta had 66 data communication lines and handled 340 to 350 transactions per second at peak times. Over 23,000 terminals were connected to the Delta data network—many of these located at the 2,800 travel agencies that were online to Delta. The response time goal was 3 seconds or less 95 percent of the time.

In addition to the reservation system, airlines also operate other online systems for sending administrative messages, tracing baggage, scheduling airplane crews, and scheduling airplane maintenance. A typical 747 flight may involve more than 29,000 transactions, of which 27,000 concern passengers. With such systems, airlines are some of the most advanced, sophisticated users of voice and data telecommunications.

Automated Teller Machine Figure 2–11 shows an ATM network in which multiple ATMs are connected to computers and several computers are connected together. Networks interconnecting computers in several banks are common throughout the world. Typically, each branch of a bank has at least one ATM, but in addition, banks may place ATMs in local shopping centers, train stations, or even large business establishments. Banks that have historically been competitors in a local area or region are now

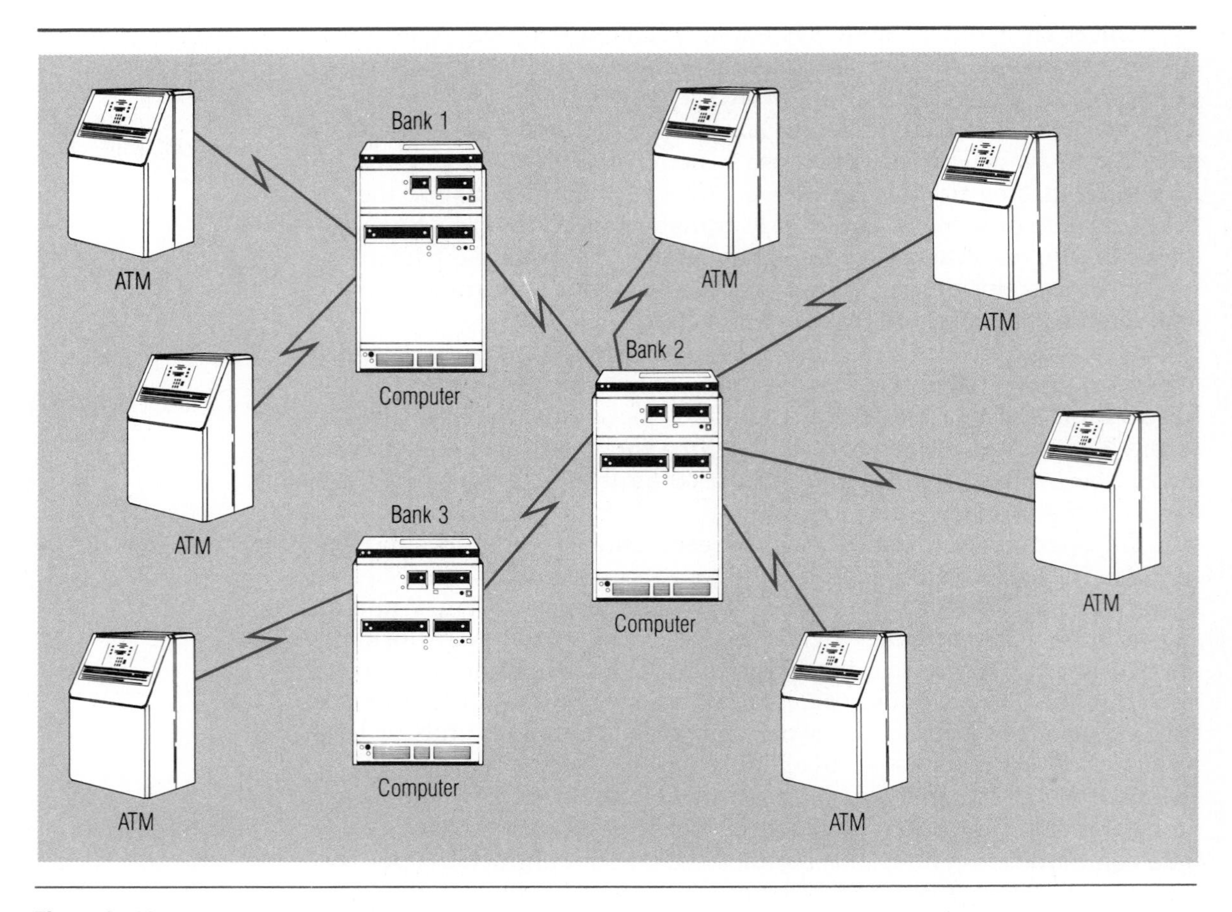

Figure 2–11
A network of ATMs from several banks.

cooperating with one another, at least to the extent of interconnecting their ATM networks.

Figure 2–11 shows that any ATM machine can make a transaction with any bank on the network. It is possible to walk up to a single ATM machine and withdraw money from an account at one bank and then put the cash back in the machine for deposit in an account at another bank.

Since ATM networks are expensive to operate, service charges may be levied for certain types of transactions to help defray the costs of the network. Customers whose transactions require access to another bank's computer or network are especially likely to be charged. In Japan, service charges also are assessed for transactions that occur outside of the bank's normal operating hours. That is, ATM transactions are free to the customer if they occur while the bank branches are open. After the branches close, the customer pays a fee for each transaction. It would seem that market competition might change this situation within a few years.

A bank customer uses an ATM. This machine can also accept bags containing the day's receipts from small businesses. (Courtesy of NCR Corporation)

Security in the ATM application is obviously very important. One concern is the physical security of the machine, since it contains hundreds of dollars in cash and checks. This problem is generally solved by keeping the money in a small vault within the machine and having the machine itself mounted in the wall of a building.

Another aspect of security is identification of the user. Bank customers are identified by issuing each user a plastic card much like a credit card. The back of the card contains a magnetic stripe on which the user's name and account number are coded. When the card is inserted in the ATM, the machine reads the characters on the magnetic stripe. If the account number is valid, the machine asks the person for his or her unique password. A computer compares the password the user enters with the stored password for that account on the computer's disk. If the two passwords match, the user is permitted to proceed with the banking transaction.

ATM applications are realtime transaction processing systems that handle a high volume of transactions. In 1986, an estimated 225 million transactions were conducted each month through 45,000 ATMs across the country. This number is remarkable, considering that the ATM technology has been available for only a little over 10 years. Although ATM operation is not yet as critical to banking as the airline reservation system is to airlines, it is obviously in the bank's best interest to keep the ATMs and network operating with high reliability, especially after hours when users can't get assistance with banking elsewhere. In addition, banks save staff costs when people use the ATM rather than a human teller. One nationwide network of ATMs experiences better than a 97 percent uptime. *Uptime* is defined as the probability that any machine on the network can contact any other machine at any time.

The number of different transactions that ATMs can perform has increased dramatically. Originally, the machines did little more than dispense cash. Now, consumers can make deposits, check their balances, get credit advances, pay bills, and transfer funds between accounts. In the next decade, it is expected that many of these systems will be expanded to new tasks, such as customer orders for checks or inquiries about loan or investment services.

Internationally, bank networks are connected together by the SWIFT network. SWIFT, an acronym that stands for the Society of Worldwide Interbank Financial Telecommunication, connects over 2,900 banks and financial institutions worldwide. The network was originally created in 1977, and in 1987 it handled over 950,000 messages. Most of these messages are interbank financial transactions or confirming messages for verification. The network is experiencing a 12 percent growth in the message traffic it handles each year and is in the midst of a complete technical overhaul to cope with the message volumes it will be required to handle in the 1990s.

Both the airline reservation and ATM applications have revolutionized the way business is done in their respective industries. In both cases, the application requires the marriage of telecommunications and computing technology. In addition, communication and cooperation between competing companies must occur. Furthermore, all of the companies must subscribe to the same standards for how transactions are processed. In addition, standards are required to allow computers in different banks, which may be from different computer vendors, to be interconnected with telecommunications. Without telecommunications,

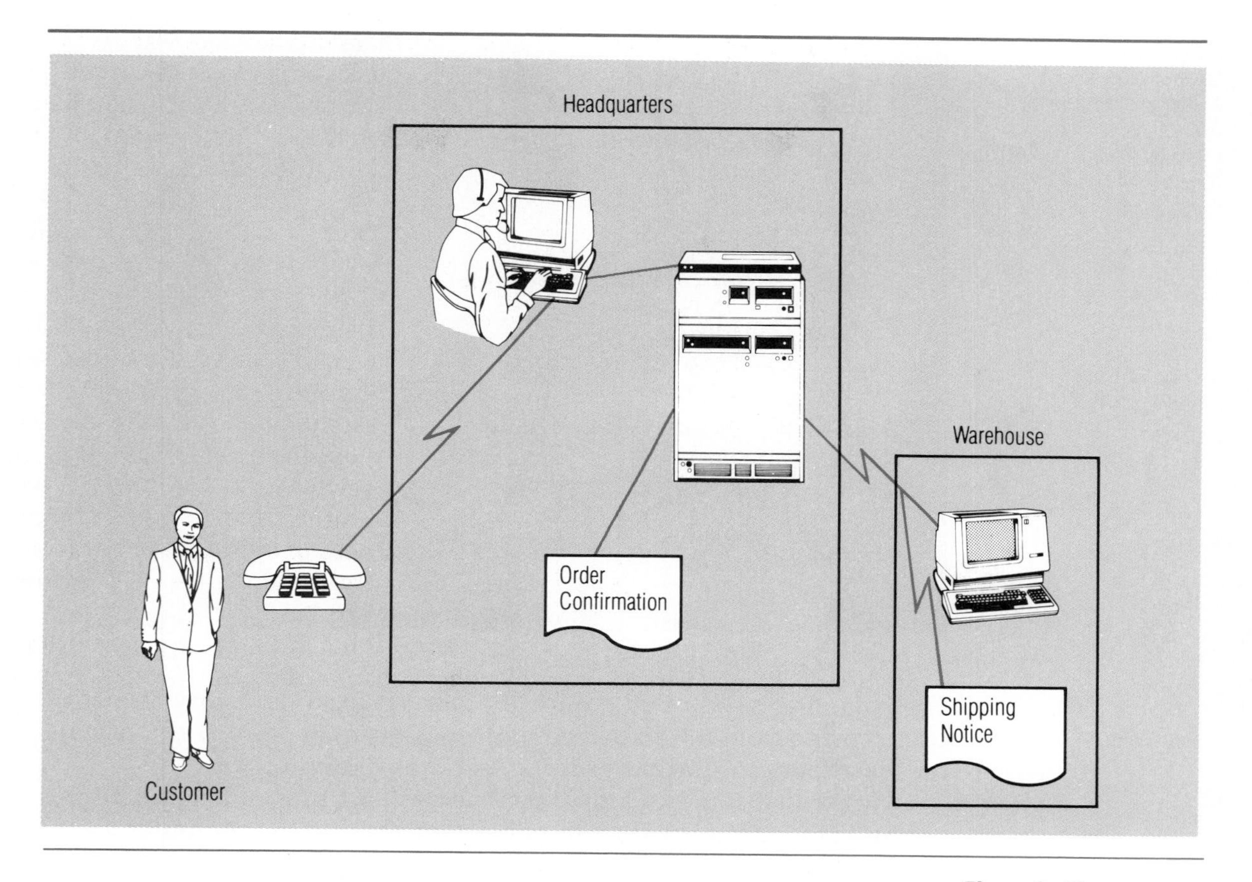

Figure 2–12
The telecommunications in an industrial sales order entry system.

these applications would not be possible, and these examples clearly illustrate how telecommunications is changing the way business is done and the structure of business itself.

Sales Order Entry Look now at another application of telecommunications in the business environment, particularly one found in a manufacturing company. Some of the products the company makes are produced and stored in a warehouse in anticipation of customer demand. Other products are produced only to fill a specific customer order.

Figure 2–12 shows a conceptual view of a sales order entry system that a manufacturing company might have. Order entry operators sitting at terminals may receive orders by mail or telephone (this example assumes a telephone order). The customer call may be handled as in the airline reservation system with an automatic call distributor routing the call to the next available order entry operator. The order entry operator keys the data required for the order into the computer. When all of the data has been entered, the operator tells the computer to process the order.

A sales order entry clerk takes an order from a customer and enters the information into the computer through the VDT. (Courtesy of Dow Corning Corporation)

The computer subtracts the quantity ordered from inventory, prints a confirmation notice to be mailed to the customer, and sends a notice to the warehouse to ship the order immediately if appropriate.

Notice that in this example there are several uses of telecommunications. The operator uses a terminal connected to the computer via a telecommunications line. The computer is also connected to a local printer that prints the order confirmation notice and a remote printer in the warehouse that prints the shipping document. There is also a terminal in the warehouse that shares the telecommunications line with the printer in a multipoint configuration. The warehouse employee uses the terminal to notify the computer when the order has been shipped. Obviously, with additional programming, the equipment used primarily for the order entry application could also send messages between the order entry clerk and the warehouse (or vice versa). The warehouse employee could also use it to request the computer balances for any inventory item.

Well-designed telecommunications systems allow terminals and printers to be used for multiple purposes. It is inefficient and unproductive to have separate terminals for each application. Each person should use a single terminal to do all of his or her work involving computers.

Point of Sale Systems in a Retail Store or Supermarket *Point of sale (POS) terminals* are used widely in large retail stores and supermarket chains. Although the technology of the terminals used in retail stores differs

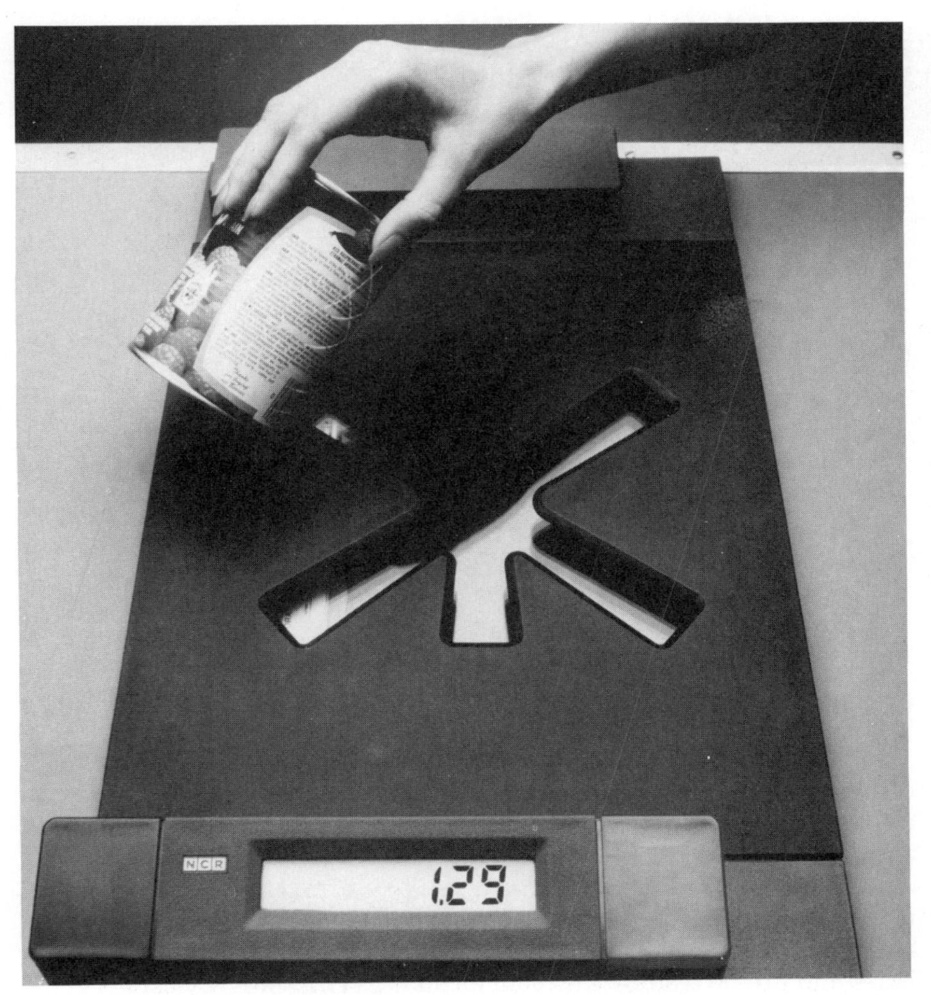

An automated checkout terminal scans the bar code on the can of raspberries. (Courtesy of NCR Corporation)

from that used in supermarkets, the use of telecommunications and the basic application are quite similar.

Several years ago, the grocery industry standardized the Universal Product Code, a bar code that can be used to uniquely identify virtually every item stocked in a supermarket. This bar code can be quickly and accurately read by a laser scanner built into the supermarket checkout counter. Good design enables the products to be read when they are passed through the laser beam at virtually any angle or speed. The supermarket checkout clerk ensures that the bar code on the item is facing in the general direction of the laser and passes the product through the beam. The laser reads the bar code and sends the product number to the store computer, which looks up the product description and price in a database. The computer transmits that information back to the terminal, which then prints the cash register tape. Occasionally, a product

A retail point of sale (POS) terminal that could be used in many different types of stores. (Courtesy of NCR Corporation)

cannot be read or does not have a bar code. In that case, the checkout clerk can key the price into the terminal. It is printed on the tape and added to the running total.

The retail store has a similar type of system. Unfortunately, products in retail stores are not universally coded in machine readable form. Some stores have merchandise tags that have a magnetic stripe or punched holes and can be read by the retail terminal, but in most cases, the clerk must manually enter the information.

The most frequent design of supermarket and retail store systems is for the store's computer to contain product and price information, which is completely self-sufficient for normal operations during store hours. Often a telecommunications link is provided to a nearby store for backup purposes. If the store's primary computer fails, the checkout terminals can continue to operate off the backup computer. These telecommunications links are illustrated in Figure 2–13.

At night a central computer at the company headquarters dials the computer at each store to collect data about the day's sales and to check inventory levels. Since the connection between the headquarters computer and the store computer is only needed for a few minutes, it would not be cost effective to have a full-time leased line connecting the two machines. In this case, the computers are connected by a standard dial-up telephone line. The call is automatically initiated by the central com-

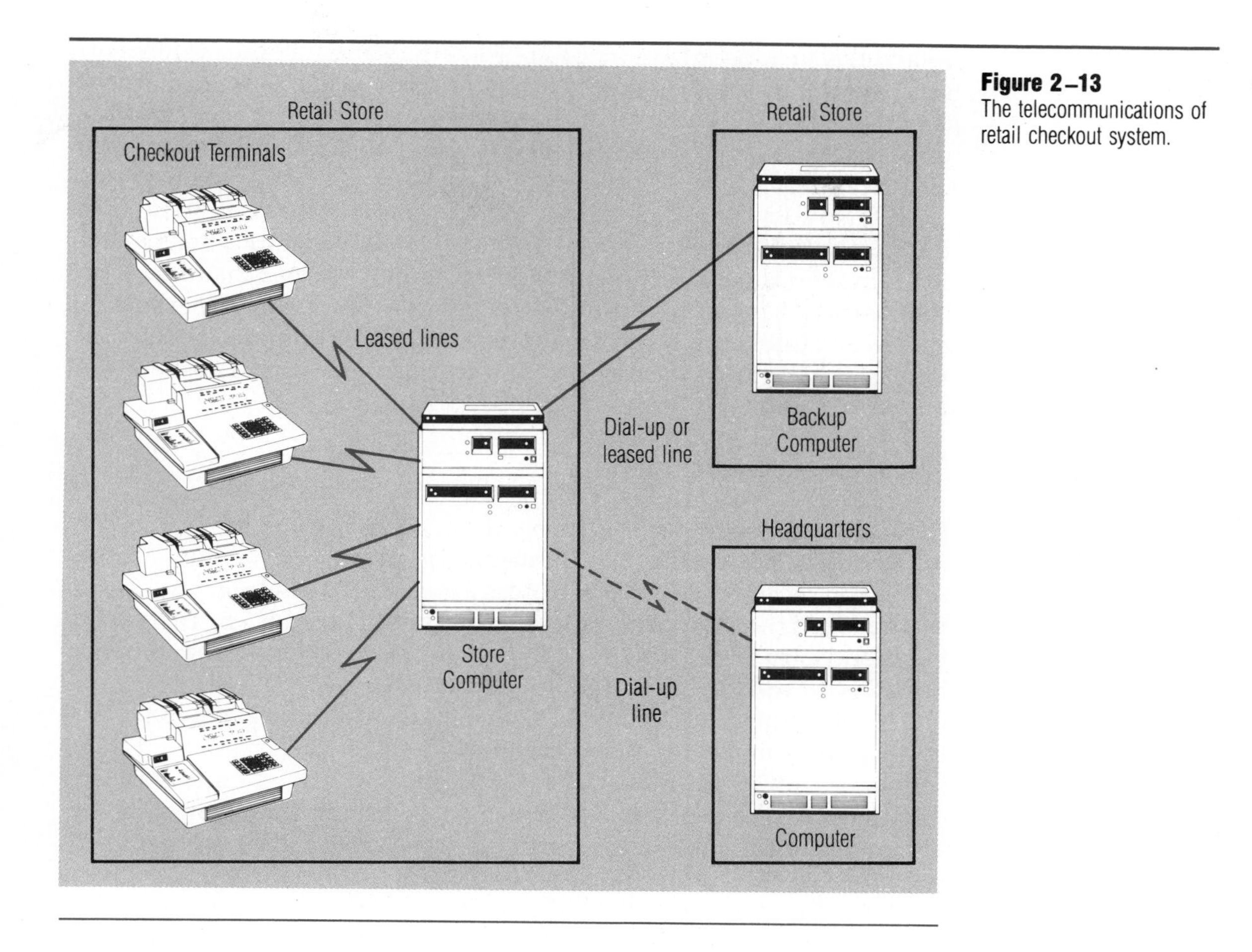

Figure 2–13
The telecommunications of retail checkout system.

puter. This call is exactly like a normal voice telephone call except that when the connection is made, data, not voice, is transmitted.

When all of the data has been transmitted and acknowledged, the connection is broken by electronically hanging up. In this way, only the time actually used for transmitting data must be paid for. After the central computer collects the data from one store, it immediately dials the next store in the chain and collects similar data and so on until all stores have been contacted. Once all of the data has been collected on the central computer, it can be processed to generate daily sales reports of various types. Also, since inventory levels were checked at each store, the computer can generate automatic restocking information.

For the customer, scanners reduce the wait in the checkout line by 25 percent, reduce cashier errors, and produce receipts showing an exact description of the products as well as the prices. Although they were largely introduced as labor saving devices, store scanners are changing the way business is done by revolutionizing inventory control. Indeed, retailing is turning into more of a science than an art because stores can see patterns of sales that were invisible a few years ago. An industry expert predicts that with all of the benefits accruing to grocers and customers alike, over 90 percent of all groceries will be scanned by the turn of the century.

Obviously, the role of telecommunications and the computer is critical in point of sale retail store systems. If a telecommunications line or the computer is not operating, customers cannot buy merchandise. Since competition is heavy in this industry and there are many supermarkets and retail stores in a given area, a failure of the computer or telecommunications system usually means that customers go elsewhere and business is lost.

Unstructured Data Applications

Electronic Mail *Electronic mail* applications, or *E-Mail* as they are sometimes called, are similar to administrative message switching applications. The purpose of E-Mail is to pass a message (mail) from one person to another. Since the content of the message is textual, in contrast to the highly structured data of typical business data processing systems, E-mail is an unstructured data application.

As illustrated in Figure 2–14, an E-Mail application requires a terminal or personal computer from which the user can access the E-Mail software running on a host computer. The software is quite sophisticated despite the fact that E-Mail is a conceptually simple application. The software provides a directory and translation facility so that users can address each other by name or at least a mnemonic code. These must then be translated to the disk address of the user's electronic mailbox.

The host computer must have adequate disk storage to store incoming messages. Users are each assigned a space on the disk known as their *electronic mailbox*. Messages are stored in the electronic mailbox until the user deletes them. The E-Mail system provides the user with exclusive access to his or her stored messages from any terminal. Although many users may check their mailboxes and delete the messages they have read on a daily (or more frequent) basis, others may leave messages on the system for days or weeks, so the disk capacity must be large enough to accommodate these differences.

electronic mailbox

E-Mail systems, such as IBM's Professional Office System (PROFS) or Digital Equipment Corp.'s All-in-One, are designed to be installed on a company's computer for its private use. PROFS offers users several options for handling incoming messages. A response can be typed for transmission back to the sender of the original message, the message can be forwarded to another person for handling, the message can be printed, or it can be stored in a log file or simply set aside for later handling.

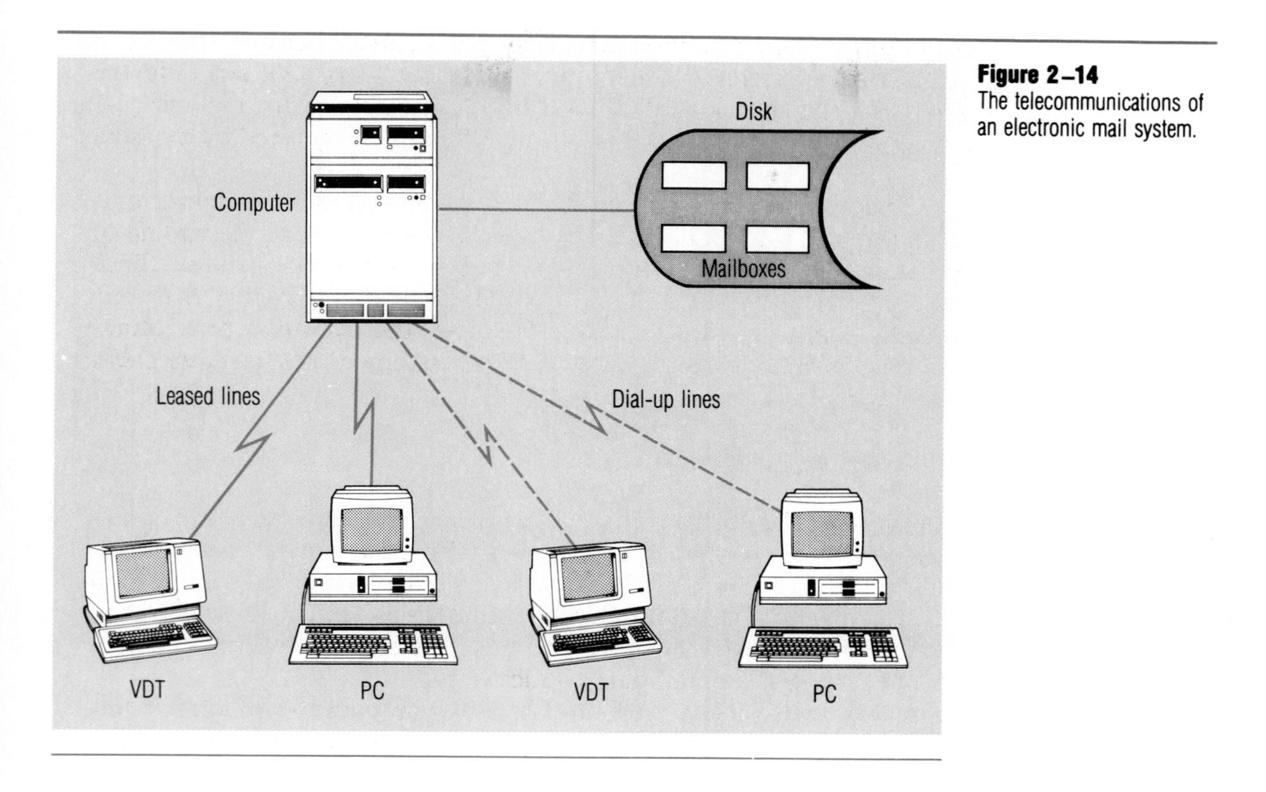

Figure 2–14
The telecommunications of an electronic mail system.

Several companies are in the business of providing public E-Mail systems. For a small fee, anyone can subscribe and be assigned an electronic mailbox. Subsequent charges vary, but they usually are based on the number of messages sent or received. Leading vendors of public E-Mail services are Tymnet's OnTyme, Western Union's Easylink, GTE's Telemail and MCI's MCI Mail. E-Mail service is also available from information service companies, such as The Source and CompuServe.

The host computer with its E-Mail software and mailboxes must be easily accessible to the users through a telecommunications network. Companies that install their own E-Mail services may also use their regular data communication network to handle E-Mail. Public E-Mail services usually are accessible by one of the public data communication networks, such as Tymnet or Telenet, and can be reached by a local telephone call in most parts of the country.

A significant difference between a modern electronic mail system and an older administrative message switching system is that the former is designed so that the user can create, format, and send the message from a computer terminal. By way of contrast, in older systems the message sender usually wrote the message by hand, gave it to a secretary who typed it, and then delivered or sent it to the communications

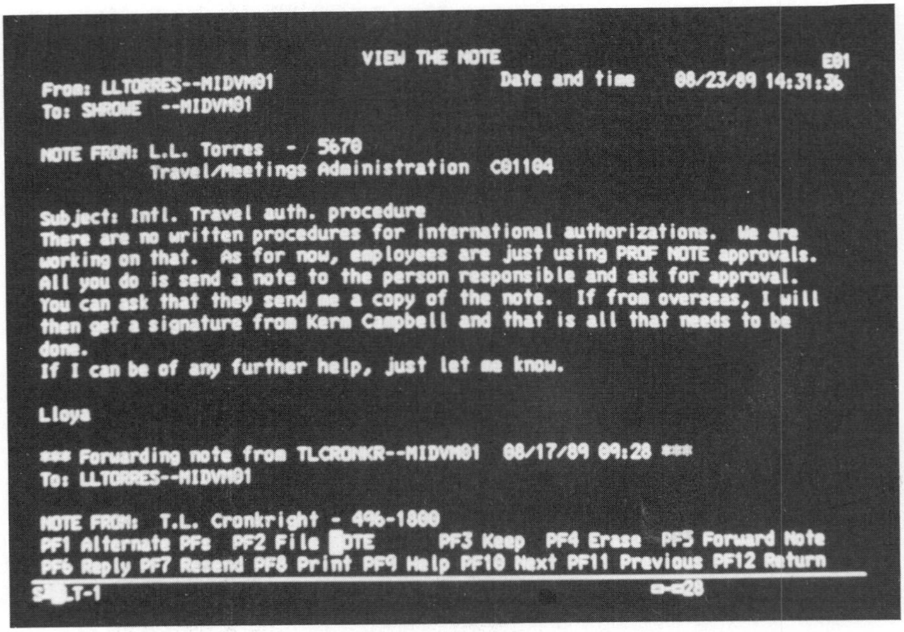

A screen from an electronic mail system.

```
                              VIEW THE NOTE                              E01
From: LLTORRES--MIDVM01                         Date and time    08/23/89 14:31:36
To: SHROWE  --MIDVM01

NOTE FROM: L.L. Torres  -  5670
           Travel/Meetings Administration  C01104

Subject: Intl. Travel auth. procedure
There are no written procedures for international authorizations.  We are
working on that.  As for now, employees are just using PROF NOTE approvals.
All you do is send a note to the person responsible and ask for approval.
You can ask that they send me a copy of the note.  If from overseas, I will
then get a signature from Kern Campbell and that is all that needs to be
done.
If I can be of any further help, just let me know.

Lloya

*** Forwarding note from TLCRONKR--MIDVM01  08/17/89 09:28 ***
To: LLTORRES--MIDVM01

NOTE FROM:  T.L. Cronkright - 496-1800
PF1 Alternate PFs  PF2 File NOTE      PF3 Keep  PF4 Erase  PF5 Forward Note
PF6 Reply PF7 Resend PF8 Print PF9 Help PF10 Next PF11 Previous PF12 Return
```

room, where it was rekeyed onto a teletypewriter or other terminal for transmission. Another distinction is that E-Mail messages are deposited in the recipient's electronic mailbox on the disk of the computer instead of automatically being sent to a destination terminal for printing.

Public E-mail services are beginning to be interconnected, allowing a subscriber in one system to be able to send messages to subscribers of other systems. An evolving international standard called X.400, which is discussed in more detail in Chapter 13, is allowing this interconnection to occur even though the E-mail networks have different technical characteristics internally. One example of this interconnection is an agreement between IBM and MCI to connect IBM's Information Network (IIN) with MCI Mail. One of the first user groups of this particular connection will be the Aerospace Industries Association, made up of twenty-eight aerospace companies. Using the connection between the IIN and E-Mail systems, employees of the aerospace companies will be able to easily exchange electronic messages with one another for the first time.

Several types of businesses or business people find public E-Mail to be very beneficial:

- independent businesspeople who need to communicate with their vendors or customers who may be located anywhere in the country are one category;
- small companies that cannot afford their own private E-Mail service are another group of public E-Mail users;

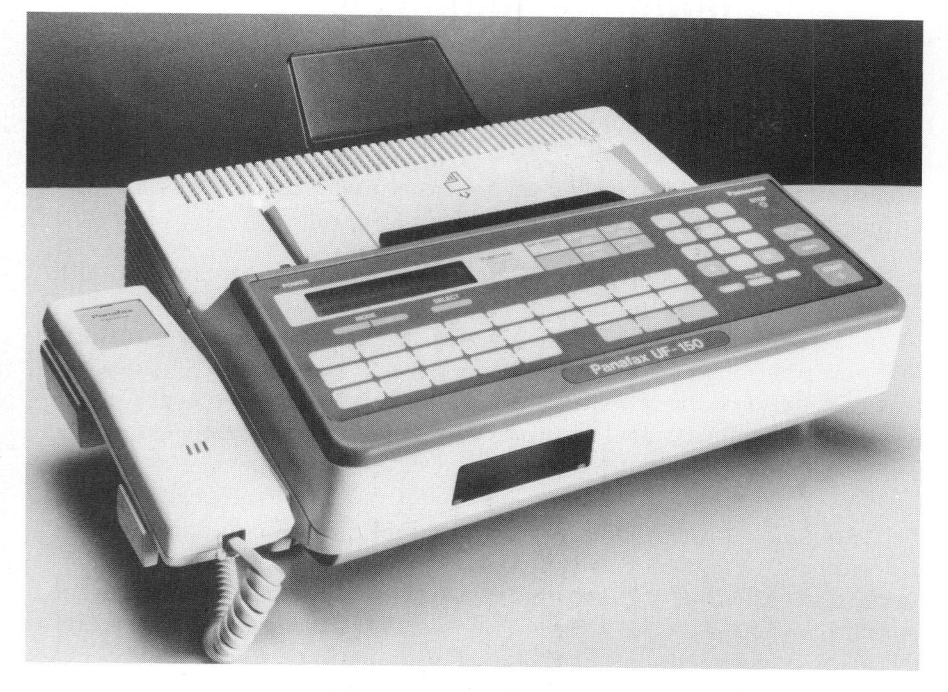

The Panafax UF-150 facsimile is small and lightweight. It can also be used as a copying machine. (Courtesy of Panasonic Corporation)

- large companies, which may have their own internal E-Mail system, may also subscribe to a public service in order to communicate with their customers, vendors, or employees;
- salespeople and other travelers, who are frequently away from their office, find public E-Mail an ideal way to keep in touch with the home office and to otherwise conduct business almost as though they were at their desk.

Image Applications

Facsimile A *facsimile (FAX)* machine scans a sheet of paper electronically and converts the light and dark areas to electrical signals, which can then be transmitted over telephone lines. At the other end, a similar machine reverses the process and produces the original image on a sheet of paper. Individual characters are not sent as such; only as contrast between light and dark. As a result, the facsimile is ideal for sending preprinted business documents or forms, as well as letters, contracts, and even photographs. Each facsimile machine, therefore, has two parts: a reader-transmitter and a receiver-printer.

Since the machine deals with light and dark areas (not letters, words, or numbers), any image, photograph, drawing, or graph can be transmitted. The facsimile machine scans back and forth across the original document in a series of scan lines. The most sophisticated facsimile machines can detect several shades of light or dark. The more shades detected, the more faithful the reproduction is at the receiving end. The speed to transmit and print an 8 1/2 × 11 document varies from 6 minutes to approximately 2 seconds, depending on the techniques used for transmitting.

Facsimile transmission has several advantages over other data communication methods in certain applications:

- if a document is already printed, it does not need to be rekeyed in order to be transmitted;
- operation of facsimile machines is very simple and requires little training;
- since the recipient receives an exact duplicate of the original image, graphs, charts, and handwritten notes can be sent as easily as typed documents.

In business, facsimile is used to transmit any document that must arrive quickly. Examples include contracts going to a law firm or a prospective client, shipping documents sent from the corporate headquarters to a warehouse, and simple engineering drawings going to a contractor or outside engineering firm. Radio stations are now accepting requests by facsimile, and some individuals have facsimile machines at home. In Japan, where addresses are complicated and streets usually are not named, facsimile is regularly used to send a map explaining how to get to one's office or home.

The Japanese have been leaders in the development of sophisticated facsimile machines primarily because of the complexity of the Kanji character set in the Japanese language and the inability (until recently) to process Kanji on computers. Because of these two factors, a large portion of the world's facsimile transmissions occur in Japan.

Videotex *Videotex* is a relatively new application in which the computer is able to store text and images in digital form and transmit them to remote terminals for display or interaction. An example of an application where videotex capability is useful is in the real estate business. A prospective home buyer can sit in a realtor's office and look at electronically reproduced photographs of homes on the market in a distant city. Since new photographs are added and deleted electronically, a very current file of available homes can be maintained, and the buyer can make a decision about houses to see before he or she travels on a house-hunting trip.

Another application of videotex is online mail order shopping. A consumer with an appropriate terminal can dial the computer of the retail store using standard telephone lines, connect the terminal to the line, and view photographs of merchandise for sale. The consumer selects the merchandise using the keyboard on the videotex terminal.

Videotex pictures are scanned, interpreted, and stored on the computer in numerical format. Obtaining good quality pictures requires a large amount of disk space for the data representing the picture. Furthermore, the large amount of data may require several minutes to be transmitted to a terminal. Many consumers don't want to wait several minutes to see a simple picture. Scanning through a catalog would be virtually impossible. Fortunately, techniques are available to reduce the amount of data that must be stored and transmitted so that high-resolution pictures can be received in a reasonable amount of time. However, compared to data and text transmission, image transmission is still voluminous.

Television Television is used in business in several ways. Among them are

- to monitor doors, parking lots, or other facilities;
- to provide information to employees;
- for video conferencing—conducting a meeting where the participants are in different locations but can see each other on television.

Security Monitoring In the security monitoring application, the television camera can be on top of a building, in a doorway or corridor,

An industrial security guard works at his VDT. The television monitors in the background are connected to cameras at various locations throughout the plant and allow the guard to watch many locations simultaneously. Other equipment monitors the use of badge readers at entrance doors. (Courtesy of Dow Corning Corporation)

or in the corner of a room viewing the room or area being monitored. The camera is unattended, and its video signal is fed to a central monitoring point such as a security desk. A video tape recorder may be connected so that any unusual activity can be recorded and then later replayed for analysis.

For security monitoring, black and white television pictures are most often used, and, in many cases, the picture is updated only every 30 to 90 seconds. This periodic refreshing of the picture is referred to as *freeze-frame television*. It requires a much slower, less expensive telecommunications line than one that must transmit *full-motion* pictures. Full-motion pictures are like those seen on commercial television stations; to achieve the motion, 30 frames are sent every second.

Providing Information to Employees Providing information to employees is another use for television in business. The information may be textual information that announces promotions, job changes, or upcoming events. An announcement might be displayed on the screen for 15 or 20 seconds and then automatically replaced with the next one. Information could also be conveyed through regular full-motion, full-color television broadcasts. The programming might include interviews with company officers, company news broadcasts, or training films.

Whether textual or full-motion, in-house television is most successful if television sets to receive the broadcasts are located in convenient locations around the office or plant. Having the sets in coffee break areas, the cafeteria, and other gathering places helps ensure that the messages are conveyed to the maximum number of people.

Video Conferencing A third use of television in business is *video conferencing*. In this application, meetings are conducted in rooms equipped with television cameras and receivers. The participants in one room can view those in the other rooms via the screen. The television signals are transmitted between the rooms on telecommunications lines.

There are actually two types of video conferencing: one-way and two-way. In one-way conferencing, a company broadcasts a program from a central location, which is received simultaneously at numerous receiving stations equipped with relatively inexpensive receivers. Some audio audience interaction is possible using telephone lines, but the picture communication is one-way. This is in contrast to two-way teleconferencing, which is more like a conventional meeting and requires cameras and transmission equipment at each end of the connection.

Another way to categorize video conferencing is whether it is full-motion or freeze-frame. In this application, the trend has been toward simple operation using freeze-frame television. In most meetings, full-motion pictures are not necessary. Once participants are in the room and seated, a picture can be transmitted to the other locations so that the participants can see who they are talking to. After that, an updated picture every 30 to 60 seconds usually is adequate. The use of freeze-frame television greatly reduces the amount of information that has to be transmitted and keeps equipment and transmission costs down.

The control panel for a multilocation video conference. (Courtesy of AT&T)

Television equipment cost and complexity vary widely. Factors that affect them are

- black and white versus color;
- full-motion versus freeze-frame;
- lighting sensitivity;
- special equipment for graphics or text transmission.

The more the video conferencing capability becomes like commercial television, the more elaborate and costly the equipment. Sending 30 frames per second obviously requires a much higher transmission speed than sending 1 frame every 30 seconds. Studios and control rooms with editing equipment are sometimes found in the corporate environment, but in most cases they are unnecessary and difficult to justify. They are used where the television broadcast has the elements of a television production.

Most video conferencing today is done from a conference room setting, not in a studio. Corporate executives don't like wearing makeup or having a crew of technicians hovering around during a meeting. Usually one camera is focused on the meeting participants, and a second camera is used to transmit pictures of transparencies, slides, or drawings that are used to illustrate points during the meeting. Ideally, the video conferencing equipment is simple enough to be operated by the meeting participants.

public video conferencing centers

Several companies, such as AT&T and Holiday Inn, have set up public centers in major cities for video conferencing. Centers are fully equipped video conferencing rooms rented by the hour. The rental fee includes the use of the room, any technicians required to operate the equipment, and transmission of the video conference by satellite to another, similarly equipped room. Using public centers is ideal for a company that occasionally needs to use video conferencing but cannot justify installing its own equipment. In addition to using the rooms for conducting meetings, some companies regularly use them to announce new products, conduct sales meetings, or communicate with customers. The conference might feature the vice president of marketing in a room in New York City and salespeople gathered in rooms in other major cities. A new product could be announced to the entire sales force simultaneously, and the two-way nature of the media would allow the salespeople to ask questions and interact with the vice president.

A number of companies have established one-way video conferencing networks. Ford Motor Co. established its network in 1985 and has since expanded it to 184 plants, district offices, and other facilities in North America. General Motors has a network that links about 100 facilities in the United States and Canada. K Mart is installing a system to link corporate headquarters with each of its more than 2,000 stores. These links enable management to reach people they were unable or unwilling to reach by traveling. IBM has used one-way video conferencing for several years for employee and customer education. The instructor works in a studio with complete graphic aids available. Students sit in classrooms at IBM facilities throughout the country and watch the instructor on large screen television. They can ask questions and respond to the instructor using audio communication.

Two-way video conferencing is still too expensive for many companies, although several—such as IBM and Hercules Chemical—have installed systems. Other companies—such as GM and Ford—have installed two-way conferencing on a limited basis, in Ford's case between its Dearborn, MI, world headquarters and offices in Dunton, England, and Cologne, West Germany. General Electric sees itself becoming a heavy user of international video conferencing and is installing a new global private network. GE will use several dozen internal video conferencing rooms to hold meetings with customers and suppliers as well as employees, regardless of where they are located. With video conferencing costs coming down and the quality coming up, it is expected that the growth of both one-way and two-way video conferencing will climb rapidly in the next five years.

Other Applications

Radio Paging *Radio paging* is the broadcast of a special radio signal that activates a small portable receiver carried by the person being paged.

Typical users of radio paging include doctors and medical technicians in hospitals, specialists in other fields who are on call and unreachable by telephone, maintenance people, security guards making their rounds, and computer programmers who are on call in case of program failures on the central computer.

In actual operation, the person who wants to make the page calls a central dispatcher who can activate the radio transmitter. Some paging systems activate the paging receiver with a tone, and the recipient knows to call a predetermined telephone number. Other systems allow the transmission of actual voice messages that tell the recipient who to call or what to do.

Both public and private paging systems exist. A public system is run by a company that specializes in providing paging services, to which anyone can subscribe by paying a monthly fee. The company provides the dispatcher and radio transmitter and the telephone number. Medium or larger companies usually run their own paging systems.

In the past, the use of radio paging systems has been limited to a range of 20 or 30 miles. Now, larger systems with broader geographic coverage are available. These systems either use repeater transmitters or relay the message to a distant transmitter on telephone lines for broadcast near the area where the recipient is known to be. With this type of relay capability, nationwide radio paging is becoming available.

SPECIAL CONSIDERATIONS OF TELECOMMUNICATIONS APPLICATIONS

Most telecommunications applications involve people. True machine-to-machine communication with no human involvement is rare. Since people are involved in the majority of telecommunications applications, a number of human factor elements must be considered in the planning or design of a telecommunications application or network.

Response Time

Matching the response time of a telecommunications system or network to the user's expectations, or vice versa, is an important human factors consideration. Of equal importance is response time consistency so that the user has the feeling that the system is always performing in the same way. We look into this issue in detail in Chapter 3.

Security

System security is another important consideration in telecommunications systems, particularly where computer data is involved. Businesses are becoming extremely sensitive about the protection of their data and are insisting that ever more stringent security measures be put in place to protect it. When a telecommunications system is being designed, security must be carefully evaluated, and security techniques appropriate for the application must be implemented. We discuss these techniques in detail in Chapter 14.

Planning for Failures

A third consideration of terminal-based systems is the special procedures required when the computer or network fails. Companies using telecommunications networks tend to become extremely dependent on the network to conduct normal business operations, yet network or computer failure occurs. The telecommunications system designer must plan for how the business will operate when the computer or network is down. Telecommunications adds an element of complexity to computer applications, particularly if long distance lines are involved, since they are subject to problems caused by severe weather, electrical interference, and a misguided bulldozer or backhoe.

failures will occur

Therefore, it can be safely assumed that someday when a user picks up the telephone or sits down at the terminal, it won't work. The outage may or may not be a problem. Some telephone calls are more important than others; some can wait or were optional in the first place. The same is true of computer applications. In the sales order entry system described earlier in this chapter, the company may not need to enter a customer order or a material today if the product is not to be shipped for several weeks or months. By contrast, if the supermarket checkout system is not operating, customers will go elsewhere to buy their groceries.

The key is planning and having thought through what will happen when (not "if") the computer or network fails. Several options are available:

- wait and do nothing except try to determine the reason for the failure and the likely amount of downtime;
- fall back to manual procedures, the type that were used before a computer and telecommunications network were in place;
- manually switch to a backup computer or telecommunications line;
- have a standby computer or telecommunications line in place ready to automatically take over in case of failure. This is called a *hot standby* system.

Each of these alternatives must be evaluated in light of the business situation. In all but the most critical applications, the usual approach is to try to bring the system back up. At the other extreme, the airlines with their reservations systems have standby computers and duplicated networks to make sure that when a failure occurs, redundant facilities are available to take over immediately without missing a reservation. Few companies would actually wait and literally "do nothing" when the computer or communication system is down.

The problem with falling back to manual procedures is that once staff becomes used to automated systems, almost nobody remembers how to do things manually. Most companies find that manual procedures are not a very acceptable backup alternative. In addition, the volume of transactions often cannot be handled manually.

The difference between having backup computers and lines and a hot standby system is that the backup computers and lines usually are used for other types of work until they are needed. When a failure of the critical computer or network occurs, work is switched off the backup computer and the application is restarted on it. Careful planning will ensure that the time required to switch to the backup computer is not excessive. With a hot standby system, the switchover takes place rapidly and automatically. Again, the speed required is very dependent on the application. Hot standby systems are expensive, and the benefits must be weighed against the costs.

planning is key

The key to handling failures is the planning itself. System outages need not be catastrophic if someone has thought through the implications of an outage and actions to be taken when the outage occurs. Many companies conclude that although it would not be convenient, the best alternative for many of their applications is to simply try to bring the system back up as quickly as possible and then catch up on the processing by having people work extra hours. Of course, in some applications, this is just not possible. More elaborate procedures are needed.

Disaster Recovery Planning

Disaster recovery planning is an extension of the discussion about planning for normal system outages. For our purposes, a *disaster* is defined as a long-term outage that cannot be quickly remedied. A fire, flood, or earthquake may be the cause, for example. No immediate repair is possible, and the computer facility and equipment are unusable.

Again, planning is the key. Having a computer in an alternate site and the ability to switch the telecommunications lines to the alternate computer is a viable alternative in many situations. In lieu of that, quickly obtaining a new computer from the vendor may be possible. Most vendors state that they will "take the next computer off the manufacturing line" to replace a computer damaged in a disaster. Of course, the computer is of no use if a suitable facility cannot be found to house it—one that has enough space, power, and air conditioning. Of equal importance is the ability to switch the telecommunications network to the new computer site, again something that may be relatively easy if it has been planned for ahead of time. Telephone companies are developing techniques and facilities to allow networks to be switched to alternate sites, a capability that is being demanded by companies that are putting disaster recovery plans in place.

mutual aid pact

Sometimes it is possible to work out a mutual aid pact with another company. Both organizations agree to back each other up in case of a disaster. They may agree to providing computer time on second or third shift in case of a disaster even if it means not running some of the low-priority processing. Computer centers are easier to back up with mutual aid pacts than are telecommunications networks. With proper planning, however, extra lines could be installed between the companies providing at least some backup transmission capability. If the two companies are in close physical proximity, they must be concerned about a disaster that would hit both of them simultaneously. An earthquake or tornado could easily hit several companies in a several mile radius and effectively neutralize any backup plans.

Several companies exist for the sole purpose of providing disaster backup recovery facilities. These companies, such as Comdisco Disaster Recovery Services, have one or more large computer centers filled with hardware ready for use in an emergency. In order to use the facilities, a business must subscribe to the service by paying a membership fee and annual dues. Then if the use of the emergency facility is necessary, a one-time activation fee must be paid as well as usage charges for the actual time the disaster site is occupied. Subscribing to one of these services is much like buying a form of insurance.

Whatever plan is developed for disaster recovery, it must be specific for different kinds of disasters. Corning Incorporated, located in Corning, New York, had a disaster plan detailing how the company would recover from a fire. In 1972, however, the town of Corning was hit by a massive flood, which literally put the Corning computer center under water. Although some of the procedures previously developed for a fire were appropriate, many were totally useless or inappropriate for problems caused by water damage.

testing is very important

Disaster recovery plans must be tested. Rarely will all of the problems of a real disaster be covered when a plan is written, and although testing does not recreate a real disaster, it does identify weaknesses in the plan. If a contract has been signed with a disaster recovery firm, that firm will assist in testing the disaster plan. The contract normally provides for a specified number of hours of test time at the disaster recovery site. Companies use this time to ensure that they can transfer their software from their mainframe computers to the backup computer and to test the telecommunications links that connect the disaster site to parts of the company not affected by the disaster.

Tests of the disaster plan can be conducted in other ways. One company's data processing manager worked with the computer vendor to develop a disaster test. Late one night, the vendor's service manager removed a critical but obscure part of the mainframe computer. When the computer operations staff could not bring the computer up the next morning, they called the vendor for service. The vendor's technicians tried for several hours but were unable to precisely diagnose the problem, so they called for help from the national support center. In the meantime, the prolonged computer outage had caused the company to activate its disaster recovery plan, and because only the data processing manager and the vendor service manager knew the real nature of the outage, the "test" was extremely realistic. It was allowed to run for over 24 hours before it was revealed to be a test. Although there was some grumbling, there was also general consensus among the computer users and management that much valuable information had been gained.

As a company becomes increasingly reliant on telecommunications and computer networks, it must constantly reassess how long it can afford to be without the systems. Getting the users involved in assessing the impact of an extended outage caused by a disaster is one way to build a case for management to support spending time and money on developing and maintaining disaster recovery procedures.

SUMMARY

This chapter has looked at a number of business applications that involve the use of telecommunications and the computer. Telecommunications and computing go hand in hand in providing modern business systems that help companies to be more effective and efficient. Simple uses of telecommunications for message switching and online data collection have given way to sophisticated, interactive applications that enable a company to do business in new ways.

The integration of these applications into the business brings with it new responsibilities. Provisions must be made for appropriate security to ensure that company data is not lost or misappropriated. Contingency planning must be done to determine how the company will operate when short or long outages occur. By now, you should have a good understanding of the many ways in which companies use telecommunications and the breadth and depth at which it can penetrate an organization's activities.

CASE STUDY

F/S Associates

F/S Associates is a small computer software development firm located in Hermosa Beach, California. The company specializes in the development of software for telephone companies to aid in the formatting and printing of telephone directories, especially the Yellow Pages. Chet Floyd, president of F/S Associates, incorporated the company in 1981 and is also the chief software developer. He uses two IBM Personal Computers for the software development work.

Several years ago, Chet established a business relationship with another company, Strategic Management Systems, a software development firm located in New Jersey. At the present time, F/S Associates is under contract to Strategic Management Systems to develop and test software for several large customers.

F/S Associates is using three different kinds of telecommunications in its business. Chet has a standard telephone, which was installed and is maintained by General Telephone Co. There are two telephone lines coming into the office so that Chet can be carrying on a telephone conversation on one line while his computer is transmitting data on the other. He says that this capability is extremely useful when he is debugging programs. It allows him to test complex programs as if he were at the customer's site. He adds, "And my availability to a project is increased while costs and lead times are reduced."

Chet Floyd, president of F/S Associates, in his office. (Courtesy of F/S Associates, Inc.)

Because F/S Associates is located on the west coast and Strategic Management Systems on the east coast, electronic mail is the second frequently used communications technique. Both companies subscribe to MCI Mail and use it almost exclusively for exchanging written communications with each other. Chet says, "When we got started with MCI Mail, we wrote our own software for our computers to automate the dialing and logon process. Now you can buy that type of software, but we're familiar with the software we wrote so we keep on using it." Chet has built commands into a batch file on his PC so that when he turns the machine on in the morning, one of the first things it asks is if he wants to check his MCI mailbox. If he responds positively, the software automatically makes the connection to MCI.

To send a message, Chet composes it offline into a file on his computer disk. He says, "MCI Mail has an editor program to help you compose and format messages, but I prefer to do it offline with the editor that I normally use for word processing."

Once the message is ready, Chet initiates the program that dials and connects to MCI Mail. When the connection is made, the mail is sent. "I use the return receipt option a lot." Chet says. "It is a no-charge option that MCI Mail offers, and by using it, I know when the mail I sent has been picked up by the recipient. That helps me to know when I might expect a response."

The third way F/S Associates uses telecommunications is for remote testing and support of software. Chet bought a software package called CloseUp, which consists of two programs called Customer and Support. With the Customer program running in an F/S Associates customer's computer and the Support program running on Chet's machine, he makes a standard dial-up telephone connection between the two computers. As the developer, Chet can then run or test his program remotely on the customer's computer. This is a big help for demonstrating the program's capabilities or in finding and correcting bugs.

Chet says that the CloseUp software also has an excellent file transfer program he uses to send new copies of the program or other data files to his customers over the telephone. The capability is also useful for exchanging updated versions of programs with Strategic Management Systems in New Jersey.

"Without telecommunications capability, it would be impossible for me to conduct business the way I am doing it," Chet says. "There would be no practical way to codevelop software with a company across the country if we couldn't send programs, data files, and messages to each other electronically. Nor could we offer support to our customers who are widely distributed geographically. We would either have to move so that we could share offices or find other business partners," he adds. "With telecommunications we're doing business in a way that lets us get the job done but still be located where we want to live."

Review Questions

1. Discuss the three ways in which telecommunications applications can be categorized.

2. Explain the following terms:
   central office
   store-and-forward
   multipoint line
   master or control station

polling
collision
transaction processing system
video conferencing
freeze-frame

3. Explain how a telephone call is made and what operations are necessary for its completion.
4. Why do the first three digits of a telephone number determine whether the call is long distance or local?
5. Why are airline reservation systems so vital to airlines?
6. Describe how a supermarket checkout system works.
7. Describe some situations where E-Mail is inadequate for communicating a message and a facsimile is required.
8. What types of businesses are not candidates to use public E-mail systems?
9. Would you rent an apartment or buy a home based on a picture you had seen on a videotex system? What other information would you require so that you could make a decision without visiting the house or apartment?
10. Identify and explain three different uses for television in business.
11. Explain the advantages and disadvantages of using video teleconferencing to conduct a business meeting.
12. Discuss the reasons why a company should have a disaster recovery plan for its telecommunications and computer systems.
13. Identify five types of "disaster" that could disable a computer-communications network.
14. Categorize the applications shown on the left into one or more of the categories shown on the right.

car rental reservation system	voice
data collection for machines at a paper mill	realtime
computer-aided drafting	store-and-forward
telemetry from a spacecraft	structured data
a facsimile transmission	unstructured data
	person-to-person
	person-to-machine
	machine-to-machine
	image

Problems and Projects

1. In order to install a video conferencing system, a company must invest money in cameras, monitors, and other equipment. What costs, if any, will be offset by the video conferencing? Is video conferencing an alternative to some other type of communication?

2. As the dean of your school, you have been asked by the president to develop a disaster recovery plan for the school's computers. Describe such a plan, focusing on the recovery procedures in case of fire. If classroom space could be found so that classes could continue, how would the computing for the students, faculty, and administration be handled?

3. Visit a supermarket that has an automatic checkout system with scanners. Observe several of the checkers for 5 minutes each and keep a count of the number of items whose bar code cannot be read and for which the data must be entered through the keyboard. What is the "read failure" rate? Do you notice any significant difference between the checkers?

4. Visit a travel agency that is connected to one of the airline reservation systems and find out how they conduct business when their computer terminal is down. How often do they experience downtime? What has happened to their telephone use since they installed the reservation terminals?

Vocabulary

online
realtime
store-and-forward
batch
remote batch
remote job entry (RJE)
local loop
central office
central office switch
toll trunk
toll office
plain old telephone service (POTS)
least cost routing
voice messaging
teletypewriter
point-to-point line
punched paper tape
torn tape message system
message center
multipoint line
control station
master station
subordinate station
slave station
polling
collision
addressing
protocol
telex
teletypewriter exchange (TWX)
teletex
timesharing
transaction processing system
automatic call distribution unit (ACD)
point of sale (POS) terminal
electronic mail (E-Mail)
electronic mailbox
facsimile (FAX)
videotex
freeze-frame television
full-motion television
video conferencing
radio paging
hot standby

References

"Airline Data Exchange Endangered," *Transnational Data and Communications Report,* February 1986, p. 5

Gifford, David, and Alfreda Spector. "The Cirrus Banking Network," *Communications of the ACM,* August 1985, p. 748.

———. "The TWA Reservation System," *Communications of the ACM,* July 1984, p. 650.

Gullo, Karen. "GE Enters The Video Age," *Information Week,* June 5, 1989, p. 15.

Haight, Timothy. "IBM, MCI Link E-Mail," *Communications Week,* June 26, 1989, p. 8.

Ratliff, Rick. "Scanners Lead the Revolution in Grocery Store Inventory," *Detroit Free Press,* October 27, 1986, p. 3D.

Rohan, Barry. "More Firms Have Vision," *Detroit Free Press,* January 18, 1987.

Roussel, Anne-Marie. "Two Years Later, SWIFT Upgrade Approved," *Communications Week,* February 27, 1989, p. 23.

Sharff, Daniel D. "For Transmitting Documents Don't Forget the Fax Process," *Information Week,* February 3, 1986, p. 48.

Wilson, John W., Zachery Schiller, Mary J. Pitzer, Russell Mitchell, and Gordon Bock. "On-Line Systems Sweep the Computer World, *Business Week,* July 14, 1986, p. 64.

Yalonis, Chris. "Communications Post Haste," *PC World,* April 1986.

CHAPTER

Internal Influences on Telecommunications in the Enterprise

OBJECTIVES

After you complete your study of this chapter, you should be able to

- describe typical user requirements for a data communication system;
- discuss the concept of real enough time;
- discuss telecommunications' ergonomic considerations;
- discuss the elements of modern quality management thinking;
- describe the issues of telecommunications security;
- discuss the meaning of cost effectiveness as it relates to telecommunications.

INTRODUCTION

The purpose of this chapter is to show how telecommunications systems, applications, and people are affected by the environment and culture of the organization in which they operate. You will see that one important factor is the degree to which the organization relies on telecommunications and how it has integrated telecommunications in its business operations. Another important influence is the requirements of the telecommunications user. The first step in defining how telecommunications can serve the organization is understanding user requirements and how they fit into the business. Finally, management expectations are an important determinant of the environment in which the telecommunications organization operates. You will see that management today is more demanding and expects more from the telecommunications function than its predecessors did.

USER REQUIREMENTS FOR COMMUNICATION SYSTEMS

Users of communication systems have certain expectations about the capabilities the system will provide and the system operations. In this section, you will examine typical user requirements for voice and data communication systems.

User Requirements for Telephone Systems

Put yourself in the position of defining requirements to the telephone company for your telephone service. You might want the system to be

- available when you need it;
- trouble-free and reliable;
- easy to use and easy to learn;
- a universal service so that you can call to and from any location;
- fast (doesn't take a long time to get a call completed);
- inexpensive.

Some people might have special requirements, such as

- amplifiers on the telephone handset if they are hard of hearing;
- easily accessible public phones if they travel;
- telephones with built-in directories and the capability to dial calls automatically if they use the telephone frequently.

If you look at the list of "typical" requirements shown, you would probably conclude that the telephone system in the United States has done a pretty good job of meeting your needs. In fact, this is the general opinion of the American public. In many respects, the telephone companies have been sensitive to user requirements and have provided new services and capabilities as needed.

In a similar way, the telecommunications department in a company must be sensitive to the requirements of the telecommunications users in the organization. The requirements for telephone service in a company are, in general, similar to those for public telephone service. However, there are usually some special requirements, such as

- multiple-line telephones so that secretaries can answer their bosses' calls;
- hold buttons so that one call can be held while a conversation is conducted with a second party;
- call transferring ability to move a call to someone else in the organization.

There may also be some other more specialized requirements, such as an automatic call distribution (ACD) service as described in the airline reservation example in Chapter 2.

Some of these capabilities have been available for a long time. Others are just becoming available and/or economical, due to advances in electronic technology. The important thing, however, is to understand what is required by the users of the telecommunications system.

User Requirements for Data Communications Systems

From a user's perspective, many of the characteristics that are desired from the telephone system are also desired from a data communication system:

- availability—the system is there and operating when it is needed;
- reliability—the system is trouble-free and does not introduce errors into the communication process;
- online and realtime—users are able to operate the system through interactive terminals;
- responsive—the system is quick enough so that it helps the user do the job and does not hinder him or her by imposing delays on the communication;
- ease of use—it is easy to accomplish the needed communication. There are dialogues that allow the user to interact with the computer;
- ergonomics—the users' workstations and terminals must provide for long periods of comfortable use;
- flexibility—the system must be easy to change.

As in the case of voice systems, some data communication users may have special requirements, such as

- terminals that can display color images;
- terminals that can display graphs;
- the ability to get hard copy output conveniently.

Let's now examine each of these requirements in detail.

Availability *Availability* is having the system or service operating when the user wants or needs to use it. Take, for example, the telephone system. We expect telephone service to be available anytime, 24 hours a day, 7 days a week, 365 days a year. What would it be like if the telephones did not operate during certain hours, say from 10 P.M. to 7 A.M., or during the lunch hour, or on holidays? In many cases, we could get used to these reduced hours of operation. We would complain, and we would certainly have to adjust some of our habits, but in most cases, the reduced hours of operation would not be critical.

There are some situations, however, where not having the telephone system available could be disastrous. What if we needed to call the fire department or the police? What if somebody needed to reach us desperately? For these cases, other communication methods would need to be developed.

variable requirements

The real requirement for availability varies by application. Many data applications in business only need to be available during business hours. For example, the order entry application discussed in Chapter 2

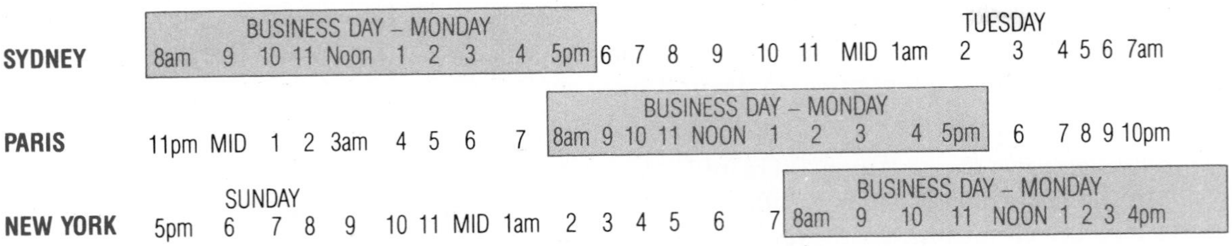

SYDNEY	[BUSINESS DAY – MONDAY: 8am 9 10 11 Noon 1 2 3 4 5pm] 6 7 8 9 10 11 MID 1am 2 (TUESDAY) 3 4 5 6 7am
PARIS	11pm MID 1 2 3am 4 5 6 7 [BUSINESS DAY – MONDAY: 8am 9 10 11 NOON 1 2 3 4 5pm] 6 7 8 9 10pm
NEW YORK	5pm (SUNDAY) 6 7 8 9 10 11 MID 1am 2 3 4 5 6 7 [BUSINESS DAY – MONDAY: 8am 9 10 11 NOON 1 2 3 4pm]

Figure 3–1
The business days in many major cities of the world have little overlap with one another. Telecommunications systems that serve international locations must have extended operating hours in order to be available during the business day in the foreign cities.

may only need to be available during business hours because customers won't be at work to place orders at other times. On the other hand, if we advertise on television at all times of the day or night and our advertisement says "call 1–800–555-XXXX. Operators are standing by to take your order," our order entry application may need to be available around the clock.

If our company does business nationwide, the hours of availability will undoubtedly have to be longer than from 8 A.M. to 5 P.M. local time. If we are located on the east coast, we will probably want to keep our systems available until 8 P.M., which is 5 P.M., the close of the business day, on the west coast. Conversely, if we are located on the west coast, we may need to start our system operating at 5 A.M. because that is 8 A.M. in the east, and our customers there are ready to do business with us.

If the communication network serves an international business, the window of availability will be longer. Figure 3–1 illustrates the situation. Allowing for an hour or two variation caused by daylight savings time, when it is 8 A.M. in Paris, it is 2 A.M. on the east coast of the United States. Therefore, the telecommunications system may have to be available from 2 A.M. until 8 P.M. eastern standard time (if it is also serving the west coast of the United States). Again, allowing for 2 or 3 hours' variation due to daylight savings time, Japan and Australia are approximately 15 hours ahead of the U.S. east coast. That means when it is 8 A.M. Monday in Sydney, it is 5 P.M. Sunday in New York. When the Australians go home at the end of their business day at 5 P.M. Monday, it is 2 A.M. Monday in the U.S. eastern time zone. If a single network served the United States, Europe, and the Far East, it would have to be available nearly 24 hours a day, 7 days a week.

Holidays are another consideration that must be studied when determining the availability requirements of a network. Even within the United States, we celebrate certain regional holidays in some parts of the country and not in others. Different customs for holidays and lunch hours also prevail. For an international network, the situation gets even more complicated. For example, Thanksgiving is celebrated in Canada but on a different day from the United States, and whereas Americans

Component	Reliability
Computer	.98
Circuit	.97
Terminal	.99

System Reliability = .98 × .97 × .99 = .941 = 94.1%

Figure 3–2
The overall reliability of a three-component, serial telecommunications system.

close businesses on July 4, no other country celebrates America's Independence Day.

Determining the real requirement for network availability is key. Whereas the public telephone system and some businesses, such as hospitals, have a requirement to be operational 24 hours a day, 7 days a week, 365 days a year, most other business organizations do not have to operate on that schedule.

Reliability *Reliability* in telecommunications is trouble-free operation. When someone uses the system, it must work. Just as we don't want our car to break down when we are on a trip, we don't want our telephone connection to break in the middle of a call. Similarly, we don't want the terminal and computer to go down when we are interacting with it.

One of the most frustrating situations for users of communication and computer systems is unpredictability. Users generally understand that systems occasionally fail, no matter how well they are designed. Once a system fails, users would generally rather have it stay out of service until it is fixed than have it come up but fail again within a short period of time.

MTBF and MTTR

A classic measure of reliability on a system is *mean time between failures (MTBF)*. The MTBF is the average time between the failures of a system. A related measure is the *mean time to repair (MTTR)*. MTTR is a measure of how long, on average, it takes to fix the problem and get the system back up after a failure has occurred. Reliability is often measured in terms of probability. If a system is 98 percent reliable, that means it is working 98 percent of the time and out of service 2 percent of the time.

Most telecommunications systems are made up of a number of components—such as the terminal, line, and computer—connected in series. This is known as a *serial system*. Each component has a certain MTBF and MTTR. To get the overall reliability of a serial system, the reliability of each of the components is multiplied together. Figure 3–2 shows an example of a three-component telecommunications system in which each component has a certain reliability. The overall resultant reliability of the system is of course *lower* than any of the components individually. You can see from the example that in order to achieve a reliability of even 95 percent, which is a typical requirement of a telecommunica-

Figure 3–3
The maximum number of minutes a system can be down in a day and still achieve a given level of reliability.

9 hour day = 540 minutes

At a Reliability of	The system would be down
.999	32 seconds
.995	2.7 minutes
.990	5.4 minutes
.98	10.8 minutes
.97	16.2 minutes
.96	21.6 minutes
.95	27.0 minutes
.90	54.0 minutes

tions system, each of the components must be considerably more reliable than .95.

The fact that the overall reliability is less than any of its components is an important concept of combinatorial probability. It is particularly relevant, since most communication systems are serial systems and have many components connected in series. To put it in other terms, if a telecommunications system is scheduled to be operational from 8 A.M. to 5 P.M. each day (9 hours), the chart in Figure 3–3 shows how many minutes the system would be down, on average, each day at different levels of system reliability. For many business applications, 20 or 30 minutes of outage each day is simply not good enough. For those applications, reliability must be greater than 95 percent or 96 percent.

Because the public telephone system is expected to be operational 24 hours a day, 7 days a week, with a very high reliability, the central office telephone equipment is designed with many redundant components. When a component failure occurs, the backup component is automatically switched into operation. With this type of failsafe design, the expected failure rate of an entire central office is approximately once in 40 years!

Most business systems don't need that kind of reliability. But again, the important thing is for the business to determine what the real reliability requirements for each of its applications or systems are and to design its systems to meet these requirements. High reliability can be achieved by designing redundancy into any system. Redundant telecommunications lines, computers, or terminals can be put in a telecommunications system to make it more failsafe. The costs of the redundant equipment must be assessed, however, and an economic analysis must be performed to determine whether the benefits of the increased reliability equal or outweigh the costs.

Online and Realtime The terms *online* and *realtime* were described briefly in Chapter 2, but some elaboration is appropriate here. Most new computers are designed for online operation through terminals. With microcomputers, the computer is built into the terminal, and with most microcomputers today, there is no capability to attach additional terminals. In any case, the microcomputer is designed to be used by one person sitting at its VDT working interactively.

If the computer can support multiple terminals, they may be directly cabled to the computer, in which case they do not use telecommunications facilities at all. Alternately, the terminals may be connected via telecommunications lines.

Realtime is a rate of response or operation fast enough to affect a course of action or a decision. The response usually is measured from the time the terminal operator presses the ENTER key, or its equivalent, on the terminal to signal the computer to perform some processing, to the time the computer delivers the first part of the output back through the network to the user's terminal. We will discuss more about response time later in the chapter.

Real Enough Time The definition of realtime says "a rate of response or operation fast enough. . . ." The question is, "What is fast enough?" The answer depends on the needs of the particular application. We can therefore think about a concept of *real enough* time. This notion suggests that the response time that is good enough for one application may not be good enough for another. For example, traditional wisdom says that 2-second response time is good enough in most applications. It isn't good enough, however, when you pick up the telephone handset on your telephone and expect to hear a dial tone by the time you get the handset to your ear. You get the telephone to your ear in less than 2 seconds, and if you don't hear a dial tone, you are annoyed and immediately begin to wonder what is wrong. Two-second response time also is not good enough for the flight controller at Cape Canaveral who is trying to destroy a rocket that is off course and headed for a populated area. The controller wants to give the command to the computer and have the rocket blown up within milliseconds to avoid a disaster. Similarly, 2-second response time is not good enough in many industrial applications, in which computers are controlling machine tools or chemical processes. In those applications, real enough time usually means something less than 1 second.

On the other hand, after you complete the dialing of your telephone call, you do not expect (nor do you probably require) instantaneous completion of the call. In that situation, real enough time is more like 10 or 15 seconds. Similarly, in the airline reservation or customer order entry application, multisecond response time is normally acceptable at the completion of a reservation or order. While the computer is busy updating files and completing the transaction, the operator has time to put papers back in a file folder, take a sip of coffee, and get ready to handle the next call.

One must look at the real needs of each application to determine what is real enough time. Indeed, one must look within the application at the various transactions or interactions that occur, since some of them may require a faster response than others.

Response Time Response time is traditionally defined as the time between pressing the ENTER key on the terminal signaling the computer that processing is needed until the first character of output is received at the terminal. Many people argue, however, that since the operator can do little when the first character is received, the time ought to be extended until the last character is received at the terminal and the operator is able to begin work again.

The overall response time that the user sees is made up of several components. In most cases, when the operator keys data into a terminal, it is stored in the terminal's memory until the ENTER key is pressed. At that time the data is transmitted over the telecommunications line to the computer. After the computer receives the data, it must process it and formulate a response. Then the response is transmitted back over the telecommunications line to the terminal, where it is displayed or printed. The time it takes to transmit the data on the telecommunications line is a function of the number of characters to be transmitted and the speed of the line.

The transmission time may be extended by delays encountered when a line is shared among several terminals. Sometimes a transaction may have to wait for another transaction to finish using the line. This wait is called *queuing*, and it can have a significant effect on the overall response time.

The processing time on the computer is a direct function of the number of instructions that must be executed to interpret the input message, process it, and formulate a message for transmission back to the terminal. This time often is extended when a computer is dealing with several terminal users concurrently. Because a computer can only process one transaction at a time, arriving transactions may encounter a *queue*, or waiting line, of other transactions waiting to be processed. Thus, if an input message from terminal A arrives while a message from terminal B is being processed, it will be placed in a queue and delayed until the processing for terminal B's transaction is complete or interrupted.

processing time

Processing time at the computer is most often calculated on an average and probabilistic basis. The usual way of stating processing time is in the form, "computer processing takes X seconds Y percent of the time." In addition to the average processing time, one is also interested in the variability. It is one thing to know that processing is completed for 95 percent of the transactions in 1 second, but what about the other 5 percent? Do those transactions take 5 seconds to process? Or 10 seconds? Or 90 seconds? Determining the averages, variances, and probabilities for a given computer system is a complicated task. It depends on a knowledge of the mathematics involved and also a knowledge of the characteristics of the workload on the computer.

Action	Seconds	Cumulative Seconds
Operator types transaction—presses ENTER key		
100-character transaction transmitted to the computer	.10	.10
Computer receives transaction—processes it	.40	.50
1000-character response transmitted to the terminal and displayed	1.04	1.54 total response time

Telecommunication Time = 1.14 seconds = 74% of total response time

Figure 3–4
A simplified response time calculation. Queuing time is not considered.

Figure 3–4 shows a typical but simplified response time calculation. No queuing has been considered. It is evident that the telecommunications network plays a very significant role in determining the overall response time that the user sees. The people who configure the computer and those who design the telecommunications network must work together to ensure that the response time requirements of the user can be met.

IBM response time study

Two studies conducted in different parts of the IBM corporation in recent years have attempted to quantify the value and economic benefit of rapid response time to users doing different kinds of work with the computer. Walter Doherty, who works at the IBM research laboratories in Yorktown Heights, New York, has said that if the response time is fast enough, the computer becomes an extension of the human brain. In reality, the person at the terminal is usually thinking several steps ahead of the work he or she is doing at any point in time and remembering those steps in the short-term memory of the brain. If response time is slow, however, the short-term memory contents get replaced as the mind wanders. When the computer responds, the person must refocus attention on the work he or she is doing with the computer.

think time

Another type of response time is *user response time*. User response time, also called *think time*, is the time that it takes the user to see what the computer displayed, interpret it, type the next transaction, and press the ENTER key. User response time is a significant part of the overall productivity of the human–machine system. Doherty says that the faster the machine responds, the faster the person will respond with the next transaction. His studies showed that the overall productivity kept climbing as the computer response time decreased to less than .5 seconds, which was the fastest computer response time that could be obtained in that particular situation.

The telecommunications network designer must look at the needs of the particular application to determine what is adequate response time or real enough time for each application. The response time required affects the telecommunications network design. Without proper design, the response time objectives will never be obtained.

Usable Dialogues As the use of terminals becomes widespread, many people who have had no previous computer experience are becoming telecommunications and computer users. The need to make the interaction with the computer easy is greater than ever before. The interaction or series of interactions between the user at a terminal and the computer is called a *dialogue*. A typical requirement of the user is that the dialogue with the computer be *user friendly*. User friendliness is an attribute of computer interactions that is difficult to describe and in many cases is strictly relative. A dialogue that may be easy for one person to use may not appear to be easy, or friendly, to another person.

It helps to break the concept of user friendliness into two subconcepts: "easy to learn," and "easy to use." Easy-to-learn interactions enable a person with little instruction or guidance to figure out how to operate the terminal or system and get the desired results. Easy-to-use interactions are ones that may take some training to learn but, once the education has been completed, are simple to perform in the desired way. The two attributes are not mutually exclusive or always mutually desired.

For example, if the dialogue between an airline reservation agent and the computer is being designed, the major concern should be for ease of use because the same set of transactions will be completed thousands of times as reservations are made. Shortcuts and flexibility are very important so that the reservation agent can jump around within the reservation process when necessary. Usually the input data is highly coded so that a minimal number of characters needs to be typed. Ease of learning the system is of secondary importance, since the operators receive some training and usually work in an environment where other, more experienced operators can provide assistance when necessary.

By contrast, the design of the dialogue for use with an automated teller machine needs to be easy to learn, since many of the users are completely inexperienced with terminals and may use the ATM infrequently. Furthermore, there is no teacher or coach standing by ready to assist, particularly in off hours when the bank is closed. Ease of use is a secondary concern. This implies that the transactions through an ATM may be somewhat rigid in sequence, content, and format, with little variation allowed. There should be an option to abort the transaction or go back to the beginning, but beyond that, the user should be taken through a straightforward series of inputs and actions to accomplish the desired results.

Ergonomics Ergonomics is another user requirement that the telecommunications designer must consider and understand. As with other requirements, the goal is to ensure that the network and all of its components meet user requirements. *Ergonomics* is, according to Webster, "the study of the problems of people in adjusting to their environment, especially the science that seeks to adapt work or working conditions to suit the worker." In recent years, as the use of computer terminals has become more widespread, ergonomic issues have received considerable attention.

The telecommunications manager or designer needs to understand the ergonomic issues and deal with them, but does not have total responsibility for resolving them. The design of the individual office or workspace and the selection of furniture is not usually within the telecommunications manager's responsibility. Often this is done by office designers or consultants who lay out the office and select the furniture to go in it. Hardware manufacturers determine the physical characteristics of their terminals, and the telecommunications department has little direct influence over the decisions made. However, equipment manufacturers have typically been sensitive to the ergonomic issues and have designed their equipment with the flexibility to meet personal preferences or legislative requirements. The specific details of workstation ergonomics are discussed in Chapter 6.

Flexibility One thing is certain about telecommunications systems and networks: They change. It is important when you plan and design a network to strive for flexibility. As companies grow, the network must handle increasing traffic volumes by adding capacity. As companies reorganize, the network may have to be reconfigured to handle new traffic patterns. As technology changes, there will be a demand from the user community to have the newest terminals and capabilities. As new uses are made of telecommunications, new capabilities, such as graphics or voice terminals, may need to be added.

Users will expect to be able to make changes to the telecommunications system and will have little empathy with technical arguments as to why changes are difficult to make. On the other hand, users will not directly ask for telecommunications flexibility, so the network designer must ask probing questions to draw the users into thinking about the future. Although not all changes can be foreseen, the network designer should work with the users' needs to anticipate as many of them as he or she can.

MANAGEMENT AND ENTERPRISE REQUIREMENTS

In any organization, the staff of various departments have certain responsibilities to perform and roles that they play. The marketing and sales departments are expected to sell products or services and to lay plans for future sales growth. The manufacturing function is responsible for producing products, but also to act as a technology center to find ways to produce the products more effectively and at a lower cost in the future. These roles can and do change over time.

Telecommunications in most companies is emerging from the back room. For years it has been buried in the depths of the company, both physically and organizationally. In the past, the mission has been to provide service at the lowest possible cost, but that usually isn't good enough anymore. Management is beginning to realize that telecommunications is vital and should become a part of the overall strategy of the organization.

In earlier chapters, we have seen several applications of telecommunications that have literally changed the way in which business was done. The airline reservation application and the use of ATMs and checkout terminals each has made a dramatic change in their respective industries. In the mid-1960s, Southern Pacific Railroad began work on the TOPS system, which automated the routing and scheduling of railroad trains and even individual railroad cars. In the insurance industry, large companies have realized that telecommunications is an effective way to communicate with field offices and agents who are not employees but independent businessmen. Insurance company networks are some of the largest installed today. Many other examples exist. Telecommunications is having a significant impact on the operations of many companies and industries.

Proactive Telecommunications Management

As the strategic importance of telecommunications is realized, the telecommunications department moves into a more visible role in the company. Management's expectations are higher. Providing telecommunications service is as important as ever, but executive management is looking for opportunities to use telecommunications as a strategic weapon to obtain competitive advantage. Managers expect telecommunications management to lead the way and identify opportunities. The emphasis is on making money with telecommunications rather than saving money. It's a subtle difference, but it has huge implications for the telecommunications department.

The classic example of strategic telecommunications is the case of American Hospital Supply (AHS). In the early 1980s, the company developed an easy-to-use order entry and inventory system for use by hospitals. American Hospital Supply installed terminals connected to the AHS network in hospitals that were its customers. Hospital personnel were able to use the system directly to track their inventory *and* to place orders for additional supplies directly from AHS. Since the system provided a service and made it easier to replenish the inventory, it became widely successful. American Hospital Supply gained a huge market share.

Proactivity in external affairs, such as getting involved in industry groups or trade organizations, is also desirable. In some cases, companies with a large stake in telecommunications are getting involved at the legislative level, attempting to ensure that any telecommunications legislation is compatible with the company's aims and objectives.

Quality Orientation

More and more companies are installing quality control programs throughout their operations. This trend began several years ago and grew out of the concern for the decline of the American automobile industry and the rise of their Japanese counterparts who were delivering high-quality, more reliable cars. The irony is that the Japanese learned many quality techniques from an American, Dr. W. Edwards Demming, who went to Japan and instilled a series of quality principles into the Japanese way of thinking about corporate operations.

Now, organizations in the United States are realizing that following the quality principles that Dr. Demming taught the Japanese makes good business sense, leads to more satisfied customers, and has a strong impact on a firm's ability to be competitive.

Modern quality thinking teaches four principles:

1. understand and meet the customer requirements;
2. aim for error-free work and "doing it right the first time";
3. manage by prevention;
4. understand and manage by the cost of quality.

Let's see how these principles can be applied to telecommunications.

Meeting the Customer's Requirements In a telecommunications network, the customers are the users of the telecommunications system. An important part of quality thinking is that the requirements are what the customer says they are. That is not to say that customers can absolutely dictate and get everything they want. It does imply however, that after joint discussion and negotiation, the ultimate decision about the requirement rests with customers.

Meeting the requirements means just what it says–*meet*. It means that the requirements should not necessarily be exceeded. For example, if 1-second response time for a certain telecommunications application meets the customer's requirements, the cost to achieve half a second may be money wasted. If 9 hour per day availability meets the requirements, it is probably not necessary or cost effective to operate the network for a longer period of time.

Aiming for Error-free Work and "Doing It Right the First Time" The principle of error-free work means that when problems occur (and they will occur) everything is done to find the cause of the problem and eliminate it . . . permanently. "Doing it right the first time" seems self-evident. But why is it that so many times in life we don't have time to do it right but always have time to do it over? Obviously, doing it right the first time is closely related to the prevention of errors. Doing it right the first time is an attitude that must be conveyed to employees.

Managing by Prevention Managing by prevention has been shown in many cases to be the least expensive and most effective way of preventing errors and problems. In product design, the sooner a design flaw is detected, the less it costs to correct it. In telecommunications, designing the network to deliver the required response time, backup, or redundancy is simpler and ultimately less costly than adding additional lines or computer processing power later to correct a response time or reliability problem. Prevention is a difficult task. It requires thought, anticipation, and careful planning. It is usually not as glamorous as firefighting, and few heroes are created, but preventing errors is the key to error-free work and improved productivity.

Managing by the Cost of Quality Understanding what it costs to provide a quality product or service is another part of the quality process. The cost of quality is composed of the cost of preventing or detecting errors plus the cost of the errors themselves that actually occur.

$$\text{Cost of quality} = \text{error costs} + \text{error prevention and detection costs}$$

In a telecommunications environment, the cost of quality is both the cost of programs put in place to prevent errors and the cost of the errors that do occur. Some of these costs are seen as losses in user productivity. An error might be the improper installation of a terminal so it fails to work. Another type of quality error occurs when response time does not meet the defined requirements.

Monitoring and measurement systems can be put in place to track quality errors. The cost of the errors can be calculated and added to the cost of prevention programs. The total of the two is the cost of telecommunications quality. With that number in hand, management can make decisions about changes to the quality program that would enhance the product or service or reduce cost without sacrificing quality.

Security

There has been a great deal of focus in recent years on computer and telecommunications security. This higher awareness and visibility have been brought about by a few well-publicized cases of computer crime, including cases where access to computers has been made by telecommunications links. This type of crime was popularized by the movie *War Games*, in which a boy with a personal computer in his bedroom was able to access computers in the Defense Department. Numerous cases have been reported where people have gained access to computers for which they were not authorized. These people epitomize the common view of the *hacker*, a term originally used to describe a technically inclined individual who enjoyed pushing computers to the limits and making them do things no one thought possible. In recent years, however, the term has come to be associated with mischievous, malevolent intention to access computers in order to change or destroy data or perform other unauthorized operations.

Obviously, management is concerned that company data be accessible and usable to those who need to use it but protected from those who don't need it. They also want to be sure the data is protected from accidental or intentional destruction and from unauthorized alteration. Specific security techniques are available to address these concerns, and they are discussed in Chapters 7 and 14. Some people would say that if security is extremely important, telecommunications probably shouldn't be designed into a system at all. It is true that the transmission of data or information outside of a company's facilities increases the chances of data being intercepted or manipulated. Most businesses, after examining the exposures and the security techniques available, conclude that the potential benefits of telecommunicating their data outweigh the risks of its being stolen or misappropriated.

Cost Effectiveness

Up to this point, we have not emphasized the importance of cost as an internal influence on the way telecommunications is managed in the enterprise. Enlightened management wants the telecommunications system to be cost effective but is not necessarily concerned that it operate at the least possible cost. Management is interested in the overall effectiveness of the money spent for a telecommunications system, in the effectiveness and efficiency of the network, and in its contribution to the overall success of the company.

The standard approach to telecommunications cost management has been to ensure that the monthly bills from the telecommunications vendors are accurate and paid on time and that the number of telecommunications lines and other equipment is kept at a minimum to provide the desired service. In the days when telecommunications was a monopoly in the United States, there were few other actions that the telecommunications manager could take.

competing services

With deregulation and the rise of competition, there are now many alternatives for most telecommunications services, and the telecommunications department has a responsibility to investigate competing alternatives. Long distance telephone service is available from several companies besides AT&T. Telephone systems may be provided by the telephone company or purchased from independent manufacturers such as Northern Telecommunications or Siemens-Rolm. When the manager compares alternatives, it is of course necessary to be sure that the services and products being compared are truly the same. This type of investigation places more of a burden on the telecommunications department; no longer can the manager just go to the telephone company and order the only product available.

renting, leasing, or buying equipment

Another cost management technique is the application of economic analysis tools to help make the financial decisions of the telecommunications department. In times past, telecommunications products could only be rented from the telephone company, and there was a fixed price. Now, in addition to selecting between competing alternatives, there are financial alternatives to evaluate—whether to rent, lease, or buy a particular product. The telecommunications manager and staff must know about and use techniques, such as discounted cash flow and the present value of money. A study of these techniques is outside the scope of this book, but they are covered in all basic finance courses. If the company has finance department specialists using and applying these tools, it is wise to get them involved to ensure that the best financial decision is made for the company.

Another set of decisions the telecommunications staff must deal with is whether to perform certain services within the department or to contract them to outsiders. Examples include publishing the organization's telephone book, equipment installation, and troubleshooting and maintenance. As usual, there are trade-offs. It may be more cost effective to have certain services provided by outside organizations. Because the company does not directly control outside maintenance people, for example, service may not be as timely as if it were done by company employees.

Most companies today do not have the staff or expertise to perform all of the service and maintenance activities. Vendors offer maintenance contracts on their equipment and have technicians with special training and testing equipment. Independent service firms, such as Sorbus or RCA, maintain many types of telecommunications equipment.

SUMMARY

Managing and operating a telecommunications department in an organization is subject to influences and pressures from telecommunications users and company management. Although not unlike the pressures that other functions in the company experience, many of these are new for the telecommunications department. Until recently, telecommunications was a back room operation that bought its products and services from a single telephone company with no competitive or cost alternatives. In that environment, decisions were simple; because there was only a limited set of products and services available, the telecommunications staff had little flexibility in the types of services they provided.

Now telecommunications is becoming recognized as a vital function within many companies. Users are more knowledgeable about telecommunications alternatives and choices, and they are demanding that their requirements for telecommunications services be given consideration. Management expects the telecommunications department to become more proactive and lead the company into opportunities where telecommunications techniques can be strategically employed to improve the overall competitiveness of the company. Not only must the telecommunications department understand its customers' requirements and select from alternative methods and products, it must also help to define new ways in which the technology can be deployed. In many ways, the telecommunications department is becoming "a business within a business," with a full set of research, marketing, financial, personnel, and operational issues to deal with.

CASE STUDY

Dow Corning Corporation

This case study is intended to describe how a company is using telecommunications. The Dow Corning case will continue throughout the book, with a different aspect of the company's telecommunications network and management being discussed at the end of most chapters. It is hoped that the student will see how many of the ideas and concepts discussed in the text apply in a real life situation.

Introduction to the Company

Dow Corning Corporation was founded in 1943 as a 50–50 joint venture of The Dow Chemical Company and Corning Glass Works. The company's business is to develop, manufacture, and market silicones and related specialty materials. Dow Corning's corporate headquarters is in Midland, Michigan, located about 120 miles north of Detroit. Midland is a town of approximately 35,000 people and is also the corporate headquarters for The Dow Chemical Company. From Midland, Dow Corning oversees a worldwide operation with 1989 annual sales of about 1.6 billion dollars. The company has sales offices in major cities throughout the world and manufacturing plants in many industrialized countries.

Since silicone products can take many forms, Dow Corning's product line is extensive. In various forms, including fluids, rubbers, and resins, silicone products function as additives to other products, as ingredients in consumer products, as processing aids, as maintenance materials, and as final products.

Dow Corning's manufacturing activities are highly integrated. Products produced by one plant may be sold to customers directly or used as chemical intermediates in other manufacturing processes. For this reason, there must be close coordination and communication between

Dow Corning Corporation's corporate headquarters in Midland, Michigan. (Courtesy of Dow Corning Corporation)

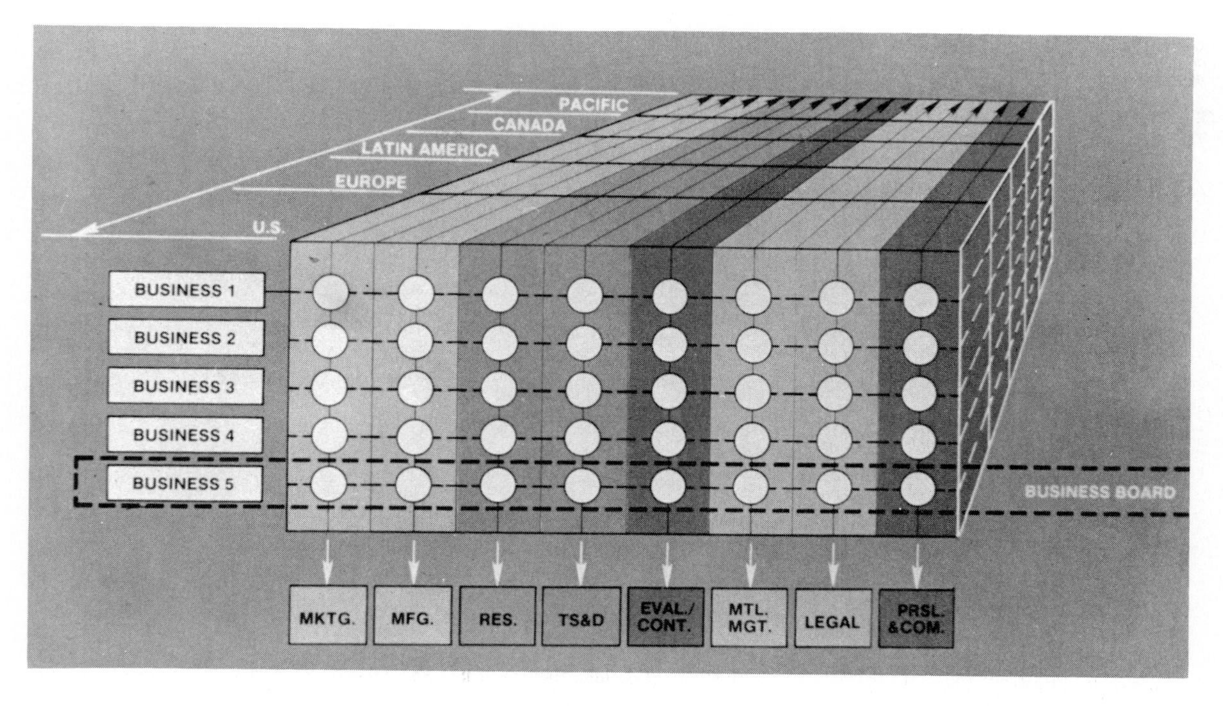

A diagram of Dow Corning's three dimensional matrix organization. (Courtesy of Dow Corning Corporation)

marketing and manufacturing to determine which products will be sold and which will be shipped to other plants for further processing. Furthermore, the plants must coordinate their activities to ensure that they have a steady supply of materials from one another.

Dow Corning's organization is a three-dimensional matrix. In one dimension, the company has the traditional functions found in most manufacturing companies. Research, manufacturing, marketing, finance, and personnel are each headed by a vice president of the company. In the second dimension, there are senior executives in charge of each major group of products, which Dow Corning calls "businesses." These business managers are responsible for the worldwide profitability of their product line. Finally, Dow Corning divides the world geographically into areas, each headed by an area manager. The United States and Inter-American area (Canada and Latin America) are headquartered in Midland. The European area headquarters is in Brussels, the Japan area headquarters is in Tokyo, and the Pacific area headquarters is in Hong Kong.

Historically, Dow Corning's management style and methods have been quite centralized. Most of the major decisions about company operations have been made at the Midland headquarters, with freedom given to the foreign areas to make operating decisions that are appropriate for their countries. Recently, a senior executive stated that, in the 1990s, Dow Corning would be "more decentralized, yet more centralized." What he meant is that as the company grows, it will be necessary to push more of the decision making lower in the organization. In addition, Dow Corning expects to experience some growth by acquiring other companies. At the same time, the executive expects that the company's financial and information systems will continue to be centralized. There is a strong feeling among Dow Corning's executive management that integrated financial and information systems are essential to maintaining the close coordination required among the company's operating units. There is also a belief that these systems provide data needed to make the appropriate decisions that enable the company to maintain its profitability.

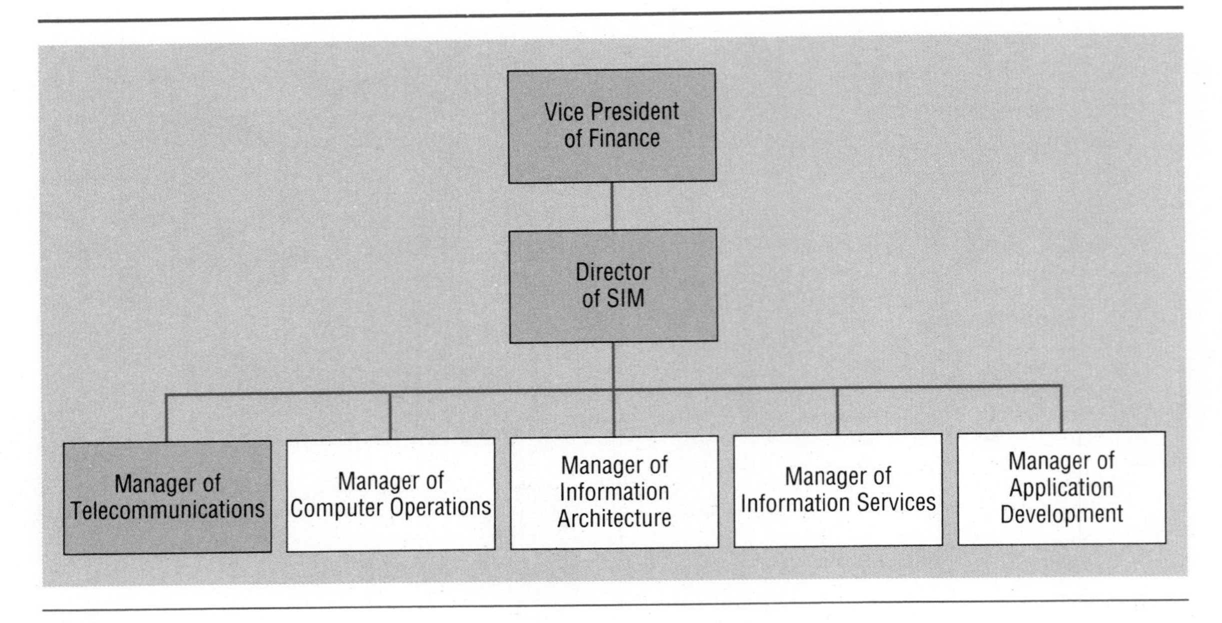

Figure 3–5
Dow Corning's SIM organization.

Telecommunications at Dow Corning

The responsibility for telecommunications in Dow Corning lies within the Systems and Information Management (SIM) department. The director of SIM reports to Dow Corning's vice president and chief financial officer. The managers of telecommunications, computer operations, information services, information systems, and information architecture report to the director of SIM, as shown in Figure 3–5.

As in the rest of the company, the staff size of the telecommunications department is kept lean. At present, it has 15 people. The department develops and operates a worldwide data communication network that connects all major Dow Corning locations to one another. The department is also responsible for the company's U.S. voice network, whereas organizations outside the United States handle their own telephone systems. There are approximately 5,000 terminals on the data network, about 3,500 of them in the United States. The reliability of the data network is about 99.5 percent during its scheduled hours of operation. One goal is to provide subsecond response time for most transactions. Although this objective is being met in many cases, constant monitoring is required to keep the performance optimized and to try to anticipate impending bottlenecks before they occur.

The data communication network is used for all types of computer-based applications. Dow Corning has integrated its order processing and inventory systems with its manufacturing planning process, and these applications are all online. Financial and personnel systems also are online and accessed by approved users throughout the company. A large electronic mail and end-user computing application runs on a separate computer from the more traditional business systems, and it shares the same telecommunications network. Most Dow Corning employees use a terminal as a routine part of doing their jobs.

Review Questions

1. What are the factors that determine the hours during which a telecommunications network must be available?
2. Compare and contrast the terms *reliability* and *availability* as they apply to a telecommunications network.
3. Explain how the reliability of a serial system is calculated.
4. Explain the difference between an *online* system and a *realtime* system.
5. Explain the concept of real enough time.
6. How is response time measured?
7. What are the factors that determine what a system's response time will be?
8. Why is queuing a factor in calculating response time?
9. Why does fast response time help the productivity of a terminal user?
10. Why is response time normally calculated on a probabilistic basis?
11. What is *think time?*
12. Compare and contrast the concepts *easy to use* and *easy to learn.*
13. Why does the telecommunications manager need to understand ergonomic issues?
14. Why is it important for a telecommunications network to be flexible?
15. Explain why it is important for a telecommunications department to be proactive and forward thinking.
16. Explain why senior management in a company is interested in cost-effective telecommunications.

Problems and Projects

1. If an ATM is 98 percent reliable, the telecommunications line to which it is connected is 96 percent reliable, and the computer to which they are attached is 99 percent reliable, what is the overall reliability of the three-component system?
2. If you were designing an "ideal" telephone system for the United States, what characteristics would it have that the present system lacks?
3. Make a chart that shows the time differences between Chicago, Brussels, Hong Kong, and Tokyo at different times of the year. (You may have to do some research to investigate the impact of daylight savings time in the various countries.)
4. Calculate the average reliability that each component of a three-component communication system would have to have so that the reliability of the total system would be 99 percent.

5. You have just been named the manager of a full-service department store, one of several in a statewide chain. The company will soon be installing its first POS terminals, which will have hand-held scanners to read merchandise tickets. You have heard that the terminals in your store will be connected to the central computer at company headquarters, but you have no other details. Next week two analysts from the corporate telecommunications department will be coming to talk to you about your requirements for the new system. In preparation for their visit, make a list of the attributes and operating characteristics that you hope the new terminals and communication network will have, and the questions you will want to ask the analyst.

Vocabulary

availability
reliability
mean time between failure (MTBF)
mean time to repair (MTTR)
serial system
real-enough time
response time
queuing
queue
user response time
think time
dialogue
user friendly
ergonomics
hacker

References

Crosby, Philip B. *Quality is Free.* New York: New American Library, Inc., 1979.

Doherty, W. J., and R. P. Kelisky. "Managing VM/CMS Systems for User Effectiveness." *IBM Systems Journal,* 1979, Vol. 18, No. 1.

"The Economic Value of Rapid Response Time." IBM brochure, number GE20-0752-0 (11-82).

Levy, Steven. *Hackers Heroes of the Computer Revolution.* New York: Anchor Press/Doubleday, 1984.

Martin, James. *Design of Man-Computer Dialogues.* Englewood Cliffs, NJ: Prentice-Hall, 1973.

CHAPTER

External Influences on Telecommunications in the Enterprise

OBJECTIVES

After studying the material in this chapter, you should be able to

- describe the three major forces at work changing the telecommunications industry;
- tell how the United States progressed from no telecommunications regulation to a highly regulated environment and then back to a middle ground;
- distinguish between the regulated and unregulated portions of the telecommunications industry;
- describe the different types of common carriers that provide communication service;
- describe the impact that technological change is having on the telecommunications industry;
- describe the impact that evolving standards are having on data communication;
- describe in general terms the ISO-OSI reference model.

INTRODUCTION

The telecommunications organization in a company is subject to many external influences. Not only is technology changing rapidly (as in the computer arena), but the telecommunications's regulatory environment is changing, particularly in the United States. Furthermore, the emergence of voice and data communications standards is affecting the telecommunications environment.

The purpose of this chapter is to look at the background, current situation, and trends of the regulatory, technology, and standards issues to further put the telecommunications world into perspective before we delve into the technical details of how telecommunications works.

THE REGULATORY ENVIRONMENT

During most of the twentieth century, the telecommunications industry in the United States has been regulated. Companies in the industry have been subject to legislation affecting the way they operate, the prices they charge, and the profits they make. In the United States, the telecommunications industry has consisted of private companies that were subject to regulation, whereas in most other countries, the telecommunications industry is operated by a government organization or bureau, such as the one that operates the post office and mail system.

In the United States, the communications regulatory situation has changed dramatically in the last several years. The communications industry has moved from a position of being almost totally regulated in 1983 to being largely deregulated and competitive today.

Why Regulation?

Why has the world's telecommunications industry been subject to heavy regulation? As the fledgling communications industry caught hold in the years around the turn of the century, it was recognized that widespread, general purpose communications could have significant national benefit if and only if they developed in a uniform, compatible way. Localized networks were not good enough. What was needed was universality, the ability of a person at any telephone to communicate with a person at any other telephone.

ensure compatibility

The only way to ensure this universality was to control what was designed and implemented and to insist that all parts of the telephone system worked in the same way. It was also felt that it was important to provide uniform communications capability to all parts of the country. People believed that the same communications services ought to be available in Billings, Montana, as in New York City. It was also important to guarantee the communications companies a fair profit for their investment and for the risk they were taking and to motivate them to stay in the communications business.

The telecommunications industry is not unique in this regard. Other industries where universality is desired or that have come to be viewed as a national resource have been regulated too, such as

- transportation;
- banking;
- insurance;
- utilities.

protect companies

Regulation can also serve to protect new companies in a fledgling industry by preventing competition and allowing them time to get their feet on the ground. This is particularly important in industries that must make heavy capital investments to build an infrastructure with which they can operate. For example, railroads spent millions of dollars to gain right-of-ways and to lay track before they could run a single train. Similarly, communications companies not only needed equipment, such as telephones, but also networks of wires and central office switching equipment.

prevent interference

In the communications industry, regulation also serves the purpose of preventing interference. What if all of the radio stations in a metropolitan area decided to broadcast on the same frequency? None of the programs would be intelligible to the listeners because they would all interfere with one another. Communications regulations allocate certain frequencies to each radio or television station and separate them so that they don't interfere with one another.

In certain industries, the regulatory concept of a *common carrier* emerged. This concept limits the number of companies that can provide key public services, such as transportation or communications. The idea is to prevent the duplication of services or the expensive, capital-intensive infrastructure, such as railroad tracks, communications transmission facilities, or pipelines necessary to provide the services. In the communications industry, the companies that provide the telecommunications networks are the common carriers, sometimes just called *carriers*. Even today their activities are regulated, in return for which they are given monopoly power within the territory or region they serve.

Milestones of Telecommunications Regulation in the United States

Figure 4–1 summarizes the important milestones of telecommunications regulation in the United States. It is important that you have some understanding of the series of events that led the country into regulation and, more recently, back to a largely deregulated status. This information is a useful aid to understanding the seemingly complex telecommunications industry.

Although AT&T had been founded to provide the nationwide network and interconnection service between Bell franchised telephone companies, many other companies sprung up to provide telephone ser-

1907	First state regulatory agencies established in New York and Wisconsin
1910	Mann-Elkins Act
1913	Kingsbury Commitment
1921	Graham Act
1934	Communications Act of 1934—FCC established
1956	AT&T Consent Decree
1968	Carterfone Decision
1969	MCI Decision
1971	Computer Inquiry I
1971	Open Skies Policy
1981	Computer Inquiry II
1982	Modified Final Judgment
1984	Divestiture
1986	Computer Inquiry III

Figure 4–1
Milestones of United States telecommunications regulation.

early competition

vice in limited markets when the original Bell patents expired in the early 1890s. As companies entered the telephone industry, competition was rampant. At one time, in the early 1890s, more than 125 companies were in operation providing duplicate telephone services. These companies naturally tended to focus their efforts in major population centers where the market was the largest while ignoring less populated areas of the country. Big cities had telephone service long before it reached small towns.

Not only was their coverage not uniform, but there was no interconnection between the various networks. Customers could place telephone calls only to other customers who were on the same telephone company's network. In many cases there was no way to call a telephone subscriber connected to a different telephone company.

Naturally, these independent companies wanted to connect to AT&T's nationwide network, but AT&T refused to make the connection. As a further defensive measure and in an attempt to maintain its unique position in the market, AT&T bought out as many of these competitors as it could.

Mann-Elkins Act Congress finally reacted to the heavy competition and, in 1910, passed the Mann-Elkins Act, which placed the activities of all telephone companies doing interstate business under control of the Interstate Commerce Commission (ICC). The main focus of this action was to install a uniform system of accounting for all telegraph and telephone common carriers. It was the first step toward regulation of the telecommunications industry.

Kingsbury Commitment The competitive pressure exerted by AT&T became burdensome to the independent companies, and in 1912 they complained to the federal government about AT&T's practices. Their major concern was the fact that AT&T was buying out the independent telephone companies and stifling competition. In early 1913, the United States Department of Justice and the ICC both informed AT&T that some of its activities appeared to be thwarting competition and were in violation of the Sherman Antitrust Act. The ICC began an investigation of AT&T's activities.

AT&T was concerned that an antitrust suit by the government could cause irreparable damage to its company and business, and so, at the end of 1913, the company agreed to

1. dispose of its stock in Western Union;
2. get approval from the ICC before acquiring any additional independent telephone companies;
3. allow the independent telephone companies to interconnect to its facilities so that they could offer nationwide telephone services to their customers.

This agreement is known as the Kingsbury Commitment after the AT&T vice president, Nathan C. Kingsbury, who authored it for AT&T. The Kingsbury Commitment is significant because it set the telecommunications system on a course toward *universal service*. For the first time it would be possible for all of the communications companies to be interconnected into one nationwide network.

universal service

Graham Act In 1921, Congress passed the Graham Act, which recognized and legitimized AT&T's monopoly. The Graham Act specifically exempted the telecommunications industry from the Sherman Antitrust Act in terms of consolidating competing companies. It also allowed the independent telephone companies to become monopolies in their own geographic territories. This ensured the establishment and building of a nationwide and nonoverlapping telephone network covering the entire country. The view of the day was that allowing these monopolies to exist was in the public's best interest.

The Communications Act of 1934 and the Federal Communications Commission (FCC) The Interstate Commerce Commission had been empowered to regulate interstate telephone services in 1910 but proved to be more interested in regulating the transportation industry than in dealing with the emerging telecommunications issues. As a result, Congress passed the Communications Act of 1934, which created a new agency, the *Federal Communications Commission (FCC)*, whose sole purpose was to regulate the telecommunications industry. The FCC, as stated in the opening paragraph of the communication act, was created

> For the purpose of regulating interstate and foreign commerce in communication by wire and radio so as to make available, so far as possible, to all the people of the United States a rapid, efficient nationwide and worldwide wire and radio and communication service with adequate facilities at reasonable charges. . . .

The formation of the FCC was the beginning of serious regulation in the communications field. The FCC is a board of commissioners appointed by the President of the United States. The commission, supported by a technical staff, has the power to regulate all interstate telecommunications facilities and services as well as international traffic within the United States. The FCC also controls the radio and television broadcasting industry and issues individual station and operator licenses.

The descriptions of all regulated telecommunications services and the rates to be charged are called *tariffs*. The common carriers file tariffs with the FCC whenever they want to offer a new interstate or international service or change the price of an existing service. The four major rate categories for tariffs are

tariffs

1. the charge for the time the communication service is used, including possible variations for different times of the day; example: direct distance dial long distance telephone charges;

2. a flat rate for full-time use of a service; example: a leased line;
3. monthly minimum charge for a basic amount of usage with additional charges when the usage exceeds the basic amount; example: WATS lines (discussed in detail in Chapter 5);
4. a charge for the amount of data sent; example: packet data transmission.

All tariffs are a matter of public record. The FCC reviews tariffs as they are submitted and ultimately approves or disapproves them.

AT&T Consent Decree In 1949, the United States Department of Justice sued AT&T for violation of the Sherman Antitrust Act. The specific charge was that the absence of competition had tended to defeat effective public regulation of the rates charged to subscribers. Of specific concern was the Western Electric Company, the Bell System's manufacturing arm, which the Department of Justice wanted separated from the rest of AT&T.

The suit was settled in 1956 by a consent decree that permitted AT&T to keep Western Electric but specifically limited the Bell System companies to the telephone business. Western Electric was prohibited from manufacturing noncommunications equipment and was ordered to limit its sales to Bell System companies. Since computing was a very young science at the time, the significance of this decision, which kept AT&T and the Bell System out of data processing activities, may not have been fully realized.

Carterfone Decision In 1968, the first chink in the monopolistic structure of the U.S. telecommunications industry appeared. The FCC decided that the Carter Electronics Corporation, a small Dallas-based company, could attach its Carterfone product to the public telephone network. The Carterfone was designed to allow the interconnection of private radio systems to the public telephone network, and Carter Electronics Corporation sought permission to sell its product. When AT&T refused to allow the Carterfone to be attached to its network, Carter Electronics sued AT&T and won.

That decision opened the door for the attachment of non-telephone company devices to the telephone network and spawned a new segment of the communications industry. After the Carterfone decision, many companies saw an opportunity to make equipment for attachment to the telephone network. This equipment provided customers with new alternatives, such as decorative telephones, private telephone systems for business, and the option to buy rather than rent telephone equipment. The new segment of the telecommunications industry that emerged has become known as the *interconnect industry*.

interconnect industry

MCI Decision The next step toward deregulation and competition was a decision the FCC made in 1969. This decision allowed Microwave Communications Incorporated (MCI) to provide intercity telecommunications links to organizations that wanted to lease them on a full-time basis for their exclusive use. This type of private line service had been offered by AT&T and the Bell System for many years, so the product wasn't new, but allowing a private company to provide network facilities was. The MCI decision required the telephone companies to interconnect the MCI lines into the public telephone network, giving their customers nationwide access.

The MCI decision opened the door to network services, and several companies quickly jumped in to provide comparable network offerings. Since they were not required to provide uniform capabilities throughout the country, the new companies tended to offer their services only on high-density routes between major cities, such as Chicago to New York or New York to Washington. Southern Pacific Communication Company, a subsidiary of Southern Pacific Railroad, was one of the first companies in the business and offered network services by taking advantage of the railroad right-of-ways it already owned. It was a relatively simple matter for Southern Pacific to run telecommunications wires alongside the tracks, and it became the first company outside of the Bell System to offer a coast-to-coast alternative service. These companies became known as *other common carriers (OCC)* or *specialized common carriers (SCC)*. (MCI was later renamed to MCI Telecommunications Corp, still commonly abbreviated as MCI.)

other common carriers

Computer Inquiry I (CI–I) In 1971, the FCC concluded a study known as *Computer Inquiry I (CI–I)* in which it examined the relationship between the telecommunications and data processing industries. At issue was the fact that although the two industries were rapidly growing closer, the data processing industry was unregulated and the telecommunications industry was regulated. An attempt was made by the FCC to try to determine which aspects of both industries should be regulated over the long term. The outcome of CI–I was that the FCC said that the data processing industry was not subject to its control.

Open Skies Policy Also in 1971, the FCC reversed a previous decision regarding communication by satellite. The resulting open skies policy declared that, with few restrictions, anyone could enter the communication satellite business. This decision opened the door to the formation of satellite transmission companies. Several existing companies, such as Western Union and RCA, soon jumped into this market, and new companies, such as Satellite Business Systems and American Satellite, were formed to take advantage of the new policy and offer satellite communication services.

Computer Inquiry II (CI–II) The year 1981 marked the conclusion of another study by the FCC to look at the relationship between data processing and telecommunications services. This study resulted in a decision known as FCC *Computer Inquiry II (CI–II)*, which stated that

1. computer companies could transmit data on an unregulated basis;
2. the Bell System was allowed to participate in the data processing market;
3. customer premise equipment and enhanced services would be deregulated and provided by fully separate subsidiaries of the carriers;
4. basic communications services would remain regulated.

CI–II provided structural safeguards against cross subsidization by requiring the carriers to have different subsidiary companies handling the enhanced and basic services.

enhanced service

In *enhanced service*, some processing of the information being transmitted occurs. The processing may be simply a conversion of the speed or the coding system, but it can be more extensive. Enhanced services are not regulated by the FCC.

basic service

In contrast with enhanced services are the *basic services*, which are tariffed and regulated. Basic service provides only the transportation of information. No processing or other change to the information is allowed. Basic services are provided by the traditional common carriers.

Because of CI–II, the *Bell Operating Companies (BOCs)*, the 22 telephone companies that were members of the Bell System, and AT&T organized new subsidiary companies to handle the unregulated products and services. AT&T chose the name American Bell for its unregulated subsidiary; however, this company name was to be short-lived.

Modified Final Judgment 1982 was a landmark year in the United States telecommunications industry. It marked the completion of an 8-year antitrust suit by the federal government against AT&T. The suit, which had originally been filed in 1974, was settled by a consent decree between the U.S. Department of Justice and AT&T.

Many in the data processing industry thought that the monopolistic nature of AT&T prior to 1982 was having a serious and detrimental impact on the U.S. telecommunications system. Problems included the lack of willingness to meet customer—particularly business customer—requirements and slow incorporation of technological change into offered products. These were the primary reasons the government brought suit against AT&T.

The original consent decree issued by the court was modified slightly by Judge Harold Green before it received final approval, and it became known as the *modified final judgment (MFJ)*. The judgment stipulated that on January 1, 1984, AT&T would divest itself of all 22 of its associated operating companies in the Bell System, such as Michigan Bell Telephone Company, Pacific Bell Telephone Company, and Southwest Bell Telephone Company. Also, the MFJ stated that the Bell name was reserved for the use of the divested operating companies. AT&T had to rename its new American Bell subsidiary, and it chose AT&T Information Systems (ATTIS) as the new name.

The restructuring that followed the modified final judgment was traumatic for AT&T, the BOCs, their employees, and customers. It meant that the operating companies were, for the first time, independent. There was a great deal of confusion about which companies would supply specific services and how they would be provided. Employees were transferred from AT&T to the operating companies and vice versa. Later in the chapter, we will look at the current state of the telecommunications industry in detail.

The changing cost structure of the local telephone companies and AT&T caused prices for products and services to be adjusted, in some cases radically. For years, the Bell system had subsidized less profitable services with revenue gained from services that were more profitable. For example, the long distance telephone rates for calls on a route between two large cities with a high volume of traffic were set high enough to cover the costs of calls between cities in remote locations or with a low volume of calls. Callers from New York to Chicago helped pay for the cost of calls between Sioux City and Fargo. The modified final judgment specifically prohibited this type of cross-subsidization and essentially stated that each service must pay for itself.

relief from CI–II

After CI–II, AT&T and the BOCs were required to offer any enhanced services through separate subsidiary companies. AT&T used ATTIS to sell the unregulated, enhanced services, whereas AT&T Communications (ATTCOM) sold the regulated, network products. The BOCs established similar subsidiaries. All of the carriers felt that the separation was unwieldy. Each subsidiary had a separate marketing effort and sales force calling on the customer, which increased the carrier's overall costs.

Almost immediately after divestiture, the carriers petitioned the FCC to permit them to sell regulated and unregulated products from a single company. In late 1986, the FCC granted AT&T "relief from CI–II," which eliminated the requirement for separate subsidiaries. The terms of the relief included new accounting and network disclosure rules to replace the structural safeguards against cross-subsidization that had been provided by CI–II. As a result of this relief, AT&T merged ATTIS and ATTCOM into a single unit, called AT&T Communications Services.

Computer Inquiry III (CI–III) *Computer Inquiry III (CI–III)* was a study conducted by the FCC to determine how and to what extent the carriers could offer enhanced services. In a decision announced in mid-1986 and reaffirmed in early 1987, the FCC ordered that the BOCs and AT&T can offer unregulated, enhanced services if the companies agree to a complex set of provisions called *open network architecture (ONA)*. The independent telephone companies, however, are exempted from the FCC order. The BOCs and AT&T must have their plans for the implementation of ONA approved by the FCC before the enhanced services can be offered. The companies submitted their first proposals in February 1988. However, because of inconsistencies in the filings, the FCC told the companies to get together, examine each others' network architectures, and make new proposals for enhanced services. The new proposals were submitted to the FCC in May 1989, and as of this writing (late 1989), the FCC has not responded.

Deregulation

In the 110 years since the telephone was invented, there has been a transition from unrestricted competition to complete monopoly powers vested in AT&T and the Bell System and now back to an intermediate ground. In the meantime, computing and data processing have come on the scene, and the nationwide telecommunications network—which was originally designed strictly for voice communication—has become heavily used for data transmission.

The FCC remains as the national regulatory agency that approves all rates and services offered on an interstate basis. Each state has a *public utility commission (PUC)*, sometimes called the *public service commission (PSC)*, which has jurisdiction over intrastate rates and services. The first PUCs were established in New York and Wisconsin in 1907.

PUC

The implication of having state-level PUCs is that it's possible for all 50 states to have different rates for the same service. Fortunately, this is usually not the case; the state PUCs do talk to one another and to some extent coordinate their activities.

The common carriers file tariffs with the appropriate PUC whenever they want to offer a new intrastate service or change the price of an existing service. The process is essentially the same one that is used for filing interstate tariffs with the FCC. Having to file 51 different tariffs (one for each state plus one to the FCC for the interstate tariff) for a nationwide service places a large burden on the common carriers.

Implications of Deregulation

The intent of deregulating a portion of the U.S. telecommunications industry was to provide better, more economical service and new, more flexible products to telecommunications customers. It was felt that competition would stimulate development of innovative products. Furthermore, it was expected that prices for communication services would drop as competition increased. A competitive marketplace was expected to force all companies to become more aggressive in developing and pricing new products and services.

For the telephone companies, this type of competition was a culture shock. In the past, being monopolies, they simply took orders for products and services. The products they chose to offer were the only products available; marketing was unnecessary.

In the new competitive environment, the telephone companies must actively market their products and services or risk losing the business. The telephone companies are free to market their unregulated products and services anywhere in the country. There are no exclusive territories anymore.

Although the impact of deregulation should be generally advantageous for the customer in the short term, the long-term results are more of an open question. Whether the traditionally excellent products and reliability that have been hallmarks of the Bell System can be maintained in the unregulated environment remains to be seen. The regulated environment of the past made it easier to build an integrated telephone system, and the U.S. telephone system is generally regarded as the world's finest.

Status of Regulation in Other Countries

In most other countries, telecommunications has been and continues to be heavily regulated. Generally, the government owns and operates the common carriers and, in some cases, the equipment manufacturers. With deregulation occurring in the United States, other countries have been watching U.S. developments closely, and a few countries have made some deregulatory moves of their own.

Generically, the regulatory agencies in most other countries are called the PTT, which stands for post, telephone, and telegraph. The name comes from the fact that in most countries the same regulatory agency has responsibility for the postal service as well as all telecommunications. A list of the regulatory agencies in selected countries is shown in Figure 4–2. The status of the telecommunications industry in several of

Country	Name of Agency
Australia	Department of Communications (DOC)
Belgium	Regie des Telegraphes et des Telephones (RTT)
Brazil	Ministry of Communications Special Secretariat for Informatics (SEI)
Canada	Provincial governments Canadian Radio-television and Telecommunications Commission (CRTC)
France	Poste Telegraphe et Telephone (PTT)
Germany	Deutsche Bundespost (DBP)
Japan	Ministry of Posts and Telecommunications
Switzerland	Poste Telegraphe et Telephone (PTT)
United Kingdom	Office of Telecommunications (OFTEL)

Figure 4–2
Regulatory agencies in selected countries.

these countries is described in the next few paragraphs. Information about other countries is presented in Appendix A.

Canada Telecommunications transmission facilities in Canada are provided by companies regulated by the government. The regulatory bodies are the provincial governments except in British Columbia, Ontario, and Quebec, where the regulation is provided by The Canadian Radio-television and Telecommunications Commission (CRTC). Services and rates are defined by tariffs, much as in the United States.

In 1931, the 10 largest Canadian telecommunications companies formed a voluntary association called Telecom Canada. This organization provides the primary nationwide voice and data communications network. Within the Telecom Canada association, one company, Bell Canada, provides over 50 percent of the telephones in the country. Competitors to Telecom Canada, such as CNCP Telecommunications, do exist, but their networks tend to be smaller and more specialized. The role of CNCP has been compared to that of Western Union in the United States.

Canada has been very aggressive in pursuing the installation of a nationwide network specifically designed to carry data. In 1972, Telecom Canada introduced the new network, called Dataroute. Since its original installation, Dataroute has been upgraded continually with new capabilities and service offerings.

United Kingdom In the United Kingdom, the Office of Telecommunications (OFTEL), a part of the Department of Trade and Industry, regulates telecommunications. The first steps toward deregulating telecommunications were taken when the government-owned British Telecommunications company became publicly owned in 1984. The ensuing years have seen many new services offered and a general upgrading of the technology of the British public telecommunications network. A competitor, Mercury Telecommunications, was even allowed to be formed. Although British Telecommunications still holds a vast share of the communications market in the United Kingdom, the presence of competition has forced it to move more quickly than it might have otherwise.

Brazil At the other extreme, Brazil has had a very tightly controlled and regulated telecommunications and data processing environment. Brazil wanted to develop its own data processing industry, ostensibly for national defense reasons, and established a Special Secretariat for Informatics (SEI) in the 1970s that regulated all data processing-related matters. SEI mandated that virtually all data processing would be done within the country on locally made computers. Several small computer manufacturers were established with heavy government subsidization in the late 1970s using technology imported from the United States, Japan, and

Germany. These companies produced small to medium sized computers but were never very successful.

Since Brazil wants all data processing done within the country, SEI and the Ministry of Telecommunications, the primary regulatory agency for telecommunications, put very tight controls on data transmission to ensure that data processing is not exported when it could be done locally. Permits had to be obtained to do any type of international data transmission from Brazil. On the other hand, telephone service, although regulated, was relatively open, and telephone traffic with other countries was freely permitted.

This strategy did not work well for Brazil, however, and changes are being made to open the data processing and telecommunications industries. SEI's authority has been downgraded, moving it from the previous role of controller/authorizer of any request relative to data processing to the new role of technical consultant. The state-owned telephone company, Embratel, reviews and approves all applications for telecommunications services. Some justification for new services is still required, but the requirements are much less restrictive than just a year or two ago.

Japan Japan's telecommunications regulation is governed by the Ministry of Posts and Telecommunications (MPT), and the agency has made aggressive moves toward deregulating the Japanese telecommunications industry. On April 1, 1985, a new Telecommunications Business Law took effect that was designed to open the communications industry to competition. The Nippon Telegraph & Telephone Company shifted from being fully government owned to a semipublic status. Three classifications were established for companies that wish to participate in the telecommunications business.

> **Type I**—telecommunications carriers must obtain "permission" from the MPT because they establish and operate telecommunications lines and have a heavy public responsibility. The rates they charge are "authorized" by the MPT;
>
> **Special Type II**—telecommunications companies must "register" with the MPT. These companies provide public services that use the transmission facilities provided by the Type I carriers. They must "notify" MPT of the rates they intend to charge;
>
> **General Type II**—telecommunications companies must "notify" the MPT of their business intent. These companies offer communication services on a private basis to companies or other organizations, and their sphere of influence is relatively limited. They are not required to inform the MPT of the rates they charge.

Until recently, Japan has been closed to the importation of telecommunications equipment made elsewhere. Understandably, this situation has angered telecommunications equipment manufacturers in the rest of the world, since Japan has been steadily increasing the amount of telecommunications equipment it exports through companies such as Nippon Electric Co. (NEC) and Toshiba.

Belgium Although it is a small country, Belgium is important because it has encouraged foreign companies to establish offices and plants in the country through liberal tax laws and other incentives. As a result, many international corporations have established their European headquarters in Belgium.

Telecommunications in Belgium is highly regulated yet relatively progressive. The government body that controls and operates telecommunications is called the Regie des Telegraphes et des Telephones (RTT) in French, or, because Belgium is officially a bilingual country, the Regie van Telefoon en Telegraaf in Flemish! The Regie, as it is most commonly known, performs the same regulatory functions as the FCC in the United States. It operates the two (French speaking and Flemish speaking) television networks and the public telephone and telex networks. The Regie also provides leased communications lines for voice or data use.

The Belgian telephone system has for years been among the world's most advanced. For example, it provided direct dialing to many foreign countries in the mid-1960s, long before that service was generally available elsewhere. The quality of telephone connections is generally excellent, yet there are some interesting quirks. In order to obtain telephone service in a home or business, it is necessary to deposit a substantial sum of money with the Regie (several hundred dollars for a residence telephone) in case the bill isn't paid. A wait of several weeks for telephone service to be installed is not uncommon.

Data transmission capabilities to and from Belgium are of good quality and readily available. Many of the international companies located in Belgium have extensive data networks radiating to other European countries.

Transnational Data Flow

A concern in the international telecommunications arena is the subject of *transnational data flow (TNDF)*. Several countries, especially in Europe, have enacted legislation to prohibit or restrict the transmission of data and information across national borders. One of their concerns has to do with national defense and the fear that national secrets will be transmitted to unfriendly countries. Another concern involves personal privacy and the fear that private data about citizens will be transmitted to foreign countries. A third concern is the export of information processing, resulting in the loss of jobs. Canada, for example, has been very concerned that because of the close proximity and the common language with the United States, U.S. companies with operations in Canada would choose to do all of their data processing in the United States, thereby limiting the growth of the Canadian information processing industry.

Multinational companies are big generators of transnational data flows. Data is transmitted across borders for processing customer-related transactions, for financial management, and for the coordination of production and inventory management. The question being raised by multinational companies is whether restrictions imposed by countries on the installation and use of international telecommunications and computer processing will create unworkable limits on the management of worldwide activities. Some examples of the types of regulation that could be damaging would be

- a tax or tariff on information transfer;
- monitoring the content of international communications;
- restricting the availability of private leased lines;
- privacy legislation mandating that no personal information be transmitted across international borders.

Several international agencies have been active in transnational data flow issues, such as the Council of Europe (COE), the Organization for Economic Cooperation and Development (OECD), and the United Nations. The OECD has developed guidelines about the protection of personal privacy when information flows across borders. These guidelines have been adopted by most of the countries of Europe and by Japan.

Individual countries have taken widely differing positions on transnational data flow issues. The U.S. position is that a free flow of information should be encouraged, and data processing and telecommunications uses should be unregulated. By contrast, the Scandinavian countries have been leaders in establishing laws to severely restrict the international transmission of personal data. The proactive countries passed laws related to the transmission of information from their countries in the early 1980s and have been getting experience with their impact in the intervening years. It is expected that a second round of legislation based on this experience will come into being in the early 1990s.

THE TELECOMMUNICATIONS INDUSTRY IN THE UNITED STATES

Since January 1, 1984, when divestiture took effect, the structure and participants in the telecommunications industry in the United States have changed dramatically. Now we will look at the structure of the industry and the types of companies that participate in it.

Common Carriers

The common carriers are the companies that provide telecommunications transmission services. These carriers are of five types:

- telephone companies;
- other common carriers (OCC);
- value-added carriers;
- satellite carriers;
- international carriers.

Some companies fall into more than one of the categories.

Telephone Companies The 22 Bell Operating Companies, which were divested from AT&T, were grouped into seven regional corporations known as *Regional Bell Operating Companies (RBOCs)*. They are

- NYNEX Corporation;
- Bell Atlantic;
- Bell South;
- American Information Technology (AMERITECH);
- Southwestern Bell Corporation;
- U.S. West;
- Pacific Telesis (PACTEL).

Figure 4–3 shows how the 22 BOCs were regrouped to form the 7 RBOCs. Each of the RBOCs is a publicly held company, and when they were originally established, the RBOCs between them had about 80 percent of the former Bell system's assets—about $17 billion each. Each RBOC started with about 90,000 employees, making them roughly the same size as GTE, and each company was in the top 50 of *Forbes* magazine's ranking of the top 500 companies, as measured by their assets.

The RBOCs together formed a company called Bell Comunications Research (Bellcore) to be their research arm. Bellcore corresponds to Bell Labs, the research arm of AT&T. Bellcore's 1989 budget was over $1 billion, which is paid almost entirely by the telephone operating companies (BOCs). Bellcore also generated $24 million in revenues in 1988 by licensing concepts and ideas it had developed.

The RBOCs provide local telephone service within the areas they serve, using the network that was in place before they were divested by AT&T. The regional companies in most cases conduct their day-to-day business through the BOCs, many of which are organized along state boundaries, such as Wisconsin Bell Telephone Company and Illinois Bell Telephone Company. The BOCs know the regulatory issues within the states they serve and can work more effectively with their respective public utility commissions than the RBOCs could.

AT&T is still the largest common carrier in the United States, even after divestiture. AT&T retained about 20 percent of the former Bell

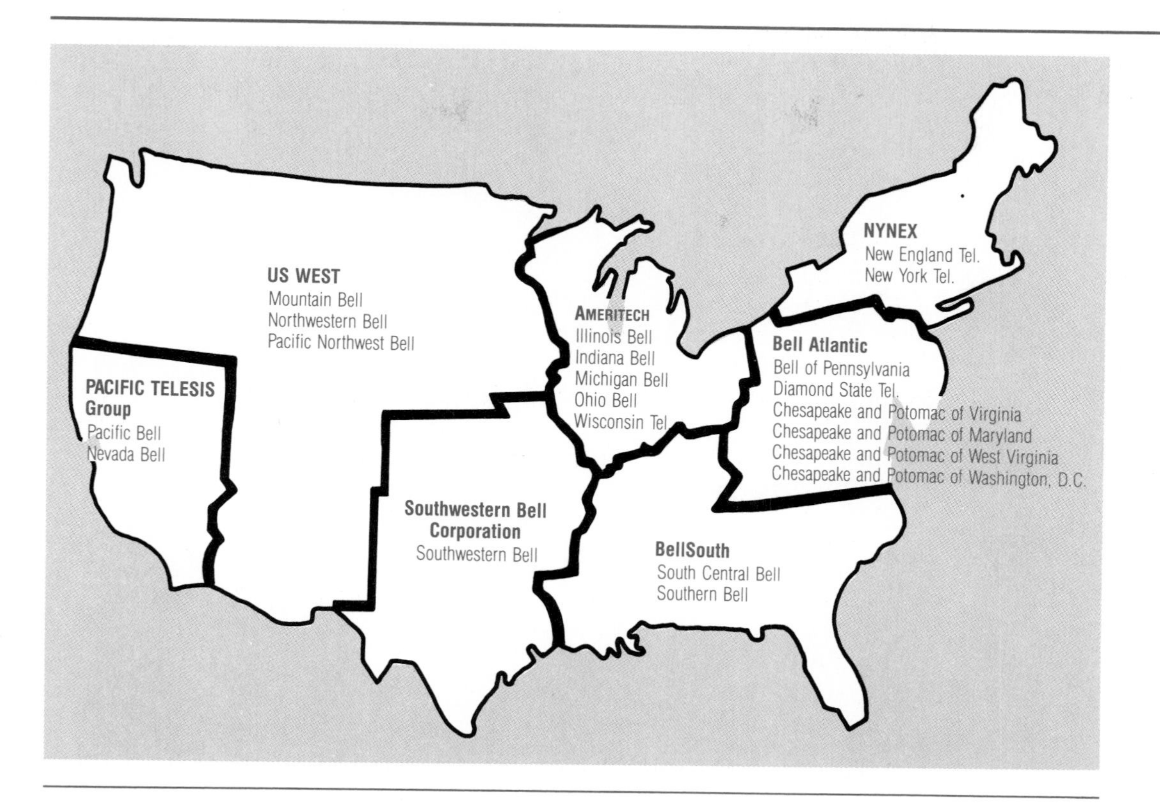

Figure 4–3
The territories served by the Regional Bell Operating Companies.

system's assets and approximately 380,000 employees, though the number of employees has been sharply reduced since that time. A division, AT&T Technologies, combined the research functions of Bell Labs and the manufacturing activities of Western Electric.

independent telephone companies

In addition to the AT&T and the BOCs, there are many independent telephone companies that were never a part of the Bell System. GTE Corporation (formerly General Telephone and Electronics) is the second largest telephone company in the United States, ranked only behind AT&T. Other companies include Continental Telephone, United Telephone, and approximately 1,375 others. Many of these are very small "Ma and Pa" companies that provide telephone service in a very limited geographic area. With modern equipment, the telephone operation is totally automatic and virtually trouble free. "Pa" can have an outside job and do any necessary repair work in the evenings, and "Ma" can type up the bills and do the accounting in her spare time. The BOCs and the independent companies together are called *local exchange carriers* or *companies (LECs)*.

As a part of divestiture, 165 local calling areas called *local access and transport areas (LATAs)* were defined within the United States. A diagram

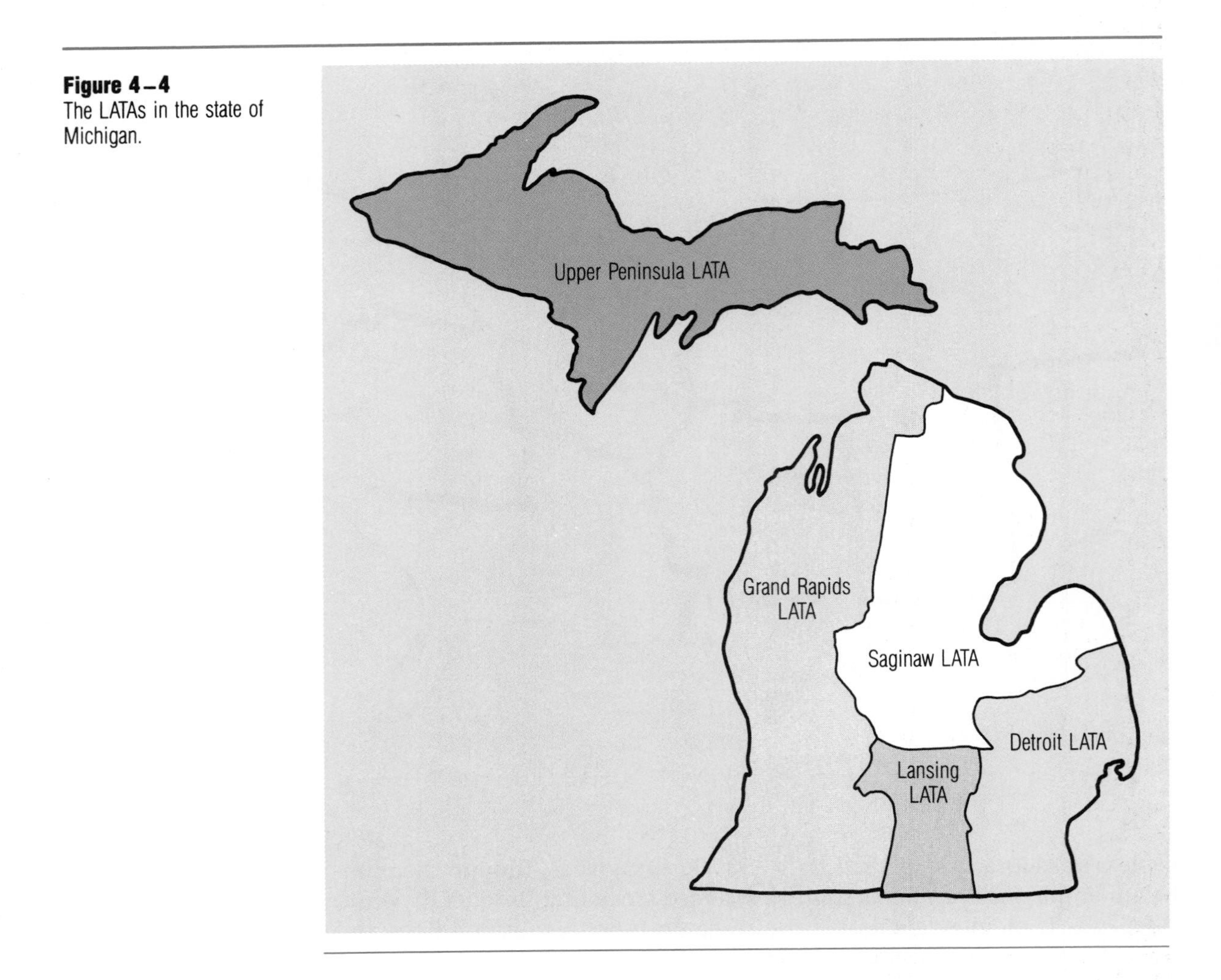

Figure 4–4
The LATAs in the state of Michigan.

LATAs

of the LATAs in the state of Michigan is shown in Figure 4–4. The LECs provide telephone service within LATAs, but inter-LATA telephone traffic must be carried by long distance carriers, such as AT&T, MCI, or U.S. Sprint. These long distance carriers are also known as *interexchange carriers (IXCs).*

equal access

One part of the MFJ was called the equal access ruling. *Equal access* means that long distance calls are no longer automatically routed to AT&T. Instead, customers have equal access to all of the long distance carriers and can choose long distance services from among them. The MFJ specified that the BOCs must provide all IXCs with access that is equal in type, quality, and price to those provided to AT&T.

points of presence

To implement equal access, one or more of the IXC's central offices in each LATA is designated as the *point of presence (POP).* The IXCs connect their long distance networks to the LEC at the POP. Unless an interexchange carrier has a POP in a LATA, it cannot offer long distance service. It is important to note that today every IXC does not have a point of presence in every LATA.

AT&T provides nationwide long distance service, using the network that it installed before divestiture. Other long distance companies, such as MCI, have been building their own networks over the past 15 to 20 years. This is a very expensive process, and even today, they must sometimes rent lines and facilities from AT&T where they don't have their own lines in place.

The responsibility of the telephone companies for lines and equipment now extends up to residential premises but not within them. Inside wiring and equipment, such as telephones, are the customer's responsibility. That is, if a problem with the wiring or telephone in the house occurs, the telephone company will fix it for a charge; if the problem is in the wiring outside the house, the telephone company will fix it at its expense. The point where the telephone company responsibility ends and the customer responsibility begins is called the *demarcation point*.

The situation is more complicated with businesses. With small business locations, the rules are the same as for residences. For large business campuses or multistory buildings, the telephone company's responsibilities frequently extend into the building to a central equipment room or even to an equipment room on each floor.

Since divestiture, the RBOCs have been very actively pursuing other business opportunities besides traditional telephone service. Most of them have set up subsidiary companies to handle unregulated products and services. For example, Bell Atlantic has gotten heavily into the maintenance of computer and telecommunications equipment. It purchased Sorbus, a large, independent computer repair firm in 1984 and has more recently purchased several other, smaller maintenance companies serving particular markets. PACTEL has gotten heavily into providing mobile cellular telephone and radio paging service in several locations around the country. NYNEX purchased IBM's retail computer stores. Several of the RBOCs are selling communications equipment and parts on a regional scale.

Other Common Carriers The other common carriers (OCCs) provide alternatives to AT&T for long distance communication. The largest companies in this group are MCI and U.S. Sprint. Many of these companies were in operation before divestiture and originally designed their networks to serve customers with high-volume data transmissions. Since divestiture, they have expanded their services to handle normal long distance telephone calls.

The OCCs provide services primarily in areas where traffic volume is high—the most profitable markets. This has allowed them to charge less for their services than the traditional carriers could. They build or lease lines that connect with the local exchange carriers at each end and take advantage of the local distribution that the LECs provide. The MFJ requires the LECs to connect the OCCs' lines into their networks to provide access to the OCCs' customers.

When divestiture took effect, it provided for the "gradual" elimination of the cross-subsidization between local and long distance telephone rates. As the cross-subsidization has disappeared, certain access charges paid by the OCCs to the local telephone companies have gone up. This has increased the OCCs' costs and forced them to raise their prices. As a result, long distance service is very competitive today, and AT&T's economies of scale are allowing them to put strong price pressure on the OCCs.

Value-Added Carriers The *value-added carriers* provide public data transmission services and are considered by the FCC to be providers of enhanced services. The largest value-added carriers are GTE Telenet, Tymnet, Inc., Graphnet, Inc., and the IBM Information Network. Value-added carriers have leased line networks that blanket the country. In some cases, they have connections to value-added carriers in other countries that provide international service. Access to the network is obtained through either a dial-up or leased line to the carrier's nearest location. Charges are based on the amount of time the user is connected to the network and the amount of data transmitted in each direction. The transmission technique most frequently used by the value-added carriers is called *packet switching*. It is discussed in detail in Chapter 11.

Satellite Carriers *Satellite carriers* offer communication through satellites. Companies offering satellite communication include RCA Americom, American Satellite Company, ITT, and AT&T. Satellites, like other communications media, can carry voice, data, television, or facsimile signals, and the companies listed sell all of these capabilities. The satellite earth station may be located at the carrier's location or on the customer's premises. If it is at the carrier's location, the signal is carried from the customer to the earth station on terrestrial lines. Satellite transmission is discussed in Chapter 9 in detail.

International Carriers *International carriers*, sometimes called *international record carriers (IRCs)*, provide communication services between countries. TRT Telecommunications, MCI International, and AT&T are the major U.S. IRCs. Teleglobe Canada (Canada), FTC Communications (France), British Telecommunications (United Kingdom), Kokusai Denshin Denwa Co. (Japan), and Overseas Telecommunications Company (Australia) are some of the major IRCs in other countries.

In general, IRCs provide international voice and data services. The data services include telex, Datel (a dial-up data service), and leased lines for data transmission.

Company	Home Country	1988 Sales Revenue from Data Communications ($ million)
Northern Telecommunications	Canada	$ 900
Alcatel	France	722
Canon	Japan	1,132
Nippon Electric Company (NEC)	Japan	964
Matsushita	Japan	929
Ricoh	Japan	850
Fujitsu	Japan	817
Toshiba	Japan	762
N.V. Philips	Netherlands	460
IBM	U.S.A.	1,600
AT&T	U.S.A.	1,250
Motorola	U.S.A.	565
Racal Electronics	U.S.A.	492
Hewlett-Packard	U.S.A.	400
Siemens	W. Germany	1,338

Figure 4–5
The largest companies in the global data communications marketplace.

The costs for international leased voice or data lines are divided into two parts. Each country charges for one half of the line; unless special arrangements are made, the customer receives a bill from the company providing each end of the line. Conceptually, on a trans-Atlantic line, it's as though each carrier billed for the line from the carrier's country to the middle of the ocean!

Unregulated Aspects of the Telecommunications Industry

There are literally hundreds of companies operating in the telecommunications industry today. Telephones, computer terminals, and private telephone systems for business use are just a small sample of the types of products that are totally unregulated. AT&T and the RBOCs have subsidiaries or marketing units to sell these types of products and are often in direct competition with one another. Trade journals abound, filled with advertisements for equipment and services. The telecommunications department in a company has a wide choice of products from which to choose.

Several multinational companies with headquarters outside the United States compete in the global telecommunications marketplace. Although naturally strong in their home country, these companies have all been working hard to expand their territory beyond national boundaries. The United States is fertile territory because of its size and the relative sophistication of its telecommunications applications. Each of these companies is actively selling its products here. A list of the major telecommunications companies is shown in Figure 4–5.

AT&T and IBM

Two of the larger participants in the telecommunications industry deserve special mention: AT&T because of its historical dominance of the telecommunications industry in the United States and ownership of the nationwide telephone network, and IBM because of its marketing and financial strength and its participation in the telecommunications industry worldwide.

AT&T's organization before divestiture and a high-level view of its current organization are shown in Figure 4–6. The operating divisions shown on the chart are divided into 19 business units, each of which handles a group of products or services. As previously mentioned, AT&T participates in both the regulated and unregulated sides of the industry. Bell Labs continues to be the research arm of AT&T and has a long

Figure 4–6
AT&T's organization before divestiture, and as of November 1, 1989.

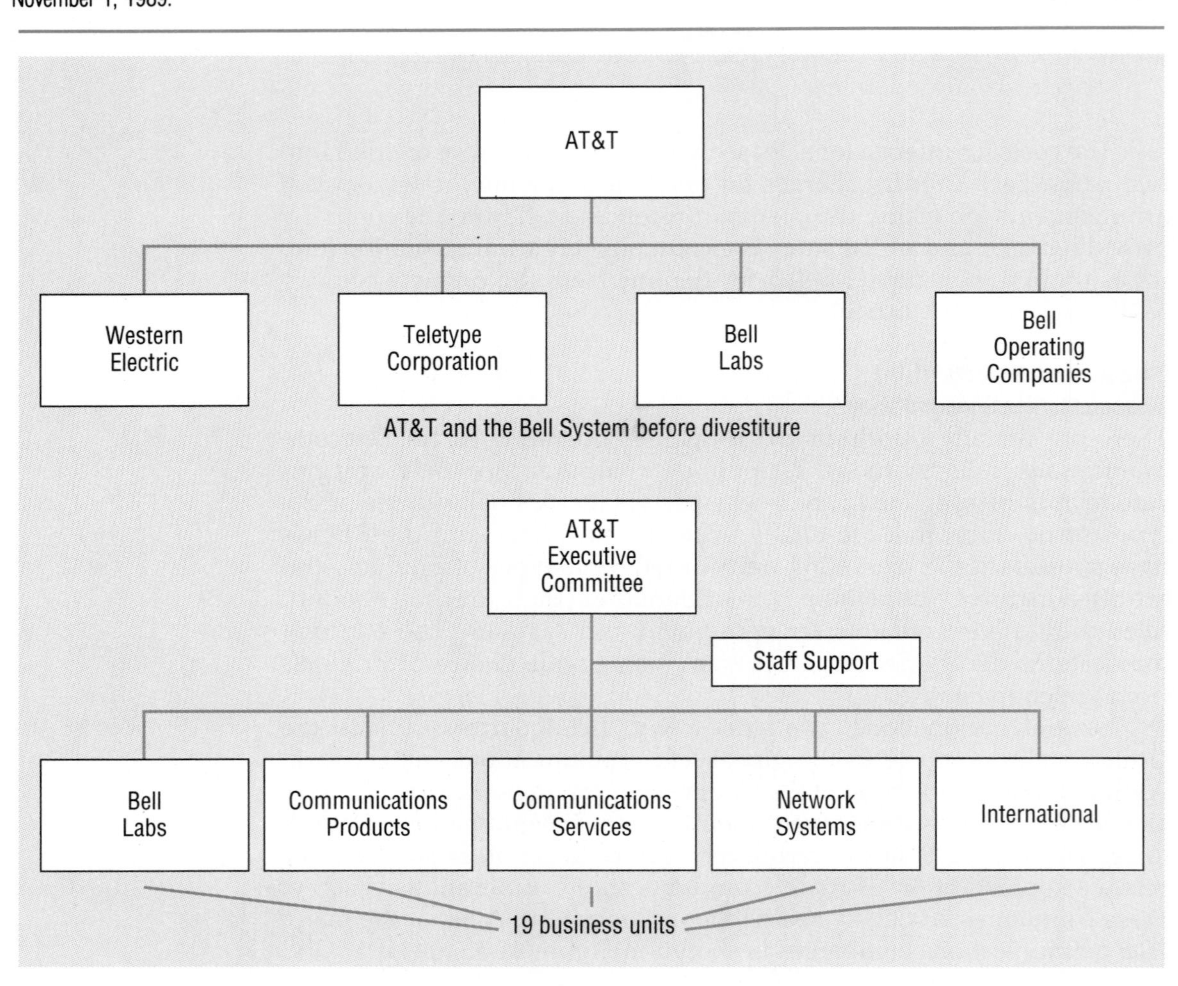

record of doing some of the finest basic research in the United States. The Communications Products Division and associated business units manufacture and sell PBXs, key systems, and computers, primarily to business customers. The Communications Services businesses manage AT&T's long distance service and private lines. Network Systems businesses engineer, manufacture, and market communications equipment and services, primarily to telephone companies. AT&T International provides products and services in foreign countries.

The focus of AT&T's expertise is the nationwide network for voice and data transmission, computerized telephone switches for central offices and private businesses, and a wide range of data communications products and services.

IBM's current communications organization is shown in Figure 4–7. The company's predominant strength over the years has been in the marketing and manufacturing of computers and related equipment. As the use of terminals grew, IBM became heavily involved in designing and manufacturing data communications equipment in order to meet its customer's requirements.

Figure 4–7
IBM's organization chart as of August 1989.

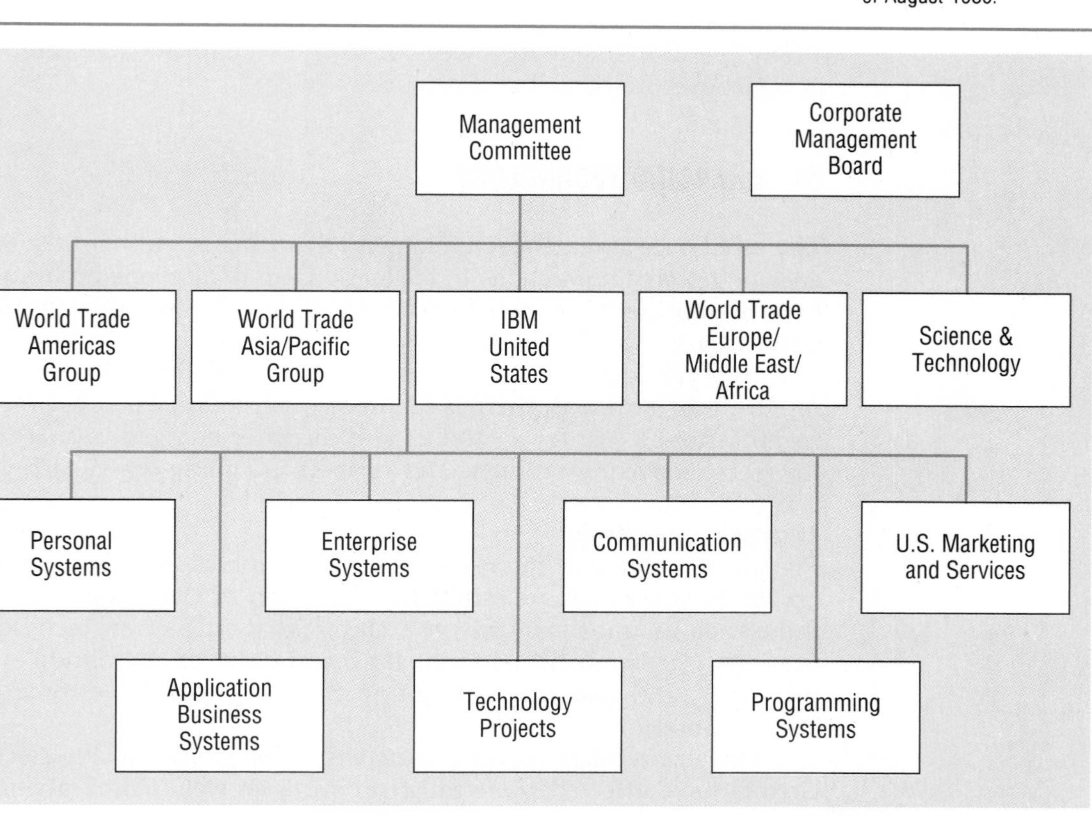

Telecommunications activity in IBM is divided among several divisions, but overall strategy is focused on the Communication Systems division. Communications Systems handles data communications product development, such as modems and terminals, and operates the IBM Information Network, a communications network that provides communications and computing services to many companies around the world. Basic communications research is conducted by the Research Division. The computer divisions have people who design and implement the telecommunications hardware within their computers. All data communications products are marketed by the marketing divisions around the world along with all of the other IBM products.

IBM had some voice communications products for sale in Europe as early as the 1960s, but they did not enter the U.S. voice market until IBM purchased Rolm Corporation in 1984. In 1989, IBM concluded an arrangement with Siemens of West Germany, whereby Siemens will take over the development of voice telecommunications products from IBM-Rolm, and IBM and Siemens will jointly sell the products under varying marketing arrangements in different parts of the world.

The general feeling is that IBM will continue to choose its markets very carefully, probably concentrating on its historical strength—the business (as opposed to consumer) marketplace. Because of its financial strength and marketing prowess, IBM will be a formidable competitor in whichever markets it enters.

■ CHANGING TECHNOLOGY

The rapid pace at which technology is advancing is another major influence on telecommunications in business. Communications problems that have no solution today may be solved by a new development or product tomorrow. Leading edge products are often obsolete within 3 or 4 years. The price performance ratio of communications equipment is improving rapidly, too. Although there is not necessarily a direct link between the manufacturer's cost for a product and the price charged to a customer, the prices for many products and services are going down each year.

Advances in Electronics

Everyone is aware of the revolution in electronics that has been going on for 30 years or more, since the invention of the transistor. Today, thousands of transistors fit on a chip less than half an inch square. Circuitry has been shrinking in size by an order of magnitude every 4 to 5 years, and costs, though not necessarily prices, have come down at a similar rate.

The size reduction and the increased capabilities of the electronic circuits have affected the size of products as well. Think about how radios and calculators have shrunk in size in the last 20 years. Size reduction of many products has its limit, however, especially where human users are involved. A calculator keyboard can only get so small before it becomes unusable. The same is true with a telephone keypad, watch, or keyboard on a terminal.

In the business world, intelligent terminals coupled with a central computer that contains massive amounts of corporate data give power, flexibility, and independence to the users. Soon it will be common to find a combined VDT and telephone providing answering machine functions, the translation of text to speech, and the capability to be driven by voice commands. All of these enhancements are aimed at increasing productivity by making it easier to do the mundane tasks and making it possible to do certain tasks that couldn't be done without computer power.

As applied to telecommunications, these intelligent workstations are only a part of the story. More intelligent, capable telephone systems located either at the telephone company's central offices or on the business premises are providing new options, such as the automatic forwarding of a call to another number and voice messaging.

EMERGING STANDARDS

A third area of change impacting the telecommunications environment is the area of standards. For years, since the inception of telephone systems, there have been few published communication standards. In early telephone history, competition in the telephone business led to proprietary systems that were incompatible with one another. These incompatibilities were resolved as the telephone network was brought together to work as a single system. When that happened, the proprietary systems had to be either changed or equipped with conversion devices to meet the standards of the network.

The standards that did exist were mostly electrical in nature, such as the voltage required to cause the telephone bell to ring, the electrical resistance or impedance of telephone lines, and the drop in signal strength permitted over various distances. These standards were established when there was a need.

In the United States, early telecommunications standards were determined and set by the predominant force in the industry, AT&T. Other countries either adopted the U.S. standards or developed their own and then used converters when calls went to international locations. Differing standards hampered the implementation of a global telephone system. International standards organizations eventually got the differences ironed out so that today it is possible to make calls anywhere.

In the data communications world, standards evolution is occurring, but today, few global standards are in place. The situation is more complicated than with voice. In addition to electrical standards, the "language" of data communications must be standardized so that terminals and computers can understand each other. The issue is further complicated by the fact that the data processing industry has been unregulated, and each company has set its own communications standards. IBM's data transmission techniques are different from and incompatible with those of Digital Equipment Corporation, which are again different from those of Honeywell or Hewlett-Packard. Although they all follow the same *electrical* connection standards so they can connect to the telephone network, this is not enough to guarantee successful communication.

It's one thing to make a connection and another to communicate. For example, in the voice world, it is possible to make a "connection" between a telephone in the United States and a telephone in Japan. If the American only speaks English and the Japanese person only speaks Japanese, "communication" will not occur. In data communication, it may be possible to connect IBM and Honeywell computers, but since they essentially speak different languages, no communication occurs.

THE ISO-OSI REFERENCE MODEL

International data communications standards are primarily following a model developed by the *International Standards Organization (ISO)*. The model is called the *Open Systems Interconnection reference model (OSI)* and was originally proposed by ISO in 1978. Since 1978, work has been underway to convert the model to a set of standards by precisely defining each part of the model (called *layers*) and the interfaces between them. It is only after standards are written and agreed to that the model can be converted to real products and the benefits of having the model and the standards can be enjoyed by users. Development of international standards is a slow process, however, and although much progress has been made, the work is still underway.

The ISO-OSI model is discussed in detail in Chapter 13. However, it is important to introduce it here and for you to gain a basic familiarity with it. The model will provide you with a framework into which you can fit much of the technical information that is presented in Chapters 6 through 12. With such a framework, you will be able more quickly to apply the information on communications hardware, networks, and protocols and make sense of what you learn.

The Seven Layers of the OSI Model

The OSI model uses a layered approach, each layer representing a component of the total process of communicating. A diagram of the seven OSI layers is shown in Figure 4–8. A brief description of each layer is given.

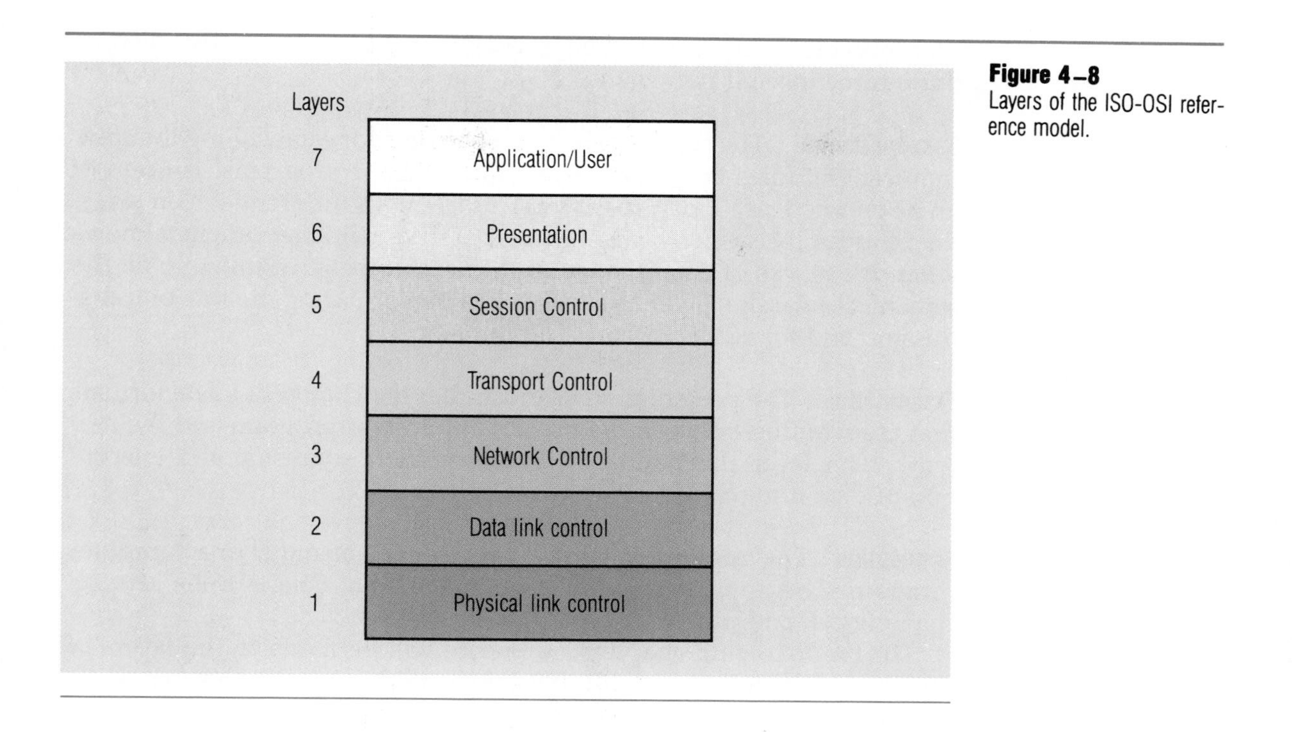

Figure 4–8
Layers of the ISO-OSI reference model.

Physical Link Control The physical link layer specifies the electrical characteristics of the communications link. It specifies how the signals are carried on the wires, what types of connectors are used, and which pins in the connectors are used for which signals.

Data Link Control The data link layer is responsible for establishing a link between two points on a network and ensuring that data is transferred between the two points successfully. This layer must include means for detecting errors in transmission and correcting errors when they do occur. The data link layer also is responsible for dividing the data into fields for transmission.

Network Control The network layer is responsible for routing a message all the way through the network from the transmitter to the receiver. If parts of a message are sent through the network on different routes, the network control layer at the receiving end is responsible for assembling the entire message before delivering it.

Transport Control The transport layer selects the route a message is to take between two points on the network. In large networks, there may be more than one route between some heavily used locations on the network. The transport layer selects the route to use for each message.

This layer also plays a role when messages are sent between two distinct but interconnected networks.

Session Control The session layer is responsible for establishing the communications rules between specific machines or programs. If users of an application use video display terminals with different screen sizes, for example, the session layer ensures that the application program knows what size blocks of data it must send to each type of terminal to fill the screen. The session layer also is responsible for pacing the rate of transmission and for certain accounting functions.

Presentation The presentation layer handles the changes of data formats that are required between the user or application program and the network. This layer also handles the compression/expansion and encryption of data if required.

Application The application layer is the user at a terminal or a computer program. The application layer determines what data is to be sent on the network and processes data that is received.

In the following chapters, reference will be made to the layers of the OSI model to help you fit that information into a framework. Coupled with this book's overall outside-in approach that leads you from familiar telecommunications applications to unfamiliar technical details, you have two excellent bases on which to build your knowledge.

■ SUMMARY

Three major forces are at work changing the environment in which telecommunications operates:

1. regulatory;
2. technological;
3. standards.

The United States has moved from a regulated, monopolistic telecommunications structure to a largely unregulated, competitive structure. Several types of common carriers provide basic and enhanced transmission services. Thousands of companies are making telecommunications products and offering services. New companies enter the marketplace every day, and many fail. The telecommunications department in a business needs to be careful to be sure that the products and services they select will be supported in the future.

AT&T, IBM, the RBOCs, and several large foreign companies have sufficient size, experience, and money to keep them in the telecommunications business. The RBOCs are particularly aggressive and are branching into many diversified product and services offerings.

Both vendors and customers are having trouble keeping up with technological change. For vendors, the product life cycle often is too short for them to recover their research and development investment. For telecommunications users, the constant implementation of change, albeit with new capabilities, features, and options, can be overwhelming and draining on company resources.

It used to be that communications equipment was installed and depreciated over a 10- or 20-year period, and the view was that it would last forever. Now, technology is bringing significant changes annually, and telecommunications equipment, technologies, and knowledge are good for just 3 to 5 years and by then become obsolete.

International communications standards are emerging that are providing a framework for the communications process. Of particular significance for data communications is the seven-layer ISO-OSI reference model, standards for which are nearing completion.

The implication for the telecommunications department in a company is that it will be working in a world of constant change. As new technologies emerge or standards come into widespread use, there will be good business reasons to take advantage of them. Taking advantage will mean changes to telecommunications hardware, software, and procedures—or all three simultaneously.

One of the key jobs of the telecommunications manager is to learn how to cope with constant change and to implement it smoothly and without disruption. Not implementing change is tantamount to putting the company at a competitive disadvantage or out of business. Implementing new products, services, or techniques carelessly is disruptive, may waste money, and can spur a loss of productivity or business effectiveness.

CASE STUDY

Impact of the Telecommunications Industry and Deregulation on Dow Corning's Telecommunications

Like most other companies, Dow Corning Corporation has felt the change that divestiture and deregulation have imposed on the vendors and the telecommunications industry. Because its location (Midland, Michigan) is not in a large metropolitan area, fewer of the small telecommunications vendors call on the company. Historically, the primary communications vendors in Midland have been Michigan Bell Telephone and AT&T. IBM has been the major vendor of computers and terminals. Both AT&T and IBM have national account teams in Midland. These teams are responsible for all of the marketing by their companies to Dow Corning throughout the United States.

When divestiture occurred on January 1, 1984, both Michigan Bell and AT&T formed new subsidiaries to handle the unregulated portions of their business. Michigan Bell pulled most of its marketing people back to Detroit and established Michigan Bell Communications to sell the unregulated equipment. AT&T's unregulated subsidiary was subsequently merged back into the parent organization so that one sales team markets all of AT&T's products and services to Dow Corning. MCI began calling on Dow Corning and has demonstrated a capability to provide communications services in the Midland area.

Dow Corning's telecommunications managers state that since divestiture, their life has become more complicated. In the past, they dealt with Michigan Bell on local communications matters and AT&T for nationwide lines and services. Since divestiture, they have had to work with both the regulated and unregulated divisions of each of those companies as well as with their many competitors. Another example of the complications imposed by divestiture is that the bills for telephone usage and other services have become more complex. This has meant that additional time is spent each month verifying the amounts.

On a more positive note, deregulation has allowed many companies to bring new telecommunications products to the market, some of which have potential benefits to Dow Corning. The telecommunications staff is always on the lookout for vendors who can provide new products while maintaining Dow Corning's requirement for excellent quality and service.

Review Questions

1. Explain why the telecommunications industry was regulated in its early years.
2. Explain the terms:

BOC	carrier
RBOC	Other Common Carrier
divestiture	international carrier

3. Describe the roles of the FCC and state PUCs.

4. Explain the four major rate categories for tariffs.
5. What was the significance of Computer Inquiry II? What impact did the relief from CI–II have on AT&T?
6. Compare and contrast the five types of communications carriers.
7. What was the significance of the Carterfone decision?
8. What was the modified final judgment? What significance did it have to AT&T?
9. Explain the difference between deregulation and divestiture.
10. What happened on January 1, 1984? How was the Bell System reorganized?
11. What is the function of a PTT?
12. What is the concern about transnational data flow?
13. What is the U.S. position on transnational data flow issues?
14. Explain the term *equal access.*
15. Why is the establishment of standards more complicated in the data communication environment than in the voice environment?
16. Explain the difference between *connection* and *communication.*
17. What is the ISO-OSI reference model? Why is it important for you to learn about it?
18. What are the functions of the physical, data link, and network layers of the OSI model?

Problems and Projects

1. Find out what percentage of the market for long distance telephone service is held by each of the companies in the industry. How has this changed in the last year? Do you think the smaller companies will be able to stay in the business and remain profitable? Why or why not?
2. Why do you think the United States is leading the world in deregulating its telecommunications industry? Are deregulation and competition really better than a regulated monopoly for the communications industry? Is deregulation leading to lower prices for users? Defend your position.
3. At divestiture, the RBOCs were all about the same size, as measured by their revenue and number of employees. Do some research and find out what has happened to their sizes in the years since divestiture occurred. Which companies have grown? Which have shrunk? Which are the most profitable today?
4. Compare your telephone bill at home for a period of time (quarter or year) with the same period last year. Separate the changes in the number of calls that were made and determine whether your rates went up, down, or stayed the same. Find out what the telephone company claims happened to the rates and compare its claim to your own experience.

5. For the sales order entry application described in Chapter 2, describe what functions would be performed by the application and presentation layers of the OSI model.

Vocabulary

common carrier
carrier
universal service
Federal Communications Commission (FCC)
tariff
interconnect industry
specialized common carrier (SCC)
other common carrier (OCC)
Computer Inquiry I (CI–I)
Computer Inquiry II (CI–II)
enhanced service
basic service
Bell Operating Company (BOC)
modified final judgment (MFJ)
Computer Inquiry III (CI–III)
open network architecture (ONA)
public utility commission (PUC)
public service commission (PSC)
post, telephone, and telegraph (PTT)
transnational data flow (TNDF)
Regional Bell Operating Company (RBOC)
local exchange carrier (LEC)
local access and transport area (LATA)
interexchange carrier (IXC)
equal access
point of presence (POP)
demarcation point
value-added carrier
satellite carrier
international carrier
international record carrier (IRC)
International Standards Organization (ISO)
open systems interconnection (OSI) reference model
layers

References

Barteé, Thomas C. (editor). *Data Communications, Networks and Systems.* Indianapolis, IN: Howard W. Sams Co., Inc., 1985.

Bielecki, John, Susan Briner, Richard Brownfield, and Steve Dimmitt. "The Challenge of Equal Access," *Telesis,* 1986, Vol. 2, p. 4.

Blyth, W. John, and Mary M. Blyth. *Telecommunications: Concepts, Development and Management.* Indianapolis, IN: The Bobbs-Merrill Company, Inc., 1985.

"Brazil's Informatics Model Challenged," *Transnational Data and Communications Report,* October 1986, p. 5.

"Computer III Pushes Regulatory Liberalization," *Transnational Data and Communications Report,* July 1986, p. 6.

Doll, Dixon. *Data Communications Facilities, Networks and Systems Design.* New York: John Wiley & Sons, 1978.

Keller, John J., and Mark Maremont. "Bob Allen is Turning AT&T Into a Live Wire," *Business Week,* November 6, 1989, p. 140.

Mercer, Robert A. "ONA: Why Should Users Care?" *Business Communications Review,* September–October 1988, p. 50.

Pasztor, Andy, and Bob Davis. "Justice Agency Urges Judge to Remove Curbs on Businesses of Ex-Bell Firms," *The Wall Street Journal,* February 3, 1987.

Petersohn, Henry H. *Executive's Guide to Data Communications in the Corporate Environment.* Englewood Cliffs, NJ: Prentice Hall, Inc., 1986.

Powell, Dave. "What You Should Know About ONA," *Networking Management,* August 1989, p. 36.

PART TWO

How Information Is Communicated

The Technical Details

Part Two describes how communications systems work. Building on the background and vocabulary from Part One, you'll be introduced here to the technical details of communications systems. Voice communication is covered first, followed by data communications.

Chapter 5 delves into the detail of telephone systems, first looking at the telephone handset, then discussing the public telephone network. Modulation and frequency division multiplexing are explained, and the technical discussion provides a good introduction to many of the data communications topics.

Chapter 6 discusses data terminals. Various types of terminals are examined, and teletypewriters and video display terminals are discussed in depth. The functions of a cluster controller are introduced, and the chapter examines various amounts of terminal intelligence and the uses to which the intelligence is put.

Chapter 7 explores how data is coded for storage in a computer and for transmission on a communications line. Concepts and techniques of data compaction and encryption also are discussed.

Chapter 8 deals with the interface of terminal to communications line, including how digital signals of the terminal are converted to the analog signals that many communications lines handle. Emphasis is placed on the functions of the modem, the device that performs the conversion.

Communications lines—the transmission facility over which all of the communications take place—are the subject of Chapter 9. The various media with which lines can be constructed are discussed, and several ways of classifying lines are explored.

Chapter 10 examines data link protocols, the rules by which data communications lines operate. Three categories of protocols are explained, and a specific protocol from each category is examined. You'll also learn about protocols for local area networks in Chapter 10.

Chapter 11 covers communications networks. You'll learn how communications lines are connected together in various ways to form networks. Networks will be classified several ways, and you'll see that the classifications frequently overlap. Local area networks are covered in detail, including the software that is required to make them operate.

Chapter 12 explores how the communications line is connected to the host computer. The front-end processor is introduced, and the implementation trade-offs among host and front-end processor, hardware, and software are discussed. The chapter also covers the functions of the software that is involved in data communications-based applications.

Finally, telecommunications architectures are described in Chapter 13. The ISO-OSI model, which was introduced in Chapter 4, as well as several vendor architectures, are studied in detail. The advantages and disadvantages of having an architecture also are discussed.

After studying Part Two, you will be conversant with all of the elements required to make a voice or data communications system operate. A good knowledge of the material will place you on a par with many communications professionals who are working in industry today.

CHAPTER

Voice Communications

OBJECTIVES

After you complete your study of this chapter, you should be able to

- describe the purpose and functions of the various parts of the telephone system;
- describe the evolution of telephone switching equipment;
- differentiate between public and private telephone switching equipment;
- describe the characteristics of analog signals;
- describe several types of telephone lines;
- explain several ways to acquire telephone service at a discount;
- describe several types of special telephone service.

INTRODUCTION

This chapter describes and discusses the three basic elements of telephone systems:

- the telephone handset;
- telephone company switching equipment;
- telephone lines.

Through your study of this chapter, you will gain a detailed understanding of the way the telephone system and telephone equipment work together to provide the reliable telephone service that we have come to expect. We will also look at the way in which voice signals are sent on telephone lines. Finally, we will look at other aspects of telephone communications, such as the special capabilities and services provided by telephone companies.

A study of voice communications is important for two reasons. First, it provides a solid foundation for studying data communications later in the book. The principles of voice communications carry over into the data world. The majority of data transmissions in the world today occur on lines and networks that were originally designed to carry voice. The data transmissions are adjusted to fit the parameters of the voice network.

The second reason is that voice communications is an important part of the overall business communications scene. Eighty percent of business telecommunications expense in most companies is for voice; only 20 percent is for data. Your knowledge of business telecommunications is not complete without a good understanding of voice communications concepts, capabilities, and products.

THE TELEPHONE SET

Primary Functions

The primary functions of the telephone set are to

- convert voice sound to electrical signals for transmission;
- convert electrical signals to sound (reception);
- provide a means to signal the telephone company that a call is to be made or a call is complete (off-hook and on-hook);
- provide a means to tell the telephone company the number the caller wants to call;
- provide a means for the telephone company to indicate that a call is coming in (ringing).

Telephone sets come in many sizes, shapes, styles, and colors. Historically, there was just one type—the traditional black desk phone—and it was available only from the telephone company. Now phones to fit any decor or color scheme are available from a variety of vendors. One entrepreneur in Texas is selling plastic disposable phones for hospitals. He suggests that to reduce the possibility of spreading germs

The Merlin family of telephones from AT&T. A telephone with the appropriate capabilities for each user can be selected for use on a modern telephone system, such as the Merlin. (Courtesy of AT&T)

The Merlin cordless telephone can connect to up to five lines through a key system. (Courtesy of AT&T)

between patients, the hospitals throw the phones—which cost under $9—away when a patient checks out of a room.

Telephones may be equipped with amplifiers to make the incoming voice signal louder for the hard of hearing. They may have loud bells or lights to indicate ringing in noisy areas. Explosion-proof telephones are available for use in chemical plants or other places where a spark might set off an explosion. Also, telephones with clocks, calendars, and elapsed timers are available.

Telephone Transmitter

Built into the handset of the telephone is the transmitter. It converts sound waves into electrical signals that can be transmitted on telephone lines. When you speak, your voice generates sound waves or vibrations that move the air in ever widening circles, as shown in Figure 5–1. Some of these sound waves (which are really variations in air pressure) flow into the telephone mouthpiece, causing a thin diaphragm in the microphone to move back and forth. The vibrations put alternately more and

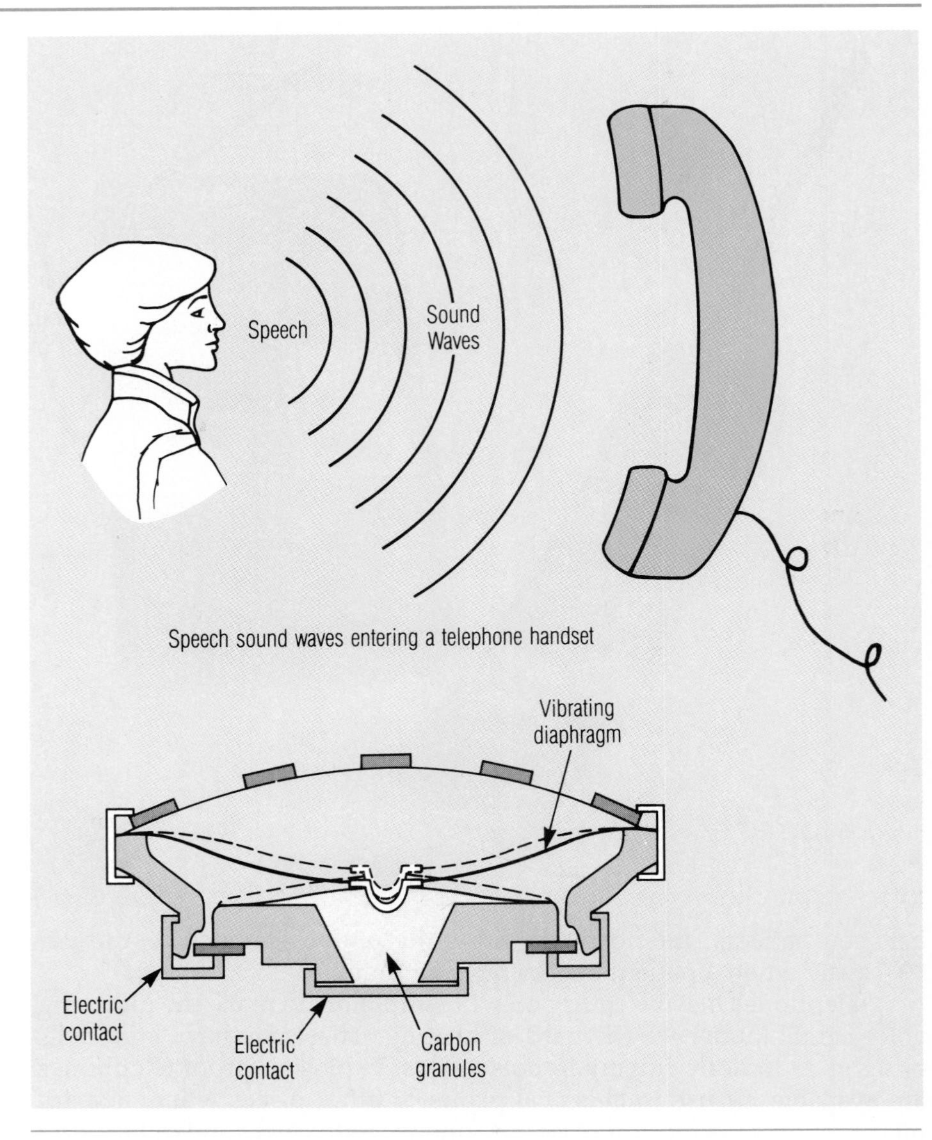

Figure 5–1
In the telephone transmitter, the sound waves of the voice cause the diaphragm to vibrate. The vibration puts varying amounts of pressure on the carbon granules, causing more and less electrical current to flow.

less pressure on carbon granules in the microphone. As more pressure is applied, the granules are packed more tightly against one another and conduct electricity better. Conversely, between the pressure waves, the granules move apart and do not conduct electricity as well. With voltage applied across the electrical contacts, this varying resistance of the carbon granules in the microphone causes a varying amount of current to flow. The varying current is an electrical representation of the sound waves generated by our voice.

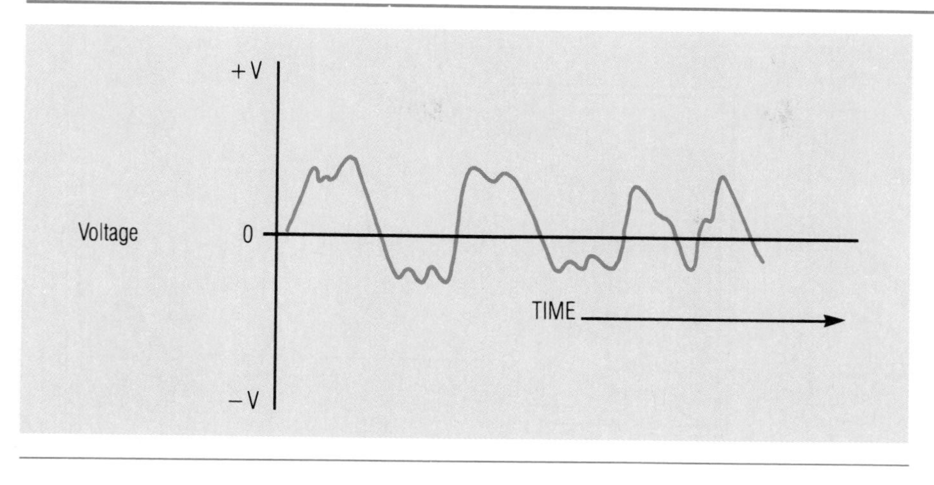

Figure 5–2
Wave of a typical voice signal.

Since the electrical signal is analogous to the sound waves, we call it an *analog* signal. If we graphed a typical voice signal or watched it on an oscilloscope, it would look like the wave shown in Figure 5–2. To be more complete, an analog signal is one that is in the form of a continuously variable physical quantity, such as voltage.

Telephone Receiver

At the other end of the telephone handset, the receiver's job is just the opposite. The receiver converts incoming electrical signals to sound waves that can be heard by the listener. Figure 5–3 is a diagram of this process. The incoming electrical signal flows through the voice coil in the receiver, and the interaction of the current in the voice coil with the magnet causes the coil to move. The coil is attached to a large diaphragm called the *speaker cone,* which vibrates back and forth, causing sound waves that flow outward from it and into the ear of the person listening.

An important aspect of the telephone and its handset is that when you talk into the transmitter, some of the electrical current generated by the microphone in the transmitter is fed into your own receiver so that you hear yourself talk. This small amount of signal is called *sidetone.* Although sidetone is not technically required for the telephone to operate properly, it is generated for human factors purposes. Without sidetone, the telephone seems dead, and people tend to talk too loudly. With too much sidetone, people talk too softly. Sidetone is a form of feedback to the person who is speaking and gives the person a reference point for determining whether he or she is speaking too loudly or too softly.

sidetone

Telephone Switchhook

The telephone *switchhook* derives its name from old style telephones that had a hook on the side on which the handset was hung. In modern

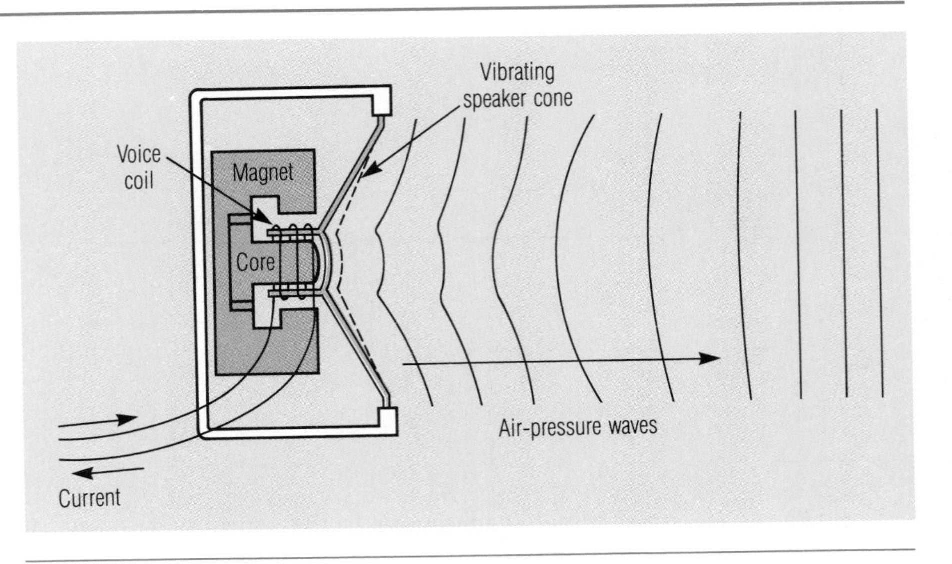

Figure 5–3
In the telephone receiver, the electrical current activates the voice coil, causing the speaker cone to move. The vibrating speaker cone creates air pressure waves that are heard as sound by the ear.

telephones, the switchhook consists of the buttons that are depressed when the handset is placed in the cradle of the telephone. These buttons might better be called just "the switch."

When a user lifts the handset, the switch is closed, and electrical current flows to the telephone company's central office. This current signals the central office that a call is about to be made. This condition is called *off-hook*. The central office equipment responds with a *dial tone*, which lets the caller know that the central office is ready to accept the call.

At the end of a call, replacing the handset in its cradle depresses the switchhook buttons, opens the circuit, and signals the central office that the call is complete. This condition is called *on-hook*. When the caller hangs up the phone, the on-hook signal causes the creation of a data record in the central office computer that is used for billing the customer at the end of the month.

"Dialing" a Number

A few years ago, when someone lifted the handset on a telephone, the response from the telephone company central office was not a dial tone but instead the voice of a human operator asking "Number please?" In those days, the caller spoke the number he or she wished to call and the operator made the connection manually. As the volume of telephone calls grew, telephone company forecasts showed that many additional operators would be required. In fact, the forecasts showed that all young women leaving high school would need to be employed as operators

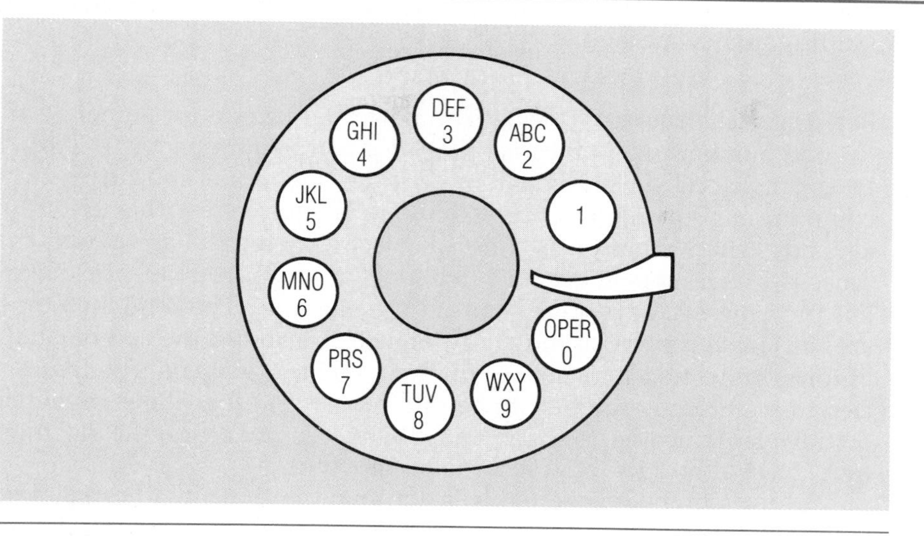

Figure 5–4
A diagram of a telephone's rotary dial.

just to handle the projected volume of telephone calls! It was evident that the process of completing the call had to be automated.

automating call connections

Rotary dials were added to telephones, which transferred the manual aspect of making the call to the caller. In more recent years, rotary dials have largely been replaced by numeric pads also known as dual-tone-multifrequency tone generators. Both the dial pulsing and dual-tone-multifrequency methods of signaling the central office the number of the telephone to be called will be discussed.

Dial Pulsing The *rotary dial* shown in Figure 5–4 generates pulses on the telephone line by opening and closing an electrical circuit when the dial is turned and released. The number of pulses is determined by how far the dial is turned. When the dial is released, the pulses are generated as a spring rotates the dial back to the resting position. The pulses are generated at the rate of 10 pulses per second. Each pulse is 1/20 of a second long, with a 1/20 of a second pause between pulses. This process is called *out-pulsing*.

The rotary dial mechanism is being replaced in some telephones today by an integrated electronic circuit. This allows the telephone to have a pushbutton key pad, but the signaling is still done by electrical pulses that are generated by the electronic circuitry. Since people can push the buttons faster than the pulses can be sent on the telephone lines, this type of telephone must be equipped with an electronic buffer to store digits that have been keyed but not yet out-pulsed.

The dial pulse technique is the most commonly used technique in the world for signaling the desired number, but it is being rapidly displaced in the United States and many other countries by the dual-tone-multifrequency method.

Dual-Tone-Multifrequency (DTMF) The newer technique for signaling the desired number to be called is called *dual-tone-multifrequency* (DTMF). It is accomplished by sending tones on the telephone line. In its most common implementation, the telephone is equipped with a 12-button keypad. When a button is pressed, electrical contacts are closed that cause two oscillators to generate two tones at specified frequencies, much like pressing two keys on a piano at the same time. The combined tones are the signal for one of the digits. Figure 5–5 shows the combinations of tones generated for each key on the telephone keypad. The frequencies were carefully selected to be different from other tones or signals on the telephone line. To be accepted by the central office, the dial tones must have a duration of at least 50 milliseconds.

Tone dialing is considerably faster than pulse dialing. For example, dialing the number zero with tones takes 50 milliseconds for the tones plus 50 milliseconds between tones for a total of 100 milliseconds or 1/10 of a second. Using the dial pulse technique at 10 pulses per second, it takes 1 second to dial the digit zero. The digit 1 takes the same length of time for either pulse or tone dialing. The average telephone number can be dialed 10 to 15 times faster with tone dialing than with rotary pulses. Tone dialing is replacing pulse dialing because it is faster and because it is generated electronically rather than mechanically and is therefore more reliable. The Bell System tradename for DTMF dialing is *Touchtone*.

Ringing

When a call has been dialed and a connection through the central office of telephone network has been made, the telephone company central office that handles the receiver's telephone indicates the incoming call to the receiving telephone by sending a ringing signal (voltage) to it. This voltage may activate a bell, an electronic ringer, a light, or other device. In the United States and Europe, the ringing signal is sent for 2 seconds with a 4-second pause between rings. Other countries have different timings.

At the same time the ringing signal is being sent to the called telephone, an audio ringing signal is sent to the calling telephone. This signal lets the caller know that the telephone company has completed the call connection process and that the called telephone is ringing. The two ringing signals are generated independently, however, and occasionally the called telephone will be answered before the ringing signal back to the caller has been generated. You may have been surprised by this situation yourself when someone you were calling seemingly answered the call "before it rang."

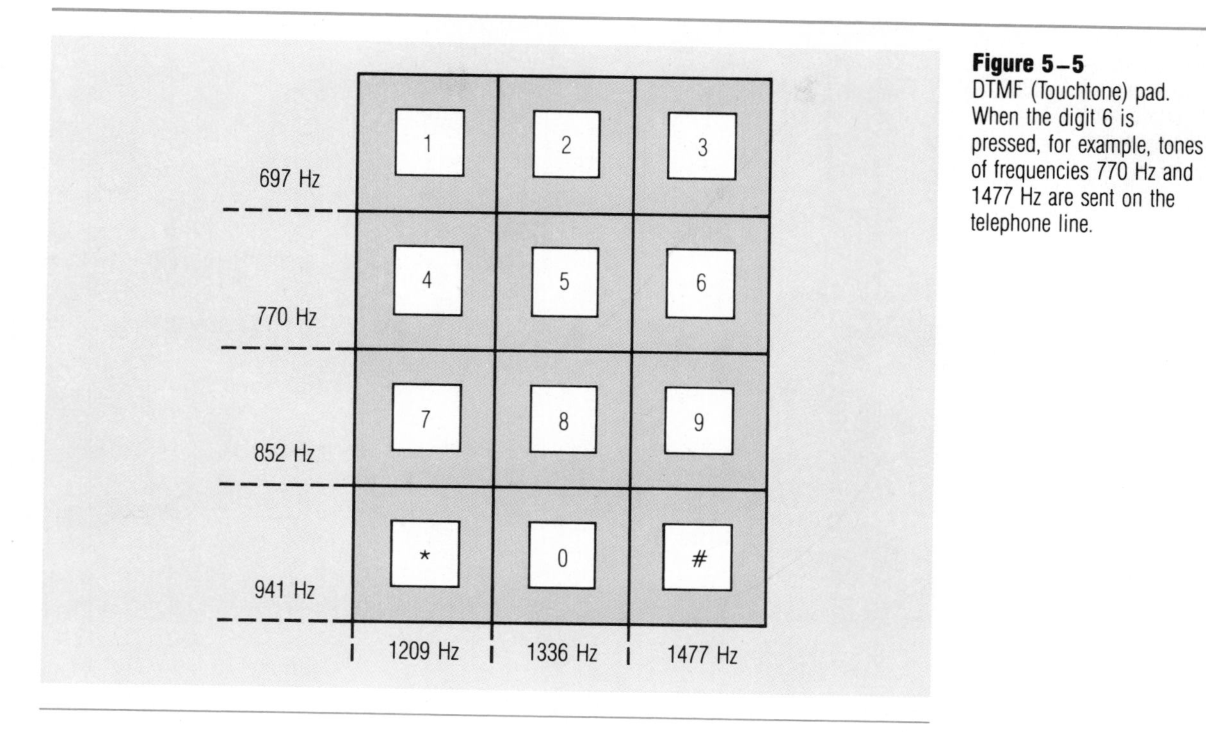

Figure 5–5
DTMF (Touchtone) pad. When the digit 6 is pressed, for example, tones of frequencies 770 Hz and 1477 Hz are sent on the telephone line.

When the called party picks up the telephone (goes off-hook), a signal is sent to the central office that tells it to stop sending the ringing signal to both the caller and receiver.

tracking call costs

In most cases it also signals the telephone company to begin charging for the call, although some carriers use other methods to determine when the call begins. Among the long distance carriers, AT&T uses this signal to time all of the long distance calls it handles. By doing so, it measures and charges only for the duration of the conversation. Some of the other carriers include the time to connect the call, which is called *call setup time*, in their charges. Still other carriers subtract a predetermined number of seconds to compensate for an average call setup time, and others measure noise on the line and begin charging for the call when talking begins.

CENTRAL OFFICE EQUIPMENT

For a short time in the early days of telephony, there was no central office or switching equipment. As described earlier, all telephones had to be connected directly to each other or they could not communicate. Clearly this situation was cumbersome and neither economical nor manageable. Switching (connecting) equipment quickly evolved. The con-

This early manual telephone switchboard was installed in New York City in 1888. (Courtesy of AT&T Bell Laboratories)

cept is quite simple; all telephones are connected to a central office switch, and the central offices are connected with one another, as illustrated in Figure 5–6.

Manual Switching

The earliest type of switching equipment had a central console where all telephone lines terminated in jacks similar to the headphone jack on a stereo receiver. When a caller went off-hook, a light on the console lit and the operator plugged her headset into the jack and talked to the caller. She got the number to be called and plugged a *patch cord*, a cable with plugs on both ends, into the caller's and receiver's jacks to connect the two. Special jacks were available so that the operator could connect her switchboard to other operator's switchboards and to special lines that went to other cities for long distance calls.

Automatic Switching

From the 1920s (when telephones with dials began to replace older style telephones) until today, the connection of telephones at the central office has been increasingly automated. The equipment used has progressed through a series of technologies from *step-by-step switches* (also known as *Strowger switches* after their inventor, Almon B. Strowger), to *crossbar switches* and *reed relays*. Each of these switches was electromechanical and, therefore, more prone to failure than today's electronic switching equipment.

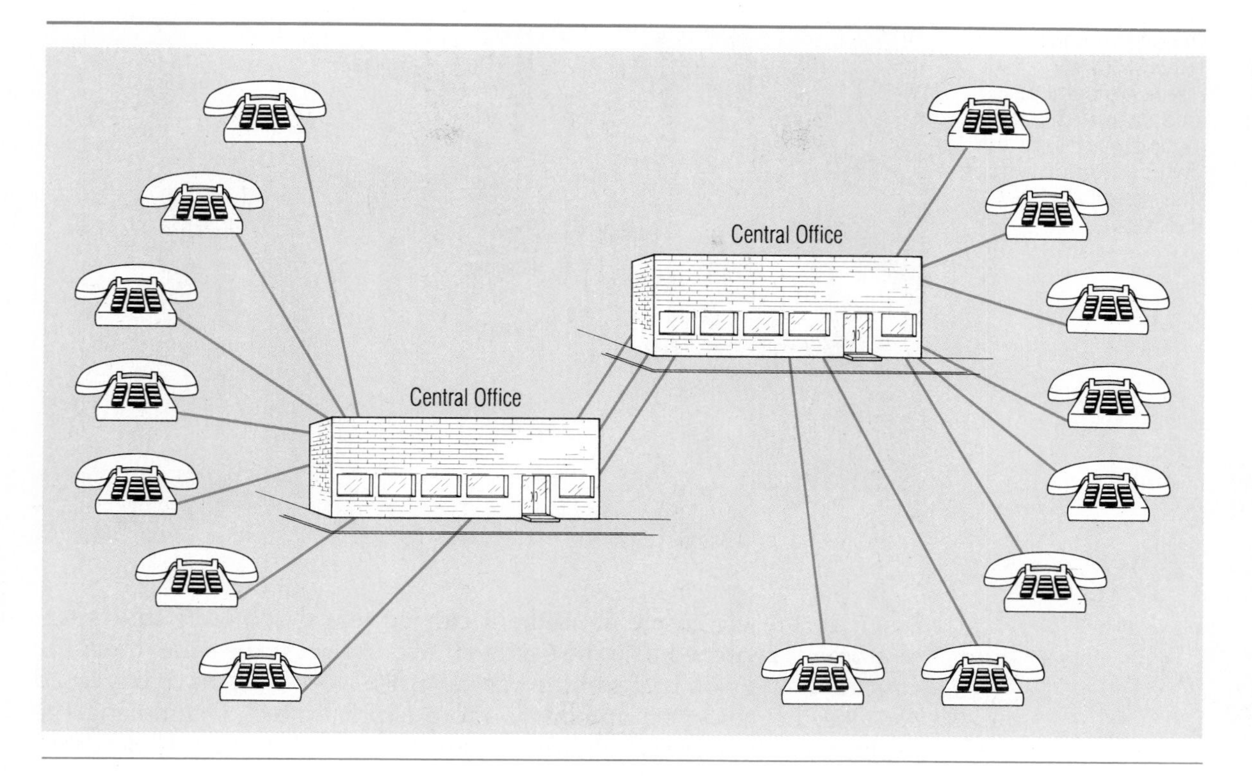

Figure 5–6
It is more efficient to connect telephones to a central office and then connect the central offices together than to connect all telephones to each other.

Electronic Switching

With the development of the transistor and the evolution of the integrated circuit, telephone switching equipment has moved rapidly from the electromechanical to the electronic age. Modern central office switches are entirely electronic, and the connection of calls is completed much more rapidly than before. Since the switches are electronic, they are much less prone to hardware failure than their electromechanical predecessors. However, since they are really programmable computers, they are susceptible to program bugs. Central office switches are designed to be highly redundant so that the failure of any portion of the equipment can be circumvented. The most sophisticated central offices are designed to have no more than 1 total failure in 40 years.

rare central office failures

Electronic central office switches handle thousands of lines. Compared to switches implemented with older technologies, such as step-by-step switches, they are physically much smaller by thousands of square feet. They are also significantly quieter because there are no relay contacts opening and closing or crossbars moving in frames.

As central offices have evolved to electronic switches, the staffing requirement in the central office has dropped dramatically. The main reason is that the newer equipment is more reliable, and fewer maintenance people are required to keep it operating. Another reason is that

staff costs reduced

A modern electronic central office switch is a specialized digital computer and is considerably more compact and reliable than earlier mechanical switches. (Courtesy of AT&T Bell Laboratories)

much of the electronic equipment can be tested remotely and spare equipment switched into the system if necessary. At the same time, the education and skill levels of the central office work force have increased because the newer equipment is more sophisticated. Diagnosing the problems that do occur requires sensitive electronic test equipment and a knowledge of how to use it to isolate the failure.

Design Considerations

Blocking When telephone facilities are designed to provide good service, designers must make a trade-off between the amount and cost of central office switching equipment and the level of service it provides. It is not necessary to include enough equipment or lines to handle simultaneous calls from every telephone connected to a central office. In normal circumstances, only a small percentage of the telephones are in use at any one time. Therefore, the central office can be designed to handle a fraction of the maximum theoretical number of simultaneous calls. If, however, all of the central office equipment is in use and one more person tries to make a call, that call is said to be *blocked* because there are no central office facilities available to handle it. The call cannot be completed even though the called number's telephone is not in use. Blocking can also occur when all of the lines connecting the central offices are in use. Locations where blocking can occur are shown in Figure 5–7. In that case, the central office handling the caller's telephone may be able to handle the call, but it is unable to make a connection to the central office handling the receiver's telephone.

In the U.S. telephone system, blocking is rare. When it does occur, the condition is signaled to the caller by either a fast busy signal or a

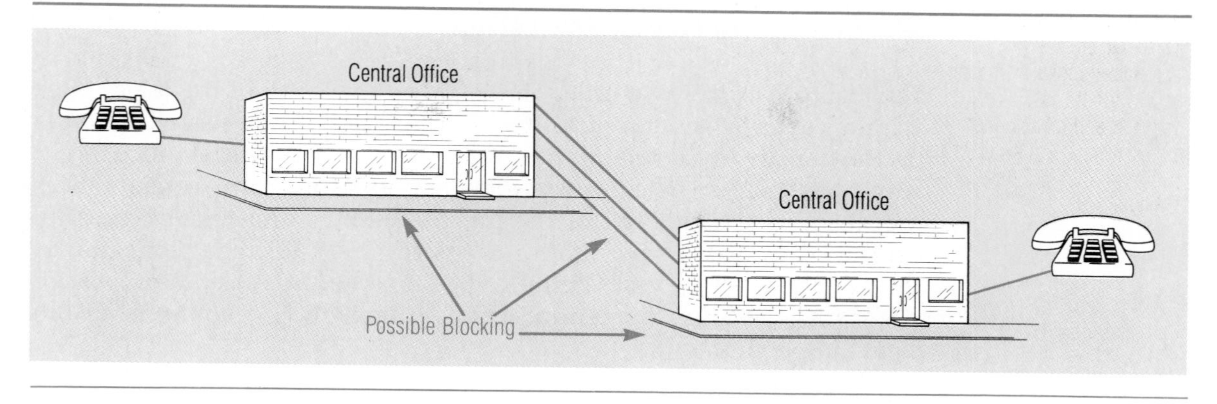

Figure 5–7
Blocking can occur at any central office or on the lines connecting them.

voice recording. The most common occurrences of blocking are when all lines are busy on major holidays, such as Christmas or Mother's Day.

The Busy Hour In deciding how much equipment and how many lines to install, the usual procedure is to study or forecast how many calls will occur in the busiest hour of the day. In businesses, the *busy hour* usually occurs between 10 and 11 A.M., and sometimes another peak occurs between 2 and 3 P.M. Obviously, if enough equipment and lines are installed so that all calls can be handled in the busy hour, there are sufficient facilities available for the calls being made during other hours of the day.

A study of the number of calls to be handled in the busy hour must also take into account their durations. Clearly, eight 3-minute calls tie up more equipment and lines than do eight 1-minute calls. Three 8-minute calls tie up a different mix of equipment than either of the other situations.

For cost reasons, the equipment and lines in a telephone system frequently are designed to handle less traffic than occurs during the busiest hour. When a detailed analysis is performed, the results often show that there is a significant savings in equipment costs by designing the facilities to handle 99, 98, or even 95 percent of the busy hour traffic. This means that 1, 2, or 5 percent of the calls will be blocked, and the caller will have to try again. For most telephone systems, some level of blockage is tolerable.

The study of telephone traffic patterns involves probability and statistics and has grown to be a very rigorous mathematical discipline, although many of the critical calculations have been reduced to tables. The *grade of service* is the proportion of blocked calls to attempted calls expressed as a percentage. If 5 calls of 100 are blocked, the grade of service would be 5/100 or 5 percent. This is designated as a P.05 grade of service. Most public telephone facilities are designed to provide at least a P.01 grade of service, meaning that only 1 call of 100 would be blocked.

grade of service

CENTRAL OFFICE ORGANIZATION

Each central office, also called an *exchange*, serves all of the telephones within a specific geographical area. The size of the area served depends on the density of the telephones. In rural areas, the central office might serve many square miles, whereas in New York City, many central offices are needed to handle all of the office buildings and residences. The central office acts as the hub for the wires and cable connecting all of the telephones it serves. Wiring comes together in a rack called a *main distribution frame* in the central office, from which it connects to the switching equipment.

Central offices are optimized to perform specific functions. Some are primarily designed to switch local telephone calls from and to businesses and residences. This type of central office is commonly known as an *end office*. The central office to which a specific telephone is connected is known as that telephone's *serving central office*. In 1985, the BOCs had approximately 9,000 end offices and controlled about 85 percent of U.S. telephone service. The other LECs also had approximately 5,000 end offices serving the remaining 15 percent of the telephones.

toll offices

Other central offices, called *toll offices* or *switching offices*, are designed primarily for forwarding long-distance calls to other parts of the country. Offices are connected in a weblike pattern that provides a high level of redundancy and alternate routing, as shown in Figure 5–8. Any central office may be connected to any other central office by direct lines, but a major consideration as to whether to connect two offices is the amount of traffic (number of telephone calls) that flows between them. In some cases, several lines are needed to carry the traffic between offices; in other cases, a direct connection is not necessary, and traffic is routed in a less direct fashion.

A call is handled at the lowest level office that can complete it or provide the required service. If a call cannot be handled by the serving central office that first received it, the call is forwarded to the most appropriate central office. The routing is determined by tables stored in the memory of the central office switch. For example, a call from a business in Dallas going to New York City might be routed from the serving central office in Dallas to a toll center serving part of Dallas. The toll center's tables might tell it that calls for New York should be routed to a certain toll office in Atlanta or, if Atlanta is busy, to Saint Louis. The objective of the network is to route the calls to the destination central office in the shortest, fastest way possible.

In AT&T's long distance network there are over 150 toll offices. AT&T uses a technique called *dynamic nonhierarchical routing (DNHR)* to route calls. Each toll office can examine up to sixteen different routes to complete a call if necessary. The goal of AT&T's network is to have a long distance call pass through no more than four toll offices. In other words, in the worst case, a call would pass from its serving office through

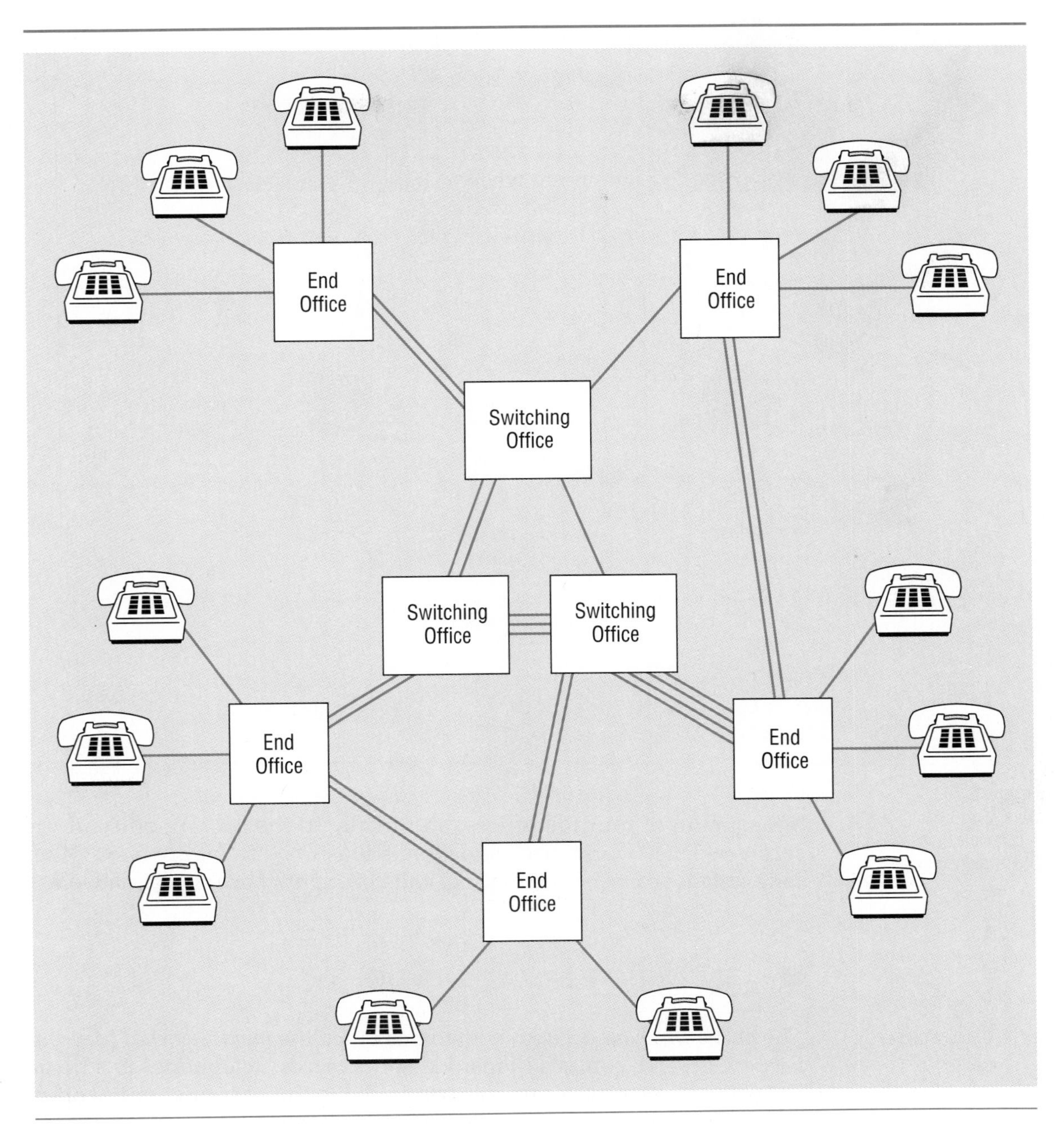

Figure 5–8
The connection of the central offices in the nationwide telephone network.

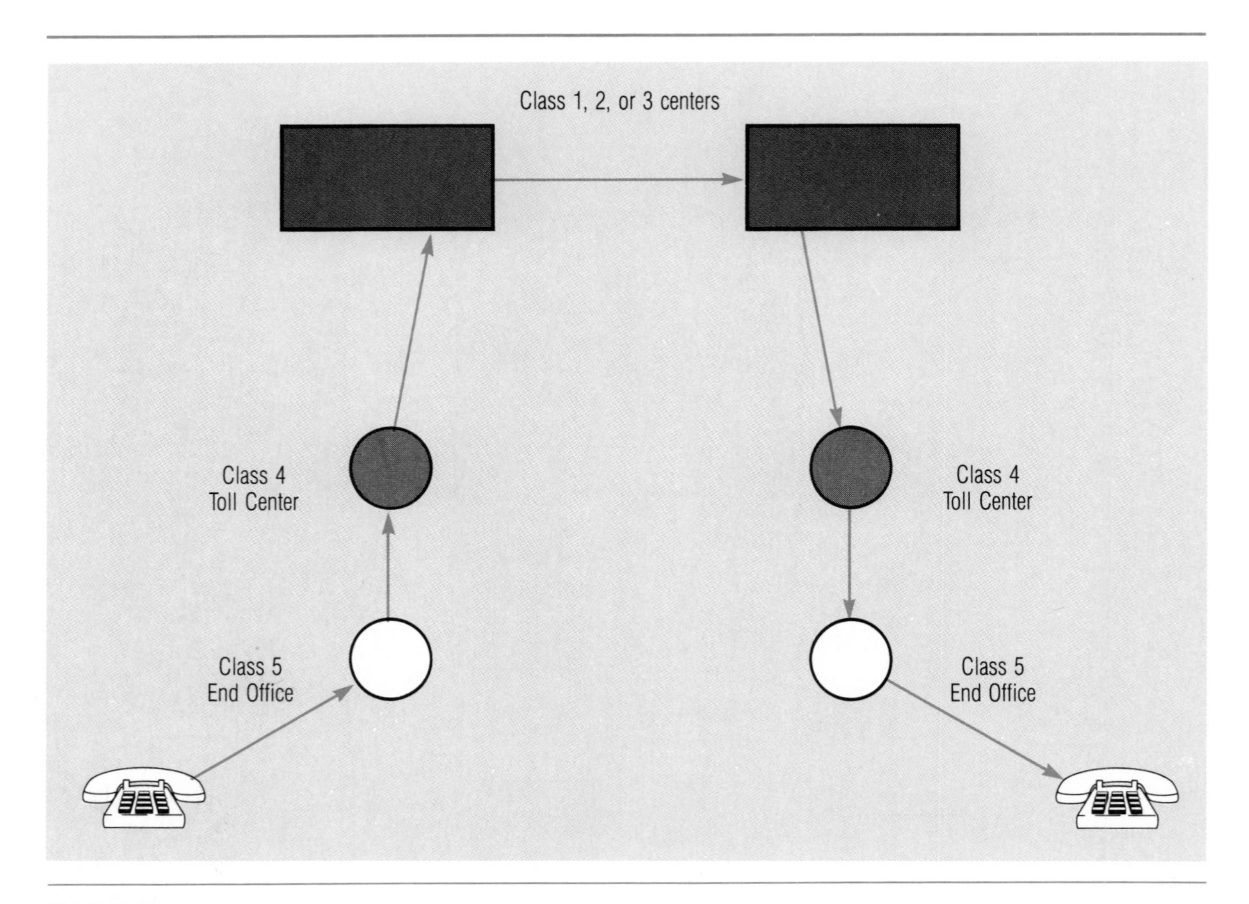

Figure 5–9
Worst case routing of a telephone call in AT&T's DNHR routing system.

a maximum of four toll offices, and finally to the serving office of the receiver. This "worst case" situation is illustrated in Figure 5–9. Other long distance carriers, such as MCI and U.S. Sprint have similar networks.

THE PUBLIC TELEPHONE NETWORK

The *public telephone network*, sometimes called the *public switched telephone network (PSTN)*, consists of many distinct pieces. Telephones in a home or business are most commonly connected to their serving central office by a pair of copper wires called the *local loop*. In telephone company jargon, these two wires are often referred to as *tip and ring*. In a residential neighborhood, the wire coming from the house, called the *drop wire*, runs to a pole (or underground equivalent) where it joins other similar wires to form a *distribution cable*. Eventually, several distribution cables join together to form a *feeder cable* that terminates at the central office, as shown in Figure 5–10. Clearly, the cable gets physically larger

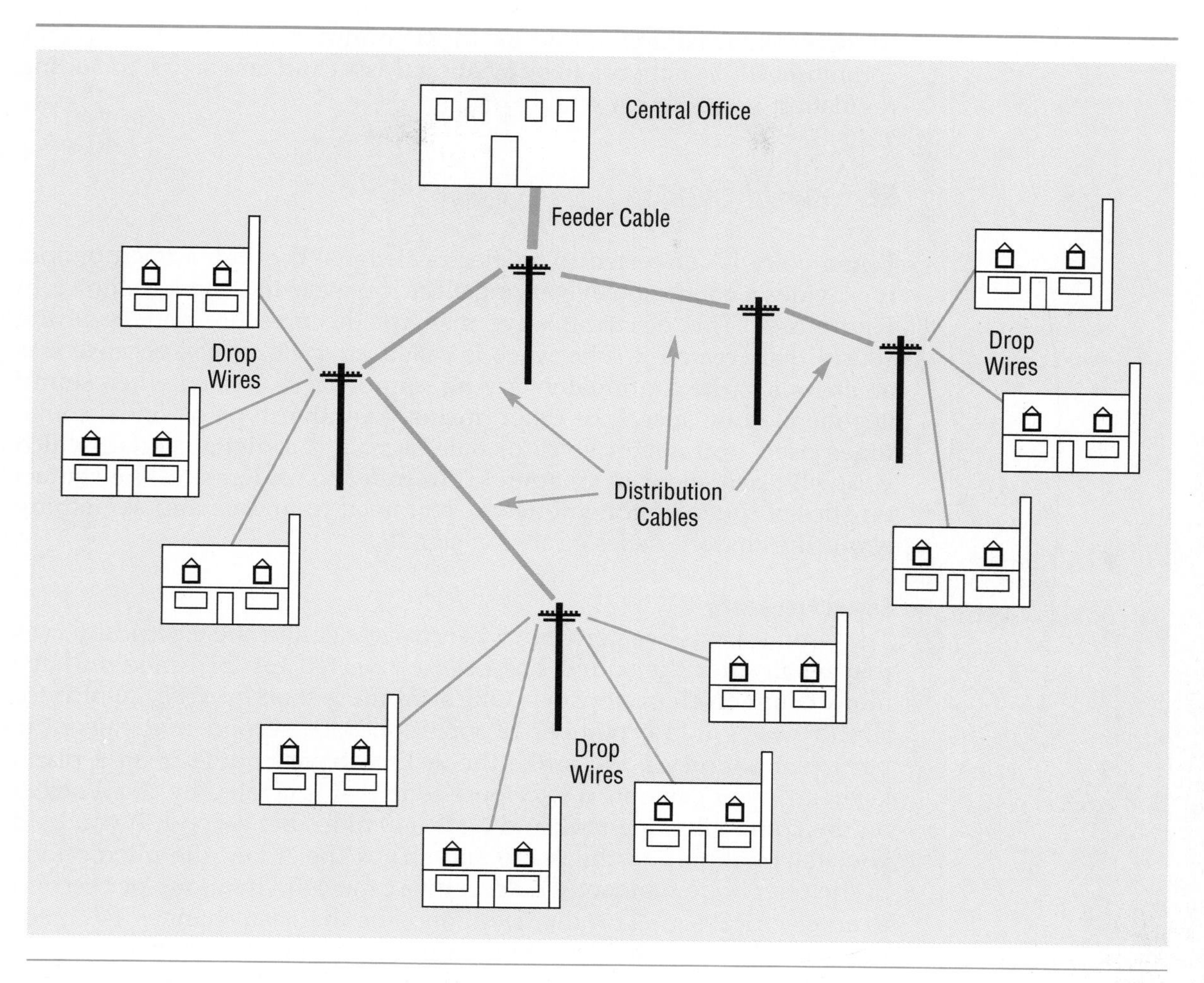

Figure 5–10
Residential telephone cabling.

the closer it gets to the central office. Cables may run above ground on poles or underground, but most new installations being made today are underground, where they are better protected and out of sight.

Central offices are connected to each other with multiple lines called *trunks*. A trunk is defined as a circuit connecting telephone switches or switching locations. (The term *circuit* is defined in Chapter 9, but in the meantime, think of a circuit and a line as being synonymous.) Functionally, a trunk and a line do the same thing—carry communications—but a trunk connects switching equipment together, whereas a line connects to a telephone, computer terminal, or other device. Trunks may be implemented with regular copper wire but are more often implemented with coaxial cable, microwave radio, or fiberoptic cable.

The PSTN, with its millions of miles of lines, handles virtually all of the voice and data communications in the world. It is diverse and highly redundant; therefore, it is impervious to massive outages. It is resilient when failures do occur. Most countries consider their public communications network to be a national asset and take steps, including regulation, to protect it.

ANALOG SIGNALS

When voice is converted to an electrical signal through a microphone, it provides a continuously varying electrical wave like the one shown in Figure 5–2. This electrical wave matches the pressure pattern of the sound that created it. The wave is called an *analog signal* because it is analogous to the continually varying sound waves created when sound is generated by speech or other means. (Another type of signal sometimes used to transmit voice or data signals is a digital signal, which you will learn about in Chapter 7.) In order to understand the characteristics of the telephone network, you need to understand something about the characteristics of analog signals.

Signal Frequency

The sound waves we generate when we speak and the electrical waves that result after the sound has been converted for transmission have many common characteristics. One attribute is their *frequency*, which for sound waves is the number of vibrations per second that cause the particular sound. If you strike the A key above middle C on a piano keyboard, you generate a very pure tone that is created by the A string on the piano vibrating back and forth 440 times per second. If you held your telephone up to the piano and struck the A key, the microphone in the telephone handset would convert the 440 vibrations per second to an electrical signal on the telephone line that also changes 440 times per second. This signal is commonly diagrammed as a *sine wave*, as shown in Figure 5–11. Each complete wave is called a *cycle*, and the frequency of the signal is the number of cycles that occur in one second. The unit of measure for frequency is the *Hertz*, abbreviated *Hz*. We say that the A key we struck on the piano generated a tone with a frequency of 440 Hz. The corresponding (analogous) electrical signal also has a frequency of 440 Hz. By way of comparison, a higher tone on the piano has a higher frequency. The A key that is an octave above A 440 has a frequency of 880 Hz. The lowest note on the normal piano keyboard has a frequency of 27.5 Hz, and the highest note has a frequency of 4,186 Hz.

sine wave

The human ear can hear sound with a range of frequencies from about 20 Hz to approximately 15,000 Hz. Between 15,000 and 20,000 Hz, most people can sense the sound but not actually hear it. Good stereo systems will reproduce sounds up to approximately 20,000 Hz and have better fidelity (sound better) than systems that do not reproduce frequencies that high. Figure 5–12 illustrates some of these frequency ranges.

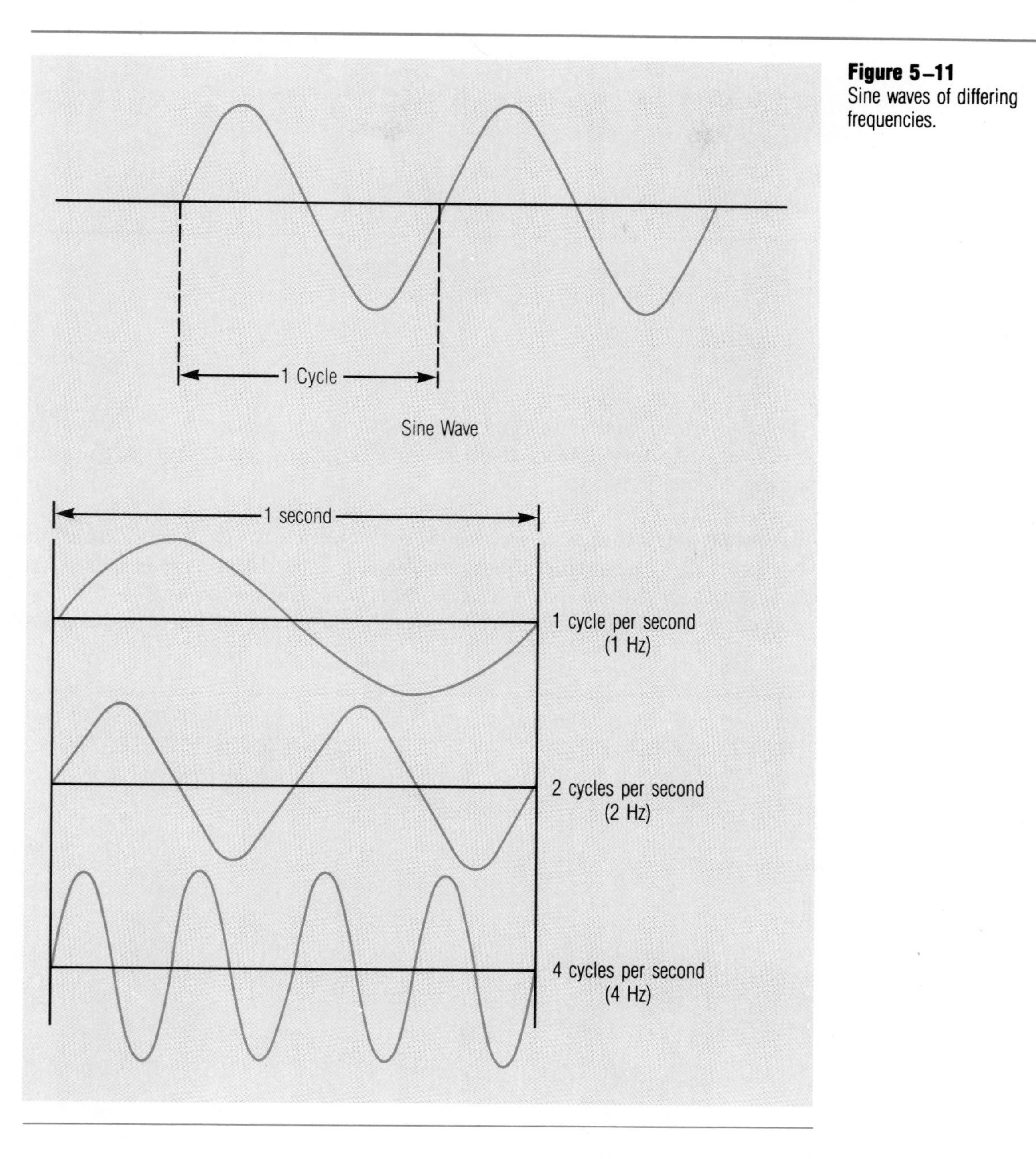

Figure 5–11
Sine waves of differing frequencies.

Sound waves, electrical waves traveling in a wire, and electromagnetic waves traveling through space, such as radio waves, have essentially the same characteristics. All are represented as sine waves whose frequency is measured in Hertz. Figure 5–13 shows the frequency spectrum, the full range of frequencies from zero Hz to several hundred thousand million Hz. When referring to very high frequencies, we commonly use the designation *kilohertz (kHz)*, *megahertz (MHz)*, and *gigahertz*

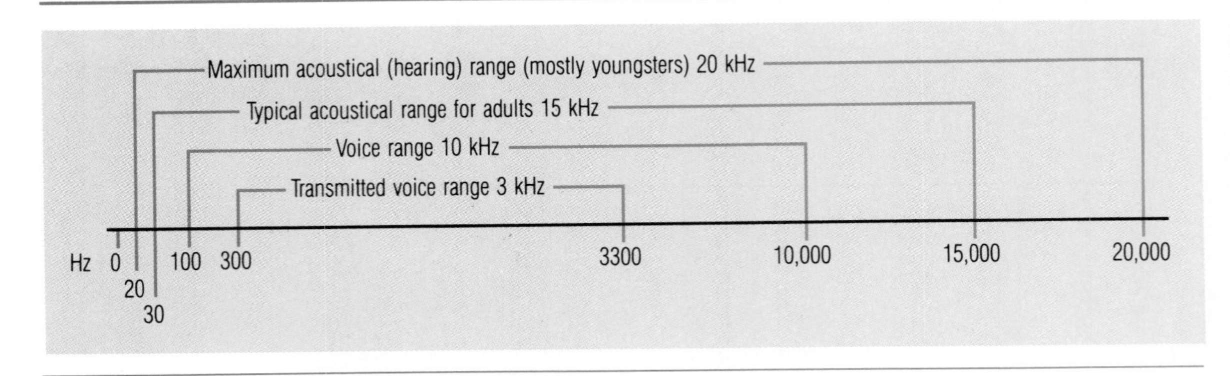

Figure 5–12
The frequency ranges of some common sounds.

(gHz) to more easily describe the frequencies. Figure 5–14 shows the full range of abbreviations used for very large and small units of measure in the scientific world.

Bandwidth Another way to look at a frequency range is the difference between the upper and lower frequency. This difference is called the *bandwidth*. In the case of a telephone signal, the bandwidth is 3,000 − 300 Hz or 2,700 Hz. Voice circuits in the telephone system are designed

Figure 5–13
The frequency spectrum showing the common names applied to certain frequency ranges.

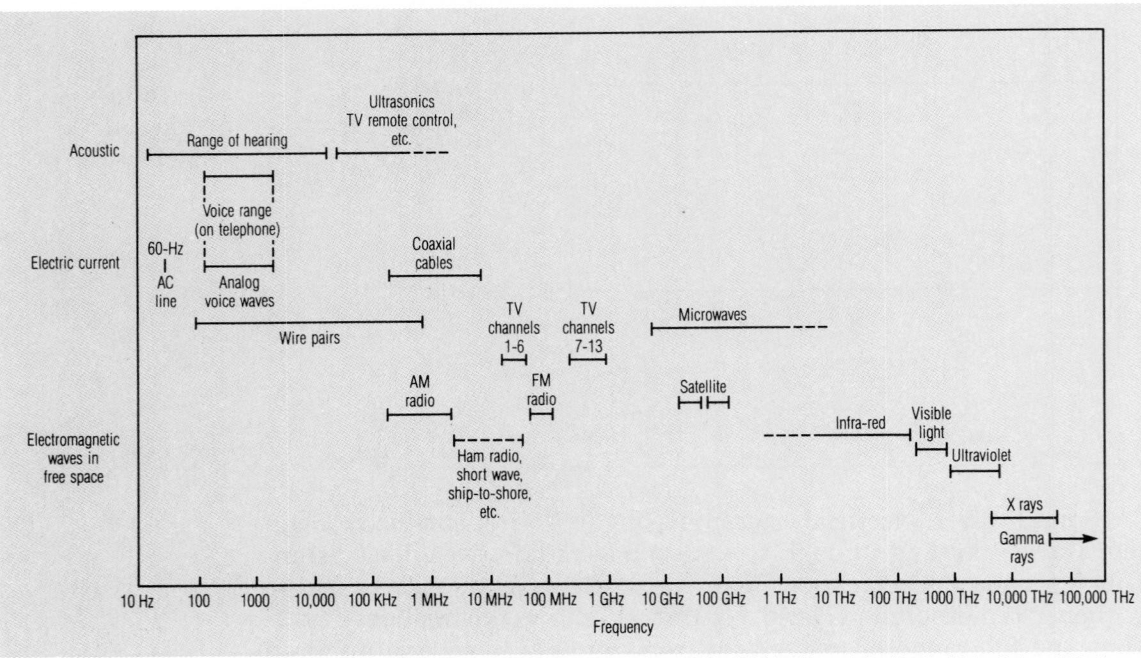

From Michael L. Gurrie/Patrick J. O'Connor, *Voice/Data Telecommunications Systems: An Introduction to Technology*, © 1986, p. 204. Adapted by permission of Prentice-Hall, Inc., Englewood Cliffs, New Jersey.

pico	trillionth	.000000000001	1×10^{-12}
nano	billionth	.000000001	1×10^{-9}
micro	millionth	.000001	1×10^{-6}
milli	thousandth	.001	1×10^{-3}
centi	hundredth	.01	1×10^{-2}
deci	tenth	.1	1×10^{-1}
deca	ten	10	1×10^{1}
centa	hundred	100	1×10^{2}
kilo	thousand	1000	1×10^{3}
mega	million	1000000	1×10^{6}
giga	billion	1000000000	1×10^{9}
tera	trillion	1000000000000	1×10^{12}

Figure 5–14
Common abbreviations for very large and very small quantities.

to handle frequencies from 0 to 4,000 Hz, as shown in Figure 5–15, but special circuitry limits the voice frequencies that can pass through it to those between 300 and 3,000 Hz. The additional space between 0 and 300 Hz and between 3,000 and 4,000 Hz is called the *guard channel* or *guard band*, and it provides a buffer area so that adjacent telephone conversations or data signals don't interfere with each other.

Signal Amplitude

Another characteristic of analog signals is their loudness, or *amplitude*. As you speak more loudly or softly into the telephone, the sound waves

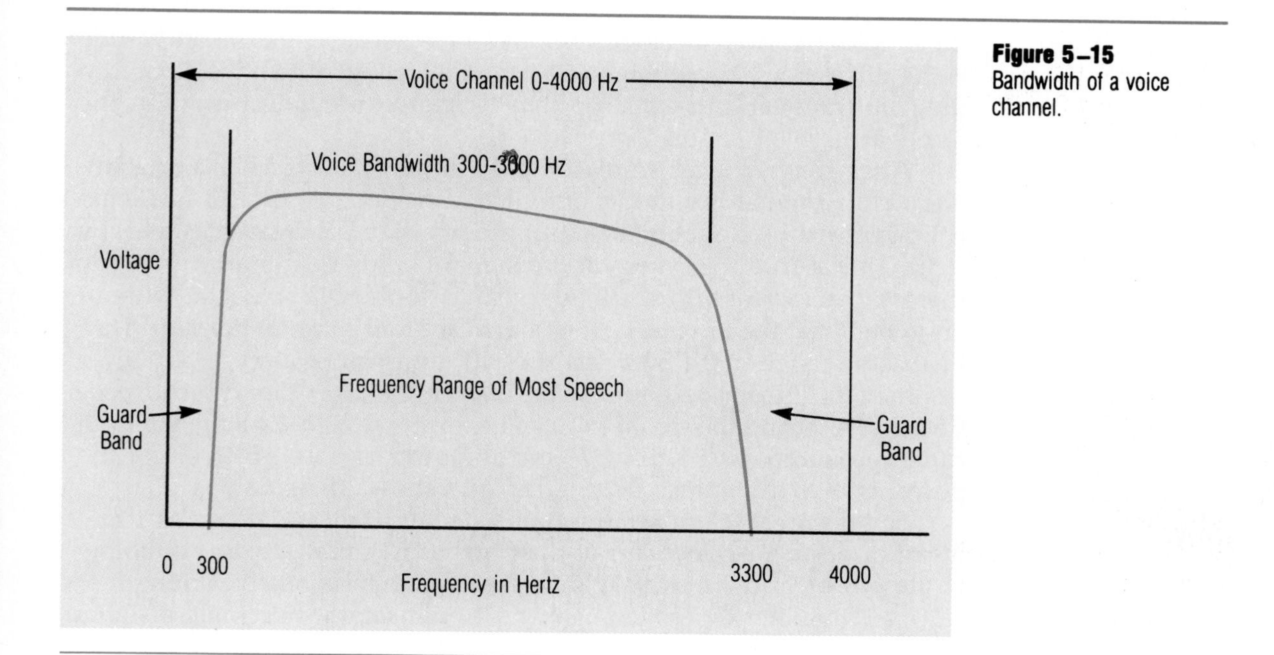

Figure 5–15
Bandwidth of a voice channel.

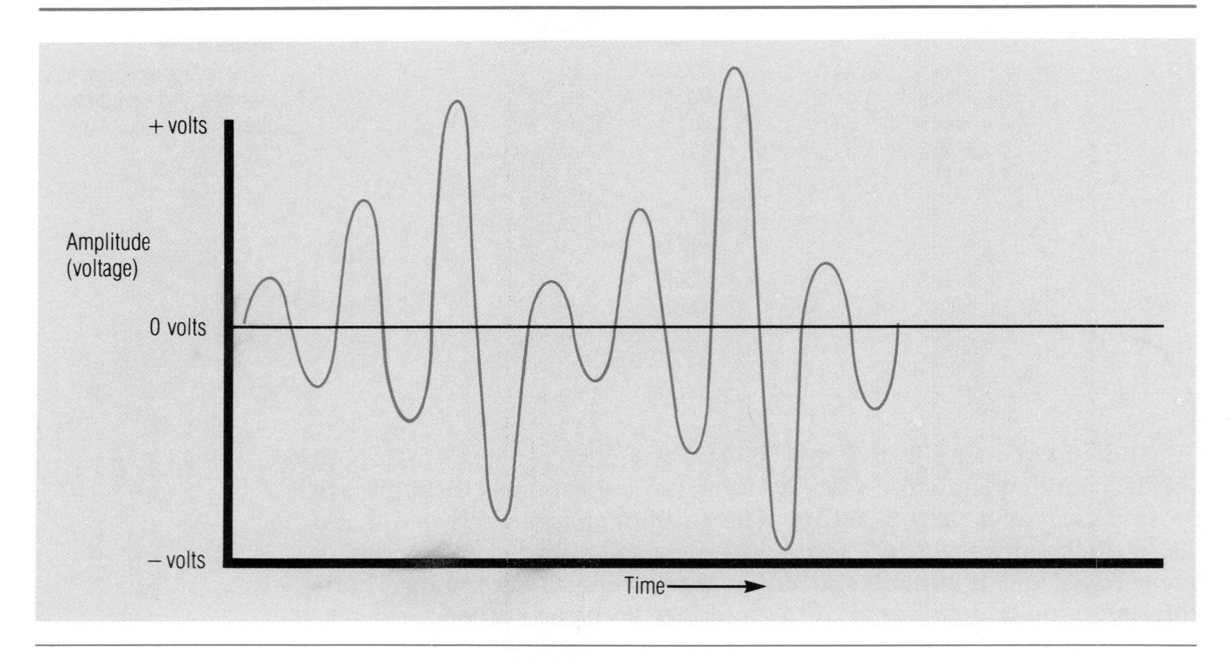

Figure 5–16
Analog wave with constant frequency and varying amplitude.

create larger and smaller electrical waves that are represented by higher peaks and valleys of the signal's voltage (shown in Figure 5–16). The amplitude of the signal is also called its *level*, and whereas with sound the amplitude relates to loudness, in an electrical signal the amplitude is the difference between its most negative voltage (the lowest point in the sine wave) and its most positive voltage (the highest point in the sine wave).

Analog signal level is measured in *decibels (dB)*, which is a logarithmic ratio of signal input and output power. Because the dB is a logarithmic measure, doubling the strength of a signal increases its level by 3 dB. This is true regardless of the signal's original strength. If we say that a signal increased by 3 dB, we mean it doubled in strength, without knowing what the original or new signal strengths are. In the same way, increasing its strength by a factor of 10 raises its level by 10 dB, by a factor of 100, 20 decibels, and so on. Working in the other direction, we find that reducing the signal to 1/2 of its former level causes the strength to be measured as −3 dB; 1/10 of the power is −10 dB; 1/100 of the power is −20 dB, and so on. Figure 5–17 shows these values.

For electrical telecommunications signals, 0 dB is defined as 1 milliwatt of power. An increase of the power to 2 milliwatts is a doubling of the power, and the signal would therefore have a relative strength of +3 dB. Doubling the power again to 4 milliwatts would yield a signal

Decibels	Relative Power
+30dB	1000
+20	100
+10	10
+3	2
0dB	1
−3dB	1/2
−10	1/10
−20	1/100
−30	1/1000

Figure 5–17
The relative power of a signal measured in decibels.

with a strength of +6 dB. The mathematical formula for the relationship between power and signal strength is

$$\text{dB} = 10 \times \log(\text{multiplier})$$

where the multiplier is the number of times the power increased. The formula works if the power decreases, too. If the power is reduced from 5 milliwatts to 2 milliwatts, the multiplier used in the formula would be .40.

Decibels and the strength of a signal are of considerable interest in telecommunications. If too much power is put on a line, a particular type of interference called *crosstalk* (which is discussed in Chapter 9) can occur. The loss of signal strength is also of interest because if a signal does not have enough strength at the receiving end of a communication path, it will be unusable. This loss, which is called *attenuation*, is measured between two points on a line, as shown in Figure 5–18. At the point where the signal is injected on the line, it has a certain strength.

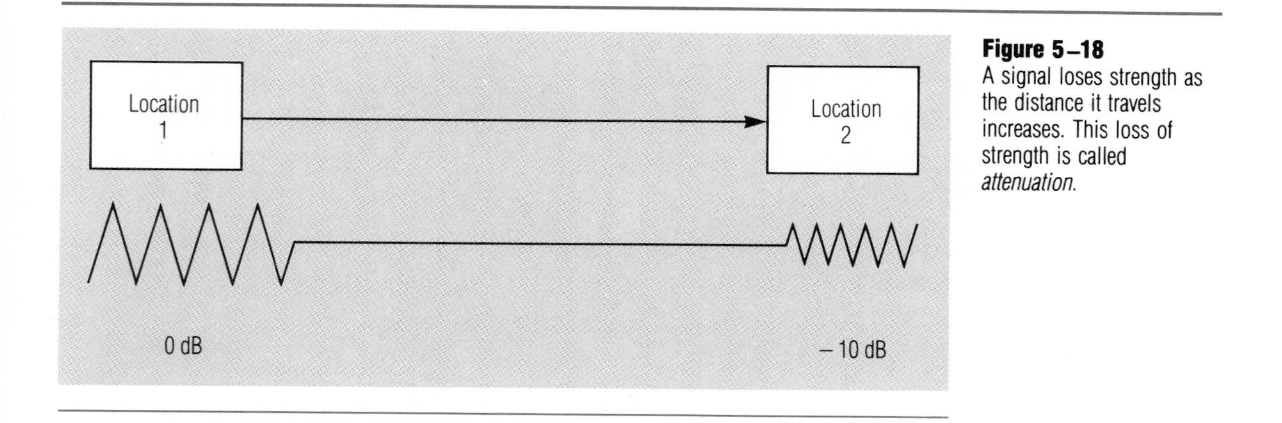

Figure 5–18
A signal loses strength as the distance it travels increases. This loss of strength is called *attenuation*.

As the signal moves away, its strength is reduced due to the attenuation of the line. The reduction is measured in decibels.

Signal Phase

A third attribute of an analog signal is called its *phase*. In contrast to frequency and amplitude, phase is harder to relate to the physical world and, therefore, is somewhat harder to understand. A signal's phase is the relative position of the sine wave measured in degrees. Sine waves can be measured in degrees, where 360° is one complete cycle of the wave. Figure 5–19 shows a sine wave that appears to break and start again, skipping a portion of the wave. This is a phase shift. Since ¼ of the wave has been skipped, it is called a 90° phase shift. Phase shifts are created and detected by electronic circuitry.

Whereas amplitude and frequency changes can be detected by the human ear, phase changes cannot, and they therefore are of little importance in voice transmission. They are very important in data transmission, however, and Chapter 8 discusses phase shifting in more detail.

ATTRIBUTES OF A VOICE SIGNAL

Whereas single tones produce clean sine waves of a specific frequency and amplitude, the human voice, music, noise, and most other sounds are made up of a large range of frequencies and amplitudes. As a result, the wave pattern is far more complex than the simple sine waves we have looked at thus far. Normal speech is made up of sounds with frequencies in the range of 100 to 6000 Hz, but most of the speech "energy" falls in the 300 to 3,000 Hz range. Although some people with high-pitched voices emit occasional sounds above 6,000 Hz, the majority

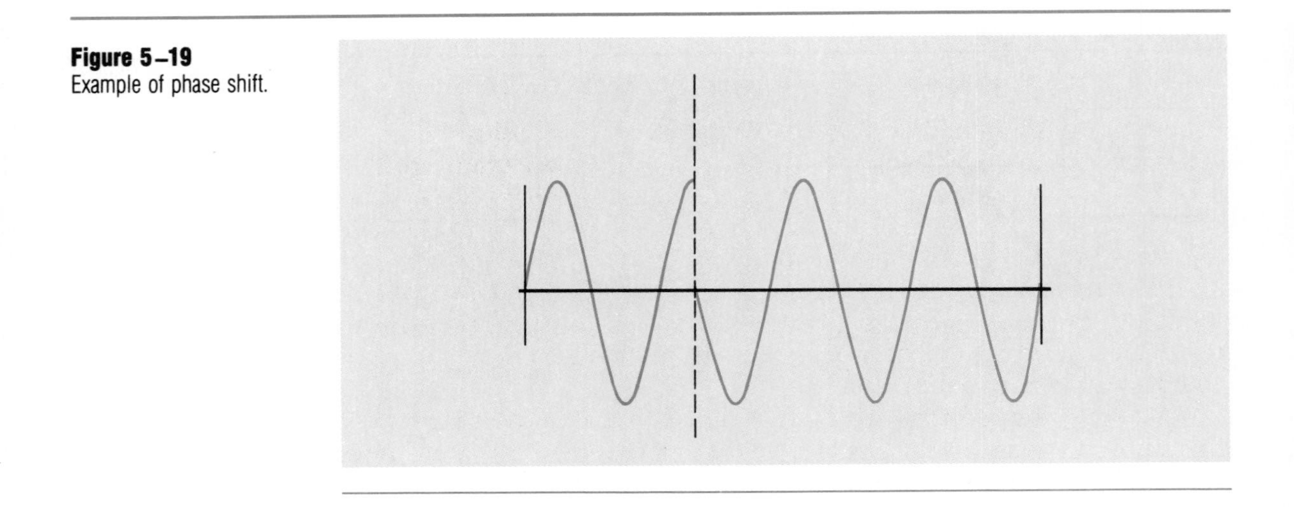

Figure 5–19
Example of phase shift.

of the sound still falls in the range of 300 to 3,000 Hz. That is why the public telephone system is designed so that all of the lines, handsets, and other components will pass voice frequencies in that range. Frequencies outside that range are filtered out by electronic circuitry and are not allowed to pass.

FREQUENCY DIVISION MULTIPLEXING (FDM)

Whereas the individual telephone circuit has a bandwidth of 4,000 Hz, the pair of wires or other media carrying it has a much higher bandwidth capacity. Twisted-pair wires have a bandwidth of approximately 1 million Hz. Dividing 4,000 Hz into 1 million Hz shows us that, at least theoretically, a standard pair of telephone wires should be able to carry approximately 250 telephone conversations. This is a very theoretical number; in practice 12 or 24 voice signals normally are carried. Naturally, the telephone company would like to take advantage of the ability to carry multiple conversations on one line, especially on trunk lines between central offices in which hundreds of telephone calls are handled simultaneously.

The technique of packing several analog signals (phone calls in this case) onto a single wire (or other media) is called *frequency division multiplexing (FDM)*. It is accomplished by translating each voice channel to a different part of the frequency spectrum that the media can carry. Using the telephone wire pair as an example, if a second voice signal could be relocated from its natural frequency of 0 to 4,000 Hz to say 4,000 to 8,000 Hz, and a third voice signal relocated to 8,000 to 12,000 Hz (as shown in Figure 5–20), many telephone conversations could be packed on one pair of wires.

MODULATION

Frequency division multiplexing is accomplished by transmitting a sine wave signal in the new frequency range in which the original signal is to be relocated. The new sine wave is called a *carrier wave*, not to be confused with a *common carrier*. The carrier wave in itself contains no information, but its attributes are changed corresponding to the information in the original signal. This change to the carrier wave is called *modulation*. Modulation converts a communication signal from one form to another, more appropriate form for transmission over a particular medium between two locations.

FM, AM, and PM

You have learned that there are three attributes of the sine waves that can be changed. If the amplitude is changed, it is called *amplitude modulation (AM)*; changing the frequency is called *frequency modulation*

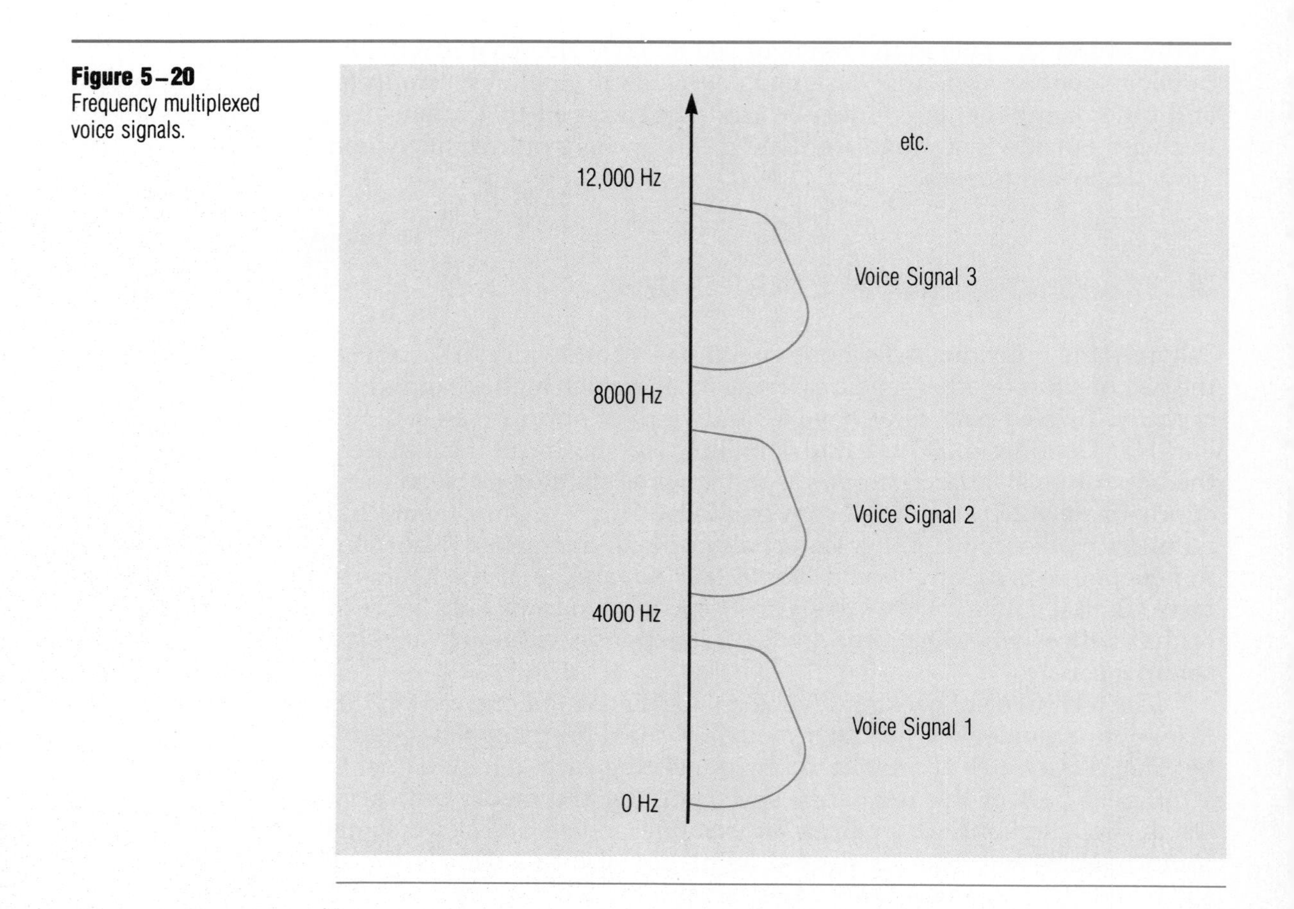

Figure 5–20
Frequency multiplexed voice signals.

(FM); and changing the phase is called *phase modulation (PM)*. Amplitude and frequency modulation are shown in Figure 5–21; phase modulation is discussed in Chapter 8. Combinations of these modulation techniques are also possible, for example, phase amplitude modulation (PAM).

Shifting the frequency of a signal to a different frequency range is one important use of modulation (we will look at the other use in Chapter 8), and the result is that the original signal is relocated to a different set of frequencies. At the receiving end, an electronic circuit called a *detector* must be able to unscramble the modulated signal and relocate it back to the original frequencies—a process called *demodulation*. In the telephone example, the modulation of the original 0–4,000 Hz voice signal occurs at the central office serving the person who is speaking, and the demodulation occurs at the serving central office near the listener.

Multiplexing equipment (multiplexers) at the central office pack groups of twelve 4 kHz voice signals into 48 kHz signals called *base groups, channel groups*, or just *groups*. Other multiplexers then pack five 48 kHz groups into *supergroups*, which have a bandwidth of 240 kHz and contain

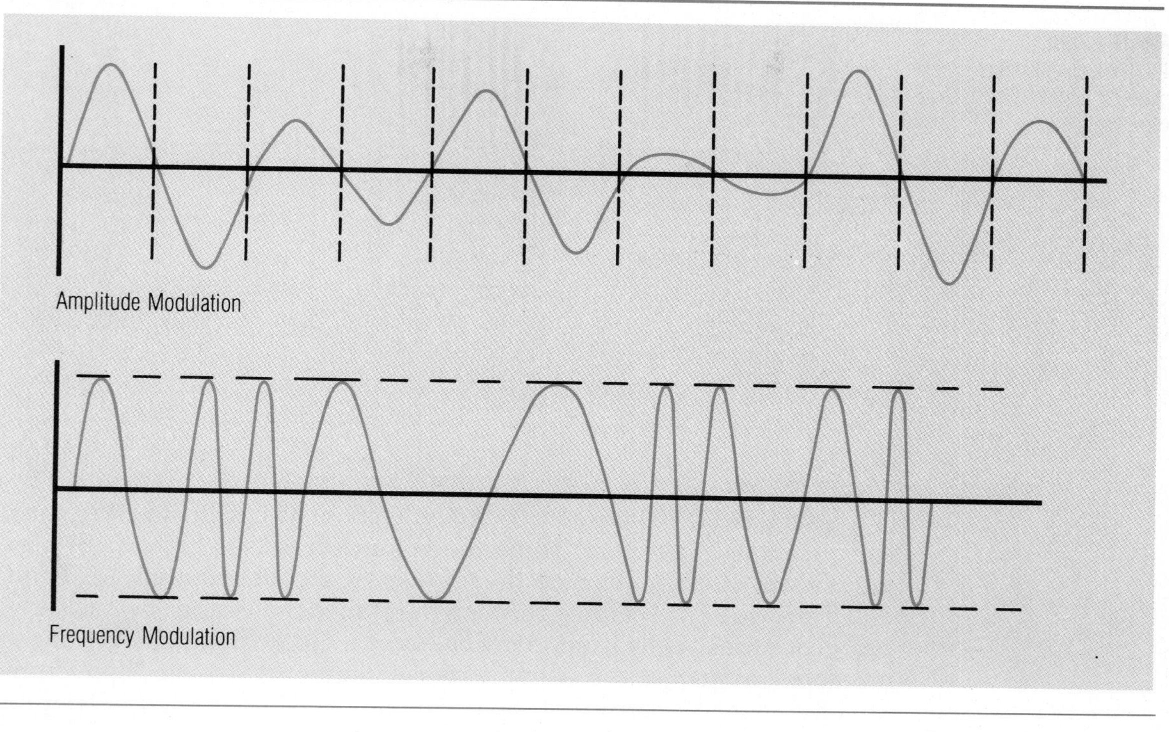

Figure 5–21
Amplitude and frequency modulation.

60 voice signals. Ten supergroups are multiplexed into a *master group* for long distance transmission. You can see from Figure 5–22 that as it travels from transmitter to receiver, the individual 4 kHz voice signal is modulated to different frequencies several times by the FDM equipment of the telephone company. As a result, lines and trunks are used efficiently between telephone company offices where traffic volume is high.

■ TASI VOICE TRANSMISSION

A technique that has long been used on undersea cables is called *time assignment speech interpolation (TASI).* TASI is another way of packing multiple voice conversations onto a single telephone line. TASI takes advantage of the fact that there are pauses in speech and approximately 10 percent of the time in a normal voice conversation, no one is speaking. TASI equipment detects when a person starts speaking and, within a few milliseconds, assigns a communication circuit to the speaker. Although a very small amount of the first syllable may be lost, it is almost undetectable by the listener. When the person stops talking, the communications circuit is taken away and assigned to another speaker. When the person starts talking again, a new circuit is assigned.

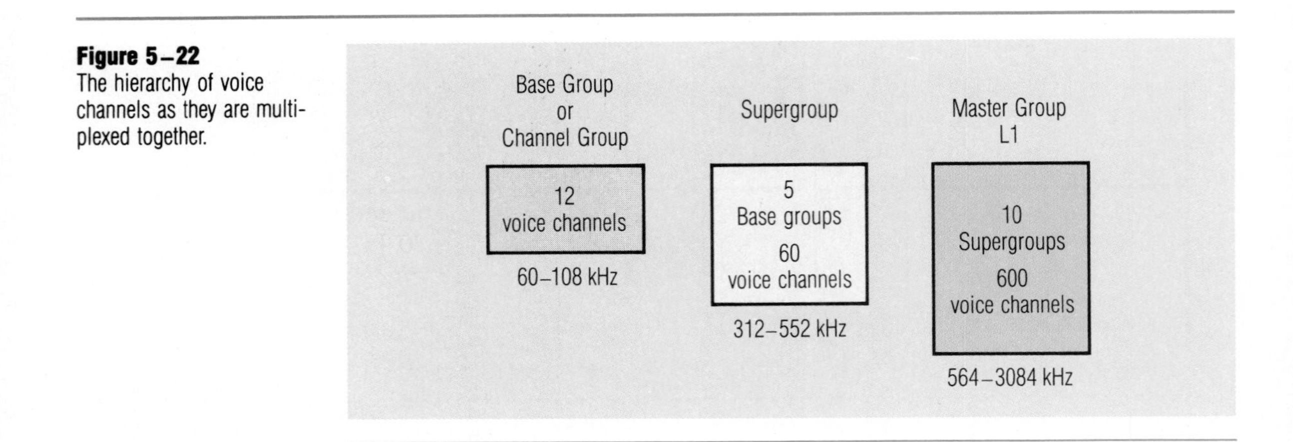

Figure 5–22
The hierarchy of voice channels as they are multiplexed together.

Using TASI, 100 talkers require only about 45 circuits to carry their conversations. Therefore, when the volume of calls is high, TASI provides a way to economize on the number of circuits required. MCI and AT&T provide TASI service between the United States and several European countries. Calls from many customers are pooled for TASI transmission. Companies can obtain a trans-Atlantic TASI voice "line" from MCI for about 60 percent of the cost of a normal, dedicated telephone line.

INTEROFFICE SIGNALING

Direct Current (DC) Signaling

In addition to carrying voice signals, the telephone lines must also carry various other kinds of signals used to set up a telephone call and indicate its status. One type of signaling occurs simply by opening or closing the electrical circuit between the telephone handset and the central office. This is called *direct current (DC) signaling*. DC signaling is used primarily between the serving central office and the customer. It is analogous to turning on a light switch that allows electrical current to flow to a light bulb. The electrical current required for DC signaling is generated by a power supply at the serving central office.

As used on the local loop, DC signaling works as follows. When the telephone handset is on-hook, the circuit is open, and no current can flow. When the handset is lifted, the circuit is closed, current flows to the central office, and it sends a dial tone (a tone signal) to the handset.

Another type of DC signaling is pulse dialing. When a digit is pulsed, the flow of current is interrupted by the pulsing mechanism (a rotary dial or its electronic equivalent), which opens and closes the circuit a certain number of times depending on which digit is being dialed.

Tone Signaling

Another type of signaling used in the telephone system is *tone signaling*. When you lift the handset, you hear a dial tone (assuming everything is working) that is a combination of a 350 Hz tone and a 440 Hz tone. On telephones with DTMF dialing, each button pressed creates a tone also made up of a combination of two frequencies. As the call is set up, you either hear a ringing signal that is a combination of 440 Hz and 480 Hz tones, a busy signal that is a combination of 480 Hz and 620 Hz tones, or a congestion signal, which means that toll trunks between central offices are busy. This is sometimes called the *fast busy* and is made up of tones with a frequency of 480 plus 620 Hz that are sent more rapidly than a normal busy signal. If you leave your telephone off-hook, you get an off-hook signal that combines tones at 1,400 Hz, 2,060 Hz, 2,450 Hz, and 2,600 Hz. This signal is much louder than the others in order to get your attention.

fast busy

Other signal tones are used between central offices. For example, the DC signal generated by a pulse dial telephone cannot be transmitted between central offices, so it is converted to tone signals at the originating central office. This conversion is an example of *E&M signaling*—a special type of signaling that takes place between switching equipment.

Notice that all of the tones mentioned so far fall in the 300 Hz to 3,000 Hz frequency range allowed for the voice signal. These are called *in-band signals*. Most of the tones that the central offices use for signaling each other also use in-band signals, but frequencies between the 3,000 Hz cutoff for the voice signal and the 4,000 Hz boundary of the telephone circuit are sometimes used. These are called *out-of-band signals*, and the most commonly used frequency is 3,700 Hz.

Common Channel Signaling

The most important signaling between central offices occurs on a special network of lines that are reserved exclusively for signaling information. This network, called the *common channel interoffice signaling system (CCIS)*, uses a set of signals called *Signaling System No. 7* or *SS7*. SS7 was first proposed by the CCITT in 1980 and updated in 1984 and 1987, and the implementation is just being completed by the LECs and IXCs in the United States.

SS7 uses separate lines to set up telephone calls from those used for the actual voice or data transmission. The advantage of using a separate signaling system is that you don't have to tie up a line until the call is actually established. Since up to 40 percent of calls that are attempted result in busy signals or no answer, SS7 saves a great deal of time on the actual voice lines.

SS7 is optimized for use in digital telecommunications networks in conjunction with intelligent, computerized switches in the central offices. It allows for database access as a part of the call setup, and that allows the telephone companies to provide certain enhanced telephone services, such as automatic callback and calling number identification. Implementing SS7 has been a big job for the telephone companies in the last few years, but the benefits to the companies and telephone users will far outweigh the costs.

TELEPHONE NUMBERING

Under the guidance of an international standards group, the *Consultative Committee on International Telegraphy and Telephony (CCITT)*, a reasonably consistent numbering plan exists for telephones around the world. This ensures that every telephone number, in its fully expanded form, is unique. According to the CCITT plan, the world is divided into nine geographic zones, which are:

1. North America;
2. Africa;
3. Europe;
4. Europe;
5. South and Central America;
6. South Pacific;
7. Union of Soviet Socialist Republics;
8. Far East;
9. Middle East and Southeast Asia.

Countries within each zone are assigned country codes beginning with the zone's digit. The countries that have, or are projected to have, the most telephones are assigned one-digit country codes; countries with the fewest number of phones are assigned three-digit country codes. In general, the form of a telephone number is as shown in Figure 5–23. The area code is sometimes shortened to one or two digits, and in some countries it is called a *routing code* or *city code*, but the results are the same.

Bellcore is the administrator of the North American Numbering Plan (NANP), which covers the United States, Canada, and some Caribbean countries. This territory has been divided into areas, each with a unique three-digit area code, as shown in Figure 5–27. Areas in close geographical proximity to one another have area codes that are quite different to avoid confusion and accidental misdialing. According to Bellcore, a numbering shortage faces North America. Until all of the telephone companies convert to 1+ dialing, in which the digit 1 must be dialed before any long distance number, there are only 160 unique area codes that can be assigned. At the present time there are only about 12 unassigned area codes. Bellcore is pressing the telephone companies to complete the conversion to 1+ dialing, but it isn't likely to occur before 1995. When it does happen, 800 area codes will be available.

Country Code	Area/City Code	Exchange Code	Subscriber Code
XXX –	N0/1X –	NXX –	XXXX

where X = 0–9 (any digit)
N = 2–9
0/1 = 0 or 1

Sample Countries and Area Codes

Country	Country Code	Area Code	Area
United States	1	212	New York City
		616	Western Michigan
Australia	61	2	Sydney
Brazil	55	11	Sao Paulo
Ireland	353	1	Dublin
		91	Galway
Japan	81	3	Tokyo
United Kingdom	44	1	London
		222	Cardiff
West Germany	49	069	Frankfurt
		89	Munich

Figure 5–23
The general form of a telephone number and some sample country and area codes of several countries.

The first three digits of a seven-digit telephone number are called the *exchange code*. Within each area code, the exchange codes are unique. Most central offices handle more than one exchange code, although some of the smaller offices only handle one. For example, a central office might handle exchange codes beginning with 631, 839, and 832.

LOCAL CALLING

Local calling is defined as telephone service within a designated *local service area*. The local service area includes telephones served by the central office and usually several other central offices nearby. Calls within a local service area are *local calls*. Local calls are charged in one of two ways. *Flat rate service* gives the user an unlimited number of local calls for a flat monthly rate. *Measured rate service* bases the charges for local calls on the number of calls, their duration, or the distance.

Local service areas frequently overlap, as shown in Figure 5–24. Mayville's local service area includes Middleburg, Freeland, and Wanigas. Bayport's is Hopedale, Freeland, and Wanigas. Freeland's local service area includes all of the communities shown.

Figure 5–24
A local service area usually includes the territory covered by several central offices.

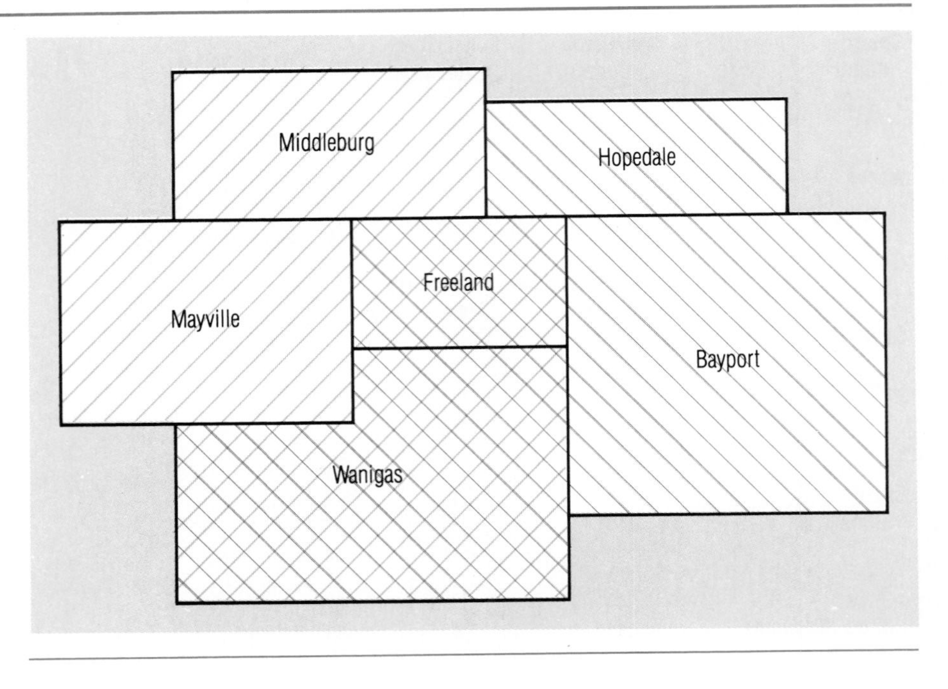

LONG DISTANCE CALLING

There are two types of *long distance calls*. Both are sometimes referred to as *direct distance dialing (DDD)*. *Toll calls* are calls outside of the local service area but within the LATA. Toll calls are handled by the local exchange carrier (telephone company), which also does the billing. The other type of long distance call is one that crosses LATA boundaries and must be handled by an interexchange carrier (IXC).

Recall from Chapter 4 that LECs provide telephone service within a service area consisting of one or more LATAs. The LEC may not provide service across LATA boundaries; inter-LATA service is provided by long distance carriers (IXCs). Thus, the LEC must pass any call that crosses a LATA boundary to an IXC, even though the call never leaves the LEC's service area. Looking at Figure 5–25, it can be seen that because the caller and answerer are in different LATAs, Michigan Bell Telephone must pass a call between Lansing and Detroit to one of the IXCs, even though Michigan Bell serves both cities. Therefore, since it crossed LATA boundaries, it is a long distance call. The customer will receive a bill from the long distance carrier that provided the inter-LATA service.

When someone calls out of a LATA, a long distance carrier's facilities are accessed by appending an access code to the front of the telephone number. Each long distance carrier has its own three-digit code. To reach a long distance carrier, the access code 10 must first be dialed,

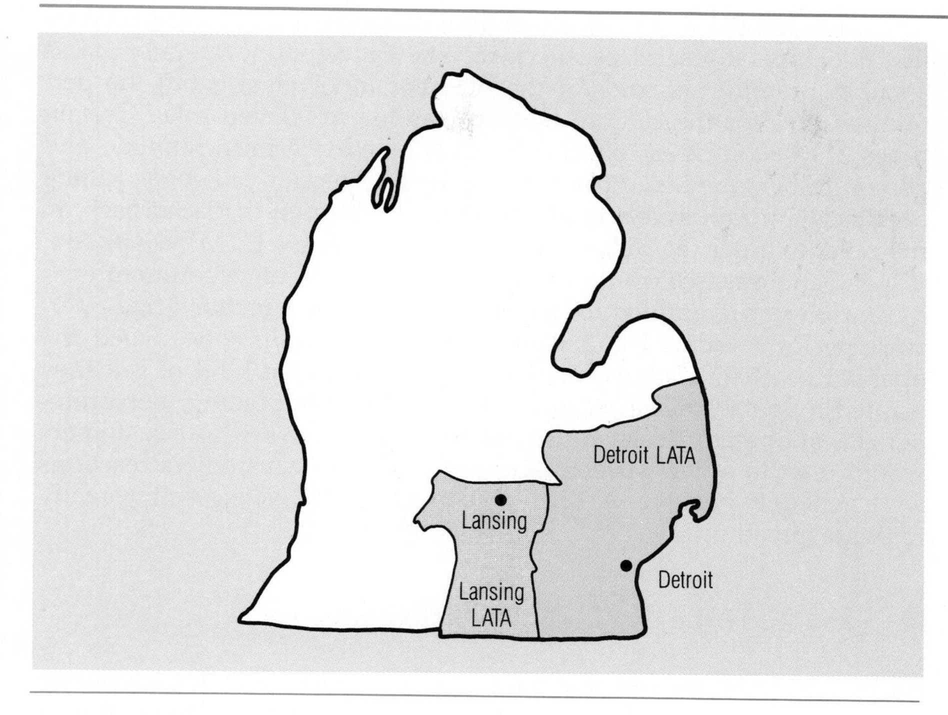

Figure 5–25
An inter-LATA call from Lansing to Detroit must be handled by an IXC, even though Michigan Bell Telephone serves both cities.

to let the central office know that a carrier's code—and not a telephone number—follows. The total sequence is

Access Code	Carrier's Code	Area Code	Telephone Number
10	XXX	N0/1X	NXX–XXXX

To reduce the number of digits that have to be dialed, each telephone customer designates one long distance carrier as its primary carrier, and that company can be reached with the single-digit access code 1. The shortened form of the long distance number is:

Prefix	Area Code	Telephone Number
1	N0/1X	NXX–XXXX

To reach a carrier other than the designated primary carrier, the caller must dial that company's prefix plus the carrier's normal three-digit code.

Telephone calls to most foreign countries can be completed automatically by dialing an international access code, 011, preceding the foreign telephone number. This is called *international direct distance dialing (IDDD).*

operator service

Although the vast majority of long distance calls are self-dialed by the caller, operator services still exist. When a telephone operator places a call, a premium is paid for the operator involvement, but the per-minute rates for the call are the same as for self-dialed calls. Certain types of calls, such as collect calls, calls billed to a third number, and person-to-person calls, must be set up by an operator. However, calling card calls are rapidly being automated. The burden is placed back on the caller to enter the calling card number through a DTMF telephone.

In San Francisco, in the fall of 1986, AT&T ran an experiment with a computer programmed to perform many of the functions that operators perform today. For 2 months, San Franciscans who dialed the operator reached a computer that asked them what kind of call they wanted to make: collect, calling card, third-number billing, person-to-person, or operator assisted. Researchers wanted to see how customers would react to a new speech recognition system. Customers reactions were generally positive. AT&T hopes that the new system will be ready for widespread use by 1990.

SPECIAL TYPES OF TELEPHONE SERVICES

For people or businesses that have special requirements for telephone calling or make large numbers of calls, telephone companies offer a number of special calling services, including discounted prices for high volume telephone use. Many of the services are specially oriented to the customer who has particular calling patterns, such as a large number of calls from a certain part of the country or the desire for the receiver to pay for a call instead of the calling party. Most of the plans are aimed at long distance calling and are, therefore, offered by the interexchange carriers.

Discounted Calling

Customers who make a large number of long distance calls are eligible for discounts from the normal charges. One category of discounting is based on call volume, measured by the amount of money spent for long distance calls each month. For example, if a caller or business has more than $100 per month in long distance calls, they may receive a 10 percent discount. If the call volume is greater than $500 per month, a discount of 20 percent may apply. Call volumes over $3,000 per month might receive a 35 percent discount, and so on. Each long distance carrier has its own plan, but all are similar. AT&T calls their plan Pro America; MCI's plan is called Prism.

Very large customers, those spending more than $10,000 per month, qualify for even higher discounts of 45 or 50 percent. Usually, however, they are required to have a direct connection to the long distance carrier's nearest point of presence (POP). This full-time line is capable of carrying at least 20 or 25 simultaneous voice calls. Long distance calls

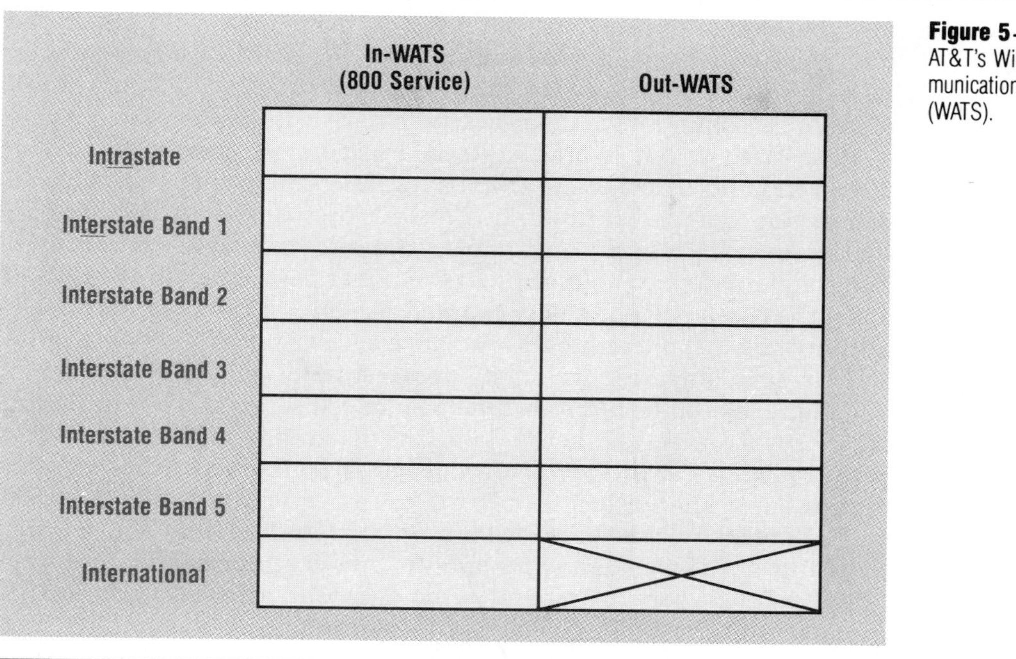

Figure 5–26
AT&T's Wide Area Telecommunications Service (WATS).

are routed through the dedicated line to the POP and from the POP on the public telephone network. For these very large customers, the cost of the special access line is usually well justified to obtain the larger discounts. This type of service is typified by AT&T's Megacom service and MCI's Prism I.

Wide Area Telecommunications Service (WATS)

Wide area telecommunications service (WATS) offers quantity discounts to the company that has a high volume of calls. WATS service can be classified three ways:

1. geographic—intrastate, interstate, international;
2. coverage—bands 1 through 5;
3. call direction—in-WATS, out-WATS.

The various services are shown in Figure 5–26. A customer purchases one or more WATS lines in any of the categories shown in order to obtain that service. For example, the customer might purchase an intrastate in-WATS line and an interstate out-WATS line for his or her business needs. A WATS line is like any other telephone line in that it can handle only one telephone call at a time. Therefore, if the traffic volume is heavy, multiple lines may be required. The customer pays a flat monthly rate to have access to a WATS line and an additional charge for each minute of use. The cost per minute is in the form of a tiered rate, that is, the more minutes that are used, the less the cost per minute. AT&T's per-minute price reductions occur after 25 and 100 hours of use per month. Other long distance carriers have different tiers.

In-WATS, which is officially called *800 service*, allows calls to be placed from a section of the country to the location that subscribes to the service. Companies typically select 800 service to provide a way for customers to call long distance to place orders or inquire about products. The company that purchases the 800 number pays for all of the calls to it. To the customer, the call is charged like a local call. Another use is to allow salespeople or others who are traveling to call the home office and have the long distance billing done centrally.

Many special features are available to 800 service users. Incoming calls may be routed to different locations depending on the time of day, or the caller may be prompted to enter a digit to indicate which of several parties he or she wishes to talk to. You have undoubtedly used 800 service to make a travel reservation, order a product from a mail order company, or call a company for specific product information.

Out-WATS service provides the opposite capability. It allows employees at the office to call out to other areas of the country at a reduced rate. Out-WATS is primarily used as a way to reduce a company's telephone bill by taking advantage of discounted pricing.

Intrastate WATS service provides either in-WATS or out-WATS service within the subscriber's home state. Interstate WATS comes into play outside of the home state. The continental United States is divided into a series of five concentric bands, with the home state at the center. Each of the bands contains approximately 20 percent of the telephones in the country. The bands are different for each state, and Figure 5–27 shows how the bands are arranged for Missouri. Band 1 for Missouri includes the states of Iowa, Nebraska, Kansas, Oklahoma, Arkansas, Tennessee, Kentucky, and Illinois. Band 2 adds the states of Wisconsin, Indiana, Mississippi, and Louisiana. Band 5 always includes the entire continental United States. Calls made on the WATS lines are priced according to the bands that they are placed to. A call to an area code in a more distant band incurs a higher rate than a call to a location in a nearby band.

To get total coverage of the United States, a company would have to purchase at least these four WATS lines:

- an intrastate in-WATS line;
- an interstate in-WATS line;
- an intrastate out-WATS line;
- an interstate out-WATS line.

Firms can also purchase 800 service from selected foreign countries to the United States. At present, AT&T offers service from nearly 20 countries, with others to be added in the near future.

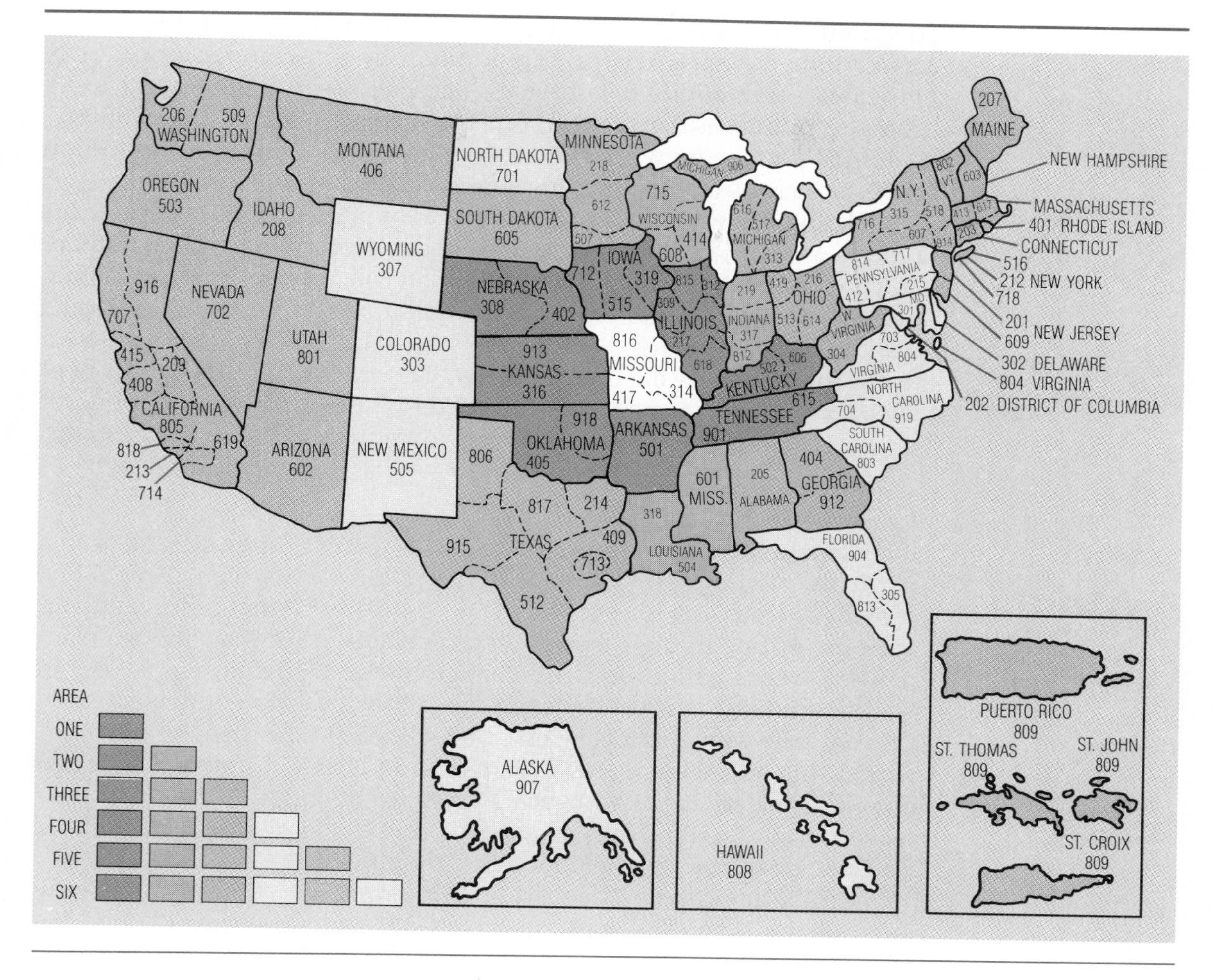

Figure 5–27
WATS bands for the state of Missouri.

WATS was originally developed and marketed by AT&T. They did not, however, register the WATS name, and, therefore, other long distance carriers use the names for their own WATS services. Typically, the cost per billed minute of WATS usage from a company such as MCI is slightly less expensive then AT&T's comparable service, but this is a rule-of-thumb, and the situation for each business customer must be examined individually.

Conceptually then, WATS lines provide another form of bulk telephone pricing or discounted calling. The WATS service is very profitable for the carriers and quite dynamic. Compared to a few years ago, WATS is much more flexible than it used to be. Further changes are expected in the next several years.

Software Defined Network

The *software defined network (SDN)* is another bulk pricing offering of AT&T designed for large companies. A business accesses SDN on either dedicated or dial lines, after which the calls are carried on the normal AT&T long distance network. In addition to a discounted price, SDN provides some additional services, such as seven digit dialing to all company locations that are connected to the network and special billing. Additionally, the network can be defined so that special authorization codes are required by individuals to make certain types of calls.

To analyze which of the many discount pricing plans is best for his or her company, the network analyst needs a knowledge of the number and duration of the telephone calls that will use the service, calling patterns by location and time of day, and information about the carrier's discount pricing plans. The carriers will help the customer analyze a firm's requirements for these services by performing a traffic study of telephone call frequency, length, and distribution patterns, hoping, of course, to influence the customer to buy more services.

Foreign Exchange (FX) Lines

If a telephone customer makes or receives many telephone calls from a particular city, a *foreign exchange (FX) line* can be installed. An FX line provides access to a remote telephone company central office so that it appears as though the subscriber has a telephone in that city. If a company in Dallas had a foreign exchange line to Houston, employees in Dallas could make Houston telephone calls at local Houston rates. Also, the company would have a Houston telephone number that, when called, would ring at the company switchboard in Dallas. Companies located outside a major city often have FX lines to the heart of the city if they make many telephone calls or want to provide a local telephone number for their customers who are located "downtown."

An FX line can handle calls in either direction, but of course only one call at a time. The subscriber pays a flat monthly rate to have the line in place, plus a per-call charge from the local carrier.

PRIVATE TELEPHONE SYSTEMS

When a business requires more than two or three telephones, it usually acquires some type of a private telephone system to provide special services and help manage the telephone traffic. As a company grows, a large number of the telephone calls that are made are intraoffice calls, from one department to another or one building to another. Without a private telephone system, each telephone would require a local loop connection to the central office, and each call would have to go through the public telephone network—even if it were destined for an office just down the hall. With a private telephone system connected as shown in Figure 5–28, the intraoffice calls can be handled internally, and only external calls must be sent to the telephone company's central office.

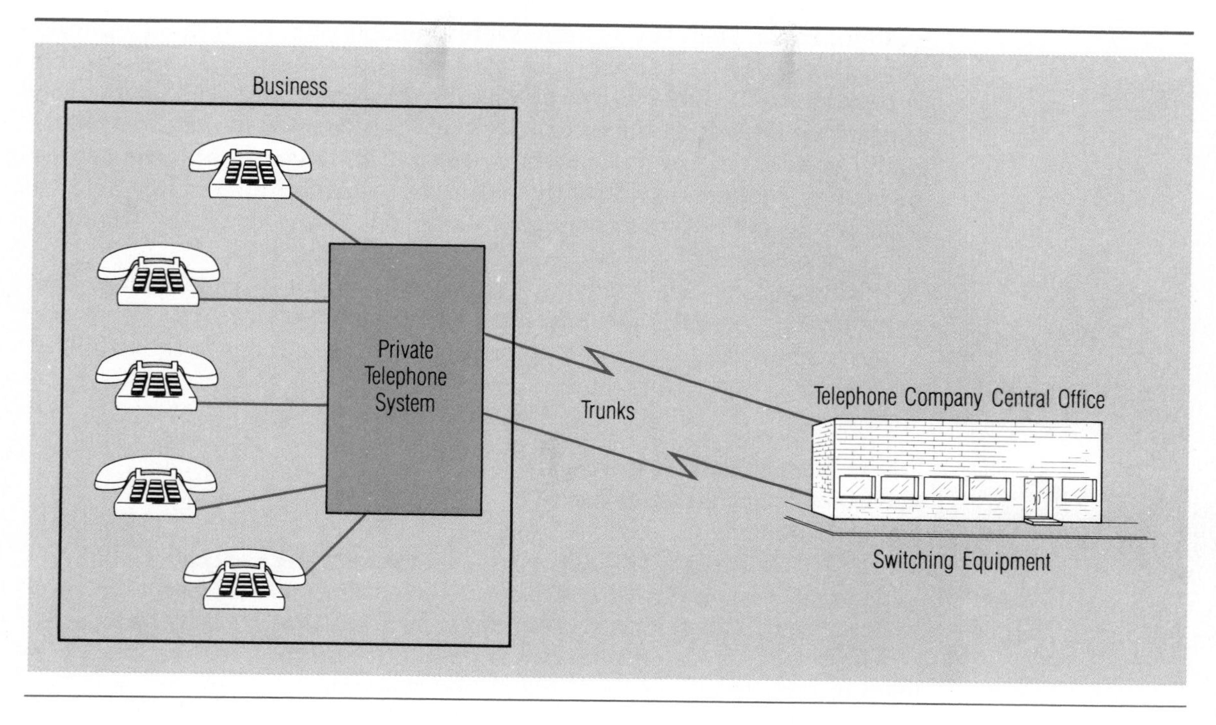

Figure 5–28
Private telephone system: Key system or PBX.

Earlier in the chapter, a *trunk* was defined as a circuit connecting telephone switches or switching locations. Our previous use of the word has been in the context of trunk lines connecting telephone company central offices switches. A private telephone system is another type of telephone switch. Therefore, the lines connecting it with the switch in the telephone company central office are also trunks. Since a high percentage of the calls in a business are internal, the number of trunks connecting the private telephone system to the central office can be substantially less than the number of telephones in the office.

Key Systems

Small private telephone systems are called *key systems*. This name is a holdover from earlier days when a telephone connected to this type of system had pushbuttons or keys that allowed a line to be selected. In a typical key system, each telephone can access two or more lines, and lamps on the telephone indicate whether each line is busy. The caller selects a line by pushing a button on the telephone to seize the line, then dials the call. The telephone usually also has a hold button so that a call on one line can be held while a second call is made or answered.

Key systems typically handle from 3 to 50 telephones. This segment of the telephone equipment market is the largest and fastest growing because of the thousands of small businesses that can use this size phone system. In the past, key systems were available only from the telephone companies, and the capabilities they offered were closely tied to the capabilities of the central office to which they were connected. Since the Carterfone Decision and particularly since deregulation in the early 1980s, many vendors of key systems have entered the market, and the equipment has become significantly more sophisticated and better able to provide capabilities independent of the telephone company's central office equipment.

Key systems are available with a wide range of capabilities. The price of the system is partly related to the number of features that the system offers. The types of features available are similar to those found on PBXs.

Private Branch Exchange/Private Automatic Branch Exchange/Computer Branch Exchange

Private branch exchange (PBX), private automatic branch exchange (PABX), and *computer branch exchange (CBX)* are terms often used interchangeably to describe private telephone switching systems that are larger and usually more sophisticated than key systems. In this book, the term *PBX* is used to refer to any of these telephone systems.

A PBX is a private telephone system designed to handle the needs of a large organization. It is the next step up from a key system in

This Rolm 9750 PBX comes in several models depending on the number of lines and telephones to be handled. The PBX is all electronic and capable of handling from 100 to 20,000 telephones. (Courtesy of Siemens-Rolm)

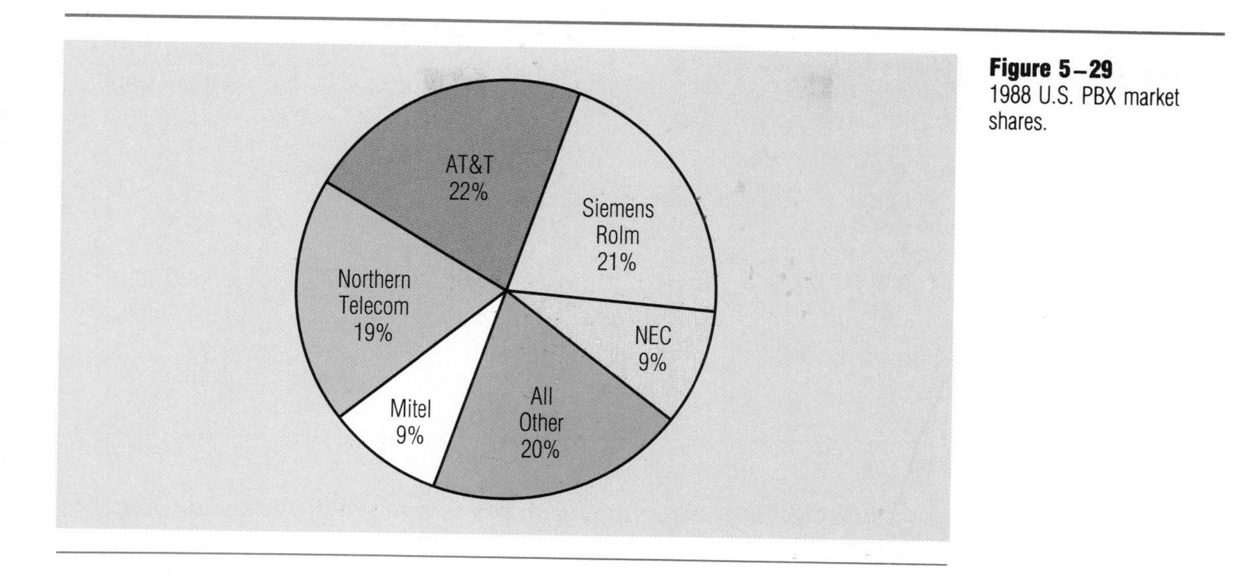

Figure 5–29
1988 U.S. PBX market shares.

capacity and the number of special features it supports. PBXs are typically designed to handle from 50 to more than 10,000 telephones. Each telephone on a PBX (or key system) is called a *station* or an *extension*. Larger sizes of PBXs are similar in capacity and capability to the switching equipment used in telephone company central offices. Since PBXs are designed for the exclusive use of one company, however, they are located on the company's premises.

The three companies with the largest share of the PBX market in the United States are AT&T with its Definity series of PBXs, Northern Telecommunications, which sells the Meridian SL-1 and SL-100 families, and Siemens-Rolm with its 9750 and Saturn PBX products. Figure 5–29 shows the division of the PBX market among the participating companies.

PBXs, like the switching equipment in central offices, are computers especially designed for handling and switching voice telephone calls. Older PBXs, like older central office switches, had mainly mechanical components, and although many of these older PBXs still exist, the new units are all electronic and programmable. That is, they require an environment suited to computers—clean and air conditioned; they can be upgraded to add more capacity; they are physically smaller than their predecessors; and they require software to operate. PBXs, like key systems, are available from the telephone companies and many other firms.

purchasing a PBX

The decision to acquire a PBX is not one to be taken lightly. The prospective purchaser needs to be sure he or she understands the firm's needs as well as the types of maintenance, training, and other support required by his or her firm as opposed to support provided by the PBX vendor. The level of support the vendor will provide is usually some-

Figure 5–30
Features of a PBX. Many of these same features are found also in key systems.

Station Features	System Features
Automatic redial	Automatic Call Distribution (ACD)
Automatic reminder	Class of service
Call forwarding	Data communication
Call park	Direct inward dialing
Call transfer	Hunt group
Call waiting	Least cost routing
Camp on	Paging
Conference calls	Pick-up group
Distinctive ringing	Ring-down station
Do not disturb	Station message detail recording/Call detail recording (SMDR/CDR)
Speed dialing	Voice messaging

what flexible. Support, along with price, are two of the major points of negotiation between the PBX vendor and a prospective customer.

A PBX gives its owner more control over its telephone system and usage than telephone systems provided by the telephone company. In addition, the PBX may provide features to improve the capability and efficiency of the telephone service that are otherwise not available. Features are usually divided into two categories: system features and station features. System features are capabilities that operate for all users of the PBX and that are in many cases transparent or unnoticeable to the user. Station features are customized to each user or telephone to provide the separate capabilities most useful to the individual. A list of system and station features is shown in Figure 5–30. The most common features are described next.

System Features PBX *system features* are available to all users. In reality, all of the features may not be activated for all users because some are designed for particular needs and may be applicable only in certain departments or parts of the company. We will look at some of these features in more detail.

Data Communication Most PBXs have the capability to handle data communications as a standard part of their hardware. With this feature, a user with a computer terminal who needs to access several computers can use the PBX as a switch. First, the user dials one computer and connects the terminal to it through the PBX, and then, when the communication to it is finished, the user can dial the number of another computer and repeat the process. The data communications feature can be used to connect the terminals in a company to computers within the company, or through outside telephone lines, the PBX can make the connection to computers in outside service bureaus or other organizations.

Direct Inward Dialing (DID) This feature gives outside callers the ability to call directly to an extension number so calls don't pass through an operator. Outside callers dial the normal seven-digit telephone number, and the call passes from the telephone company central office through the PBX and directly to an individual's telephone. Without the DID feature, all incoming calls pass through an operator, who makes the connection to the desired extension.

Hunt Group The hunt group is another method of distributing calls to one of several individuals in a predetermined sequence. Hunt groups are often set up for departments in which several individuals can handle incoming calls but where there is a definite preferred sequence. When a call comes in, it is passed to the first extension in the hunt group. If that extension is busy or is not answered, the call is passed to the second extension, and so on throughout the entire group.

Least Cost Routing The least cost routing feature attempts to place outgoing long distance calls on the line over which the call can be completed at the least cost. For example, if a PBX had foreign exchange lines and WATS lines connected to it, the least cost routing feature would use a table stored in the memory of the PBX to determine which line should be used to place a particular call. If the preferred type of line is not available, the second choice is used, and so on. When all alternatives have been tried, the call is sent out on the standard long distance facilities as a regular DDD call (the most expensive alternative). The least cost routing facility usually provides statistics showing the number of calls that went out on each type of line, as well as the number of calls that had to overflow to a more expensive alternative. Using these statistics, the network analyst can determine whether more lines of a particular type are required.

Pick-Up Groups Pick-up groups allow any member of the group to answer an incoming call. For example, in a group of programmers that are connected on a pick-up group, if a ringing telephone does not get answered, any member of the group may pick up his or her own phone, key in the pick-up code (usually an asterisk) and have the call transferred to his or her telephone so that it can be answered. The idea is that incoming calls will be answered by someone who does a similar kind of work and who can potentially help the caller.

Station Message Detail Recording (SMDR) *Station message detail recording (SMDR),* sometimes called *call detail recording (CDR),* is the feature of the PBX that records statistics about all calls placed through the system. The data recorded includes at least the calling and called extension numbers, the time of the call, and its duration. It may also include other statistics, such as the user's class of service or the line numbers used. The statistics usually are accumulated on a magnetic tape or disk attached to the PBX. Once the data is captured, it can be used for recording and billing purposes. Usually the data is taken off the PBX and transferred to another computer that performs the reporting and billing functions. Some PBXs can transmit the call detail data over a communications line to another location for processing.

In addition to providing the source data for telephone billing, the SMDR feature provides data to network analysts. Using this data, analysts can examine calling patterns and traffic volumes and make decisions about whether the mix of telephone lines is adequate to provide the required grade of telephone service to the company.

Station Features *Station features* are activated by a PBX system user. Whereas the features may be made available to individuals or groups of people, it takes some action by the individual to use the feature.

Automatic Reminder The automatic reminder feature lets a user tell the PBX to call him or her back at a specified time. Its most common use is in a hotel for wake-up calls. You tell the hotel operator when you want to be called, and he or she instructs the PBX to ring your telephone at the specified time. Often the PBX can play a prerecorded message when it makes the automatic reminder call, such as "Good morning, it is 7:00 A.M."

Call Forwarding Call forwarding lets calls for one extension ring at another extension. If a person is going to be out of the office, he or she can activate call forwarding to have all of the calls ring at the secretary's desk or at whatever extension he or she will be. Call forwarding can also be set to forward the call if the extension is busy. A person might have calls forwarded to different extensions depending on which condition (such as ring with no answer, or a busy signal) occurs.

Call Transfer The call transfer feature allows calls to be transferred to another extension. In the blind transfer, a transfer code and the extension number are keyed in. When the person who was originally called hangs up the telephone, the call is transferred. In the consultation transfer, keying in the transfer code places the caller on hold. The new extension is dialed, and when it is answered, the person originally called and the person to whom the call is being transferred can converse. When the original answerer hangs up, the transfer is completed.

Call Waiting The call waiting feature indicates an incoming call while a call is in progress. The second call is indicated by a tone or lamp, and the parties in conversation can decide whether to take the second call.

Camp On When you place a call and the number you called is busy, the camp on feature lets you tell the system to call you back when the number is free. The PBX tests the extension that was called and, when it is free, calls you back. If you answer your telephone, it automatically redials your call for you. This keeps you from having to continually redial a busy number yourself.

Distinctive Ringing With the distinctive ringing feature, a telephone may have different ringing signals for calls from within the com-

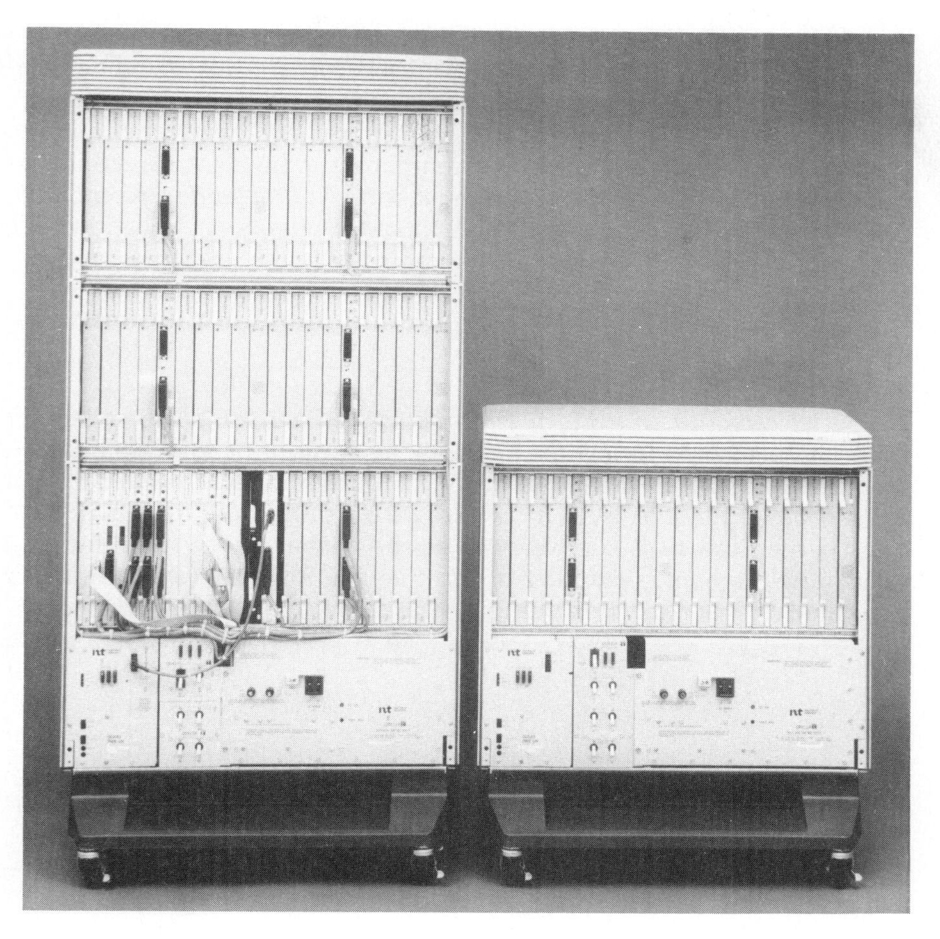

This Northern Telecom PBX is configured for 600 telephone lines. Line cards can be added or removed for smaller or larger configurations. (Courtesy of Northern Telecom, Inc.)

pany and outside calls. Other distinctive rings may be available for emergency calls or trouble calls.

Do Not Disturb The do not disturb feature may be implemented in several ways. In its simplest form, it gives the caller a busy signal even though the extension called is not in use. Another implementation causes the caller to receive a distinctive busy signal that indicates the person being called has his or her telephone in do not disturb mode. A third implementation signals the PBX to automatically forward calls to another extension.

Speed Dialing Speed dialing allows frequently called numbers to be stored in the PBX's memory and then accessed with a shorter set of digits. For example, a PBX may allow every user to store 100 ten-digit telephone numbers. Each one can then be accessed by dialing a speed dialing code followed by a two-digit number, 00–99. A common use for speed dialing is to store the numbers of other company locations or frequently called customers or suppliers.

Several of the features just described are becoming available on the public telephone system. These features become available when the serving central office is upgraded to an electronic switch, which is a big brother of the PBX. Call waiting, call forwarding, and conference calling are available from most telephone companies today. Since there is an extra charge for these features, you must notify your telephone company if you wish to have them activated for your telephone. In addition, automatic redial is a feature available on many home telephones today.

DATA PBXs

A special category of PBXs is the *data PBX*. Data PBXs are switches especially designed for switching data; they do not handle voice calls. They connect many terminals to many computers on an as-needed basis. A user at a terminal dials, or has a hardwired connection, to a data PBX. The user indicates the computer he or she wishes to be connected to, usually with a name or acronym, and the data PBX makes the connection. If all of the connection points on the computer are busy, some data

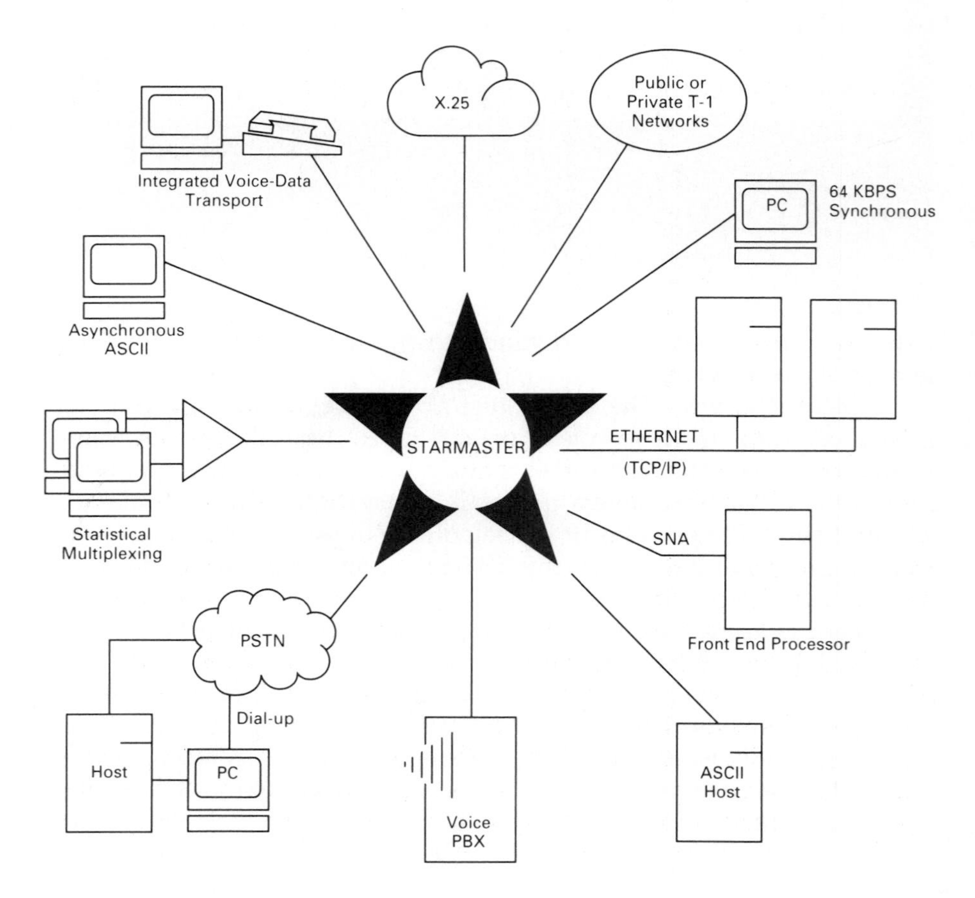

This diagram shows the capabilities of Gandalf's most advanced data PBX, the Starmaster Hybrid Networking System. (Courtesy of Gandalf Technologies, Inc.)

PBX	Centrex
Ultimate control	Control shared with telephone company
Not regulated	Regulated by state PUC
Total ability to manage	Management shared with telephone company
Total responsibility to manage	Telephone company is primarily responsible
Requires capital to buy the system	No capital required
User/vendor must provide service	Telephone company provides all service
Usually less redundancy is built-in	More redundant hardware may mean higher reliability
Growth capability depends on the inherent design of the PBX	Essentially unlimited growth capability

Figure 5–31
Factors to be considered in selection of a PBX versus a Centrex system.

PBXs will hold the call in a queue and call the user back when the connection can be made. When the user no longer needs the connection, it is broken, and the hardware is available to another user.

Data PBXs are less expensive than PBXs designed to handle both voice and data. They can add an element of security to data connections by requiring a password before the connection to the computer is made. Data PBXs can also perform many of the conversions that data transmissions often require. These conversions will be examined in detail in succeeding chapters, but suffice it to say for now that they involve the code, speed, and format of the transmitted data.

Data PBXs are offered by a number of companies. Among the largest vendors are Gandalf, Develcon, and Micom.

CENTREX SERVICE

Centrex service provides a telephone service much like that of a PBX, but the telephone companies use equipment located at their central office, not at the customer site. The central office equipment may be dedicated to the customer, but more likely it is the same equipment used to provide normal public telephone switching functions. It is important to understand that the Centrex service is not precisely defined or standardized, even though the name is used nationwide. Since each telephone company can offer Centrex service and several different types of central office equipment are used, the Centrex offering is a combination of the technical capabilities of the central office equipment and the features that the telephone company decides to make available to its customers. Centrex service is regulated by the state public utilities commission; a PBX is a private system and unregulated.

You might wonder why a business would consider acquiring a PBX if telephone company-provided Centrex service would give the same service. There are definitely pros and cons to both types of systems, as shown in Figure 5–31. A PBX gives the company ultimate control of its telephone system and total ability to manage it. On the other hand, it requires the company to make the capital investment in the PBX equipment and to provide space, power, and air conditioning as well as skilled technical people to operate it. By way of contrast, when a Centrex system is used, the telephone company manages and services the equipment, and the customer pays the telephone company a service fee.

PRIVATE VOICE NETWORKS

When a company has several locations, each with its own PBX or key system, it is often desirable to tie the locations together with telephone lines. By renting lines to connect the PBXs or key systems, a private network can be built that saves money compared to the cost of making standard long distance calls. In addition, it simplifies the dialing and, in many cases, can appear to the telephone user as a single, integrated, private telephone system.

Tie Lines

Leased lines that connect the private PBXs or key systems are called *tie lines* or, more properly (because they connect switching equipment), *tie trunks*. They are acquired from the common carrier or can be installed privately if the locations are in close proximity. If leased from the telephone company, a fixed price is paid for full-time, 7-day, 24-hour use.

With tie trunks in place, a telephone user at one location dials an access code to access the trunk and PBX at the other end. Usually, a second dial tone is heard from the remote PBX, and then the extension number of the called party is dialed. If a speed dialing feature is installed, this process may be simplified.

Private Networks for Large Organizations

When an organization has a very large volume of calls between its locations, it is economical to build a private network. Only the largest companies can afford to install the lines and equipment that are necessary, but under the right circumstances, private networks are very cost effective. The largest private networks rival the networks of small telephone companies in size.

ETN

One type of private network, often called an *electronic tandem network (ETN)*, is served out of PBXs on the customer premises. Special software in each PBX on the network contains tables that determine how each long distance call will be routed. Some calls may be sent directly to the destination PBX, whereas others may be routed through intermediate PBXs before reaching the final destination. This type of network is similar to the public long distance networks in its capability. There are approximately 1,000 such networks installed in large organizations.

Another type of network has similar capability but uses switching equipment in the carriers' offices. Often, this equipment is dedicated to a specific customer, such as the federal government or one of its branches. AT&T calls this type of network *enhanced private switched communication service (EPSCS)*.

EPSCS

VOICE MESSAGING

Voice messaging provides an electronic voice mailbox where callers can leave messages for other people. Voice messaging is the voice equivalent of E-Mail systems discussed in Chapter 2. Voice messaging systems may be a system feature built into a PBX or provided as a separate piece of hardware. Some companies make voice messaging units that may be attached to a wide variety of PBXs or that work with Centrex systems. It is estimated that at the end of 1986 there were 4,000 voice messaging systems installed in the United States. VMX, Inc., and Siemens-Rolm are the market leaders, each with about 20 percent of the market. A number of other companies—including Wang and AT&T—divide the rest.

Whether it is a feature on a PBX or a stand-alone system, the voice messaging system can be used by callers inside a company or calling in from the outside. In addition to simply providing a voice mailbox capability, a voice messaging system provides other functions as well. It allows voice messages to be sent to several people at the same time. For example, a manager could communicate to everyone in a department and be sure that they all heard the same message said the same way. Comments may be added to voice messages, and the original message with the comments can be forwarded to another party. Voice messages may be put in the system with instructions to place them in voice mailboxes at a later time.

Using this capability, a department manager could, before leaving on a trip on Tuesday afternoon, put a voice message in the system for all of his employees, announcing a job change or new product. He could tell the system not to put the message in his employees' mailboxes until Thursday morning if it were important that the announcement not be made until a certain time.

A voice messaging system is not just a simple tape recorder. It converts an analog voice signal to a digital signal and stores it on a magnetic disk for later recall. When a person calls into his or her mailbox, the digital signal is converted back to an analog signal, and the voice message is spoken to the recipient.

Voice messaging capability is useful in a variety of situations. Research has shown that up to 83 percent of all business telephone calls do not reach the called party on the first attempt. An important advantage of voice messaging is that the calling and called parties do not need to be

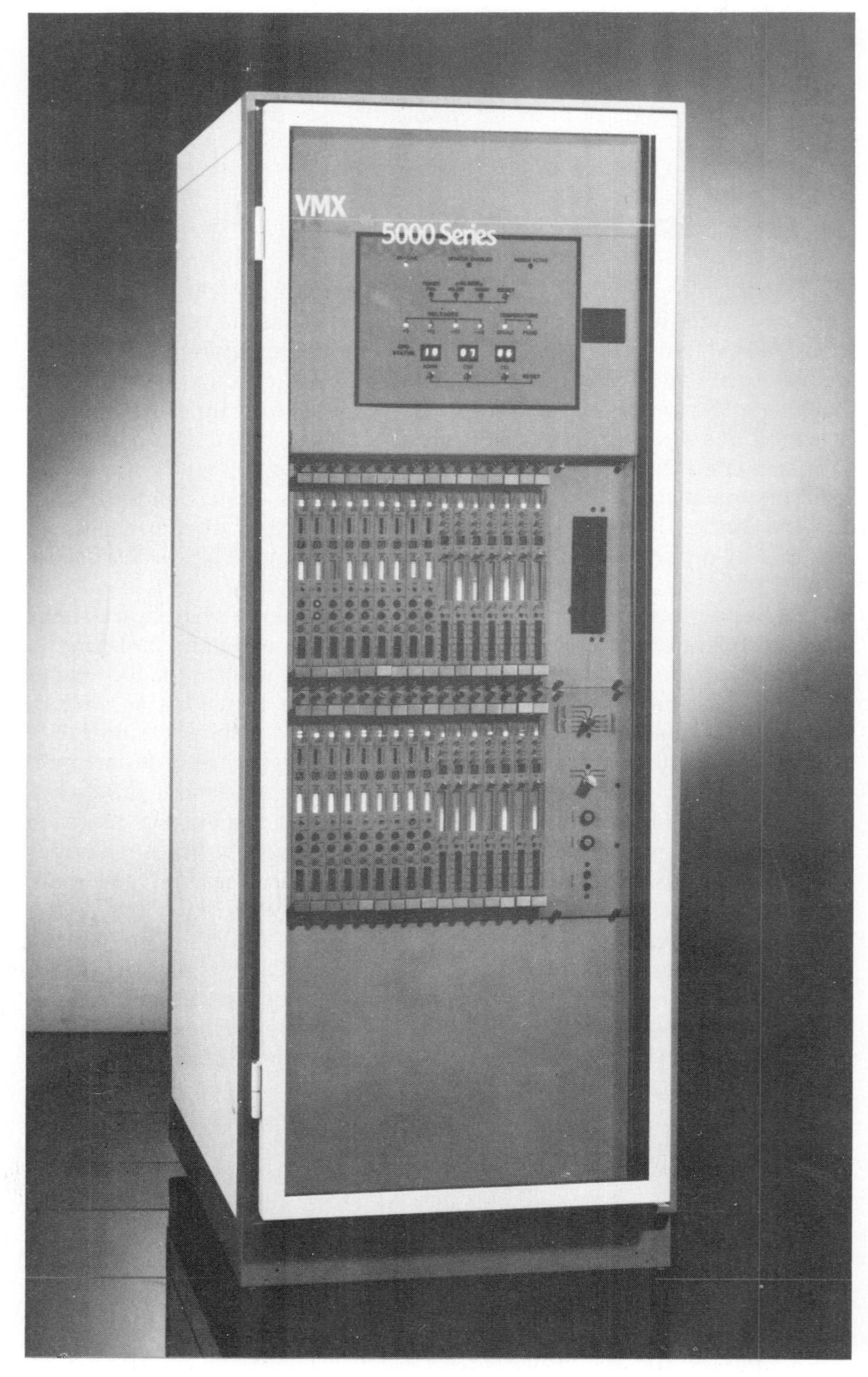

The VMX 5000 voice messaging system includes many advanced features. It can be configured to support several thousand users. (Courtesy of VMX, Inc.)

present at the same time. Voice messaging allows the caller to easily leave a detailed message. If the caller can go into detail and ask a question or explain a situation, it is possible to avoid the need for a call back in 50 percent of the cases. For these situations, telephone tag can be minimized or avoided altogether. Some people expect that electronic conversing, where the two people never actually talk to one another, will become an accepted form of intracompany communication.

Another use for voice messaging is in minimizing the problems caused by time zone differences. Callers on the east coast can leave messages for business associates on the west coast before they get to work. Conversely, west coast callers can leave messages after east coast workers have gone home for the day.

Voice messaging is also useful for salespeople and marketing people who do a lot of traveling. They can call into the voice messaging system from any tone dialing telephone and have the system play back all of the messages stored for them. Then they can respond to the messages by sending a reply, forwarding the message to another person for handling, or sending new messages.

Audiotex

A relatively new development is called *audiotex*. Audiotex combines a voice messaging system with the capability to access a database on a computer. Using this type of system, a person can call Dow Jones & Co.'s Dowphone service and get the latest stock prices, dividend information, and financial headlines spoken over the telephone. To listen to stock quotes, callers key in identification numbers from their printed directories. For example, keying in 4205 brings information about Boeing Co., 5918 about Zenith Electronics Corporation, and so on.

TWA uses audiotex for crew scheduling. A crew member calls an 800 number and enters an identification code. The voice messaging machine passes an inquiry to a computer, which looks up the crew member in a database and lists any assignments for the person. The information is passed back to the voice message machine, which formulates a response and delivers a voice message to the caller. The most useful attribute of audiotex systems is the way in which voice messaging and computer technology have been combined to deliver unique, sometimes complex messages in spoken form to a caller.

OTHER TELEPHONE SERVICES

Audio Teleconferencing

Another type of telephone service that some businesses find useful is audio teleconferencing. Usually used in a conference room, an audio teleconferencing setup has an omnidirectional microphone in the center of the table and a speaker, both of which are connected to the telephone.

An audio teleconference in progress. The tall thin microphone on the table is omnidirectional and can easily pick up voices from anywhere in the room.

Participants sit around the table, and one participant makes a standard telephone call to an individual or to another similarly equipped conference room. All the people in the room can hear what is said at the other end of the connection on the speaker, and the microphone picks up everything that is said and transmits it over the telephone lines. In some audio teleconferencing systems, multiple rooms can be connected together so that several groups of people can converse. About the only special requirement—in addition to the microphones and speakers—is the human factor consideration of taking turns and being polite.

Businesses have found audio teleconferencing useful for allowing a group of people to converse with a specialist or expert at a remote location, for having status meetings between groups of people at diverse locations, and for holding meetings where the participants are in many locations. In some cases, it can be a substitute for travel. In one recent audio teleconference, a company conducted a sales meeting. Eleven locations, each with approximately 20 participants, were connected together in a large audio teleconference.

Marine and Aeronautical Telephone Service

In both the marine and aeronautical telephone service, radio communication is established between a transceiver on a boat or airplane and a land-based transmitter and receiver that are connected to the public telephone system. The land-based equipment is usually voice-activated so that it switches automatically from transmit to receive as the parties at either end talk. In most cases, an operator must initially establish the call and monitor its progress to ensure that the radio transmitter and receiver are working properly.

Aeronautical telephone service has the special problem that because the airplane is moving so fast, it quickly gets out of the range of the transmitter. Unless a means exists to pass a call from one transmitter to another as the plane moves, the call is terminated as the plane goes out of range. A small system of coordinated ground-based equipment has been established along some major air corridors. The transmitters and receivers are close enough together so that if a plane goes out of range of one, it is within range of another, and the call can be passed with no interruption.

Some commercial airlines are today offering telephone service on some of their flights. In the most common implementation, the passenger goes to the telephone and inserts a credit card. This unlocks the handset, which is cordless, and the passenger takes it back to his or her seat. The number is dialed, and the call is broadcast from the airplane to the nearest ground receiver. When the connection is made, the call progresses like any other telephone call. Upon completion of the call, the passenger returns the handset to the telephone and retrieves his or her credit card. Billing is done through the credit card. There is a basic flat rate for the call plus a per-minute charge.

Cellular Telephone Service

On a more limited geographic scope, moving automobiles suffer the same problem as airplanes. When they move out of range of the transmitter/receiver, the connection is lost. Traditionally, cities have had a single powerful radio transmitter and receiver for automobile telephone service. Each telephone call required a separate frequency, and although a city like Chicago had approximately 2,000 frequencies available, there was a demand for more.

The problem of insufficient frequencies became so universally acute in large cities that the whole concept of mobile telephone service was reworked in the early 1980s to take advantage of modern computer technology. Cities were divided into cells (shown in Figure 5–32), and low-powered radio equipment controlled by a central computer was installed in each cell. Because the transmitters use low power, the same frequencies can be reused in nearby cells without interference. Thus if a city has 10 cells, 5 to 8 times as many channels are available for telephone conversations as were available in the previous system.

This type of system is called a *cellular telephone system*; its basic technology was developed at the Bell Laboratories in the early 1960s. Access to the system is made from a vehicle equipped with a telephone connected to a two-way radio. The radio equipment usually is installed in

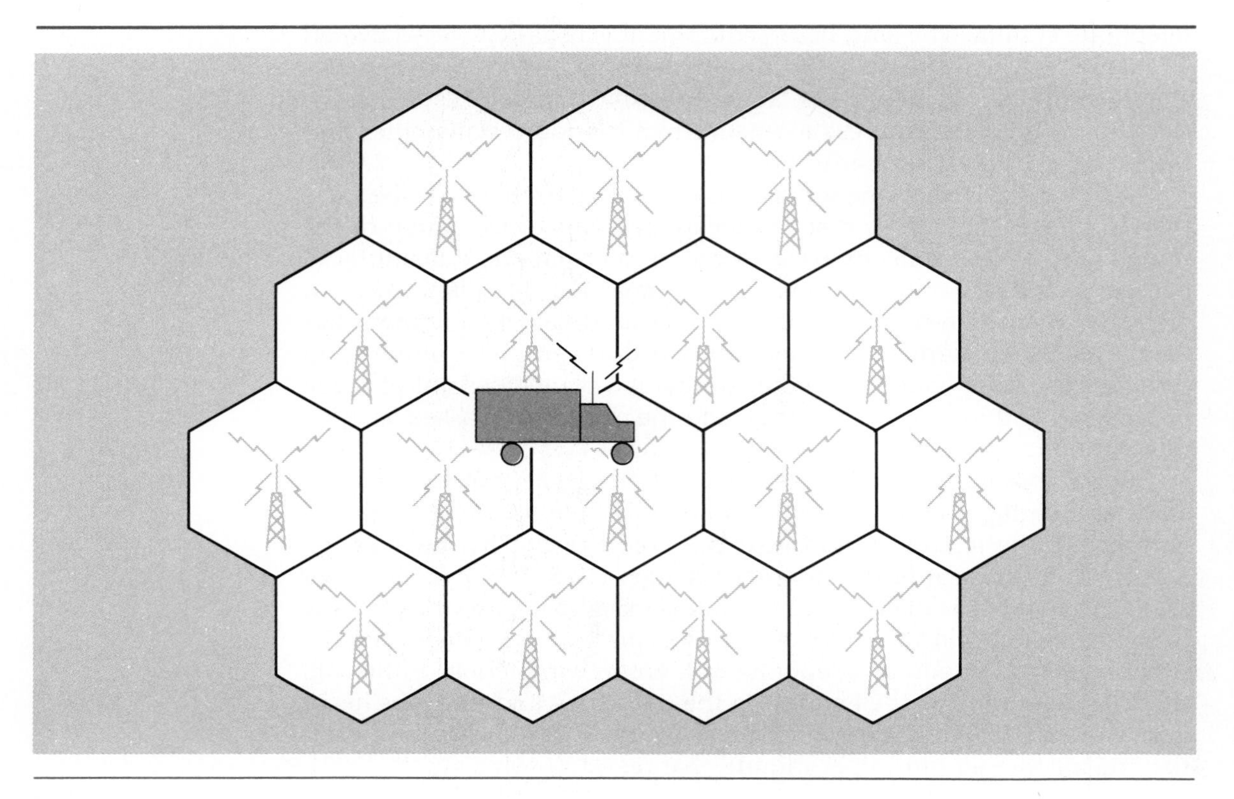

Figure 5–32
The layout of a cellular telephone system.

the trunk for security and aesthetic purposes, leaving only the telephone in the interior of the vehicle. When a call is initiated the cell transmitter assigns the telephone call to a voice channel. Calls destined for a car telephone go to all the cell transmitters, which broadcast the car's telephone number. When the unit in the car detects its number, it sends an acknowledgment, and the nearest cell transmitter takes over and assigns a voice channel.

Mobile telephone calls are monitored every few seconds by a central computer. When the signal from the car starts getting weak, the computer determines which cell to switch the call to and instantaneously switches it to the new cell's radio equipment. The switch occurs so fast that it is unnoticeable by the individuals who are conversing. (It does present problems if data is being transmitted on the cellular connection, however.) One of the key attributes of cellular telephone systems is that they operate completely automatically. No operator is required to set up, monitor, or switch calls from one cell to another.

A growing number of people are using cellular telephone service for facsimile and data transmissions. Data transmission can be nearly 100 percent reliable if the proper error checking circuitry is built into the equipment. One ambulance company is using a cellar telephone

Cellular telephones can be useful in case of emergencies as well as for normal business or personal conversations. The handset shown in this photograph is connected to the main electronics unit mounted inside the car. (Courtesy of Tandy Corporation)

system to transmit data about patients to the hospital while the ambulance is enroute. Contractors are sending facsimiles of drawings from the office to the job site by cellular phone. And the Chemical Manufacturers Association maintains a database of information about hazardous chemicals that can be transmitted via a cellular telephone system to the site of a chemical spill.

Since cellular telephones transmit calls by radio, there has been some concern about eavesdropping. Cellular calls can be picked up by some radio scanners as far as 5 to 10 miles away from the transmitter. A federal law makes cellular telephone eavesdropping punishable by a $500 fine, but critics say is impossible to enforce. Users need to be aware that calls on cellular telephones are not as secure as calls placed on conventional telephones.

Cellular telephone service has been a big commercial success. Although cellular service is more expensive than a regular telephone line, it can be a boon to any company that makes its living from customer service. It also is widely used for business and pleasure by people who spend large amounts of time commuting or tied up in traffic.

In late 1988 there were more than 1.7 million cellular telephones, nearly four times the total in late 1986. There are expected to be more than 4 million units in use by the early 1990s. The cost of cellular telephone equipment has been coming down; good systems today cost from $500 to $1,000, about 60 percent of the cost just 2 years ago. Usage costs also have been coming down. Most cellular carriers charge a flat monthly fee ranging from $25 to $35, plus 25 to 45 cents per minute. This works out to $125 to $150 per month for an average subscriber. In general, call quality is good, though users in some cities, such as Los Angeles, suffer from interference due to overuse of the available frequencies.

Cordless Telephones

Cordless telephones are an even more limited distance form of mobile telephone service. The cordless telephone is a small, battery-operated, low-power transmitter/receiver designed to work within a few hundred feet of a base unit transmitter/receiver that connects to the telephone line. The transmission between the two units is a frequency modulated carrier. Some cordless telephones have the dial or keypad on the base unit; others have it on the portable unit. If it is on the portable unit, the dial pulses or tones must be modulated onto the carrier when a number is dialed.

Usually, portable telephones are designed so that they operate in a standby mode, conserving battery power but able to detect incoming calls by a buzz when the base unit sends a ringing signal. When the user wants to answer or place a call, he or she first switches the transmitter on and then continues with the call. Most units are designed so that the handheld telephone will recharge its batteries automatically when placed in the base unit.

THE VOICE NETWORK USED FOR DATA TRANSMISSION

Succeeding chapters discuss the subject of data communications. Data communications is a much more recent development than voice communications. Much of the ubiquitous nationwide public telephone network that we take for granted is, to all appearances, an analog network designed and optimized to handle voice communications. Data communications has certain unique attributes and different requirements, but these have been made to fit the parameters of the voice network. Of the 300 million telephone calls that take place in a day, several million are data exchanges that routinely take place over the voice telephone network. You will see how this is done in the following chapters.

SUMMARY

This chapter presented the various components of the voice telephone system. It looked at the telephone instrument itself, the switching equipment located in the central office, and the hierarchy of central offices that compose the public telephone network. You have also studied how voice signals are converted to electrical signals, the characteristics of those signals, and the way in which they are modified (modulated) for transmission on the telephone network. In addition, you studied private telephone systems and their features, systems like those found in most businesses today. Finally, you were introduced to a number of the services offered by the telephone companies.

Many of the concepts and facilities presented in this chapter will come up in later chapters that discuss data communications. By understanding the vocabulary and subject material presented in this chapter, you will have a good foundation for the chapters that follow.

CASE STUDY

Dow Corning's Telephone System

Dow Corning's main telephone system in Midland is a Centrex system provided by Michigan Bell Telephone. There are approximately 2,500 telephones in the Midland area attached to the Centrex. The system has direct inward dialing and provides four-digit dialing between stations. The DID feature allows the company to provide good telephone service with only two central telephone operators. These operators also serve as receptionists for the Dow Corning Center headquarters complex.

Each telephone user (or department) pays for telephones and long distance calls separately. The company produces an internal telephone bill each month from magnetic data tapes provided by Michigan Bell, AT&T, and some internal tables of information that Dow Corning maintains. All costs of operating the telephone system are charged out to the telephone users. Several rates for various types of long distance calls are calculated, and these are applied to the actual calls made as reported by the telephone company. In 1988, the company purchased its telephone handsets after an economic evaluation showed that substantial savings would accrue from this decision.

Users in the Midland area make long distance telephone calls through a network of out-WATS lines. Least cost routing is provided by the Centrex system. To make a long distance call, a telephone user simply dials the area code and telephone number; the Centrex system determines the most economical way to route the call. If all of the WATS lines are busy, the call overflows to normal lines.

Dow Corning's plants and sales offices in the United States have a variety of equipment suited to their requirements. Some of the plants have PBXs, whereas others have key systems.

Dow Corning's telephone operators also serve as receptionists for the corporate headquarters. (Courtesy of Dow Corning Corporation)

Many of the sales offices are quite small, so key systems serve their needs adequately. In general, the plants have installed PBX equipment. Some of the plants have acquired their own WATS or tie lines to reduce their long distance telephone costs.

Dow Corning conducted a study in 1984 to determine whether a more centralized approach to telephone management was appropriate. One conclusion of the study was that over time it would be possible and desirable for the company to build an integrated nationwide telephone network. The second conclusion of the study was that it made sense to position the company to integrate its voice and data networks when and where it was appropriate.

After looking at the product offerings of several vendors, the study team recommended that when telephone equipment is replaced, preference should be given to Rolm PBXs or key systems. There was a strong feeling that since Rolm had recently been acquired by IBM, there was a better chance of being able to integrate a Rolm telephone system with the IBM computers and data network than if another supplier's telephone system were selected. Of course, this recommendation was subject to review from time to time.

Between 1986 and 1988, Rolm PBXs were installed in two manufacturing plants, the office in Toronto, Canada, an independent business unit in Midland, several of the larger sales offices in the U.S., and the central order processing (COP) department in Midland.

The PBX that serves the COP group was installed to handle the special needs of that department. COP enters and processes all orders for Dow Corning products from customers throughout the United States. The department has its own in-WATS network for customer calls. This network passes through an ACD system, which routes each incoming call to the appropriate customer service representative. There is also an automated voice response unit that helps when the customer isn't sure which customer service representative to speak with. The voice response unit provides information to the customer and allows him or her to key digits into the telephone

A Dow Corning customer service representative takes an order. She is using a headset connected to her Rolm telephone and a flat panel video display terminal that lets her be logged on to four different applications simultaneously. (Courtesy of Dow Corning Corporation)

to direct the call to the appropriate person. The COP PBX is connected to the Midland area Centrex system. All of the telephones on the Rolm system have four-digit telephone numbers beginning with 7 and can be accessed by Centrex users the same way they access any other telephones.

In 1988, a study was performed to investigate the feasibility of replacing the Centrex system in Midland with a large Rolm PBX. About the time the study team was ready to make its recommendations, IBM announced that they were selling most of their investment in Rolm to Siemens, a large German telecommunications company. Because of the significant nature of this corporate change of direction, Dow Corning decided to upgrade its Centrex system rather than install a Rolm PBX. The new Centrex system, which costs less than the old Centrex, is based on digital switching equipment at the Michigan Bell central office and provides most of the same capabilities as a PBX.

Connected to the Centrex system is a voice messaging system provided by VMX, Inc. Voice messaging was first investigated in 1982 and seemed to have applicability, particularly for sales people who travel a great deal. In 1983, Dow Corning began using voice messaging on a public service bureau system and then installed its own voice messaging computer in 1984. Today there are a large number of voice messaging users, mainly in marketing, sales, and technical service. The costs of the system are divided among the users, who each receive a monthly bill for their voice messaging use. The VMX system has proven to be very reliable, and the users are extremely happy with the capability it provides.

A final capability of the Midland area telephone system is its leased tie lines connecting the Centrex in Midland to a number of the outlying plants and offices. Of particular interest is the tie line that connects to the PBX in the Brussels, Belgium, office. This tie line operates completely automatically. A user in Midland dials a three-digit access code. If the line is not busy, he or she receives a dial tone from the Brussels PBX. At that point, the user dials the four-digit extension number of the telephone in Brussels. No telephone operators at either location have to handle the call. Making a call from Brussels to Midland is similar. This line, which uses TASI technology, is provided by MCI and costs Dow Corning about $5,000 per month. The line saves the company money compared to the cost of the long distance telephone calls that would be made between Midland and Brussels if the line was not in place.

Review Questions

1. List the five functions of the telephone set.
2. Compare and contrast dial pulsing with the DTMF technique for dialing a telephone number.
3. If the telephone system is designed for a P.03 grade of service, how many calls would you expect to be blocked for every 500 calls placed?
4. Distinguish between a telephone line and a trunk.
5. In normal speech, some sounds above 3,000 Hz are generated. What happens to these frequencies when they are sent through the public telephone network?
6. Explain the term *modulation*. For what is it used?

7. Is the electrical representation of the voice signal that is transmitted between the home and the telephone company central office modulated? Explain your answer.
8. What are the three attributes of a sine wave?
9. Explain what a foreign exchange line does.
10. Compare and contrast key systems and PBXs.
11. Distinguish between drop wires, local loops, distribution cables, and feeder cables.
12. What is the bandwidth of the AM radio broadcasting band of frequencies in the United States? The FM radio band? (If necessary, do some research at the library.)
13. An analog signal is fed into one end of a twisted pair of wires. At the other end, the relative strength of the signal is measured as −10 dB. How much has the power of the signal dropped as it travelled through the wire?
14. List an example of an electrical wave with a frequency of 60 Hz, a radio wave with a frequency of 640 kHz, a radio wave with a frequency of 102 mHz, and sound wave with a frequency of 880 Hz.
15. What are the attributes of a tone used for tone signaling in the telephone system?
16. What is the importance of Signaling System 7?
17. Explain the difference between LATAs and area codes. Can an LEC carry a telephone call between two telephones whose telephone numbers have different area codes? If so, under what conditions?
18. Explain the ways in which WATS service may be categorized.
19. What factors must the network analyst consider when selecting the type of long distance service that is best for his or her company? How can he or she get help in analyzing the data and making the decision?
20. Compare and contrast a voice messaging system with an ordinary telephone answering machine.
21. Explain how a cellular telephone system works.

Problems and Projects

1. Would a company ever want to acquire band 2 and band 5 WATS lines at the same time? Under what circumstances?
2. A company with 300 employees in Texas wants to acquire a private telephone system. Would you suggest a key system or a PBX? Why?
3. A company headquartered in St. Louis, Missouri, wants to provide the ability for its customers in Iowa and Illinois to call its headquarters toll free. Would the company be better off considering WATS lines or foreign exchange lines to provide this service? Why?

4. Visit a local hotel/motel and find out what type of telephone system it uses. Does the system have the automatic reminder feature for wake-up calls? If not, how are wake-up calls handled? Are there any special features on the hotel's telephone system that weren't described in the text? If so, what do they do?
5. Visit a local small business that has a key system installed. How many telephones and outside lines will the system handle? How much expansion capability does it have? Did the company purchase the system, or is the firm leasing/renting it? What features does the key system have? (Use the features described for PBXs as a checklist.)
6. Pick one of the following services to investigate:

   cellular telephone system voice messaging system audio teleconferencing system

7. Find a company in your area or do research at the library and write a two-page report describing the capabilities, features, and shortcomings of the system you have chosen.

Vocabulary

analog
sidetone
switchhook
off-hook
dial tone
on-hook
dial pulsing
rotary dial
out-pulsing
dual-tone-multifrequency (DTMF)
Touchtone
call setup time
patch cord
step-by-step switch
Strowger switch
crossbar switch
reed relay
blocked, blocking
busy hour
grade of service
exchange
main distribution frame
class 5 central office
end office
serving central office
toll office
dynamic nonhierarchical routing
public telephone network
public switched telephone network (PSTN)
local loop
tip and ring
drop wire
distribution cable
feeder cable
trunk
analog signal
frequency
sine wave
Hertz (Hz)
kilohertz (kHz)
megahertz (MHz)
gigahertz (gHz)
bandwidth
guard channel
guard band
amplitude
level
decibel (dB)
attenuation
phase
frequency division multiplexing (FDM)
carrier wave
modulation
amplitude modulation (AM)
frequency modulation (FM)
phase modulation (PM)
demodulation
multiplexer
base groups
channel groups
groups
supergroups
master groups
time assignment speech interpolation (TASI)
direct current (DC) signaling
tone signaling
E&M signaling
in-band signals
out-of-band signals
common channel interoffice signaling system (CCIS)
Signalling System 7 (SS7)
Consultative Committee on International Telegraphy and Telephony (CCITT)
exchange code

- local calling
- local service area
- local call
- flat rate service
- measured rate service
- long distance
- direct distance dialing (DDD)
- toll calls
- inter-LATA calls
- international direct distance dialing (IDDD)
- Wide Area Telecommunications Service (WATS)
- in-WATS
- 800 service
- out-WATS
- software defined network (SDN)
- foreign exchange line
- key systems
- private branch exchange (PBX)
- private automatic branch exchange (PABX)
- computer branch exchange (CBX)
- system features
- station message detail recording (SMDR)
- call detail recording (CDR)
- station features
- data PBX
- Centrex
- tie lines
- tie trunks
- electronic tandem network (ETN)
- enhanced private switched communication service (EPSCS)
- voice messaging
- audiotex
- cellular telephone system

References

Benkoil, Dorian. "Data PBXs, What, Why and When?" *Teleconnect*, December 1986, p. 67.

Berney, Karen. "Mobile Phones' Business Appeal," *Nation's Business*, August 1986, p. 38.

Blythe, W. John, and Mary M. Blythe. *Telecommunications: Concepts, Development, and Management.* Indianapolis, IN: The Bobbs-Merrill Company, Inc., 1985.

Briere, Daniel. "The Numbers Game," *Network World,* June 19, 1989.

Brodsky, Ira. "Dialing Data Via Mobile Telephones," *Data Communications,* October 1989, p. 119.

Bruckman, Sally. "Company Uses Audioconferencing to Conduct Its Sales Meetings at 11 Locations Simultaneously," *Communications News,* February 1987, p. 37.

"Centrex CPE: Taking a Bite Out of Big Bad Centrex," *Teleconnect,* December 1986, p. 104.

Cooper, Ray J. "A Dozen Area Codes Remain. The Czar of Numbering Needs a Break," *Data Communications,* November 1986, p. 107.

Davis, Bob. "Eavesdropping Looms as a Problem As Cellular-Telephone Use Widens," *The Wall Street Journal,* October 29, 1986.

Doll, Dixon R. *Data Communications: Facilities, Networks, and Systems Design.* New York: John Wiley & Sons, 1978.

Falk, Howard. *Microcomputer Communications in Business.* Radnor, PA: Chilton Book Company, 1984.

Fike, John L., and George E. Friend. *Understanding Telephone Electronics.* Dallas: Texas Instruments, 1984.

Finnigan, Paul F., and James G. Meade. "Audiotex: The Telephone As Data-Access Equipment," *Data Communications,* November 1986, p. 141.

Fitzgerald, Jerry. *Business Data Communications: Basic Concepts, Security, and Design,* 2nd ed. New York: John Wiley & Sons, 1984.

Gantz, John. "SS#7: Its Promise and Pitfalls," *Telecommunication Products + Technology,* September 1988.

"Going the Distance," *Teleconnect,* August 1989, p. 127.

Gurrie, Michael L., and Patrick J. O'Connor. *Voice/Data Telecommunications Systems.* Englewood Cliffs, NJ: Prentice-Hall, Inc., 1986.

Hafner, Katherine. "Hello Voice Mail, Goodbye Message Slips," *Business Week,* June 16, 1986, p. 80D.

Hapgood, Fred. "At 411 It's Simply a Matter of Keeping in Tune with the Numbers," *Smithsonian,* November 1986, p. 67.

Jesitus, John. "Using Cellular for Data Links," *Communications News,* June 1989.

Keller, John J., Frances Seghers, and Mark Ivey. "AT&T Is Eating 'em Alive," *Business Week,* February 16, 1987, p. 28.

Knight, Fred S. "Cellular Hits the Big Times," *Business Communications Review,* August 1989.

Levine, Jonathan B., and John J. Keller. "Craig McCaw's High Risk Empire," *Business Week,* December 5, 1988.

Levine, Richard O. "Extending Networks with Signaling System 7," *Business Communications Review,* September–October 1988.

Parkinson, Richard. "Straight Talk on Digital Communications," *Information Systems Management,* Vol. 4, No. 1, Winter 1987, p. 18.

Rappaport, David M. "Voice Mail: Key Tool or Costly Toy?" *Data Communications,* October 1986, p. 153.

Ratcliff, Rick. "Cellular Phones Have Big Numbers," *The Detroit Free Press,* October 14, 1988.

Rey, R.F. (editor). *Engineering and Operations in the Bell System,* 2nd ed. Murray Hill, NJ: AT&T Bell Laboratories, 1983.

Roberts, Johnnie L. "Cellular Phones Take Off—And Become an Integral Part of People's Work Lives," *Wall Street Journal,* July 24, 1986.

Roehr, Walter. "Knocking on Users' Doors: Signaling System 7," *Data Communications,* February 1989.

Stamper, David A. *Business Data Communications,* 2nd ed. San Francisco: The Benjamin/Cummings Publishing Company, 1986, 1989.

Stix, Gary. "Office Network Update," *Computer Decisions,* August 27, 1985, p. 66.

"Telephone Headsets That Don't Give Germs a Chance," *Business Week,* December 15, 1986, p. 100K.

"This Long-Distance Operator Is a Machine," *Business Week,* December 22, 1986, p. 46H.

"What is a Key System?" *Teleconnect,* February 1987, p. 55.

"Who Switches Data along with Voice? PBX Users, Increasingly," *Data Communications,* February 1987, p. 77.

Stallings, William. "Demystifying SS7 Architecture," *Telecommunications,* March 1988.

CHAPTER

Data Terminals

OBJECTIVES

After you complete your study of this chapter, you should be able to

- classify terminals several different ways;
- explain the capabilities and characteristics of several types of terminals;
- discuss the functions of a cluster control unit;
- describe uses for several specialized types of terminals;
- distinguish among intelligent terminals, smart terminals, and dumb terminals and describe the characteristics of each;
- explain the criteria used when terminals are selected for a particular application.

INTRODUCTION

This chapter begins a look at the details of data communication systems by studying the part with which you are most likely to be familiar—the data communications terminal. The study of data communications in this book takes an outside-in approach. The discussion begins at the terminal and, through the next five chapters, works its way in to the computer to which the terminal is attached. For most students, this approach will go from familiar territory—the terminal—through the unfamiliar and new technical aspects of data communications, and then back to more familiar ground at the computer. Coupled with the structure provided by the ISO-OSI model that was introduced in Chapter 4, you have two ways in which to categorize and classify your new knowledge.

Although you may have studied the material about terminals in a data processing class, the information will serve as a quick review of the different types of terminals in common use. The chapter also highlights the telecommunications implications of the various types of terminals. It may also serve to fill in the gaps in your knowledge of data communication terminals and how they work.

DEFINITIONS

A *terminal* is an input/output device that may be attached to a computer via direct cable connection or via a communications line. The terminal may be dependent on the computer for computational power and/or for data. If the terminal is directly connected to the computer by a cable, telecommunications may not be involved. Hardwired terminals may use signaling methods other than those commonly used in telecommunications.

A VDT or other device does not need to be connected to the computer full-time to qualify as a terminal. A personal computer is an example of a device that has its own intelligence and may operate independently of a host computer a good deal of the time. At other times, however, a connection may be made to allow the personal computer to be operated as a terminal attached to a host computer.

data terminal equipment

data circuit-terminating equipment

In communications terminology, both the terminal and the computer to which it is attached are properly known as *data terminal equipment (DTE)*. Data terminal equipment operates internally in digital format. Its output signal is a series of digital pulses. *Data circuit-terminating equipment (DCE)*, which you will study in Chapter 8, provides the interface between data terminal equipment and the communications line. As shown in Figure 6–1, when analog lines are used, DCE provides the translation between the digital format of the DTE and the analog format of the transmission line. Even when digital communications lines are used, DCE is required between the DTE and the line.

TERMINAL CLASSIFICATION

Data terminals can be classified into categories, but the classification scheme may not be definitive because the categories have wide overlap. Because of the overlap, any classification of terminals is questionable.

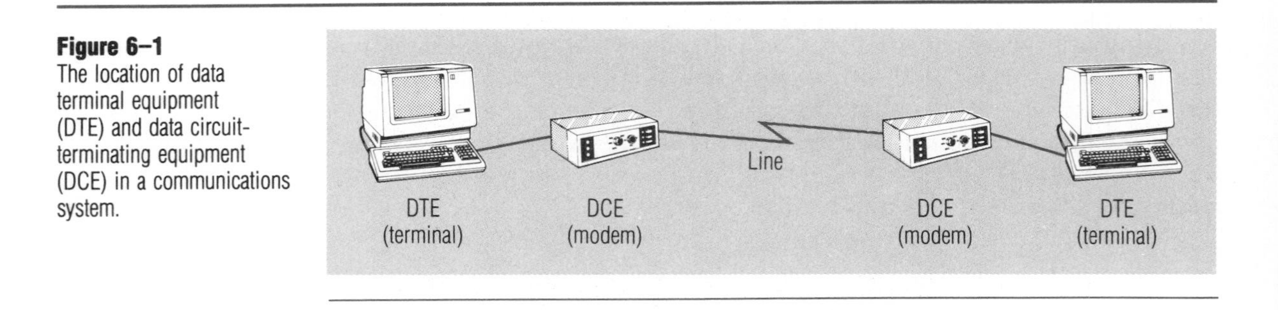

Figure 6–1
The location of data terminal equipment (DTE) and data circuit-terminating equipment (DCE) in a communications system.

Nonetheless, there are several categories that will serve us well for descriptive and discussion purposes. They are

- teletypewriters;
- video display terminals (VDTs);
- industry-oriented transaction terminals;
- intelligent terminals;
- remote job entry terminals;
- specialized terminals.

You can see the overlap already. Personal computers are intelligent devices, but they certainly use video display terminal technology; video display terminals are used in many transaction-oriented applications.

Teletypewriter Terminal

The teletypewriter was the primary terminal in use before 1970. The simplest way to think of a teletypewriter terminal is as a typewriter with additional electronics and other features added for communications purposes. Indeed, some early terminals were just that—a typewriter equipped for communication. Most teletypewriter terminals in use today were originally designed for communications and will not operate as ordinary typewriters.

The teletypewriter is equipped with a keyboard and a mechanism for continuous feeding of paper. The keyboard of a teletypewriter ter-

A contemporary teletypewriter. (Courtesy of Dow Corning Corporation)

minal is similar to that of a typewriter, although the keys are usually laid out somewhat differently. The alphabetic and numeric keys are usually in standard locations. However, the location of the punctuation and other special characters may vary. In addition, there are special keys unique to computer terminals that can send special signals to the communications equipment or the host computer. Examples of these are the ESC key, the BREAK key, and the ENTER (RETURN) key. There are many ways to construct a keyboard, but the one that is usually preferred by people with formal typing training is the keyboard with distinct keys and a touch and feel like that of a good electric typewriter.

The printing mechanism on the teletypewriter terminal varies in how it forms the characters on the paper. In early teletypewriters, each character was mounted on a print arm that struck the paper through a ribbon much like a typewriter. The IBM Selectric mechanism, in which the characters are on a golf ball-like device that rotates and twists to properly position a character before striking the paper, has been used in many terminals. Others use a matrix of wires that push the ribbon against the paper and form characters by a series of closely spaced dots. Others employ heat as well as a matrix of wires to cause a chemical reaction in specially treated paper to form characters. This type of printing mechanism does not require a ribbon.

RO devices

Some teletypewriters do not have keyboards and are designated as *receive-only (RO)* devices. They are used where no input to the computer is desired or necessary, for example, in newspaper offices that receive weather reports, stock quotations, or news broadcasts. Without a keyboard, RO devices cannot be used to enter data into computers.

KSR devices

Teletypewriters with keyboards but with no paper tape equipment are called *keyboard-send-receive (KSR)* devices. KSRs still are used frequently as data terminals, although they are being replaced by VDTs that are quieter, less mechanical, and therefore more reliable.

ASR devices

Teletypewriters with paper tape readers and punches are called *automatic-send-receive (ASR)* devices. They were very widely used in the 1950s and 1960s when torn tape message switching systems were in widespread use; they are not common today.

Most teletypewriter terminals have a switch that dictates whether pressing a key on the keyboard will cause a character to print on the printer. In some communications systems, direct printing from the keyboard is disabled. All characters typed on the keyboard are sent on the communications line to the computer, which echoes them back to the printer. This type of echoing is one way of showing whether the computer correctly received the character that was keyed. Although it is not foolproof, the technique is used in many communications systems.

Teletypewriters and other terminals operate in one of two ways: *unbuffered* or *buffered*. In an unbuffered terminal, a character is transmitted to the computer as soon as someone presses a key on the keyboard. In the buffered terminal, the keyed characters are stored in an internal storage area or buffer until a special key, such as the RETURN or ENTER key, is pressed. Then all of the characters stored in the buffer are transmitted to the host computer in one operation. With a buffered terminal, users can correct typing mistakes before data is sent to the computer. This type is the most common terminal today. Buffering is not unique to teletypewriters; it is found in most other types of terminals as well.

Some other characteristics of teletypewriter terminals are

- slow speed—seldom more than 15 characters per second;
- very mechanical—therefore not as reliable as all-electronic devices;
- appropriate where hard copy is required;
- often connected to the computer only by dialing up.

Video Display Terminal (VDT)

The names *video display terminal (VDT)* and *cathode ray tube terminal (CRT)* are often used interchangeably. Not all video display terminals use cathode ray tube technology, however, so *video display terminal* is more general and appropriate. Other technologies used for VDTs are light emitting diodes (LEDs), gas plasma, and electroluminescent displays. These technologies yield a display that is much flatter and takes up less space on a desk, hence the generic name, *flat panel displays.*

Other characteristics of VDTs are

- all electronic—highly reliable;
- buffered (usually)—data is stored until the user presses the ENTER key;
- capable of very high speed display of data;
- various types range in capability from simple, unintelligent "glass teletypewriter" to very intelligent, programmable.

The Screen The VDT usually contains a cathode ray tube on which an electron beam causes phosphors to glow, forming the desired letters, numbers, special characters, or other patterns. Three electron beams and different types of phosphorus are used to create a color image, in the same way that a color television picture is formed. A VDT screen usually measures 12 inches, 14 inches, or 15 inches diagonally.

Characters are formed when the electron beam energizes selected dots of phosphorus within a matrix. Typical matrix sizes are 5 × 7, 7 × 9, and 8 × 10 dots. The characters are formed so that there is at least one unused row of dots around them to provide spacing between characters. This concept is illustrated in Figure 6–2. The VDT with spaces for 80 characters across the screen and 24 or 25 rows down has evolved as a de facto standard, although screens with more columns and rows also are available.

all-points-addressable VDTs

Most alphanumeric screens allow screen positions to be addressed at the character level. For example, a program could instruct the com-

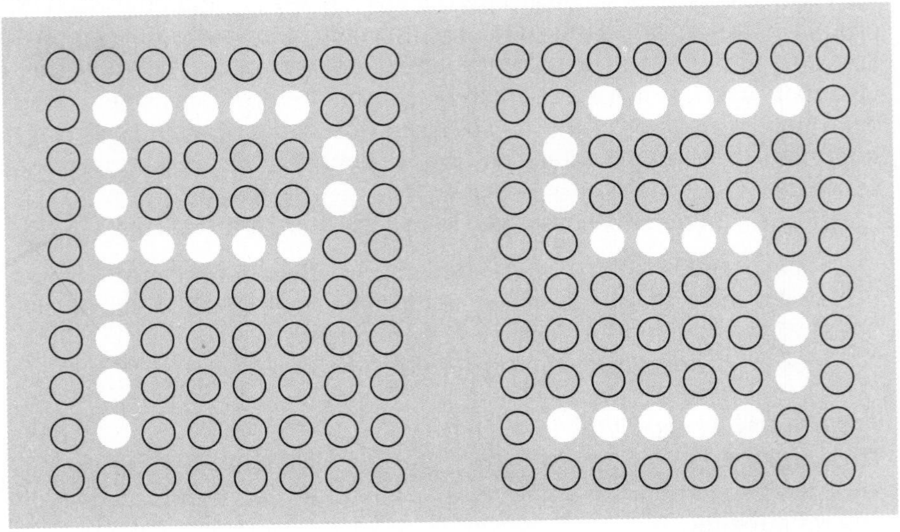

Figure 6–2
Characters formed by a dot matrix.

puter to place a character *c* in column 14, row 7, or the digit 3 in column 71, row 19. Other VDTs are *all-points-addressable (APA)* and allow each individual dot on the screen to be controlled. These dots are called *picture elements*, commonly known as *pixels* or *pels*, and they can be turned on or off or, on a color VDT, set to a specific color under the control of a program in the computer. All-points-addressable VDTs normally are used for graphics applications.

The screen of a VDT also displays a special place-marking character called a *cursor*, which indicates where the next character from the computer or keyboard will be displayed. In some cases, the cursor can be made to blink or display at a higher intensity than other characters on the screen so it can be seen easily. The keyboard of a VDT has special keys, called *cursor control keys*, for moving the cursor. These are often designated by arrows that point up, down, left, and right and that move the cursor in the indicated direction. Moving the cursor with the cursor control keys is not the same as moving it with the SPACE BAR or BACKSPACE key. The SPACE BAR inserts a space character, which will be transmitted to the computer, in the data stream. The BACKSPACE key removes a character; using a cursor control key has no effect on the data stream.

VDTs have several methods of highlighting characters for easy identification by the user. One technique is called *intensifying*, in which a character (or any collection of dots) on the screen is made brighter than the other characters around it. Some VDTs have several brightness levels. Another technique is to cause a character to *blink* by varying the intensity at which it is displayed several times each second. A third technique is called *reverse video,* which reverses the character and background colors. If, for example, the normal display shows green characters on a black background, a reverse video character would be black on a green background. A fourth technique for highlighting characters is with the use of *color,* for example, displaying most of the characters in green and certain characters, such as error messages, in red.

When teletypewriter terminals receive output from a computer, they print it, one line at a time. There is no inherent concept of a page, except as a collection of lines. This is known as *line-by-line mode.* VDTs can operate in a similar way, receiving and displaying one or a few lines at a time. When a VDT operates in line-by-line mode, the new lines normally appear at the bottom of the screen. All other lines are moved up, and the top line disappears off the top of the screen. This is much like what the user sees as printed teletypewriter output moves up and eventually disappears over the back of the terminal.

VDTs can also operate in *page* or *formatted mode.* There are several ways this is implemented. It is simplest to think of the screen of the VDT as a "page," consisting of a specific number of lines, each with a set number of characters. Programmatically, output can be placed anywhere on the screen, and if the communications lines are fast enough, the entire screen is displayed to the user at the same time. Page mode

The IBM 3192 VDT is a low-cost terminal designed for general use. The 3192 can display eight colors and graphics. (Courtesy of IBM Corporation)

The Memorex Telex 1192 series of video display terminals offers several models which display data in both monochrome and color. (Courtesy of Memorex Telex Corporation)

allows the screen to be laid out like a paper form with headings, field identifiers, and fields to be filled in, as shown in Figure 6–3. The operator may be required to fill in certain fields, whereas others are optional. After the operator types certain information on the form, a material's lot number, for example, the computer may respond by filling in other fields, such as the product's name and status.

Figure 6–3
Page mode allows the VDT to be laid out like a paper form.

QUALITY ASSURANCE SYSTEM 10/11/90
LOT NUMBER TEST RESULTS 10:57

LOT NUMBER: FA128639

ITEM ID	LOT NUMBER	QA STATUS	STATUS DATE	SAMPLE TYPE	TEST RESULTS
1959042	FA128639	APPROVED	20SEP90	1A	1:56.2 2:OK 3:7.1cm 4:HIGH AVERAGE
2876691	BQ212598	APPROVED	29SEP90	2	1:41.0 2:OK 3:4.5cm 4:OKAY
1464552	RM213356	REJECTED	30SEP90	1A	1:55.1 2:LOW 3:1.1cm 4:HIGH

This keyboard for an IBM PS/2 personal computer has a numeric keypad at the right side. Twelve program function keys are at the top. (Courtesy of IBM Corporation)

The Keyboard VDT keyboards, like teletypewriter keyboards, are similar to standard typewriter keyboards. The extra keys for cursor control already have been mentioned. In addition, there are usually special function keys, sometimes called *program function keys,* that direct the computer to perform actions predetermined by the software being run. Some keyboards have a *numeric keypad,* that is like a 10-key calculator keyboard. The numeric keypad is useful if a lot of numeric data must be entered, and its keys are in addition to the regular numeric keys on the keyboard.

Two other useful keys on a typical VDT keyboard are the INSERT and DELETE keys that the operator uses to insert or delete characters. When the cursor is moved under a character and DELETE is pressed, the character disappears from the screen, and all characters to the right of it on the line are shifted one position to the left to fill the gap. The INSERT key works just the opposite; it allows characters to be inserted on a line, with all following characters shifted to the right.

Some terminals have specialized keyboards that usually are designed to help the operator perform his or her job more quickly and efficiently. The terminals in McDonald's restaurants are one example. They have keys for the clerk to press to indicate the item the customer is ordering and the quantity. The keys are actually labeled "Hamburger," "Cheeseburger," "Large Fries," "Shake," and so on. Other specialized keyboards are used on terminals in laboratories, plants, banks, and so on.

Another type of specialized keyboard is one used in countries that use alphabets made up of other than Roman letters. In Japan, China, Korea, and the Arab countries, for example, the alphabets and characters are much different from the ones we use in western countries, and keyboards must be designed to handle those characters. Minor modifications are required in almost every country so that the keyboard has keys for national characters, such as those with an umlaut in German-speaking countries and accented characters where French and Spanish are spoken.

touch-sensitive screen

Alternative Input Mechanisms In addition to keyboards, some VDTs have other ways of controlling the movement of the cursor on the screen or entering data. The screens of some VDTs are *touch-sensitive*. If the computer displays a question on the screen with a list of possible answers the user can indicate a choice by touching the appropriate answer. Touch-sensitive screens detect the location of the finger using either a photosensitive or resistive technique. With the photosensitive technique, a matrix of infrared light sources is mounted on two sides of the screen, and photo sensors are mounted on the opposite sides. Touching the screen interrupts the light beams, and the coordinates of the area touched can be detected. The resistive technique uses two conductive layers on the screen separated by a small distance. When the screen is touched, the connection between the two conductors is made at the point of touch, and this location can be sensed.

Touch screens have been successful in applications where people unfamiliar with data processing must use the terminal because these screens are easy to use and require no training. One example is a VDT at an information booth in a fair or exposition. The general public may use the VDT to obtain information about the location of exhibits or other facilities. Obviously there is no opportunity to train the users, many of whom may never have used a VDT before.

Other input mechanisms useful in certain applications are the *mouse*, *joystick*, *trackball*, and *lightpen*. The mouse, joystick, and trackball provide alternative means for moving the cursor on the screen and have found wide use in graphics applications and electronic arcade games. The lightpen is a pointing device used to mark a spot on the screen or make a selection from a list of displayed items.

VDT Selection Criteria When you select a VDT, consider the following factors:

- Does the face of the screen have a nonglare surface to prevent reflection from nearby light?
- Can the screen be tilted or swiveled to a comfortable viewing position?
- On a monochrome VDT, is the character display in one of the two standard colors (green on a black background or amber on a black background)?
- On a color VDT, can the default background, and character colors be selected by the user?
- Is the screen image flicker free? (Flickering screen images cause eye strain.)
- Is the size of the screen appropriate for the application?
- Is the resolution of the characters fine enough to minimize eyestrain?
- Is the VDT programmable?
- Does it have graphics capability?

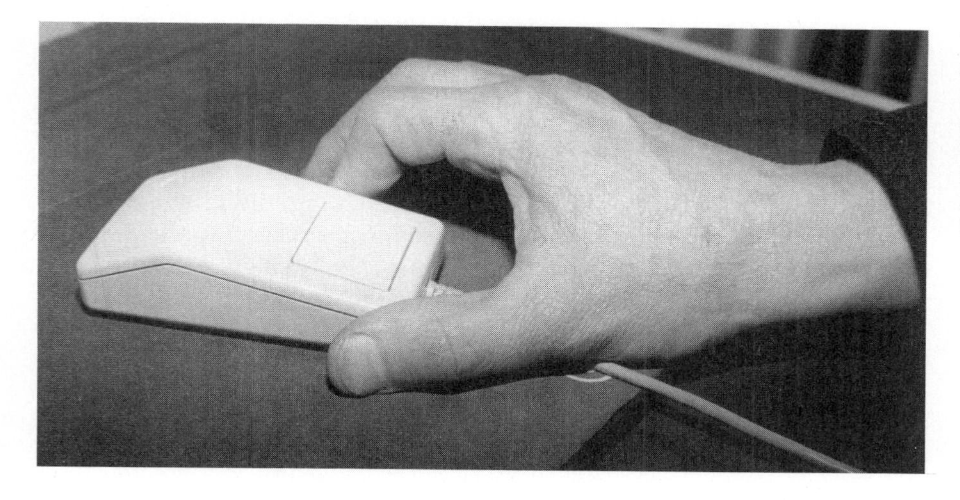
This mouse can be moved around on the desk top to control the location of the cursor on the screen of a VDT. The buttons are used to send signals to the computer. (Courtesy of H. G. Haigney)

Engineering Graphics Terminals

Engineering graphics terminals are large, all-points-addressable VDTs with very high resolution. A screen with 1,024 × 1,024 or 2,048 × 2,048 dots across its face is typical for this type of terminal. Engineering graphics VDTs are used for engineering design and drafting, the application being called *computer aided design* or *computer assisted drafting (CAD)*. The terminals usually contain their own microcomputers to perform specialized calculations to enlarge or reduce a drawing or to rotate the viewpoint of a drawing. Telecommunications for this type of terminal is used

This user is holding a lightpen, which is used for selecting items displayed on the screen of the VDT. (Courtesy of IBM Corporation)

for downloading a drawing from the host computer to the terminal and uploading it back to the computer for storage after it has been modified.

Engineering graphics terminals frequently place a heavy load on a communications line. The APA characteristic of the terminals, coupled with the large screen size and the complexity of the drawings, dictate that there are many bits or characters to be exchanged with a host computer. The amount of communication is inversely related to the amount of intelligence in the terminal. If the terminal is intelligent, it may only need to communicate with the host to retrieve and store drawings. All other work can be carried out on the terminal. If the terminal is unintelligent, it relies on the host to assist with every change made to the drawing, and the communication is very frequent and lengthy. Most CAD terminals fall somewhere between the two extremes, but the direction is clearly toward more intelligence in the terminal.

Industry-Oriented Transaction Terminals

Industry-oriented transaction terminals are specifically designed for the efficient processing of online transactions in a certain industry. Common examples of these types of terminals are the automated teller machines and supermarket checkout terminals mentioned in previous chapters. Others include the terminals used by bank tellers. Many of these have special printers for recording changes in passbooks. Point of sale terminals read the data on merchandise tags in retail stores. Badge readers are used for time and attendance reporting in factories. These terminals are all designed around the requirements of a particular type of business transaction. They are designed to be easy to use, even by untrained operators or laypeople. Since the market for these specialized terminals

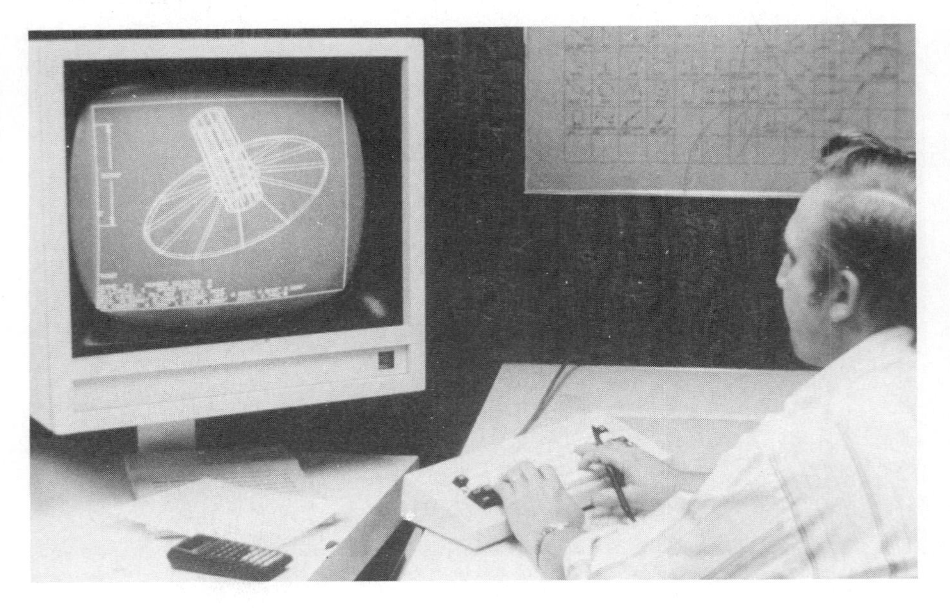

The engineering grapics terminal displays high resolution graphics that can be enlarged to show tiny details of the drawing.

usually is relatively small, the price is often higher than that for a more general terminal with the same electronic components and similar packaging.

Remote Job Entry (RJE) Terminals

Remote job entry (RJE) terminals, as the name implies, were originally used for the remote submission of batch programs or jobs to run on a host computer and for receiving and printing the output of those jobs. The classic RJE terminal consists of a reader for decks of punched cards containing the control statements or programs to be run on the computer and a medium- to high-speed printer for printing the results of the jobs' execution. To the host computer, the job appears to have been submitted from a card reader directly attached to the computer in the computer room, and the output appears to be printed on a computer room printer. The fact that telecommunications is involved is masked from the computer and from much of its operating system software.

With the obsolescence and decline in the use of punched cards and the widespread availability of interactive VDTs, work for the computer is usually submitted online with a few commands rather than with a deck of punched cards. As a result, the classic RJE terminal is seldom seen, although the high-speed remote printer is still in wide use. Printers of 100 to 1,000 lines per minute are most common.

Facsimile Machines

Facsimile machines (FAX) used as specialized computer terminals have some of the same characteristics as OCR machines. Instead of recording

This stockbroker's terminal has additional keys on the keyboard to handle the specialized requirements of the broker. Other terminals are simpler and are designed for direct use by the broker's clients. (Courtesy of K. B. Kaplan)

Figure 6–4
Facsimile machines are grouped according to their technology and transmission speed.

Group I	Internally analog Transmits a page in about 6 minutes Rarely sold anymore
Group II	Usually internally analog Transmits a page in about 3 minutes Still sold because it is relatively inexpensive
Group III	Internally digital Fully automated operation Transmits a page in about 30 seconds Much better copy quality than Group I or II Most popular today
Group IV	Internally digital Transmits a page in 2 seconds or less if the telecommunications line is fast enough Letter quality copy Expensive Not sold widely

and coding individual characters, however, facsimile machines read and code patterns of light and dark areas on a sheet of paper as was discussed in Chapter 2.

types of FAX machines

Facsimile machines are divided into four groups acccording to their technology and speed (Figure 6–4). Group I and II machines are internally analog in operation, and the transmission of an 8 ½ × 11 inch document takes several minutes. Group III and IV machines are internally digital and transmit documents in less than a minute. The speed with which Group IV machines can transmit documents is primarily limited by the speed of the communications line to which they are attached. On a digital line that transmits data at 56,000 or 64,000 bits per second, a Group IV facsimile can send a document in about two seconds.

The digital facsimile machines code a spot on the document (a pixel) as either a 1 or 0 depending on the amount of light that is reflected from it. The receiving machine produces a corresponding black or white pixel on the output document. The resolution or quality of the output depends on how many lines per inch are scanned by the transmitter—in other words, how many samples are taken. Of course, the more samples taken, the more data that must be transmitted, but some digital facsimiles compress the 1s and 0s before transmitting them.

The signal from a digital facsimile can be read into a computer and stored, since it is made up of bits. If an appropriate all-points-addressable VDT is available, the image can be displayed. Conversely, a digitized image can be sent from a computer to a digital facsimile machine to produce hard copy output. Hardware boards for personal computers are widely available to provide this capability. Facsimiles are to OCR terminals as all-points-addressable VDTs are to alphanumeric VDTs. In both cases, one is designed to handle only alphanumeric data, and the other is designed to handle images.

Facsimile machines used as terminals find application where documents must be read into a computer and stored and/or transmitted to another location. They place a significant load on a communications line, however. An 8 ½ × 11 inch page, when scanned by a facsimile, requires that about 40,000 8-bit bytes of data be transmitted. This can be reduced by compression or by scanning the document at a lower resolution. However, the received image quality may suffer.

Specialized Terminals

Specialized terminals are not designed for a specific application or industry. Their use is more limited than that of a standard VDT or teletypewriter, but for certain applications they are extremely useful or even indispensable.

Telephone One important type of terminal in this category is the standard 12-key pushbutton telephone, which can be used to send digits and 2 "special" characters to a computer. The telephone is used in banking and other applications where the input is all numeric. A common use is for entering an account number, customer number, or product number used as the key to some information stored on a computer's

This FAX35 facsimile machine can transmit a page of data in 15 seconds. It contains a 30-page memory for sequential broadcasting and a 124-station autodialer. (Courtesy of Ricoh Corporation)

disk. The computer looks up and reports particular information regarding the item identified by that key.

When the telephone provides the input, the companion output unit is often provided by an *audio response unit* on the computer that can form sounds from digital data stored on disk and "speak" the response back to the user. Audio response units produced in recent years can speak at varying rates of speed and with a pitch and inflection that is appropriate to the words being spoken. The use of an audio response is an example of the marriage of voice and data communications technology.

Optical Recognition Another type of device is the *optical recognition* terminal, which can detect individual data items or characters, and convert them into a code for transmission to a computer. This type of terminal uses a photo cell to sense areas of light or dark on paper or other media.

bar code readers

optical character recognition

There are several types of optical character recognition terminals. *Bar code readers* scan bars printed on merchandise or tags. The bars on grocery products are one example, but the use of bar coding is spreading to many industries. *Optical character recognition (OCR)* terminals detect and read individual characters of data. One type reads the characters on a typewritten page; another type can read hand-written data if it is

The Memorex integrated terminal is a combination VDT and telephone. It can also be equipped with a printer to print images of the screen, and a lightpen. (Courtesy of Memorex Telex Corporation)

clearly written. A simpler OCR device detects only marks on a page and is used to read survey forms, answer sheets from examinations, and medical questionnaires.

Integrated Terminals An integrated terminal is more commonly known as an *integrated video display terminal (IVDT)*. The integration comes from the combination of a VDT terminal with a telephone. IVDTs have been available since the early 1980s, but they have not seen widespread use. Most IVDTs integrate a telephone with a personal computer so the terminal is intelligent. One simple application of an IVDT is to store a list of frequently called telephone numbers on the IVDT's disk. The list is called to the screen, the entry for the person to be called is selected, and the touch of a button causes the IVDT to dial the telephone number. For the individual who accesses many different computers via dial-up telephone calls, the IVDT could be set up to automatically place the call and send the commands to log the user on to the computer. Both of these applications can be accomplished in other ways with less expensive equipment, however, which is the primary reason the terminals have not become popular. Nonetheless, many vendors have IVDT products in their product line, and competitive pressure keeps them there.

Other Terminals Many other types of terminals are used for simple tasks, such as counting, weighing, measuring, and reporting results to the computer. On an assembly line, for example, a simple photoelectric cell may detect the passage of each item manufactured and report the event via communications lines to a computer. In a chemical plant, a device may measure the flow rate of a liquid through a pipe (or its temperature or pressure) and report this information to a computer. With advances in microprocessor technology, many devices that previously were unable to communicate are being given communication capability.

■ CLUSTER CONTROL UNITS

Cluster control units (CCU) are used with some types of terminals as a way of sharing some of the expensive electronic components needed to support the advanced features of the terminals. The IBM 3270 family of terminals is typical: up to 32 VDTs may be attached to one CCU, and the control unit is attached to the communications line. This is illustrated in Figure 6–5.

CCUs are normally programmable devices, although frequently the programming is provided on diskettes by the manufacturer and cannot be changed by the user. CCUs may contain buffers for the terminals that can be shared. They may also perform code conversion and do error checking to ensure that data is received from the communications line correctly. Another use for the cluster control unit is to allow one or

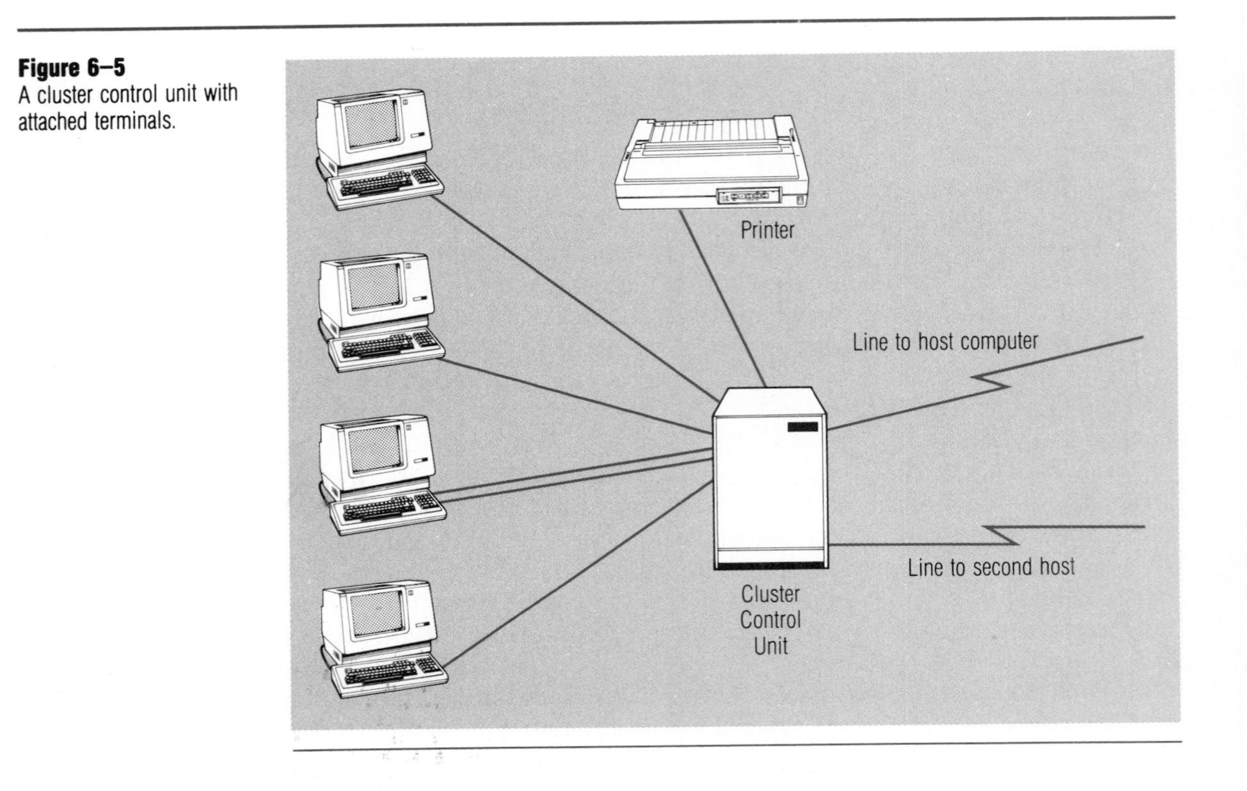

Figure 6–5
A cluster control unit with attached terminals.

more printers to be shared among the terminals attached to it. A special key on the keyboard of each of the attached VDTs allows the image on the screen of the VDT to be printed immediately on a printer connected to the control unit without sending any data to or from the host computer.

Some CCUs can be attached to more than one communications line, allowing a terminal to access more than one computer. This is valuable to users who access multiple computers. With some CCUs, the user can be "logged on" to more than one computer simultaneously. The VDT screen displays the information from one computer, but the user can switch to view information from the other computer using a key (or combination of keys) that is frequently called the *hot key.* Some CCUs allow a user to have multiple connections to one computer. Using this capability, the user can be logged on to more than one program on a single host computer and can switch back and forth between applications using the hot key. This capability is called *multiple sessions.*

Cluster control units can have different degrees of intelligence ranging from simple buffering and translation facilities to being programmable computers in themselves. One example of the latter is a store controller/cluster controller in a point of sale terminal system.

This IBM 3174 cluster control unit can attach up to 16 terminals or printers. Other models can attach up to 32 devices. The cluster control unit connects via a leased communication to a host computer. (Courtesy of IBM Corporation)

TERMINAL INTELLIGENCE LEVELS

With the reduction in the cost of integrated circuit chips, the desktop personal computer of today has as much power as a large mainframe computer had 20 years ago. If we were to classify all terminals on an intelligence scale, we would find that there is a range or continuum from fully intelligent, general purpose computers used as terminals to completely unintelligent ("dumb") terminals. Some people classify terminals into three categories: *intelligent terminals, smart terminals*, and *dumb terminals*. However they are classified, today's terminals offer a wide range of capabilities.

Intelligent Terminals

The more intelligent a terminal is, the more able it is to participate in processing data. A fully intelligent terminal is in itself a general purpose computer. If a terminal can be programmed, some of the processing tasks can be performed locally on the terminal, giving a degree of independence from the host. When processing is shared between a host computer and an intelligent terminal, it is called *distributed processing* or *distributed data processing (DDP)*. For many reasons, distributed processing is widely used in businesses. Computing tasks can be performed on the computer most suitable for the task. Screen formatting and data editing can be done by the intelligent terminal. On the other hand, where access to large databases or extensive computation is required, the power of the host processor can be brought into play. Using several computers also minimizes the risk of a massive computer failure, since users are not dependent on a single mainframe to do their processing.

When properly programmed, the intelligent terminal may be able to operate for long periods of time when there are failures in the data communications system or host computer. The terminal continues to operate without accessing the host and saves the results of its work. When the host computer becomes available, the terminal transmits the results of the processing it has performed.

Terminal intelligence is often present in less obvious ways. For example, a considerable amount of built-in intelligence is required for a supermarket terminal to read and interpret the bar code on a peanut butter jar passed through a laser beam at practically any angle or speed. Considerable intelligence also is required by bank or credit card terminals to read and interpret the magnetic stripe on a bank card, even though the stripe might have been damaged by abrasion from being carried in a person's wallet. The movement toward intelligent terminals is inevitable, depending primarily on their cost effectiveness.

Smart Terminals

Smart terminals are not programmable, but they have memory that can be loaded with information. Their memory may be loaded by a transmission from a host computer, by keying data from an attached keyboard, or by reading it through some other attached device, such as a diskette reader.

One use for the smart terminal's memory is to store constant data, such as formats, that the operator uses repetitively. An order entry operator's terminal might be loaded with the format of the order form. This would save continually transmitting it from the host to the terminal each time the operator wanted to use the form. The operator would call up the blank format from the memory of the terminal to the display screen. The blanks in the format would be filled in with the data for the order. When the operator had checked the data, he or she would press the ENTER key. Usually only the data the operator entered, but not the order format, would be transmitted to the host, another saving of transmission time.

The smart terminal has a certain amount of independence from the host. The operator works with a format, completing information and making corrections until he or she is satisfied that the form is complete and correct. Only when the operator presses ENTER is there an interaction with the computer. Unlike the intelligent terminal, the smart terminal usually does not have the capability to save the data from an order if the communications line or host computer is down.

When a smart terminal is connected to a cluster control unit, some of its "intelligence" may be provided by the CCU. For example, the CCU might provide the storage for the formats, either in its memory or on a disk.

Dumb Terminals

The advantages of an unintelligent or "dumb" terminal are simplicity and low cost. The terminal is totally dependent on the host for all processing capability and has either no storage or only limited, special purpose storage for terminal control functions. Dumb terminals are extremely popular, however, because of their low cost and the fact that they are supported for communications by virtually every type of host computer.

PERSONAL COMPUTERS USED AS TERMINALS

Personal computers (PCs), often also called *microcomputers,* are becoming very widely used as terminals on communications networks. In some applications, they are the terminal of choice because of their intelligence and ability to participate in a distributed processing system. Software that is more commonly found in the personal computer environment, such as word processing and spreadsheet programs, may be used to manipulate data before it is sent to a mainframe for further processing, or data from the mainframe may be sent to the personal computer for processing or formatting before it is presented to the user.

Personal computers may be connected to networks and mainframe computers at several levels of sophistication. Frequently, they are connected so that they act like dumb terminals because this is the simplest, least expensive type of connection that can be made. In this case, they emulate (act like) a terminal that the host computer recognizes. Terminal emulation is accomplished by special hardware in the personal computer or by a program called a *terminal emulation program.* In either case, the host thinks it is working with a standard terminal because the hardware or software in the PC responds just like the terminal would. The most commonly emulated terminals are the Digital Equipment Corporation (DEC) VT-100 and VT-220 and the IBM 3101 and 3270.

Terminal emulation is a simple, inexpensive way to use a personal computer to communicate, but when a personal computer emulates a dumb terminal, the power of the personal computer is not used to its best advantage. The host doesn't know it is communicating with another computer so it cannot tap the intelligence that resides in the PC. More sophisticated programming is required to allow the two computers to communicate in a distributed processing fashion or as peers. One simple enhancement is to add software that allows the personal computer and the mainframe to transfer files back and forth. Either computer can create a file using its own software, and then the operator can have it sent to the other computer using the SEND or RECEIVE command of the file transfer software. This type of file transfer is called *uploading* or *downloading* depending on one's point of view. Often, file transfer soft-

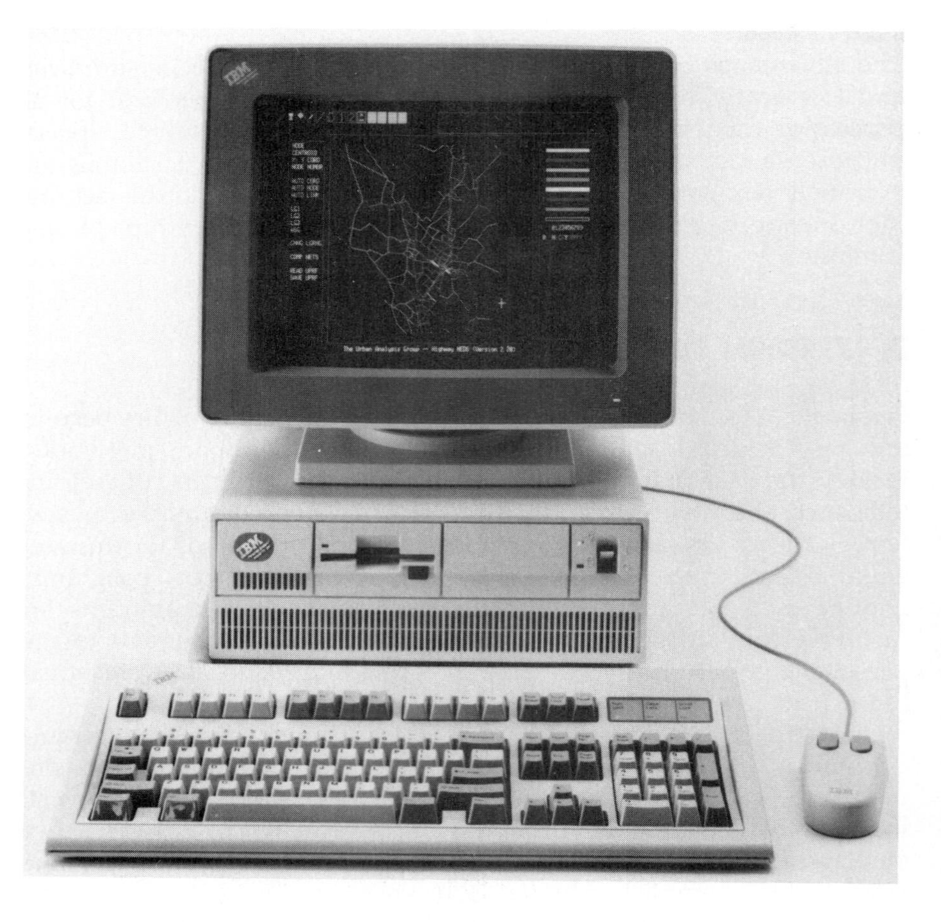

This IBM PS/2 model 70 contains a powerful processor and comes with a minimum of 1 megabyte of memory. (Courtesy of IBM Corporation)

ware also includes the capability for a printer attached to the personal computer to receive and print data directly from the mainframe.

More sophisticated software for distributed processing allows the host and personal computers, or two personal computers, to interact through a communications network in a more realtime fashion. Often called *peer-to-peer* capability, such software allows a user or program at one computer to request data from another computer and have it sent automatically. In another situation, a user on a personal computer might view data stored on another computer as a simple extension of the disk capacity on his or her own PC. This level of sophistication requires considerably more elaborate software and more powerful computers than are required for terminal emulation. Of course, the cost increases too.

Portable personal computers are often used by sales representatives or other travelers in two modes. They use the machines as standalone computers during the day to perform processing independent of the host. At night, when the representative returns to his or her hotel room, the machines are used as terminals, connected via ordinary telephone lines to a host computer at company or division headquarters. The results of the day's activities are transmitted to the host, and messages or electronic mail are received from it.

WORKSTATION ERGONOMICS

In recent years, with the increased use of terminals in the workplace, a greater emphasis has been placed on workstation ergonomics, and much has been published about the total workstation environment for terminals. The *workstation* is the place where a person sits or stands to do work. It contains the working surface, terminal, chair, and any other equipment or supplies the person needs to do his or her job. It is the place where the computer meets the user. Having information about workstation design is useful for telecommunications people because they are often called on to advise others in the company about terminal use and environment. A drawing and some information about an ergonomically designed workstation are shown in Figure 6–6.

No matter how perfectly the workstation is set up, it is important to take a periodic break from concentrating on the VDT. There is a real incidence of headaches, fatigue, muscle aches, and temporary eyestrain

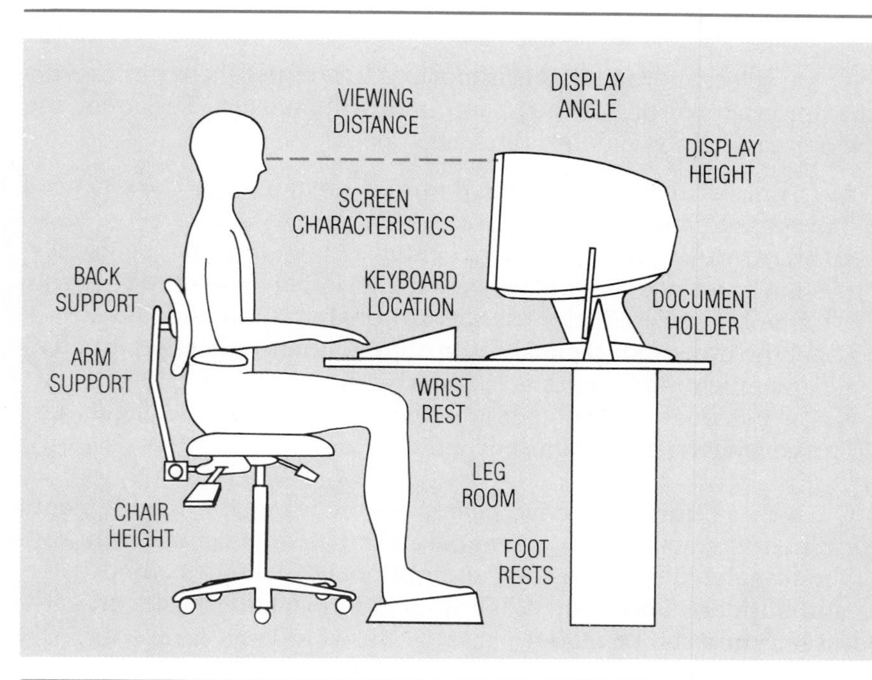

Figure 6–6 Ergonomically designed furniture allows for natural movement and allows workers to change position to prevent fatigue and stiffness.

Considerations for Ergonomically Designed Workstations

1. The VDT screen should have a non-glare screen that can be tilted back 10 to 20 degrees.
2. The screen of the terminal should be about 18 inches from the operator's eyes.
3. The top of the screen should be no higher than eye level.
4. The chair should have a seat that adjusts up and down and a backrest that adjusts separately. When seated the operator's feet should rest flat on the floor or on a footrest.
5. The keyboard should be attached to the rest of the terminal by a cable so it can be moved on the work surface. The keyboard should be positioned so that the operator's lower arms are parallel to the floor and the upper arms are perpendicular to it.
6. Nearby blinds should be closed and other sources of glare eliminated. Small adjustable task lights are usually preferable to overhead lights.

resulting from prolonged use of VDTs under poor conditions. None of these problems are permanent, and they are minimized when VDT workstations, environment, and job structure are properly designed.

THE VDT AND HEALTH CONCERNS

Much has been written about potential health hazards stemming from working with VDTs over long periods. The traditional questions are:

- Is there any radiation hazard in working with the VDT?
- Will it affect my eyesight?
- What is the most comfortable method of working with the VDT?
- How can I relieve the stress associated with constant use of a VDT?

Many health-related concerns about working with VDTs involve the radiation emitted from the display tube. Numerous studies by universities and government agencies worldwide have shown that the amounts of radiation emitted from the terminal are well below established safety standards and, in most cases, are nearly undetectable. Other studies have shown no significant association between VDT use and vision defects or eye abnormalities.

TERMINAL SELECTION

When you select a terminal, it is important to identify the requirements of the applications for which the terminal will be used. The following questions can help you determine users' needs:

- Is a special application-oriented terminal required, or can a general purpose terminal be used?
- Is hardcopy required, or can a video display terminal be used?
- What types of operators will use the terminal? Will they be highly trained or novices who use the terminal only occasionally?
- Are the proposed terminals compatible with the existing computer?
- How much intelligence is required in the terminal?
- Is a personal computer required, or can a less intelligent, less expensive terminal be used?

Once these questions have been answered and the requirements determined, the next step is to compare alternative terminals from several vendors, looking at required and optional capabilities, vendor service and support, and cost. When all this information has been gathered, a selection can be made.

These workstations are ergonomically designed to provide the operator with comfort, proper lighting, and functionality. (Courtesy of Steelcase)

On a pragmatic basis, the type of terminal selected may be essentially predetermined by the type of computer to which it will be attached or by other terminals that the company has on hand. In many cases, a simple approach may be to assume that the terminal is going to be a standard VDT unless it is shown why some other type terminal is required. VDTs represent a good mix of capability and reliability. They are effective in most applications and are available today at a very low cost.

TYPICAL TERMINAL SCENARIO

This discussion is intended to give a practical example of how terminals are connected to a host computer in typical businesses throughout the world. Although hundreds of options are available, there are in fact certain configurations that are more widespread than others.

The VDT and personal computer are the most widely used terminals by far. In some companies, a VDT can be found on nearly every desk. The IBM 3270 series of VDTs is in widespread use in part because approximately 80 percent of the mainframe computers in use are made by IBM or are compatible with its machines. The IBM 3270 series includes terminals with screens that can display 24 lines of 80 characters and terminals with other screen sizes as well. Both monochrome and color terminals are available in the family.

The 3270 VDT is attached to a cluster controller, and the cluster controller is attached via a communications line to a central computer. The communication between the cluster controller and the computer most frequently operates at a speed of 9,600 bits per second, although higher and lower speeds are used as well. Response time from the computer to the terminal in most companies averages between 1 and 5 seconds, but there is wide variability.

Also attached to the cluster controller is a printer used for local screen printing. It is shared by all of the terminals on the control unit, and although it is convenient to the users, it is not located in an individual's office. Normally, the printer is located in a central location so it can be easily accessed by all of the users.

In smaller businesses, VDTs are also in widespread use, but they are typically directly attached to minicomputers and in close proximity to them. Transmission speeds of 1,200 to 9,600 bits per second are common. Temporary dial-up connections are used by employees who need occasional access to the computer. Color and graphics terminals are less likely to be found, although they are available for use with computers of all sizes.

SUMMARY

A wide variety of terminals are available today, with a range of capabilities and prices. The trend is toward more intelligent terminals as the cost of memory and integrated circuitry drops. Intelligent terminals are more flexible and provide the capability, through customized programming, to reduce the amount of data transmitted through the network. They also can share the data processing workload with a host computer. Ultimately, however, terminal selection must be based on the needs of the user.

CASE STUDY

Evolution of Data Terminal Purchases at Dow Corning

The use of data terminals in Dow Corning started in 1953 when the first teletypewriters were installed for message communications between the corporate headquarters and several of the larger sales offices. These teletypewriters were not connected to a computer but simply connected with one another on point-to-point and multipoint lines. In 1968, this torn tape message switching system was replaced with a computer controlled message switching system using similar teletypewriters. Also in 1968, the first remote job entry terminals were installed, which allowed users in several of the outlying plants to submit data or jobs to run on the mainframe computer in Midland.

Early in 1969, the first VDTs were installed. The original two VDTs were used at the helpdesk to monitor and control the communications network. By 1970, 10 additional VDTs had been installed in the order processing department in conjunction with a new sales order entry application. There were just over 200 terminals installed in the company by 1980. By then, they were located in many departments, but only a few people in each department were users of the computer applications. By 1989, there were over 3,000 VDTs installed in nearly all Dow Corning locations around the world. Their use had become common among all levels of employees and required for most jobs.

Most of the terminals are VDTs from the IBM 3270 family or plug compatible equipment from Memorex or Telex. Users are free to select the specific type of terminal they need from a list of approved equipment. The list is updated approximately semiannually by a committee made up of SIM people and users. Users pay for their own terminals, and most are purchased when they are acquired.

Dow Corning's torn tape message switching system was controlled from this location. Telex machines can be seen at the right. (Courtesy of Dow Corning Corporation)

A terminal user at Dow Corning is scheduling a meeting using one of the central computers. (Courtesy of Dow Corning Corporation)

In 1983, the first personal computers were installed in the company, and their growth has been rapid. The company began aggressively managing the acquisition of personal computers in 1984, when their use started to grow rapidly. A personal computer policy was established, and a list of PCs approved for use within the company was published. Like similar lists in other companies, Dow Corning's was strongly oriented toward IBM PCs and compatible units. One of the early concerns was that PCs be configured so that they could eventually be connected to the company's telecommunications network. By early 1987, there were approximately 800 personal computers installed, and at the end of 1989, there were about 2,000. Approximately 95 percent of them were connected to the telecommunications network and used as intelligent terminals.

The personal computer is used in many Dow Corning departments for spreadsheets, word processing, and project management. (Courtesy of Dow Corning Corporation)

The telecommunications department negotiated an arrangement with a Dow Corning subsidiary company, Site Services, Inc. (SSI), to handle terminal and PC installation and service. SSI receives terminal and personal computer equipment from vendors, tests it to be sure it is operating properly, installs software if required, and sets the equipment up in the user's office. SSI also provides maintenance services for terminals and personal computers and keeps a small inventory of terminals available for immediate delivery or as "loaners."

Other types of terminals are used in the manufacturing and research functions of the company. Dow Corning uses Digital Equipment Corporation (DEC) and Hewlett Packard (HP) computers in its plants and laboratories, and, therefore, the terminals connected to those computers were made by DEC and HP. Interconnection of the IBM, DEC, and HP computers and communications equipment is occurring as manufacturing and laboratory systems are integrated with traditional business applications that run on the IBM mainframe computers.

Many users perform remote job entry. In most cases, they submit their jobs from a VDT terminal. The output of the job, usually a report, frequently is directed to a nearby printer, but in many cases, the report is stored on the disk of the computer and viewed online. This has the advantage of saving the time and cost of actually printing the report, particularly if it will be used infrequently.

The use of facsimilie machines has grown rapidly in the last several years. Users have found them invaluable for rapidly exchanging printed documents that are not stored in a computer. There are over 120 facsimile machines used in the United States and more in the foreign locations. Group III machines predominate, although there are a few of the older, slower Group II machines in use. Recently, a few users have experimented with circuit cards that allow their personal computer to send and receive documents to and from facsimile machines.

Review Questions

1. Distinguish between data terminal equipment and data circuit-terminating equipment.
2. What are some desirable attributes of a terminal that will be used by the general public on an occasional basis?
3. How may the higher cost of an intelligent terminal be justified compared to a standard VDT?
4. What ergonomic factors must be considered when you select a terminal?
5. What are the advantages and disadvantages of a buffered terminal compared to one that is unbuffered?
6. Under what circumstances would an APA terminal be used?
7. Describe several methods of highlighting information displayed on a VDT.
8. What is the purpose of a program function key on a terminal's keyboard?
9. Describe the factors to consider when you select a VDT.

10. For what applications is a mouse most useful?

11. How is remote job entry usually performed today?

12. Distinguish between facsimile machines classified as Group I and those classified as Group IV.

13. In order to display an image that has been read into a computer from a facsimile machine, an __________ VDT is needed.

14. List the primary purposes of the cluster control unit.

15. Distinguish between intelligent, smart, and dumb terminals.

16. What is distributed processing? What are its benefits?

17. Describe three levels of sophistication that may be employed when connecting a personal computer to a host computer.

Problems and Projects

1. Visit a company that has a large number of terminals installed. Find out what portion of the terminals are "dumb," "smart," and "intelligent." Does the company make a strong distinction in usage between the different intelligence of its terminals? Do programmers design computer applications to take advantage of the capabilities of intelligent terminals? Does the company forecast that the number of intelligent terminals will grow in the future? If so, how fast? Will intelligent terminals ever totally replace dumb and smart terminals in the company?

2. Do you expect the use of terminals in most businesses will grow to the point where there is one terminal for every employee? What are some examples where this might not be the case? Can you think of a situation where there might be more than one terminal for every employee?

3. Vendors of color terminals claim that color improves the productivity of the operator. How can a terminal improve productivity? What effect does color have? If color does, in fact, improve productivity, think of some applications where it would be especially useful.

4. Think of some applications that require a user to have a terminal that prints all of the interactions with a computer.

5. Visit a stockbroker and use one of the special "application-oriented" terminals to display current stock prices. How does the special design of the terminal make it easier to get the information? How much more difficult would it be to get stock prices if a general purpose VDT were used instead? Talk to the broker and find out what other special capabilities the terminals have.

6. Do some research on the latest findings about the potential health hazards of radiation from VDTs.

Vocabulary

all except peer-to-peer

terminal
data terminal equipment (DTE)
data circuit-terminating equipment (DCE)
receive-only (RO)
keyboard-send-receive (KSR)
automatic-send-receive (ASR)
unbuffered
buffered
video display terminal (VDT)
cathode ray tube terminal (CRT)
flat panel display
all-points-addressable (APA)
picture element (pixel, pel)
cursor
intensifying
blinking
reverse video
line-by-line mode
page mode
formatted mode
program function keys
numeric keypad
touch-sensitive
mouse
joystick
trackball
lightpen
computer aided design (CAD)
computer assisted drafting (CAD)
facsimile machine
remote job entry terminal (RJE)
audio response unit
optical recognition
bar code reader
optical character recognition
integrated video display terminal (IVDT)
cluster control unit
hot key
multiple sessions
intelligent terminal
distributed processing
smart terminal
dumb terminal
microcomputers
terminal emulation program
uploading
downloading
peer-to-peer
workstation

References

Blyth, W. John, and Mary M. Blyth. *Telecommunications: Concepts, Development, and Management.* Indianapolis, IN: The Bobbs-Merrill Company, Inc., 1985.

Brown, Bob. "Group IV Fax Grows Slowly but Steadily," *Network World,* August 7, 1989.

Cambridge, Zada B. "Making VDT Work for You Can Cut Job Aches, Pains," *The Saginaw News,* September 7, 1986.

Corson, Richard G. "VDTs—New Evidence Indicates Helpfulness over Harmfulness," *Data Management,* December 1986, p. 24.

Falk, Howard. *Microcomputer Communications in Business.* Radnor, PA: Chilton Book Company, 1984.

Fitzgerald, Jerry. *Business Data Communications: Basic Concepts, Security, and Design,* 2nd ed. New York: John Wiley & Sons, 1984, 1988.

Friend, George E., and John L. Fike. *Understanding Data Communications.* Dallas: Texas Instruments, Inc., 1984.

Martin, James. *Design of Man–Computer Dialogues.* Englewood Cliffs, NJ: Prentice-Hall, Inc., 1973.

Robins, Marc. "The Executive Workstation: A Product for the Niches," *Teleconnect,* January 1987, p. 98.

Sexton, Don. "Fax: Faster, Cheaper, Better," *Teleconnect,* July 1989.

Stark, Thomas H. "Facsimile Equipment: Faster, Better, Cheaper," *Telecommunication Products + Technology,* December 1986, p. 43.

Coding and Digitizing

OBJECTIVES

After you complete your study of this chapter, you should be able to

- define what a code is;
- describe several different coding systems;
- describe three different types of characters that compose a code's character set;
- discuss typical functions of control characters;
- describe a method of error checking that is built into a code;
- explain the purpose, pros, and cons of encryption;
- describe several ways in which analog signals are digitized;
- describe the advantages of digital data transmission.

INTRODUCTION

In Chapter 5, we saw that the human voice, when it is converted to an analog electrical signal, presents a complicated pattern to the medium for transmission. This chapter looks at how, through coding, data can be converted to a simple, two-state, digital signal and transmitted in that format. The digitization of analog voice signals also is explained.

This chapter begins the bridging process between your knowledge of analog voice signals and how they are transmitted to the digitized world of data. The material's concepts will be used throughout the rest of the book and are important to understanding the technical material in succeeding chapters.

TWO-STATE PHENOMENA

In introductory computer classes, you learned that many natural and physical phenomena are two-state systems. A baseball runner is out or safe; a basketball shot is made or missed; a light bulb is on or off; an electrical circuit is open or closed resulting in a flow of electrical current or no flow. No *ifs, ands,* or *maybes*—it's one way or the other.

We refer to these two states as *binary states*. They can be represented using the binary digits 1 and 0. The term *binary digit* is abbreviated *bit*. When bits are used to represent the settings, we say that information

Figure 7–1
The Morse code.

has been coded. A 1 bit means the runner is safe, a 0 bit means he is out; a 1 could mean the basket is good or the current is flowing. Note that the 0 bit presents information just as the 1 bit does. It does not mean "nothing."

CODING

A *code* is a predetermined set of symbols that have specific meanings. The key point is that the meanings are predefined; the sender and receiver must agree on the set of symbols and their meanings if the receiver is to make sense out of information that is sent. For data communications purposes, codes are assigned to individual characters. Characters are letters of the alphabet, digits, punctuation marks, and other special symbols, such as the dollar sign, asterisk, and equals sign.

A character code you may have heard of is the Morse code, shown in Figure 7–1. It is a binary code because it uses two code elements, a dot and a dash. Not all of the characters, however, have the same number of code elements. The code was structured so that the most frequently used characters require the fewest number of dots and dashes. In contrast, less frequently used characters are represented by more dots and dashes. The pitch (frequency) and volume (amplitude) of the coded signal are irrelevant to the meaning of the Morse code.

Morse code was designed for human use, and that is why the spaces between the letters and words have meaning. The translation of letters and numbers into Morse code is done by the human operator sending a message. Decoding from Morse code back into letters and numbers is done by the operator at the receiving end. With humans at both ends of the transmission, it is also not important that all characters consist of the same number of dots and dashes or that the dots and dashes be perfectly formed and exactly consistent. Although officially the ratio of length between a dot and a dash is 1:3, when a human is sending the Morse code using a telegraph key, this ratio varies widely. Expert telegraph operators can tell who is sending code just by listening to the way the characters sound.

MACHINE CODES

When two machines, such as a computer and a terminal, are used in communications, some of the attributes of the Morse code, or any other code systems designed for human use, are not desirable. It is much easier for a machine to process a code if the code has the following attributes:

- it is a true binary or two-state code;
- all of the characters have the same number of bits;
- all of the bits are perfectly formed;
- all of the bits are of the same duration.

A binary code works well for machines communicating by electrical means because 0 bits and 1 bits can be represented by a current flow that is on or off.

For transmission efficiency, it is ideal to have a coding system that uses a minimum number of bits to represent each character. How many bits are needed? With one bit we can have two states: 1 or 0. With two bits, there can be four combinations; 00, 01, 10, 11. With three bits, there can be eight combinations; 000, 001, 010, 011, 100, 101, 110, 111. Do you see a pattern? Each bit that is added doubles the number of combinations. If each combination of bits represents a character, then with three bits and eight combinations, eight characters can be represented. Mathematically, the number of unique combinations is expressed as 2^n where n is the number of bits.

Figure 7–2 shows a table of the powers of 2. From it you can see that using 5 bits gives 32 combinations, and with 6 bits there are 64

$2^0 = 1$
$2^1 = 2$
$2^2 = 4$
$2^3 = 8$
$2^4 = 16$
$2^5 = 32$
$2^6 = 64$
$2^7 = 128$
$2^8 = 256$
$2^9 = 512$
$2^{10} = 1024$

Figure 7–2
Powers of two.

possibilities. The number of possible combinations or characters in a coding system is called *code points*.

character assignments

A code normally has unique groups of bits assigned to represent the various characters in a code. These unique sequences of bits are called *character assignments*. Different codes, even though they may have the same number of bits, use different character assignments. Character assignments must be made for three different types of characters: *alphanumeric characters, format effector characters*, and *control characters*. The alphanumeric characters are the letters, numerals, and symbols, such as punctuation marks and dollar signs. They are also referred to as *graphic characters* because they can be displayed on a terminal screen or printed on paper. Format effector characters control the positioning of information on a terminal screen or paper. Included in this group are tabs, backspaces, carriage returns, and line feeds. Control characters can be divided into two subgroups. Device control characters control hardware connected to a data processing or telecommunications system. A device control character might instruct a printer to skip to the top of the page of paper. Another device control character might change the color on a VDT screen. Transmission control characters control the telecommunications system and provide functions, such as identifying the beginning and end of the transmission or acknowledging that data has been correctly received.

graphic characters

■ PARITY CHECKING

Many coding systems use an extra bit called a *parity bit*. The parity bit is added to each character representation for checking purposes. The parity bit is added so that the total number of 1 bits in the character will be an even (or in some cases, odd) number. From Figure 7–3 it can be seen that if the representation for the character R is 1010010, the number of 1 bits is 3, which is an odd number. If an even parity system is being used, which is the most common, a parity bit of 1 would be added to

Figure 7–3
If *even* parity is being used, the parity bit is set to 1 when necessary to make the total number of 1 bits in the character an *even* number. If *odd* parity is used, the parity bit is set to 1 when necessary to make the total number of 1 bits an *odd* number.

The 7-bit ASCII code for the letter *R* (no parity bit)	**1010010**
• with even parity	**10100101**
• with odd parity	**10100100**
The 7-bit ASCII code for the letter *S* (no parity bit)	**1010011**
• with even parity	**10100110**
• with odd parity	**10100111**

the character so that its complete representation would be 10100101. If odd parity were being used, a 0 parity bit would be added. Assume the representation for the letter S is 1010011. The number of 1 bits is already an even number. With an even parity system a parity bit of 0 would be added, giving a complete representation of 10100110. With odd parity, a 1 bit would be added.

When a character representation is transmitted on a communication line and the transmitter and receiver have agreed to use even parity, the receiving machine checks to see whether there is an even number of 1 bits. If not, the machine detects the error and takes appropriate action. The type of action taken is discussed in detail in Chapter 9.

ESCAPE MECHANISMS

Most coding systems include a technique called an *escape mechanism*. One of the code points is assigned a special meaning, often called the escape or ESC character. When the ESC character is sent as part of the data, it means that the characters that follow are to be interpreted as having an alternate meaning. The concept is similar to the use of the SHIFT key on a typewriter. When pressed, shift changes the meanings of the other keys on the keyboard. Lowercase letters become uppercase letters, numbers become punctuation marks, and so forth.

In some systems, the escape character changes the meanings of all characters that follow it until a second escape character is sent. In other codes, the escape character changes the meaning of only the single character that immediately follows it. Different systems use escape characters in different ways. It is also possible that several escape characters exist in a code. ESC1 might cause one meaning to be assigned to all of the following characters, whereas ESC2 might assign a different meaning.

The necessity to support escape codes complicates the design of equipment that is to code and decode the data. There is a trade-off between using escape codes to effectively obtain additional code points and adding another bit to the coding system to double the number of code points. If only a few more code points are needed, using escape codes may be an effective way to obtain them. Most coding systems in use today use 7 or 8 bits to represent each character but still include an escape mechanism to provide alternate meanings if required.

SPECIFIC CODES

Many different codes are used in telecommunications systems. One of the earliest was the original Baudot code invented by Emil Baudot, a Frenchman. His work produced a code structure that was eventually standardized as CCITT International Alphabet No. 1, but that code is not in use today. The name *Baudot code* has remained, however, and it is commonly but incorrectly applied to a code that was developed by Donald Murray. The Murray code was standardized as CCITT International Alphabet No. 2, and it is still in use. Since the name *Baudot code* is still so widely used, we will use it in this book to refer to the CCITT International Alphabet No. 2.

Baudot Code

The Baudot code uses 5 bits to represent a character and has no parity bit. With 5 bits, there are 32 unique code points, which are not enough to represent the alphabet, numerals, and punctuation marks. The escape mechanism used in the Baudot code is to assign two characters a unique function called the *letters shift* and *figures shift*. When a figures shift character is sent, all of the characters that follow it are treated as uppercase characters until a letters shift character is sent. Similarly, all of the characters following a letters shift are treated as lowercase characters until a figures shift is sent. The shift characters nearly double the possible characters that can be assigned. Since some characters must be recognized in either shift (examples are the letters shift and figures shift characters themselves and the carriage return and line feed control characters), the Baudot code actually has only 58 unique characters.

Figure 7–4 shows the Baudot code. The CCITT definition allows the uppercase characters to be interpreted differently depending on the country and to some extent the application. Figure 7–4 shows several different interpretations of the code points for different applications.

The Baudot code was originally developed for the French telegraph service and is still used today in telegraph, teletypewriter, and telex communications. Most teletypewriters built before 1965 were designed around the use of the Baudot code. The mechanical design of the machine assumed that the 5-bit Baudot code would be used. Changing to another code would have meant completely redesigning the equipment.

American Standard Code for Information Interchange (ASCII)

The *American Standard Code for Information Interchange (ASCII)* grew out of work done by the American National Standards Institute (ANSI) and is the most widely used code in computers and telecommunications networks today. ASCII is a 7-bit code and therefore has 2^7 or 128 unique code points, as shown in Figure 7–5. The way to read the ASCII code chart is to use the bits over the column head (high order bits) followed by the bits on the left side of the rows (low order bits). Thus, the ASCII code for the capital letter *P* is 1010000, and the code for the lowercase letter *s* is 1110011. An extended version of ASCII adds an eighth bit for parity, but that version is not in wide use.

The ASCII code has uppercase and lowercase letters, digits, punctuation, and a large set of control characters that are discussed later.

● Denotes positive current

Start	1	2	3	4	5	Stop	Lowercase	Uppercase: CCITT standard international telegraph alphabet No.2	Uppercase: U.S.A. teletype commercial keyboard	Uppercase: AT & T fractions keyboard	Uppercase: Weather keyboard
	●	●				●	A	–	–	–	↑
	●			●	●	●	B	?	?	$\frac{5}{8}$	⊕
		●	●	●		●	C	:	:	$\frac{1}{8}$	○
	●			●		●	D	Who are you?	$	$	↗
	●					●	E	3	3	3	3
	●		●	●		●	F	Note 1	!	$\frac{1}{4}$	→
		●		●	●	●	G	Note 1	&	&	↘
			●		●	●	H	Note 1	#		↓
		●	●			●	I	8	8	8	8
	●	●		●		●	J	Bell	Bell	'	↙
	●	●	●	●		●	K	(	(	$\frac{1}{2}$	←
		●			●	●	L	)	)	$\frac{3}{4}$	↖
			●	●	●	●	M	.	.	.	.
			●	●		●	N	,	,	$\frac{7}{8}$	⦷
				●	●	●	O	9	9	9	9
		●	●		●	●	P	0	0	0	ø
	●	●	●		●	●	Q	1	1	1	1
		●		●		●	R	4	4	4	4
	●		●			●	S	,	,	Bell	Bell
					●	●	T	5	5	5	5
	●	●	●			●	U	7	7	7	7
		●	●	●	●	●	V	=	;	$\frac{3}{8}$	⦶
	●	●			●	●	W	2	2	2	2
	●		●	●	●	●	X	/	/	/	/
	●		●		●	●	Y	6	6	6	6
	●				●	●	Z	+	"	"	+
						●	Blank				–
	●	●	●	●	●	●	Letters shift				↓
	●	●		●	●	●	Figures shift				↑
			●			●	Space				■
				●		●	Carriage return				<
		●				●	Line feed				≡

Note 1: Not allocated internationally; available to each country for internal use.

Figure 7–4
Baudot code.

Figure 7–5
ASCII code.

LAST FOUR-BIT POSITIONS (Bits 4,3,2,1)	FIRST THREE BIT POSITIONS (Bits 7,6,5)							
	000	001	010	011	100	101	110	111
0000	NUL	DLE	SP	0	@	P	`	p
0001	SOH	DC1	!	1	A	Q	a	q
0010	STX	DC2	"	2	B	R	b	r
0011	ETX	DC3	#	3	C	S	c	s
0100	EOT	DC4	$	4	D	T	d	t
0101	ENQ	NAK	%	5	E	U	e	u
0110	ACK	SYN	&	6	F	V	f	v
0111	BEL	ETB	'	7	G	W	g	w
1000	BS	CAN	(	8	H	X	h	x
1001	HT	EM	)	9	I	Y	i	y
1010	LF	SUB	*	:	J	Z	j	z
1011	VT	ESC	+	;	K	[	k	{
1100	FF	FS	,	<	L	\	l	|
1101	CR	GS	—	=	M	]	m	}
1110	SO	RS	•	>	N	∧	n	~
1111	SI	US	/	?	O	–	o	DEL

There are 96 graphic (printable) characters and 32 nonprintable control characters. Notice that the difference between the code for uppercase and lowercase letters is just 1 bit. Also, the last 4 bits of the code for the digits is their binary value. For example, the code for the digit 2 is 0110010. The last 4 bits, 0010, are the binary representation of the digit 2. Similarly, the last 4 bits of the digit 9, 1001, are the binary representation for the digit 9. These attributes are very useful when a computer manipulates data. The ASCII code is also designed for easy sorting by computer. Sorting by the binary value of the code yields a sequence that is meaningful to humans.

The international equivalent of the ASCII code is the CCITT International Alphabet No. 5. ASCII also is widely used in many countries.

Extended Binary Coded Decimal Interchange Code (EBCDIC)

The *Extended Binary Coded Decimal Interchange Code (EBCDIC)* was developed by IBM. It is very widely used in IBM mainframe computers and terminals. EBCDIC is an 8-bit code that has the 256 code points shown in Figure 7–6. This chart is read like the ASCII chart, described previously. The bits from the top of each column are combined with the bits

<table>
<tr><td>Bits 4,5,6,7</td><td colspan="4">00</td><td colspan="4">01</td><td colspan="4">10</td><td colspan="4">11</td><td>Bits 0,1</td></tr>
<tr><td></td><td>00</td><td>01</td><td>10</td><td>11</td><td>00</td><td>01</td><td>10</td><td>11</td><td>00</td><td>01</td><td>10</td><td>11</td><td>00</td><td>01</td><td>10</td><td>11</td><td>Bits 2,3</td></tr>
<tr><td>0000</td><td>NUL</td><td>DLE</td><td></td><td></td><td>SP</td><td>&</td><td>–</td><td></td><td></td><td></td><td></td><td></td><td></td><td></td><td></td><td>0</td><td></td></tr>
<tr><td>0001</td><td>SOH</td><td>SBA</td><td></td><td></td><td></td><td></td><td>/</td><td></td><td>a</td><td>j</td><td></td><td></td><td>A</td><td>J</td><td></td><td>1</td><td></td></tr>
<tr><td>0010</td><td>STX</td><td>EUA</td><td></td><td>SYN</td><td></td><td></td><td></td><td></td><td>b</td><td>k</td><td>s</td><td></td><td>B</td><td>K</td><td>S</td><td>2</td><td></td></tr>
<tr><td>0011</td><td>ETX</td><td>IC</td><td></td><td></td><td></td><td></td><td></td><td></td><td>c</td><td>l</td><td>t</td><td></td><td>C</td><td>L</td><td>T</td><td>3</td><td></td></tr>
<tr><td>0100</td><td></td><td></td><td></td><td></td><td></td><td></td><td></td><td></td><td>d</td><td>m</td><td>u</td><td></td><td>D</td><td>M</td><td>U</td><td>4</td><td></td></tr>
<tr><td>0101</td><td>PT</td><td>NL</td><td></td><td></td><td></td><td></td><td></td><td></td><td>e</td><td>n</td><td>v</td><td></td><td>E</td><td>N</td><td>V</td><td>5</td><td></td></tr>
<tr><td>0110</td><td></td><td></td><td>ETB</td><td></td><td></td><td></td><td></td><td></td><td>f</td><td>o</td><td>w</td><td></td><td>F</td><td>O</td><td>W</td><td>6</td><td></td></tr>
<tr><td>0111</td><td></td><td></td><td>ESC</td><td>EOT</td><td></td><td></td><td></td><td></td><td>g</td><td>p</td><td>x</td><td></td><td>G</td><td>P</td><td>X</td><td>7</td><td></td></tr>
<tr><td>1000</td><td></td><td></td><td></td><td></td><td></td><td></td><td></td><td></td><td>h</td><td>q</td><td>y</td><td></td><td>H</td><td>Q</td><td>Y</td><td>8</td><td></td></tr>
<tr><td>1001</td><td></td><td>EM</td><td></td><td></td><td></td><td></td><td></td><td></td><td>i</td><td>r</td><td>z</td><td></td><td>I</td><td>R</td><td>Z</td><td>9</td><td></td></tr>
<tr><td>1010</td><td></td><td></td><td></td><td></td><td>¢</td><td>!</td><td>|</td><td>:</td><td></td><td></td><td></td><td></td><td></td><td></td><td></td><td></td><td></td></tr>
<tr><td>1011</td><td></td><td></td><td></td><td></td><td>.</td><td>$</td><td>,</td><td>#</td><td></td><td></td><td></td><td></td><td></td><td></td><td></td><td></td><td></td></tr>
<tr><td>1100</td><td></td><td>DUP</td><td></td><td>RA</td><td>⟨</td><td>.</td><td>%</td><td>@</td><td></td><td></td><td></td><td></td><td></td><td></td><td></td><td></td><td></td></tr>
<tr><td>1101</td><td></td><td>SF</td><td>ENQ</td><td>NAK</td><td>(</td><td>)</td><td>—</td><td>'</td><td></td><td></td><td></td><td></td><td></td><td></td><td></td><td></td><td></td></tr>
<tr><td>1110</td><td></td><td>FM</td><td>ACK</td><td></td><td>+</td><td>;</td><td>⟩</td><td>=</td><td></td><td></td><td></td><td></td><td></td><td></td><td></td><td></td><td></td></tr>
<tr><td>1111</td><td></td><td>ITB</td><td></td><td>SUB</td><td>|</td><td>—</td><td>?</td><td>"</td><td></td><td></td><td></td><td></td><td></td><td></td><td></td><td></td><td></td></tr>
</table>

Figure 7–6
EBCDIC code.

at the left side of each row. Thus the EBCDIC code for the capital letter *P* is 11010111, and the code for lowercase letter *s* is 10100010.

Note the differences in the bit assignments between the ASCII and EBCDIC codes. The ASCII bits are numbered 7654321, whereas the EBCDIC bits are numbered 01234567. When data is translated from one coding system to the other, as is done frequently in data communications applications, it is important to be alert to these differing bit assignments to avoid incorrect translation.

Other Coding Systems

Many other coding systems have been invented and used for data processing or communication purposes. However, most of them are not in wide use today. The *Binary Coded Decimal (BCD) code* is a 6-bit code that has 64 character combinations. It was used in the early days of data processing and grew out of the Hollerith code used to code data stored on punched cards. The BCD code was not standardized, and therefore one manufacturer's version was entirely different from another's.

BCD code

An *N-out-of-M code* is a system that helps detect the loss of the bit settings for a small number of bits. In this type of code, M bits are used

N-out-of-M codes

to transmit each character, and N of these bits must be 1s. Therefore, the receiver has a way of detecting if certain characters are received incorrectly. IBM developed a 4-out-of-8 code in which 8 bits were used to represent each character. Four of the bits in each character were always 1s and 4 bits were always 0s. Although we won't go through the mathematics, the number of valid code points in the 4-out-of-8 code is 70. This limited the usefulness of the code. Although a 4-out-of-8 code detects errors more effectively than the use of a single parity bit does, it is not as effective as some of the other techniques used for error detection discussed in Chapter 9.

CONTROL CHARACTERS

In both the ASCII and EBCDIC code tables, you have seen that there are control characters—32 in ASCII and 27 in EBCDIC. These characters are used to control data transmission and the terminal devices connected to the lines. Some of the more common control characters and their meanings are as follows.

Transmission Control Characters

SOH Start of Header is used at the beginning of the message's header. The header is a series of characters that indicates the address of the receiver and/or routing information for the message

STX Start of Text is a control character that precedes the main body or text of a message

ETX End of Text is used to terminate the text or main body of a message

EOT End of Transmission is used to indicate the end of a transmission that may have contained more than one set of headings and text

ACK Acknowledge is a control code sent by the receiver of a transmission as a positive acknowledgment to the sender

NAK Negative Acknowledgment is transmitted by the receiver as a negative response to the sender

NUL The Null character in itself has no meaning. It is used to continue sending valid characters down the line for timing or other purposes. Note that the null character is not the same as the blank character. Blank has a different, unique binary code

Device Control Characters

BEL Bell is used to sound an audible alarm, turn on a light, or otherwise indicate a need for operator attention or intervention at the receiving end

DC1
DC2
DC3
DC4 These four device control characters are used in various ways by different communication systems. Most commonly, DC3 is also designated as the X-OFF character sent by a receiver to a transmitter when the receiver cannot accept any more data. When it is ready to receive again, the receiver sends the DC1 character, also known as X-ON

Format Effector Control Characters

CR Carriage Return indicates to the receiver that the printing element should be moved to the first position of the current print line

LF Line Feed indicates to the receiving device that it should move to the next print line on a printer or the next line on a display terminal

HT Horizontal Tabulation moves the print element to the next predetermined (tab) position on a print line or display screen

VT Vertical Tabulation moves the print element to the next predetermined vertical position on the print line or display screen

CODE EFFICIENCY

Code efficiency is a measure of how few bits are used to convey the meaning of a character accurately. Efficiency of coding is important because an efficient code minimizes the number of bits that are transmitted on expensive communication facilities. If only numeric data is to be transmitted, a 4-bit code would be significantly more efficient than a 7- or 8-bit code. All of the digits can be represented with 4 bits, and the total data transmission would occur twice as fast as if an 8-bit code were used. Most data communications systems transmit only alphanumeric data and can therefore easily get by with a 7-bit code.

Another aspect of coding efficiency is the number of bits in a character that actually convey information. Bits that are used to determine the code points in a code are *information bits*. Any other bits are called *noninformation bits*. Parity bits, for example, convey no additional information and are not a part of the original data. They are, therefore, noninformation bits.

Code efficiency is defined as the number of information bits divided by the total number of bits in a character. If an 8-bit code that includes one parity bit is being used, the code efficiency is calculated as $7/8$ = .875, or 87.5 percent. Codes that have more than one parity bit or in which only certain bit combinations are valid are less efficient than that. In addition, we will see that as data is moved to communications lines for transmission, other bits and even whole characters may be added for checking purposes. Because of checking that occurs during transmission (discussed in Chapter 9), the commonly used 7-bit version of the ASCII code and the EBCDIC code were designed with no inherent checking. They are dependent on other means to ensure that data is received accurately. Hence, both coding schemes have 100 percent code efficiency.

CODE CONVERSION

In most data communications systems, code conversion occurs from one coding system to another. Even though ASCII is the most widely used communications code, it is not universal. Although nearly all personal computers use ASCII for both internal and communications purposes, most large IBM data processing equipment stores data in EBCDIC. Probably the most common conversion today is from ASCII to EBCDIC and back, which must be done whenever personal computers, which operate in ASCII, are connected to mainframe computers, which operate in EBCDIC. Data must be converted twice: from ASCII to EBCDIC when it arrives at the mainframe and from EBCDIC to ASCII when it is sent to the personal computer.

Code conversion is conceptually quite simple and the type of task that a computer can perform readily. Usually, a table containing the target codes is stored in the memory of the computer. The binary value of the incoming character is used as an index into the table, and the target character is picked up as shown in Figure 7–7. The process gets more complicated when one code is converted to another code that uses a smaller number of bits. For example, when EBCDIC, with 256 unique characters, is converted to ASCII, with only 128 characters, many characters cannot be converted. Sometimes the ESC character must be brought into play so that one EBCDIC character translates into two ASCII characters—ESC and another character.

DATA COMPRESSION/COMPACTION

Compression or *compaction* is the process of reducing the number of bits used to represent a character, or shortening the number of characters before they are transmitted. The reasons for compressing data are to save storage space on the transmitting or receiving device or to save transmission time so that the message arrives faster and costs less to

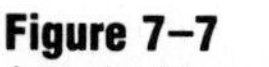

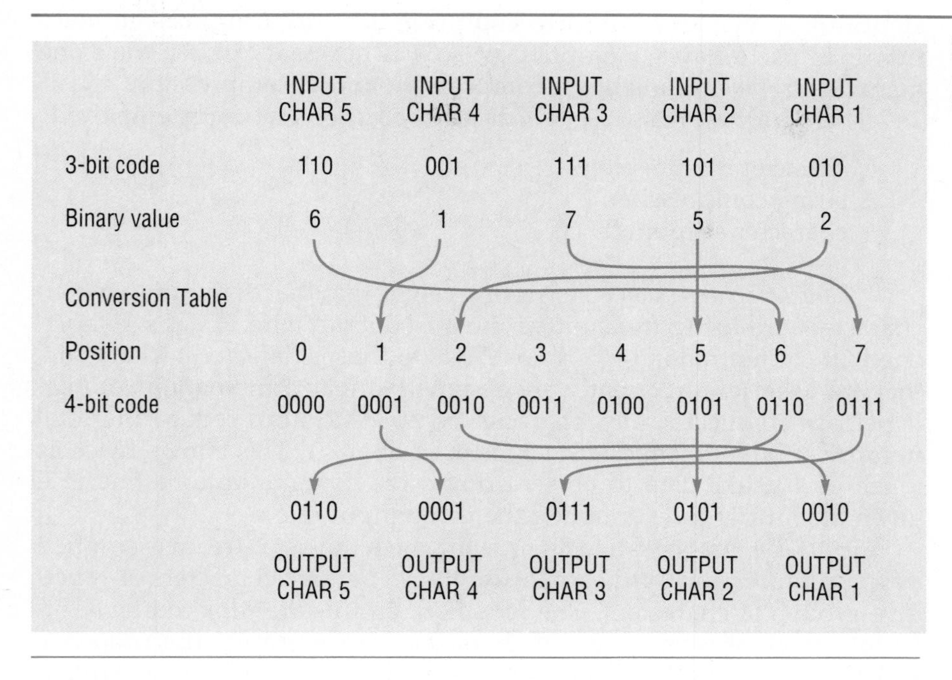

Figure 7–7
A method for converting a 3-bit code to a 4-bit code.

send. The results of successful compression are an apparent increase in transmission throughput and a reduction of storage or transmission cost.

In a typical application, a data compression device is employed at both ends of a communications line, as shown in Figure 7–8. At the transmitting end, the data is compressed using a set of mathematical rules called an *algorithm*, or a combination of algorithms. At the receiving end, a compatible device decompresses the data using the same

Figure 7–8
The locations of data compression devices on a telecommunications circuit.

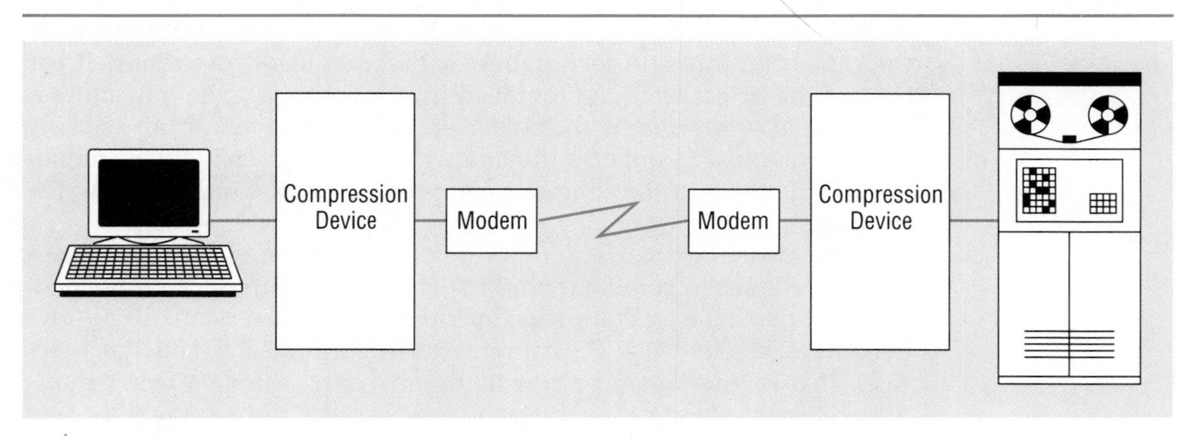

algorithm(s) that were used to compress it. Most compression algorithms in use today are proprietary, so it is necessary to use the same manufacturer's equipment for compression and decompression.

There are three major types of data compression that can be employed:

- character compression;
- string compression;
- character stripping.

character compression

Character compression consists of an algorithm that determines which characters are being transmitted most frequently and assigns a shortened bit configuration of perhaps 2 bits to those characters. Characters that are used less frequently are assigned longer combinations of bits. If performed successfully, character compression can reduce the total number of bits transmitted by a factor of about 2. This allows twice as much of a user's data to be sent down the communications line in a given amount of time doubling the throughput rate.

string compression

String compression is a technique in which the data stream is scanned by the compression hardware looking for repetitive characters or repetitive groups of characters. The repetitive groups are replaced by a different, much shorter group of characters. For example, the sequence XXXXXXXXXX might become: *control character, 1, 0, X*. This is shorthand notation that says that the letter X is to be repeated 10 times. This shorthand reduces the original 10 characters to 4, which is only 40 percent of the original number of characters. If a file of names and addresses was being transmitted and many of the addresses contained the same city name, many characters and much transmission time would be saved by substituting a 1- or 2-character sequence for the city name every time it appeared in the file. The shortened form would be transmitted, and at the receiving end, the special character sequence would be replaced by the original word.

character stripping

Character stripping removes the leading and trailing control characters from a message and adds them back at the receiving end. Although in itself this technique may seem rudimentary, when combined with the other compression techniques, it becomes very important. If the control characters were not removed from a message to be transmitted, they would be viewed by the character compression algorithm as being among the most frequently transmitted characters. This would reduce the effectiveness of the character compression on the main body of the message.

The most effective data compression devices use all three compression techniques in combination. First character stripping is performed, followed by character compression and string compression. To achieve maximum effectiveness, the compression algorithms must also be adaptive. That is, they must be constantly analyzing the data being transmitted and updating their internal tables of the most frequently sent characters or groups in order to assign the fewest bits to them.

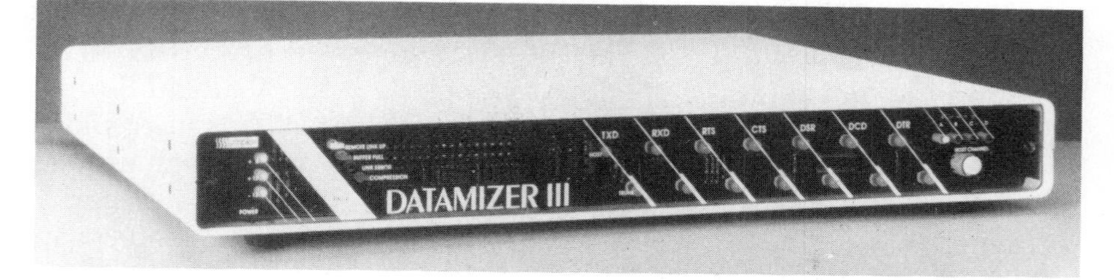

This Datamizer III data compression unit from Symplex is a sophisticated device that will compress data up to four times. (Courtesy of Symplex, Inc.)

Good data compression algorithms can reduce the data transmitted in many business applications by a factor of 4:1. The resulting savings in the cost of storage or transmission time must be weighed against the cost of the data compression equipment. For some companies, the most important benefit of data compression is the higher throughput rate, which may translate into improved response time for the users of an interactive application.

ENCRYPTION

Another reason for manipulating the data stream before transmission is to encrypt the data to keep it private or secret. For many years, encryption was rarely used except by the government, but it is now gaining wider use in industry as the concern for data security grows. Financial institutions are also big users of encryption techniques to conceal account numbers and amounts in the data they transmit.

Encryption is the transformation of the data from the meaningful code that is normally transmitted to a meaningless sequence of digits and letters that must be decrypted before it becomes meaningful again. Simple encryption schemes use character substitutions. For example, a *B* may be substituted for an *A*, a *C* for a *B*, and so forth. Unfortunately, human analysts and computers can rapidly break this type of code. Therefore, more sophisticated techniques, which take advantage of the fact that machine codes are binary digits and can be manipulated mathematically, are used. The mathematical manipulation of bits is especially feasible when computers can be used to do the transformation.

Encryption is of particular interest in data communications because of the relative vulnerability of transmission over telephone lines. The environment is different from the computer room where data is transferred over cables that run a short distance and are often heavily shielded. Data transmitted using communications lines travels relatively long distances on various media and is essentially unprotected. Voice and data transmission are both subject to tapping or other unauthorized reception.

Modern encryption technology uses a mathematical algorithm and a key that is provided by the user. The encryption algorithm may be in the public domain, but the encrypted data can still be private because of the key. The most widely used algorithm in the United States is the *data encryption standard (DES)* algorithm developed in the mid-1970s by IBM and the federal government and adopted by the National Bureau of Standards and Technology (formerly the National Bureau of Standards). The DES is approved for the government for encrypting unclassified data, at least until January 1992. In 1988, the National Security Agency (NSA) stopped using DES for classified security needs and began using new encrypting algorithms that NSA controls. The new algorithms were developed under the Commercial Communications Security Endorsement Program and carry a secret classification.

data encryption standard

The DES algorithm encrypts blocks of 64 bits using a 64-bit key. Of the 64 bits in the key, only 56 are used, but this still yields 2^{56} or more than 72 quadrillion possibilities! The output of the algorithm is a string of random bits that are transmitted. At the receiving end, the reverse process, *decryption*, occurs, using the same 64-bit key.

decryption

Since the receiving end must know which key the transmitting end used, methods must be put in place to get the key from the transmitting end to the receiving end and protect its confidentiality. This takes a combination of technical and management techniques. In many situations, managing the keys is a high cost overhead to the encryption. For example, couriers may have to be employed to deliver the keys, which are stored on magnetic tape. The couriers have to be trusted, and the tapes with the keys may need to be stored in secure boxes. The possibility that the keys could be disclosed is high. Technologies are becoming available that alleviate the situation by allowing keys to be transmitted electronically or by employing combinations of keys.

Encryption and decryption can be performed by hardware or software. The trade-off is that hardware is faster, but software is more flexible and easier to change. Hardware circuit chips that implement the DES algorithm are available and can be put into computers or terminals to implement a hardware solution. Software is also available to perform the encryption function. Its speed is acceptable in applications when only moderate amounts of data have to be encrypted.

In the voice world, we can identify a person we are talking to by the sound of his or her voice. In most situations, this is adequate. Voice encryption devices, commonly called *scramblers*, are available for voice transmissions. They make the voice transmission unintelligible to anyone without a descrambler, effectively rendering wiretapping useless. Scramblers are used to some extent in the government and defense department but are not widely used in industry.

scramblers

The decision to encrypt data must be made carefully. The cost of the encryption hardware or software can be determined easily. The time that it takes a computer to encrypt or decrypt the data if software is used can be calculated and a monetary value placed on it. One must also calculate the throughput delays that occur during the encryption/decryption process and determine whether they are significant. Finally,

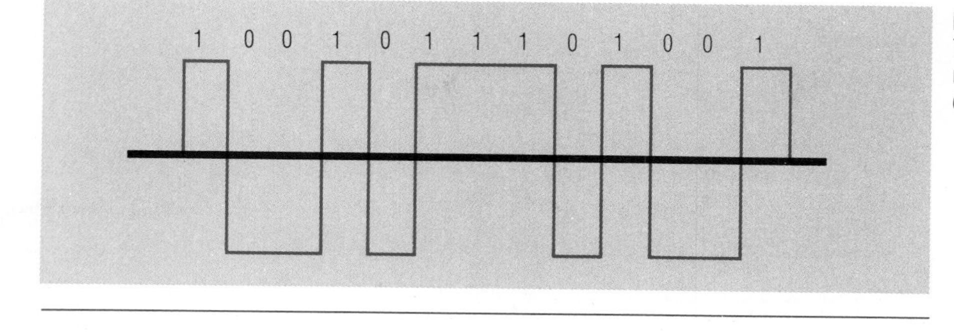

Figure 7–9
The pulses of a digital signal. Each pulse represents one bit of data.

the administrative or management costs of managing the keys, keeping them secure, and changing them regularly must be considered.

DIGITIZATION

Once data has been coded, it is in a form that is easily transmitted through communications systems. The binary representation of the data can be thought of as a series of 1 bits, represented by positive pulses on a communications line, and 0 bits, represented by negative pulses, as shown in Figure 7–9. This is called a *digital signal*. Clearly this digital pulse stream is much simpler than the analog wave form of the typical voice signal, and therefore transmitting it ought to be simpler. Unfortunately, the world's telecommunications networks were originally designed to handle analog signals. Therefore, in most cases the digital pulse stream must be converted to an analog form that is compatible with such networks. That conversion is the subject of Chapter 8.

Although most communications today are analog, there are advantages to transmitting digitally. These advantages include

- higher line capacity;
- better use of existing media;
- better quality voice transmission;
- less peripheral equipment;
- less expensive to install and maintain;
- integration of voice, data and image transmission;
- makes data communications easier.

In the future the pendulum will swing. There will be more digital communications, and the major conversion effort will be to change analog signals to digital form. The techniques to convert analog signals to digital signals have been used by the telephone companies internally for years. These techniques are now being more widely used as digital transmission capability extends to user premises.

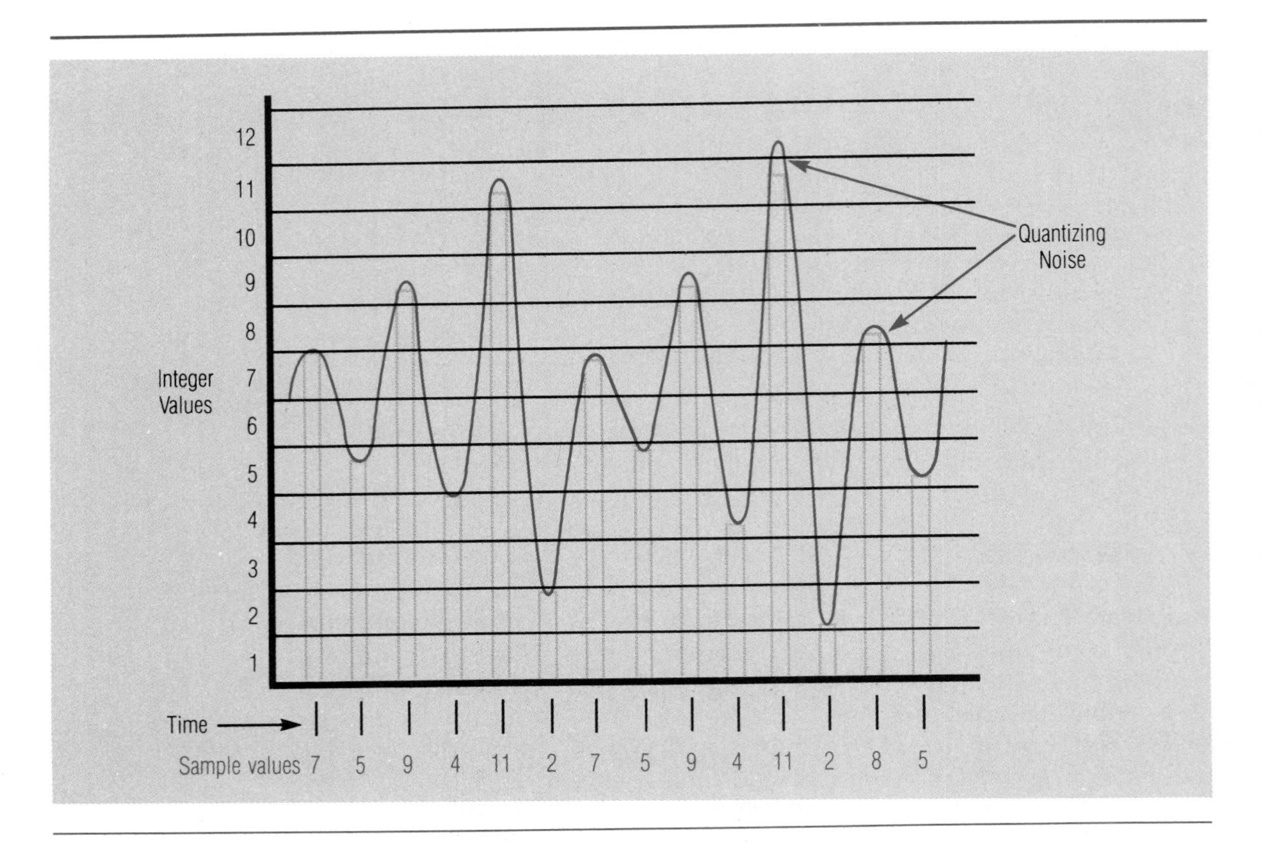

Figure 7–10
Quantization of an analog voice signal.

DIGITIZING ANALOG SIGNALS

You saw in Chapter 5 that when a user speaks into the microphone of a telephone handset, the resulting signal takes the shape of a continuously varying wave—an analog signal. If we look at the wave electronically and measure its height (voltage) at specific points in time, we obtain a series of voltages with numeric values. These values can be represented in binary form and transmitted as a series of bits. This is the way an analog signal is *digitized.*

A/D converter

Figure 7–10 shows that at every unit of time the height of the curve is measured by a special instrument called an *analog-to-digital (A/D) converter*. An A/D converter is essentially a digital voltmeter that can take many readings per second. Instead of reading actual voltages, however, the A/D converter uses a scale of integer values. In effect, this scale of integers is superimposed over the voltages so that the different heights of the curve (voltages) can be represented as integers. This is important because integers can be converted accurately to binary numbers, whereas fractional numbers cannot. The process of approximating the actual analog signal value to the predetermined integer steps is called *quantization*. It is similar to rounding. The difference between the exact height of the curve when the sample is taken and the nearest integer value is called *quantizing noise* or *digitizing distortion*.

Although the diagram in Figure 7–10 only shows 12 values, there are typically 256 different integer values that can be obtained each time a sample is taken. If there are 256 different values, the integer is converted to an 8 bit binary number; there is no room for a parity bit. The 8-bits are then transmitted over the communication line. At the next unit of time, another sample is taken, and the process is repeated.

At the receiving end, the process is reversed. The bit stream is divided into 8-bit groups. The groups are interpreted as 8-bit binary numbers, are converted to voltages by a *digital-to-analog (D/A) converter*, and the original wave form is reproduced. The more frequently that samples are taken, the more accurately the original wave form can be reproduced.

The existing worldwide standard for digitizing voice signals is called *pulse code modulation (PCM)*. PCM uses 256 integer values and samples the signal 8,000 times per second. Since 8 bits are used for each sample, the effective data rate is 8,000 times 8, or 64,000 *bits per second (bps)*. This technique for digitizing voice has been used by the telephone companies internally since the early 1960s, especially on long distance communications lines.

pulse code modulation

Recent developments have shown that good voice quality can be maintained at a lower sample rate of 4,000 samples per second. This means that the effective data rate is lowered to 4,000 times 8 bits or 32,000 bps. As the sample rate is slowed, the effect is similar to compressing the bandwidth on an analog signal. If the sample rate is slowed to say 2,000 or 1,000 samples per second, voice quality is lost, and the reconstructed voice at the receiving end loses its distinguishing characteristics, though it is still understandable. At even lower sample rates, the reconstructed voice is unintelligible.

A more common technique for reducing the number of bits that must be transmitted is to keep the sample rate at 8,000 per second but reduce the number of bits transmitted for each sample. Since voice signals do not change amplitude very rapidly, adjacent samples usually are not very different in value from one another. Thus, if only the "difference" in the integer value of the samples is sent, it can be represented with 4 bits. This is illustrated in Figure 7–11 and is called *adaptive differential pulse code modulation (ADPCM)*. ADPCM was adopted by the CCITT in 1985 as the recommended method for digitizing voice at 32,000 bps. ADPCM may become the next worldwide standard for digitizing voice because it is similar to PCM.

adaptive differential pulse code modulation

Another method of digitizing voice is called *delta modulation*. Delta modulation compares the analog signal level to the last sample taken. If the new value is greater than the previous sample, a 1 bit is sent; if

delta modulation

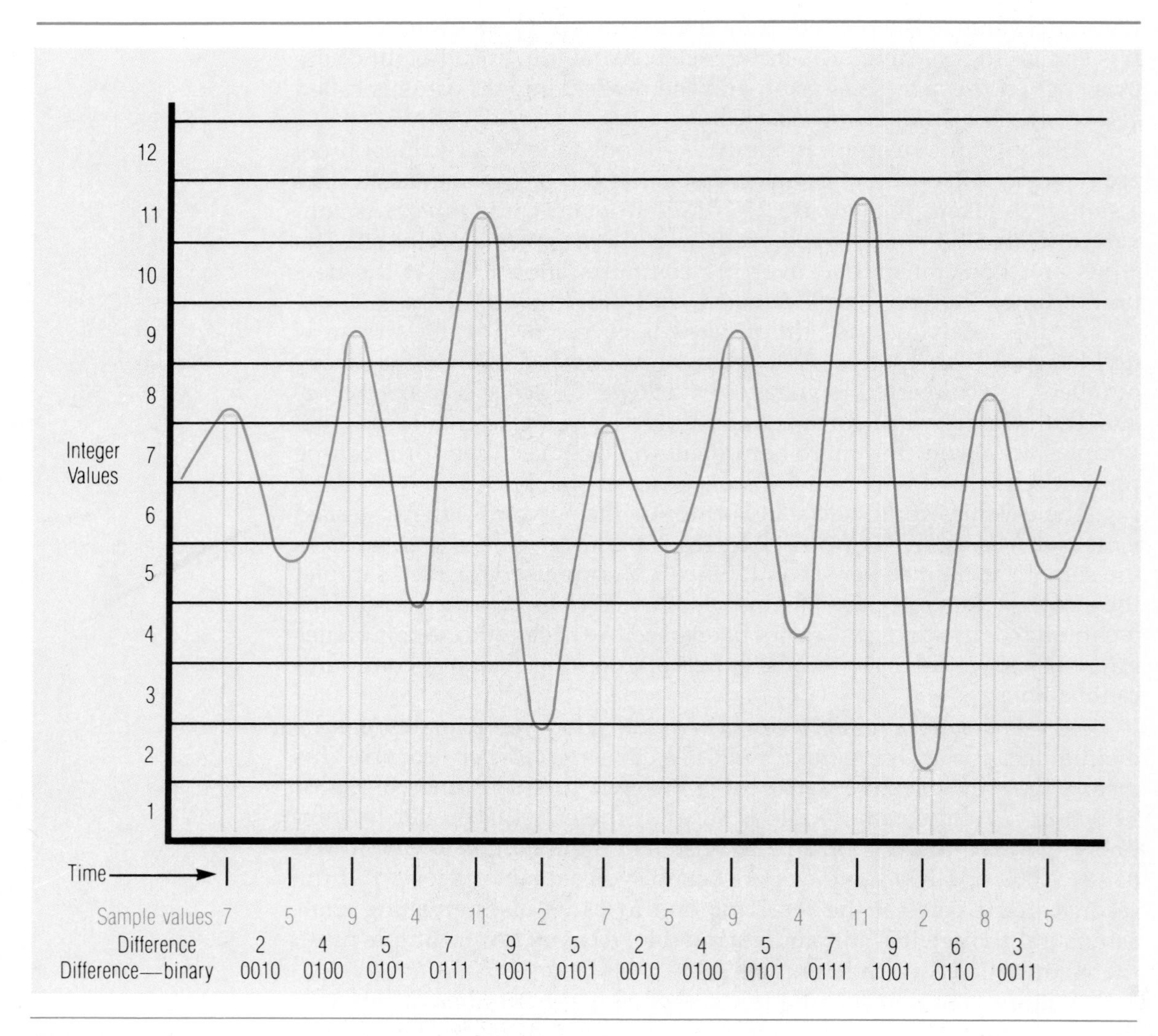

Figure 7–11
ADPCM codes the difference in signal strength in 4 bits each time a sample is taken.

the new value is less than the previous sample, a 0 bit is sent. To maintain good voice quality, delta modulation samples the analog signal 32,000 times per second. Since one bit is sent for each sample, the data rate is 32,000 bps, the same as ADPCM.

linear predictive coding

Research is continuing in ways to reduce the data transmission rate required to send digitized voice signals while still maintaining acceptable quality. One new technique is called *linear predictive coding (LPC)*. LPC samples less often than the other techniques and predicts the direction that the analog signal will take. Using LPC, AT&T Bell Labs can demonstrate high-quality voice at 9,600 bps. Rates as low as 2,400 bps can be used, but the voice quality suffers. Researchers believe that a new generation of electronic circuit chips will make the 2,400 bps voice quality acceptable.

Recently, analog-to-digital converters that are small and cheap enough to fit in a telephone handset have become available so that it is technically possible and affordable to transmit digitized voice directly from the telephone. Today, this capability is found frequently in private PBX systems.

When analog voice data is digitized, the resulting bit stream is the same as a bit stream of coded data. If someone were monitoring a digital communications line and knew nothing about what was being transmitted, it would be impossible to tell whether the bits on the line represented digitized voice, coded data, or an image from a digital facsimile machine. In fact, if digital multiplexing equipment had been employed, the bit stream could be a combination of all three. It is, therefore, important that the receiver of a digital transmission have some information about how to interpret the incoming bits.

SUMMARY

This chapter looked at various alternatives for coding information. For machine communications, a binary code in which each character contains the same number of elements is desirable. The Baudot code was one of the earliest codes designed for machine-to-machine communication, but it was limited by a small number of unique combinations because it was only a 5-bit code. Most common today are the ASCII and EBCDIC codes, which contain 7 and 8 bits, respectively. Although ASCII is the most widely used code, EBCDIC is important because it is used by IBM in most of its products.

Compaction and encryption are ways in which streams of coded characters are manipulated before being transmitted in order to eliminate redundancy or to provide privacy. The technique of digitizing analog signals is also growing in popularity. Of significance is the fact that digitized voice signals look the same as any other digital signal and can be handled by similar equipment.

Review Questions

1. Describe several attributes that are important in a code to be used for machine-to-machine transmission.
2. What is a *code point*?

3. What are the advantages and disadvantages of an 8-bit code compared to a 5-bit code?
4. What is the minimum number of bits that would be needed to encode only alphabetic data? How many bits are needed if numeric data also is included?
5. Name the three different groups of character assignments that are made in a coding system.
6. Using the ASCII table in the text, convert your name to bits, 7-bits per character. Use upper and lowercase characters as you would to write your name. Now, assuming even parity, add the parity bit to each coded character in your name.
7. Explain the meaning of the following control characters:
   SOH STX EOT ACK NAK CR LF
8. A certain coding scheme contains 5 information bits and 2 noninformation bits for each character. Calculate the code efficiency of this code.
9. Describe three different kinds of data compression techniques.
10. Explain the trade-offs between encrypting data using hardware and software.
11. Describe the advantages of transmitting data digitally.
12. Explain why, using PCM, it takes 64,000 bps to transmit a voice signal. How does ADPCM cut the data rate in half?
13. Describe the function of an A/D converter.
14. Explain quantization.
15. Why is there interest in transmitting digital voice signals at slower data rates?

Problems and Projects

1. Amateur radio operators have modified computers to automatically send Morse code when a key on the keyboard of the computer is pressed. Describe some of the difficulties that would be encountered by a computer at the receiving end in interpreting the transmission for display on the screen.
2. Not very many words in any language have long strings of repetitive characters in them. When then is data compression most useful?
3. Think of some applications where voice quality does not have to be as good as it is on the telephone system in order to be acceptable.
4. Develop a simple encryption technique to encrypt and decrypt alphanumeric data. Using your technique, encrypt this problem. If you have computer experience, write a program to encrypt data using data using the algorithm you developed. Compare the speed of your program to performing the encryption manually.

Vocabulary

- binary digit
- bit
- code
- code points
- character assignments
- alphanumeric characters
- format effector characters
- control characters
- graphic characters
- parity bit
- escape mechanism
- Baudot code
- letters shift
- figures shift
- American Standard Code for Information Interchange (ASCII)
- Extended Binary Coded Decimal Interchange Code (EBCDIC)
- Binary Coded Decimal (BCD) code
- code efficiency
- information bits
- noninformation bits
- compaction
- compression
- character compression
- string compression
- character stripping
- encryption
- algorithm
- data encryption standard (DES)
- decryption
- scrambler
- digital signal
- digitized
- analog-to-digital (A/D) converter
- quantization
- quantizing noise
- digitizing distortion
- digital-to-analog (D/A) converter
- pulse code modulation (PCM)
- bits per second (bps)
- adaptive differential pulse code modulation (ADPCM)
- delta modulation
- linear predictive coding (LPC)

References

Brostoff, George. "Utilizing Data Compression for Network Optimization," *Telecommunications,* June 1988, p. 64.

Davenport, William P. *Modern Data Communication: Concepts, Language, and Media.* Rochelle Park, NY: Hayden Book Company, Inc., 1971.

Falk, Howard. *Microcomputer Communications in Business.* Radnor, PA: Chilton Book Company, 1984.

Friend, George E., John L. Fike, H. Charles Baker, and John C. Bellamy. *Understanding Data Communications.* Dallas: Texas Instruments, 1984.

Gurrie, Michael L., and Patrick J. O'Connor. *Voice/Data Telecommunications Systems.* Englewood Cliffs, NJ: Prentice-Hall, Inc., 1986.

Kerr, Susan. "A Secret No More," *Datamation,* July 1, 1989, p. 53.

Leddy, Donald, and Lori Rubin. "Here's How Pulse Code Modulation (PCM) Works in Converting Analog Voice Signals to Digital," *Communications News,* October 1986, p. 59.

Powell, Dave. "The Hidden Benefits of Data Compression," *Networking Management,* October 1989, p. 46.

Schindler, Paul E. Jr. "AT&T Bell Labs Technique Can Digitize Voice Signals," *Information Week,* August 19, 1985, p. 23.

Wlosinski, Larry G. "A Fundamental Approach to the Basics of Networking," *Data Management,* September 1986, p. 13.

CHAPTER

Data Transmission and Modems

OBJECTIVES

After you complete your study of this chapter, you should be able to

- describe what is meant by the signaling rate of a circuit;
- describe the speed of a circuit measured in bits per second;
- describe three modes of data transmission;
- distinguish between asynchronous and synchronous transmission;
- explain how a modem works;
- describe several types of modulation used by modems;
- describe several types of interface between a terminal and the modem;
- describe several different types of modems and modem features.

INTRODUCTION

In this chapter, the way in which a terminal is connected to a communications circuit through a modem is described. In order to gain the proper depth of understanding about modems and their capabilities, however, you'll look first at the different modes of communication. This is necessary because modems are designed with specific capabilities, depending on the mode of data transmission that will be used and the speed at which the data is to be transmitted. In this chapter, continue thinking of a communications "circuit" and "line" as being the same. Although this is not technically true (and it will be clarified in Chapter 9), the distinction is not important here.

The material in this chapter and in Chapter 9 describes the first layer of the ISO-OSI model that was introduced in Chapter 4. Layer 1, the physical layer, provides the physical path through which communications signals flow. In this chapter, you will study the way in which signals are converted and sent on the communications line. Although the material is somewhat technical, many basics of data transmission are introduced and explained. With the understanding of data transmission from this material, you will be well prepared for the material about data circuits, protocols, and networks presented in later chapters.

CIRCUIT SIGNALING RATE

Chapter 5 explained the concept of an analog circuit's bandwidth, which is the difference between the highest and lowest frequencies that the circuit can carry. You also saw that the standard telephone circuit has a bandwidth of 4,000 Hz. In 1928, Harry Nyquist of Bell Labs showed that the maximum signaling rate that can be achieved on a noiseless communication channel is 2B, where B is the bandwidth measured in Hz. The *signaling rate* is defined as the number of times per second that the signal on the circuit changes, whether in amplitude, frequency, or phase. The signaling rate is measured in *baud*. Thus, if the frequency, amplitude, or phase of a signal is changed 600 times per second, it is said to be signaling at 600 baud.

baud

Nyquist's work suggests that on a circuit with a 4,000 Hz bandwidth, the maximum theoretical achievable signaling rate is 2 × 4,000 Hz, or 8,000 baud. In practice, the signaling rate that actually can be achieved is significantly less than the theoretical maximum because in the real world, noise and other transmission impairments occur on every circuit. Furthermore, the time available to detect the signal changes at the receiving end becomes very small as the speed increases. At 2,400 baud, for example, the signal changes 2,400 times per second; therefore, the receiver must detect the signal change in 1/2400 of a second, or 416.5 microseconds.

CIRCUIT SPEED

Circuit speed is defined as the number of bits that a circuit can carry in one second. It is measured in *bits per second (bps)*, a term introduced in Chapter 7. The abbreviation bps is often incorrectly used interchangeably with the term *baud*. If only one bit is sent with each signal change on the circuit, the baud rate and the bps rate are the same, assuming all of the bits are of the same length. Today, however, sophisticated techniques allow more information to be encoded in each signal change. Therefore, the baud rate and bps rate of most data transmissions normally are quite different.

bits per second

Suppose there were four unique signal changes on a circuit—that is, four unique changes of amplitude, frequency, or phase. With four possibilities, each of the four signal changes could represent 2 bits of information. For example, if a circuit could transmit four unique frequencies, one frequency could represent the bit combination 00, a second could represent 01, the third could represent 10, and the fourth frequency could represent the bit combination 11. This is shown in Figure 8–1. When 2 bits of information are coded into one signal change, they are called *dibits*.

Frequency	Dibit
f1	00
f2	01
f3	10
f4	11

Frequency	Tribit
f1	000
f2	001
f3	010
f4	011
f5	100
f6	101
f7	110
f8	111

Figure 8–1
If four different frequencies are used, each can represent 2 bits ($2^2 = 4$). If eight frequencies are used, each can represent 3 bits ($2^3 = 8$).

What if eight different frequencies were used? Then each frequency could represent 3 bits, called *tribits*, also shown in Figure 8–1. Using such techniques, dibits, tribits, and even *quadbits* can be encoded in each signaling change. Thus, circuit speeds can be achieved, as measured in bps, which are several times greater than the circuit's signaling rate as measured in bauds. When the signaling rate is 2,400 baud, it is possible to send 2,400, 4,800, 7,200, or even 9,600 bps, depending on how many bits are coded in one baud.

On a data communications circuit, the DTE at both ends of the circuit must send and receive bits at the same rate. Therefore, the bit rate of the circuit is the maximum bit rate that can be sent between the equipment. Of course, if the DTE is not generating data at the speed the circuit can handle, the actual bit rate sent will be less.

We have seen that bauds and bps are units of measure for different characteristics of a communication circuit. It is surprising that many people who work in the communications industry do not understand the difference between these two concepts and think that the terms are interchangeable. In fact, the only time that the baud rate and bps rate of a circuit are the same is when each signal change on the circuit indicates 1 bit. Although this is often the case in slow-speed transmissions, it is still incorrect to use the terms interchangeably.

MODES OF TRANSMISSION

There are many ways in which data transmissions can be classified. Three important ways are according to the

- data flow;
- type of physical connection;
- timing.

We now look at each of these in detail.

Data Flow

Simplex Transmission *Simplex transmission,* as shown in Figure 8–2, is data transmission in only one direction on a communications line. No transmission in the opposite direction is possible. Although simplex transmission is not what we usually first think of when we discuss data transmission, it is more common than you might imagine. In businesses, simplex transmission is used for monitors and alarms where a signal is sent from a sensor back to a central monitoring or control point. A similar application occurs in hospitals, where sensors on the patients send signals to the nurses' stations. Television and radio are other examples of simplex transmission.

Half-Duplex Transmission *Half-duplex transmission (HDX)* is transmission in either direction on a circuit but only in one direction at a time. It is the most common form of data transmission, and it is commonly used in data processing applications where, for example, an inquiry is sent to the computer and then a response is sent back on the same circuit to the terminal. CB radio is another example of half-duplex transmission.

Full-Duplex Transmission *Full-duplex transmission (FDX)* is data transmission in both directions simultaneously on the circuit. It requires more intelligence at both ends of the circuit to keep track of the two data streams. One place where FDX transmission is used is between computers. Computers have the intelligence and the speed to perform the

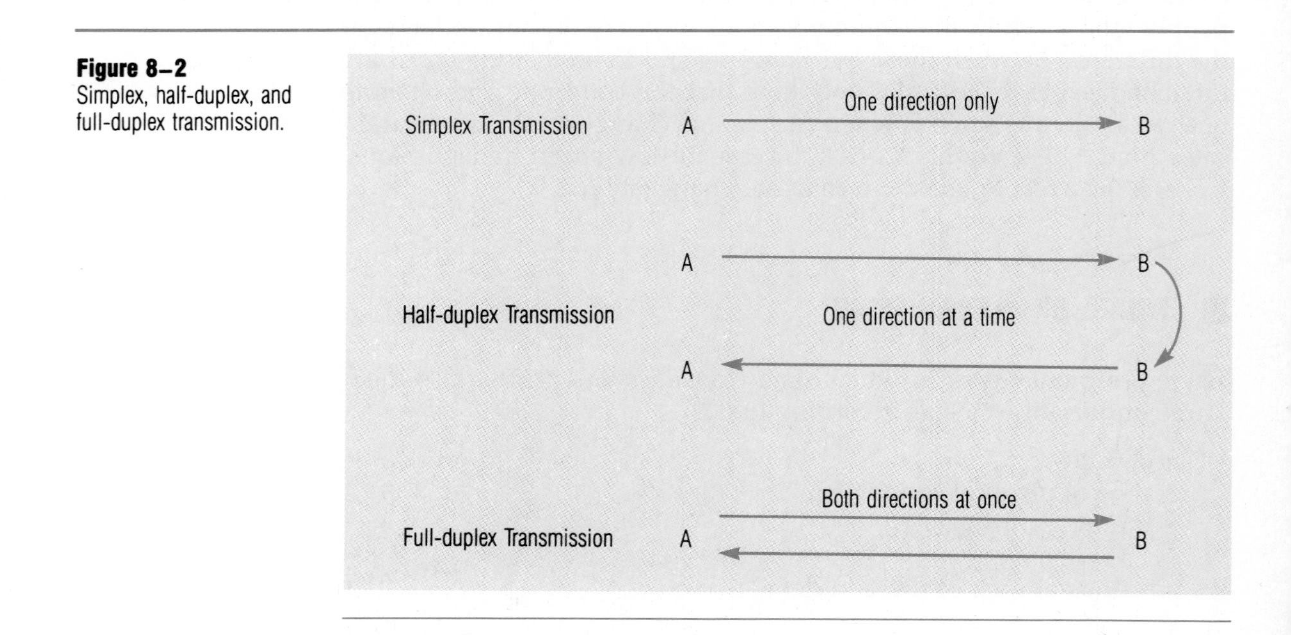

Figure 8–2
Simplex, half-duplex, and full-duplex transmission.

necessary line control functions and to take advantage of the speed that FDX transmission offers. As circuit chips get smaller and less expensive, however, other communications equipment is becoming more capable of handling the complexities of full-duplex transmission, and its throughput advantages are now starting to be realized more widely.

Type of Physical Connection

Parallel Transmission When terminals are in close proximity to the computer, they often are connected via a direct cable that has one wire for each bit in a character of the data code being used by the terminal. That is, if a 7-bit coding scheme, such as ASCII, is being used, the cable would have seven wires, as shown in Figure 8–3. There also would be several additional wires for timing, checking, and control. With multiple wires, all the bits of a character can be transferred between the terminals and computer at once. This is called *parallel transmission*. It is extremely fast, but because of the number of wires involved, it is also expensive and not practical over long distances.

Serial Transmission As distance increases, *serial transmission*, in which the bits of each character are sent down a communications line one after another, or serially, is used. This type of transmission is illustrated in Figure 8–3 and is used for all telecommunications applications we will

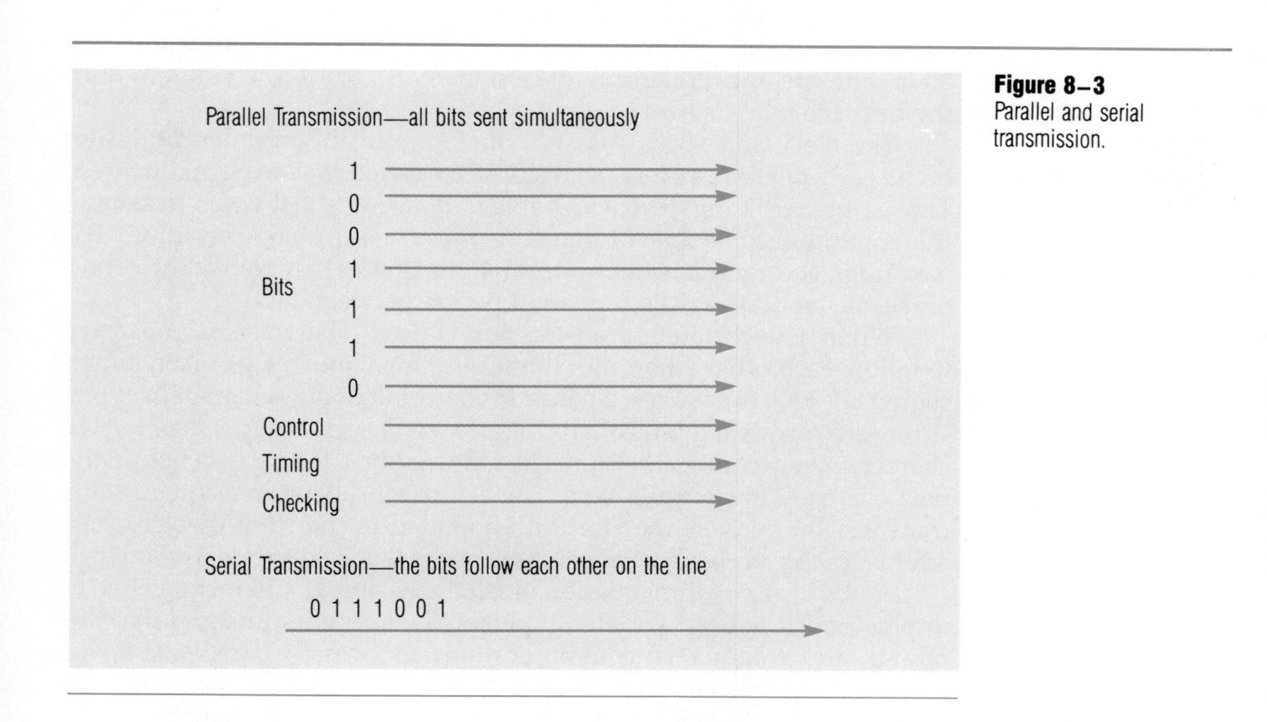

Figure 8–3 Parallel and serial transmission.

study. There is a trade-off, however. With serial transmission, the transmitter and receiver are more complicated because they have to decompose a character and serialize the bits for transmission and reconstruct the bits into a character at the receiving end.

Timing

Asynchronous Transmission *Asynchronous transmission,* sometimes abbreviated *asynch,* is a transmission technique in which each character sent on a communications line is preceded by an extra bit called a *start bit* and followed by one or more extra bits called *stop bits.* It is sometimes called *start/stop transmission.*

start bits and stop bits

Asynchronous transmission originated with the early mechanical teletypewriter terminals. For those teletypewriters, the start bit gave the mechanism in the receiving terminal time to start rotating and get up to speed and ready to receive the rest of the bits that made up the character. The stop bit gave the rotating mechanism time to get back to a known position and ready to receive the next character. Long ago, the convention was established that an idle line is one on which the signal for a 1 bit, also called the *mark signal,* is being sent. The opposite condition, the signal for a zero bit, is called the *space.*

mark and space signals

To start the mechanism of the receiver, the line is brought from the idle, or mark, state to the 0, or space, state for one bit time, thus creating the start bit. For the next 7 bit times (when a 7-bit code is being used), the signal is changed between the 0 and 1 bits to represent the character being transmitted. Then the stop bit(s), a 1 bit, is sent. After the stop bit is sent, the line remains in idle mode (mark, or 1-bit, condition) until the next character is sent.

The mark and space signals, coupled with the mechanical design of the early teletypewriters, allowed *bit synchronization* to be maintained. That is, the receiving device was able to determine just when to sample the communication line to detect a pulse or no-pulse condition. For maximum accuracy, it is important that the line be sampled at the center of the bit, not during the transition period between bits.

bit synchronization

When asynchronous transmission is used, characters do not have to follow each other along the communications line in a precisely timed sequence. This is easy to visualize if you think about an amateur typist sitting at a terminal typing characters somewhat erratically. Ten or twelve characters may come rather quickly, followed by a long pause before the next character or group is sent. This is not a problem in asynchronous transmission because the start bit appended to the front of each character gives the receiving terminal some notice that a character is following.

Asynchronous transmission is relatively simple and inexpensive to implement. It is used widely by personal computers, inexpensive terminals, and commercial communications services. A penalty in terms of *transmission efficiency* is paid, however, because at least 2 extra bits are added to each character transmitted. The exact penalty depends on the number of bits that make up the character. The table below shows the transmission efficiency calculation when the Baudot, ASCII, and EBCDIC codes are used.

Code	Number of bits in code	Start/ Stop bits	Total bits	Transmission efficiency
Baudot	5	2	7	5/7 or 71%
ASCII	7	2	9	7/9 or 77.7%
EBCDIC	8	2	10	8/10 or 80%

When the ASCII code is transmitted with an extra parity bit added for checking purposes, its transmission efficiency is 8/10 or 80 percent, like that of EBCDIC.

Asynchronous transmission does not require maintenance of precise synchronization between the transmitter and receiver for an extended period of time. When a start bit is sensed, the receiver knows that the next *n* bits (where *n* depends on the code being used) on the line make up a character. This is called *character synchronization*. After receiving a stop bit, the receiver simply waits for next start bit. The only synchronization that has to occur is during the transmission of the 5 to 8 bits that make up a character so that the receiver looks at or samples the line at the right time. The sampling rate depends on the line speed, a rate that is predetermined and known by both the transmitter and receiving terminals. Without some form of character synchronization, a receiver might not know which was the first bit of a character, and the character would be misinterpreted.

character synchronization

Asynchronous transmission is used where equipment cost must be kept low. The transmission inefficiencies of asynchronous transmission were a big disadvantage when line speeds were primarily 300 bps or less. Now that data can be transmitted asynchronously at up to 19,200 bps, concern about the inefficiency is heard less frequently.

Synchronous Transmission The asynchronous transmission method is adequate for slow-speed transmission, but it is relatively inefficient because of the extra start and stop bits transmitted with each character. For higher speeds, a more efficient technique is desirable. This need is satisfied by the *synchronous transmission* technique. When synchronous transmission is used, bit synchronization is maintained by clock circuitry in the transmitter and in the receiver. The timing generated by the transmitter's clock is sent along with the data so that the receiver can keep its clock synchronized with that of the transmitter throughout a long transmission.

With synchronous transmission, data characters usually are sent in large groups called *blocks*. The blocks contain special synchronization characters, with a unique bit pattern. They are inserted at the beginning and sometimes in the middle of each block. The *synchronization characters* perform a function similar to that of the start bit in asynchronous transmission. When the receiver sees the synchronization character, it knows that the next bit will be the first bit of a character, thus maintaining character synchronization.

Synchronous communication has one to four synchronizing characters for each block of data. Asynchronous communication has two synchronizing bits for each character. If the ASCII code is used and 250 characters are to be sent, the number of bits sent by each transmission method is:

Asynchronous

250 characters × (7 data + 2 start/stop bits per character) = 2,250 bits

Synchronous

(250 data characters + 4 synchronizing characters) × 7 bits per character = 1,778 bits

In this example, the synchronous transmission technique sends 21 percent fewer bits. It is, therefore, 21 percent more efficient than asynchronous communication in this example. The efficiency advantage of synchronous transmission improves as longer blocks of data are transmitted.

Because it has a higher transmission efficiency, especially when large blocks of data are being transmitted, synchronous communication is preferred over asynchronous data transmission in many business applications. Transmission between computers is almost exclusively done synchronously. The transmissions of RJE terminals, sending large amounts of input or receiving printed reports, is another typical synchronous application. Many interactive display terminals, such as the IBM 3270 series, also use synchronous transmission.

DIGITAL SIGNALS

Chapter 5 looked at the characteristics of a typical analog signal, one generated by a human voice speaking into a telephone. That type of signal has a very complex wave form. In contrast, the signal generated by a terminal or other DTE is very simple, since it is made up of pulses for the binary digits 1 and 0, which represent the coded data to be transmitted. This type of signal is called a *digital signal*.

There are several forms of digital signals, as shown in Figure 8–4. *Unipolar* signals are those in which a 1 bit is represented by a positive voltage pulse and a 0 bit by no voltage. *Bipolar, nonreturn-to-zero (NRZ)* signals have the 1 bits represented by a positive voltage and the 0 bits represented by a negative voltage. *Bipolar, return-to-zero* signals are similar to NRZ signals, but the pulses are shorter, and the voltage always returns to 0 between pulses. Unipolar signals rarely are used today. Bipolar signaling has the clear advantage of making the distinction between the 0 bit and a no-signal condition, a distinction useful in troubleshooting when problems occur.

When a digital signal can be transmitted directly on a circuit designed for digital transmission, a relatively simple connection between the DTE

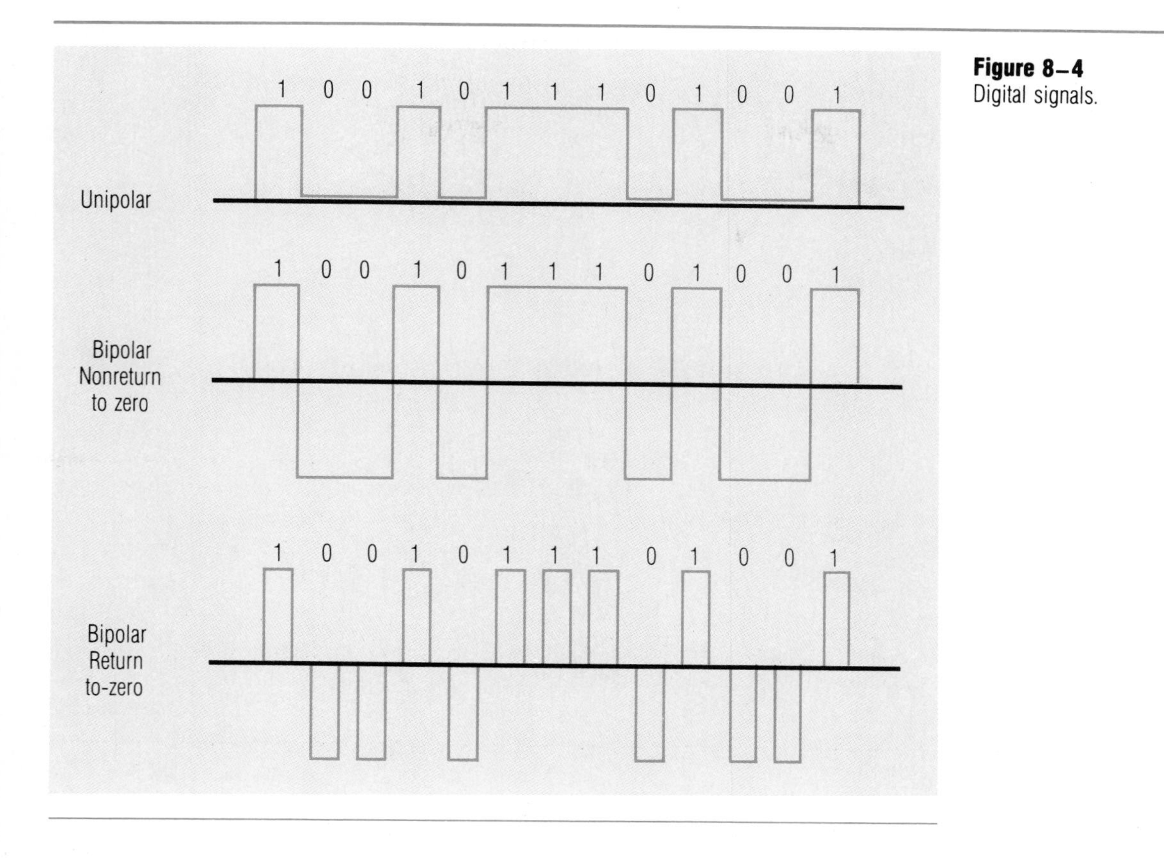

Figure 8–4
Digital signals.

and the circuit can be used. However, most of the available communications transmission facilities in place today were originally designed for transmitting analog voice signals. Therefore, before transmitting on an analog line, the user must insert a digital-to-analog signal converter between the DTE and the line to change the signal from digital to analog form.

MODEMS

A *modem* (from MOdulation and DEModulation) is a specialized analog-to-digital and digital-to-analog signal converter that works by modulating a signal onto a carrier wave and demodulating it at the other end. A modem is connected in the communications network between the data terminal equipment and the transmission network, as shown in Figure 8–5. Modems are an example of data circuit-terminating equipment (DCE), a term that was introduced in Chapter 6. Since the Carterfone Decision, many companies have entered the marketplace, and the modem market is very competitive.

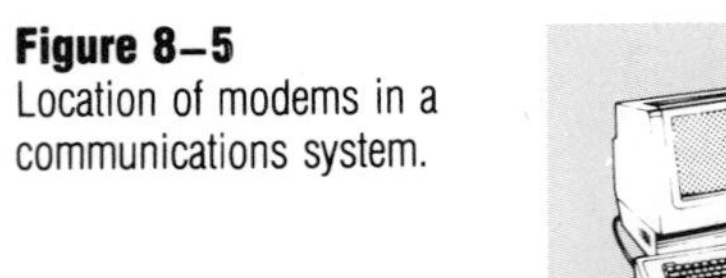

Figure 8–5
Location of modems in a communications system.

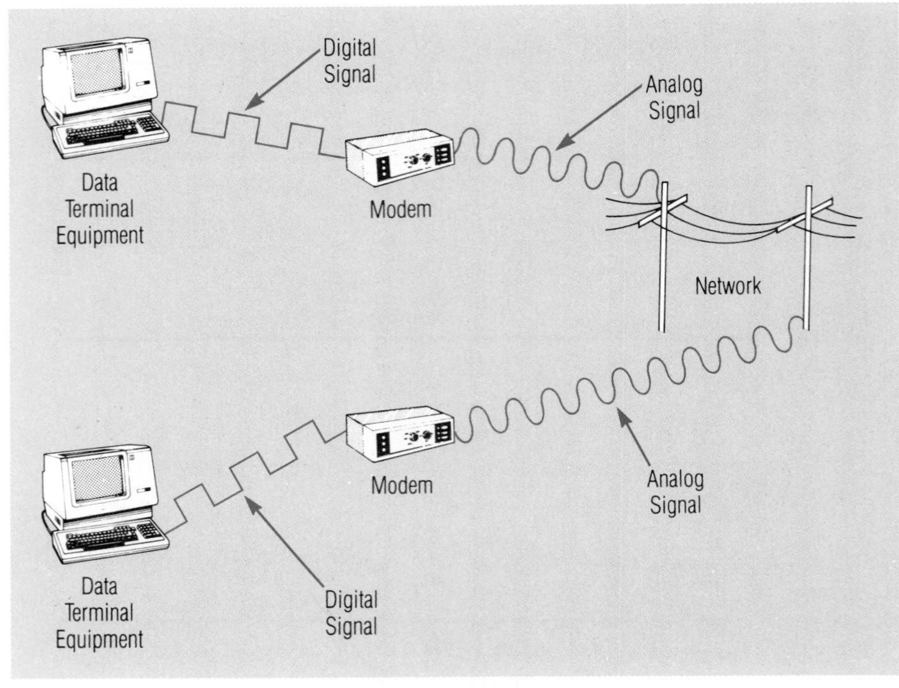

How Modems Work

In its simplest form, a modem senses the signal from the data terminal equipment (DTE). When it senses a 0 bit, it sends a signal with certain attributes (amplitude, frequency, and phase). When it senses a 1 bit, it sends a different signal, with at least one of the attributes changed. For example, when frequency modulation is used, a 0 bit from the DTE causes the modem to turn on an oscillator that sends an analog wave of a specific frequency on the telephone line. When a 1 bit is sent, the modem turns on another oscillator, which generates a wave with a different frequency. At the receiving end the process is reversed. The waves are converted to a digital signal (electrical pulses) representing the original 0 and 1 bits. This specific type of frequency modulation is called *frequency shift keying (FSK).*

frequency shift keying

We will look at the operational details of a particular modem, the Bell Type 103. It is used for asynchronous transmission at 300 bps and is a very popular modem for slow-speed transmission.

The Bell Type 103 modem uses FSK modulation. When two modems of this type communicate, one is designated as the "originate" modem and the other is designated as the "answer" modem. The originate modem transmits the 0 bits (spaces) at 1,070 Hz and the 1 bits (marks) at 1,270 Hz. The answer modem uses 2,025 Hz for spaces and 2,225 Hz for marks. These frequencies are well within the range of human hearing, and each signal sounds like a musical tone. In practice, the tones change so fast that if you listen to a telephone line when this type of modem is transmitting, the signals produce a warbling sound.

Looking at the signal diagrammatically in Figure 8–6, we can see where all of the frequencies fit within the bandwidth of a standard telephone line. This allows the transmissions from both modems to occur simultaneously, which is full-duplex transmission. Figure 8–7 shows how the wave form of the originate modem looks as it shifts back and forth from one frequency to another.

HIGHER SPEED TRANSMISSION

Although the FSK modulation technique can be used for speeds higher than 300 bps, the full-duplex transmission is sacrificed. This is because FSK modulation requires a bandwidth of approximately 1.5 times the baud rate. At 1,200 baud, 1,800 Hz would be required for transmission in each direction, giving a total required bandwidth of 3,600 Hz. This is more than the 3,000 Hz available on a telephone circuit. Therefore, when FSK is used at 1,200 baud, the transmission is limited to half-duplex.

Another practical difficulty is that as the speed increases, the frequency changes occur so fast that it becomes difficult for the receiver to

Figure 8–6
Frequencies used by a Bell Type 103 modem.

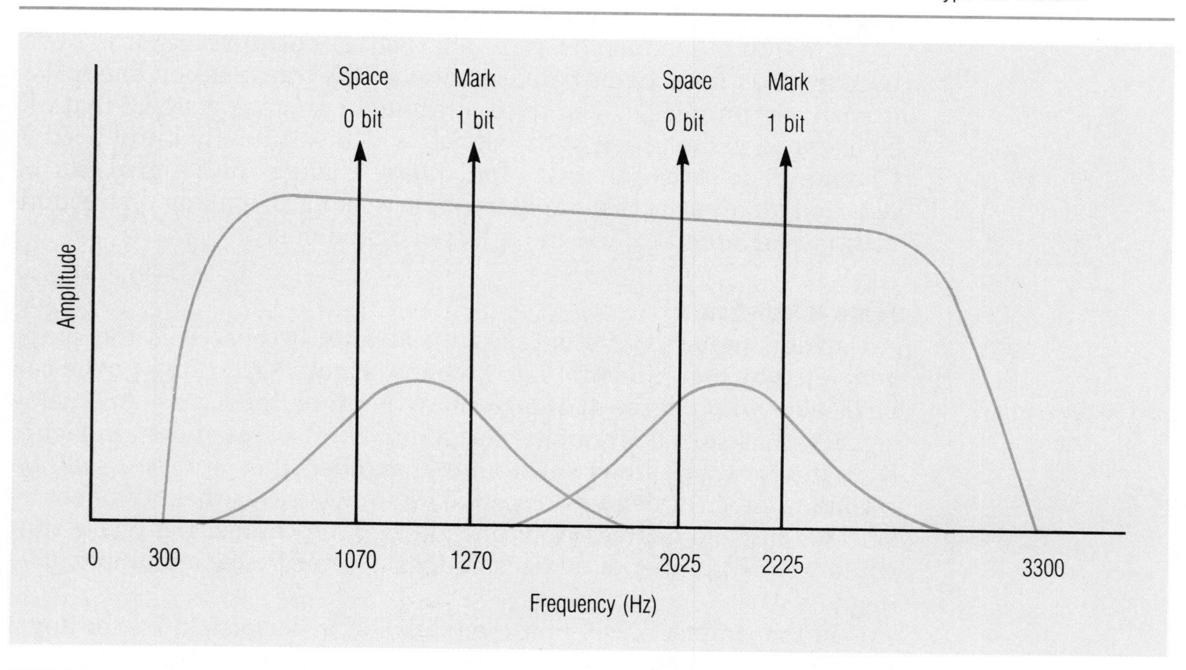

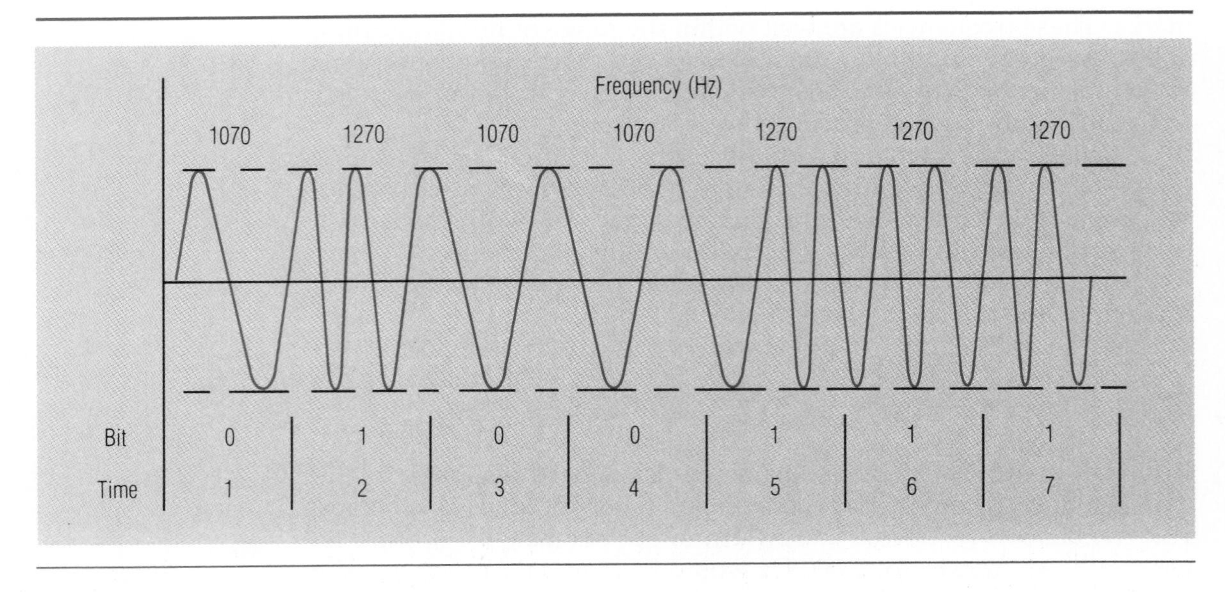

Figure 8–7
Frequency modulation in a Bell Type 103 modem.

detect them. At 1,200 baud, each tone is transmitted for only 1/1200 of a second, or .0833 seconds. At higher speeds, the duration would be shorter. The problem is solved at 1,200 baud by widening the difference between the frequencies. With the greater difference, the modem more quickly can tell one frequency from the other. The Bell 202 modem uses frequencies of 1,200 Hz for a 0 bit and 2,200 Hz for a 1 bit.

Amplitude modulation is an alternative, but in reality it is never used by itself in modems because noise on the transmission line makes it relatively unreliable. The third attribute of an analog signal that can be changed is the phase of the signal, which was briefly introduced in Chapter 5. It turns out that rapid phase changes of a signal can be detected more easily than rapid frequency changes, making phase modulation best suited for use in high-speed modems.

Phase Modulation

An analog signal's wave is called a *sine wave* because it is the shape generated by the geometric sine function. Figure 8–8 shows how it can be labeled with degree markings at any point on the X axis. An analog signal's phase can be thought of as a timing offset, as shown in Figure 8–9. If two waves are offset from one another, they are *phase shifted* a certain number of degrees. Figure 8–9 shows waves that are offset by 90, 180, and 270 degrees from one another. Of course, the phase shift could be any number of degrees; it does not need to be a multiple of 90 degrees.

In the simplest case, phase modulation is performed by shifting a sine wave 180 degrees whenever the digital bit stream changes from 0 to 1. It would shift 180 degrees again when the signal changed from 1

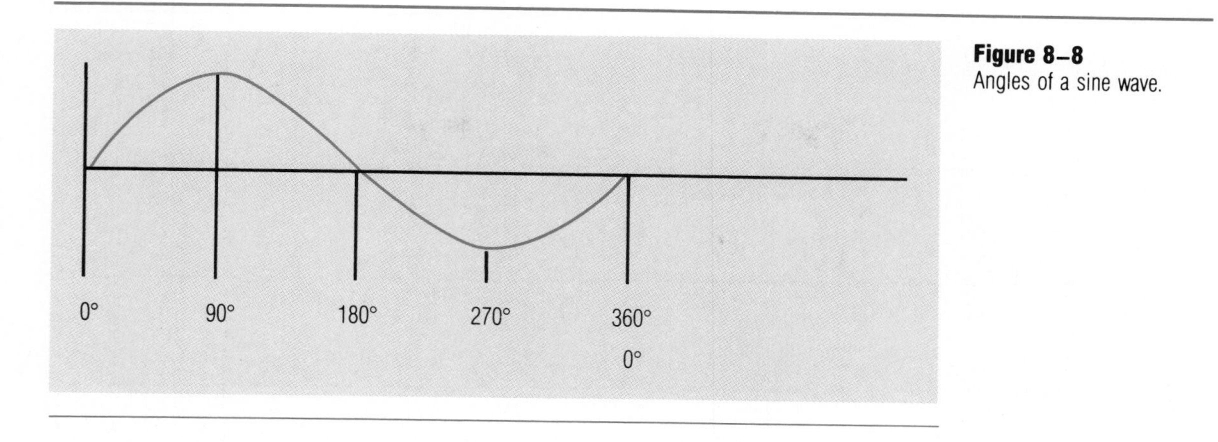

Figure 8–8
Angles of a sine wave.

to 0. The wave would look like Figure 8–10. This type of modulation is called *phase shift keying (PSK)*. The more usual type of phase shift modulation of this type is *differential phase shift keying (DPSK)*. In DPSK the phase is shifted each time a 1 bit is transmitted; otherwise the phase remains the same. This is illustrated in Figure 8–11. Notice that the

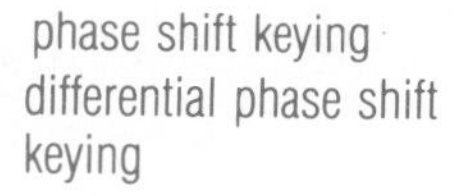
phase shift keying
differential phase shift keying

Figure 8–9
Phase shifts.

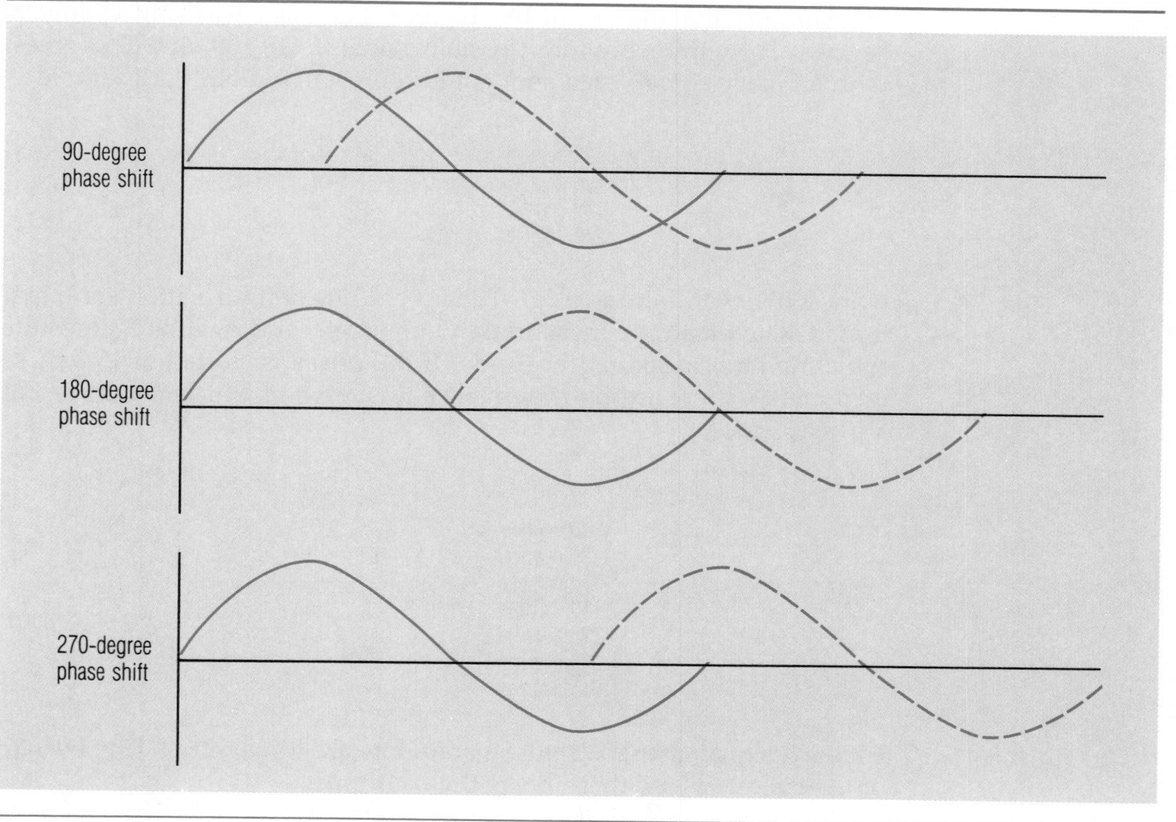

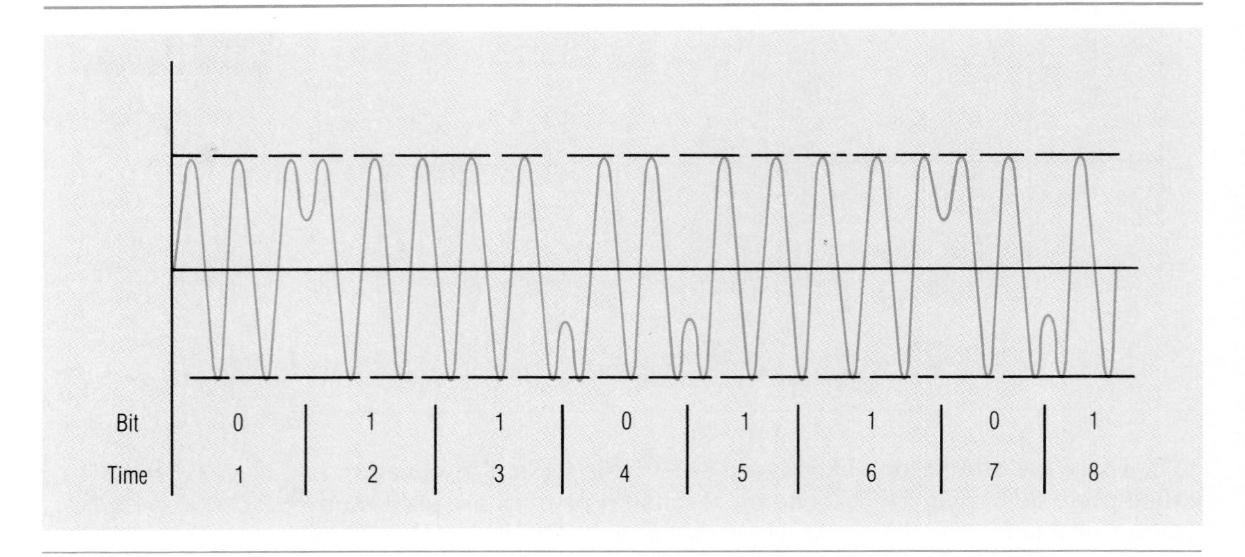

Figure 8–10
Phase shift keying (PSK).

wave form is the same in Figures 8–10 and 8–11, but the bits represented are different.

Suppose that instead of 180 degrees, the phase was shifted only 90 degrees. Now there are four possible shifts, 0, 90, 180, or 270 degrees. With a binary signal, each shift could represent a dibit. For example

Phase Shift	Dibits
0 degrees	00
90 degrees	01
180 degrees	10
270 degrees	11

Now, whenever there is a signal change, 2 bits of information are transmitted. The data rate, measured in bits per second, will be 2 times the signaling rate, measured in bauds. If the phase is shifted in 45-degree increments, there are eight possible shifts, each of which can represent a tribit.

Phase Shift	Tribits
0 degrees	000
45 degrees	001
90 degrees	010
135 degrees	011
180 degrees	100
225 degrees	101
270 degrees	110
315 degrees	111

For each signal change, 3 information bits are transmitted. The bit rate of the circuit will be three times the baud rate.

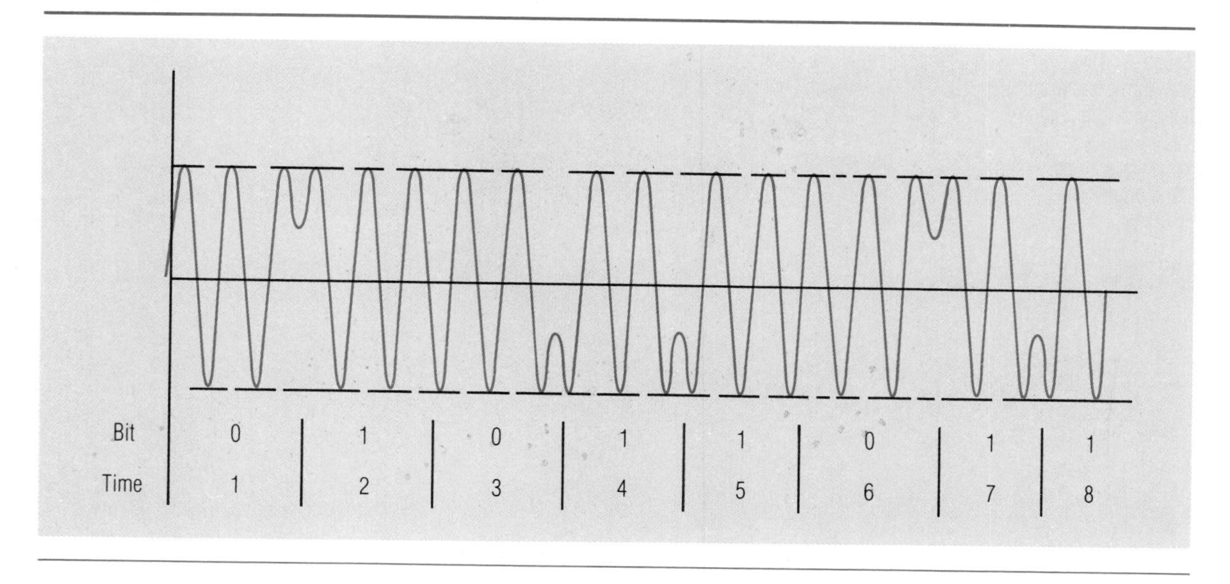

Figure 8–11
Differential phase shift keying (DPSK).

One very popular modem that uses a form of DPSK is the Bell 212A modem. This modem has a combination of capabilities. It operates asynchronously at 300 bps using FSK to be compatible with the Bell 103 modems. The 212A also operates asynchronously or synchronously at 1,200 bps, using a special form of DPSK. The CCITT V.22 specification is functionally identical to the Bell 212A characteristics, although variations of the V.22 standard have emerged in recent years.

Quadrature Amplitude Modulation (QAM)

If the phase is shifted less than 45 degrees, the receiver has a very difficult time detecting it. To get to higher speeds, modem designers have developed modulation techniques that use a combination of phase and amplitude modulation. The objective is to get 16 distinct combinations of phase and amplitude. This allows 4 bits, a quadbit, to be represented by each combination. There are several techniques by which this has been achieved, but all are called *quadrature amplitude modulation (QAM)*. The most common technique is shown in Figure 8–12. Note that there are 8 phase changes and 4 amplitudes in use, but not all combinations are valid. Also note that the quad bits are not assigned in a neat, logical sequence. The reasons are very technical but have to do with ensuring that the receiver detects as many transmission errors as possible.

The most popular modems using QAM modulation have been the Bell 209 and newer 2096 modems, which transmit data at 9,600 bps on leased lines. Both of these units signal at 2,400 baud and transmit full-duplex.

Figure 8–12
An example of the phase changes and amplitudes used in one type of modem that uses quadrature amplitude modulation.

Phase Change (degrees)	Relative Amplitude	Quadbit
0	3	0001
0	5	1001
45	$\sqrt{2}$	0000
45	$3\sqrt{2}$	1000
90	3	0010
90	5	1010
135	$\sqrt{2}$	0011
135	$3\sqrt{2}$	1011
180	3	0111
180	5	1111
225	$\sqrt{2}$	0110
225	$3\sqrt{2}$	1110
270	3	0100
270	5	1100
315	$\sqrt{2}$	0101
315	$3\sqrt{2}$	1101

Trellis Code Modulation (TCM)

As the speed of a transmission is pushed above 9,600 bps, the techniques to achieve error-free transmission become very sophisticated. As the bit rate increases, noise and other impairments of the communications line have a more significant effect. The following table shows the number of bits and characters that are affected by a noise pulse on the communications line that lasts just .1 seconds.

Speed (bps)	Bits Affected	8-Bit Characters Affected
300	30	3+
1200	120	15
2400	240	30
4800	480	60
9600	960	120
19200	1920	240

Trellis code modulation (TCM) is a specialized form of QAM that codes the data so that many bit combinations are invalid. If the receiver detects an invalid bit combination, it can determine what the valid combination should have been. Trellis coding allows the transmission of 6, 7, or 8 bits per baud. A 2,400 baud signaling rate yields speeds of 14,400 bps, 16,800 bps, and 19,200 bps on leased lines. On dial-up lines, the current maximum speed is 9,600 bps.

As modulation techniques get more sophisticated, the amount of signal processing that the modem must do increases significantly. Indeed, many of the more sophisticated modems today depend on powerful circuit chips called *signal processors* to code the data for transmission and decode it at the receiving end.

A significant use of TCM has been to provide 9,600 bps transmission on dial-up telephone lines. The CCITT published standard V.32, which was unlike any existing Bell standard. V.32 calls for 9,600 bps, full-duplex operation using TCM, and an echo cancellation technique. Echo cancellation permits high-speed signals traveling in opposite directions to exist on the same dial-up circuit at the same time in the same frequency band. (Echoes on communication lines are discussed in Chapter 9.) V.32 modems have become quite popular in the last 2 years because of their high transmission speeds and falling prices. A typical V.32 modem today costs between $1,000 and $2,000, and the prices are still falling.

Modems that operate at higher speeds of 14,400, 16,800, and 19,200 bps also are available. The CCITT V.33 standard applies to modems that operate at 14,400 bps on leased lines. Modems that operate at 14,400 bps on dial-up lines and modems that operate at the higher speed use vendor-proprietary techniques and as yet are not standardized by the CCITT. Some modems achieve high throughput rates by compressing the data before it is transmitted. Work is underway to standardize compression algorithms, and it is expected that modems with compression units will be able to achieve throughputs of 20,000 to 40,000 bps on analog lines in the next few years. For higher speeds, digital lines will be used.

MODEM CLASSIFICATION

Modems usually are classified according to the type and speed of transmission they are designed to handle. Thus, we find modems designed for 300 bps, asynchronous, full-duplex transmission, such as the Bell Type 103 we looked at earlier. Others are designed for 2,400 bps half-duplex synchronous transmission, 9,600 bps, half-duplex synchronous transmission, and so on.

When a modem is designed for half-duplex transmission, additional circuitry is added to perform a process called *line turnaround*. Line turnaround occurs when one modem stops transmitting and becomes the receiver, and the receiving modem becomes the transmitter. The modems exchange synchronization signals to ensure that they are ready to operate in the new mode, and then transmission begins again in the opposite direction. Line turnaround can occur very frequently, and the speed of this process is an important factor in overall line throughput. Commonly, it takes 50 to 200 milliseconds for a pair of modems to turn the line around. This may not seem like a lot of time, but it can represent a significant portion of total line time.

line turnaround

Modems for Asynchronous Transmission

The earlier discussion about asynchronous transmission and the Bell 103 modem covered most of the specifics of asynchronous modem operation. As the speed increases above 300 bps, more sophisticated modulation techniques are used. Modems are available to communicate

The AT&T 4024 modem transmits data asynchronously or synchronously at 300, 1,200, or 2,400 bps. Transmissions can be either full-duplex or half-duplex. (Courtesy of AT&T)

asynchronously at 9,600 bps on normal dial-up telephone lines and at higher speeds on permanent leased lines. Costs for asynchronous modems range from about $100 for a unit that will send data at 300 bps to $3,000 for a modem that yields error-free transmission at 9,600 bps on a dial-up line.

Modems for Synchronous Transmission

Modems used for synchronous transmission are more complicated in several ways than those used for asynchronous transmission. A clock circuit must be provided to time the release of bits to the communication line. The clock is usually a crystal controlled oscillator with a very high degree of accuracy. Synchronous transmission is primarily used at higher

This 3480 modem from AT&T-Paradyne is a part of a series of modems that transmit data synchronously at speeds from 2,400 bps to 19,200 bps. (Courtesy of AT&T-Paradyne Corporation)

transmission speeds of 2,400 bps and up, and the modulation techniques (such as QAM) are more sophisticated. Thus, the signal processing circuitry is more complex. Furthermore, at higher transmission speeds, minor aberrations in the communications line can cause transmission errors if the modem's signal detection circuitry is unsophisticated.

Figure 8–13 shows a general block diagram of a synchronous modem with its major components. The power supply takes standard 110 volt AC power and converts it to the lower voltages needed for internal operation.

In the transmitter, a modulator and digital-to-analog converter convert the digital bits (pulses) from the DTE to a modulated analog signal appropriate to the type of modulation being used. The output is a sine wave of the proper amplitude, frequency, and phase to represent the digital input signal.

equalization

Equalizer circuitry compensates for variability in the actual transmission line used. Although telephone circuits have certain standard parameters and specifications, there is a range of acceptable characteristics. As a modem's speed increases, it becomes critical for the modem to detect and compensate for the exact parameters of the particular line being used. *Fixed equalizers* in the transmitter assume that a certain average set of parameters exists, and they shape the transmitted wave accordingly.

More sophisticated *adaptive equalizers* are used in the receiving section of modems operating above 2,400 bps. A standard, training sig-

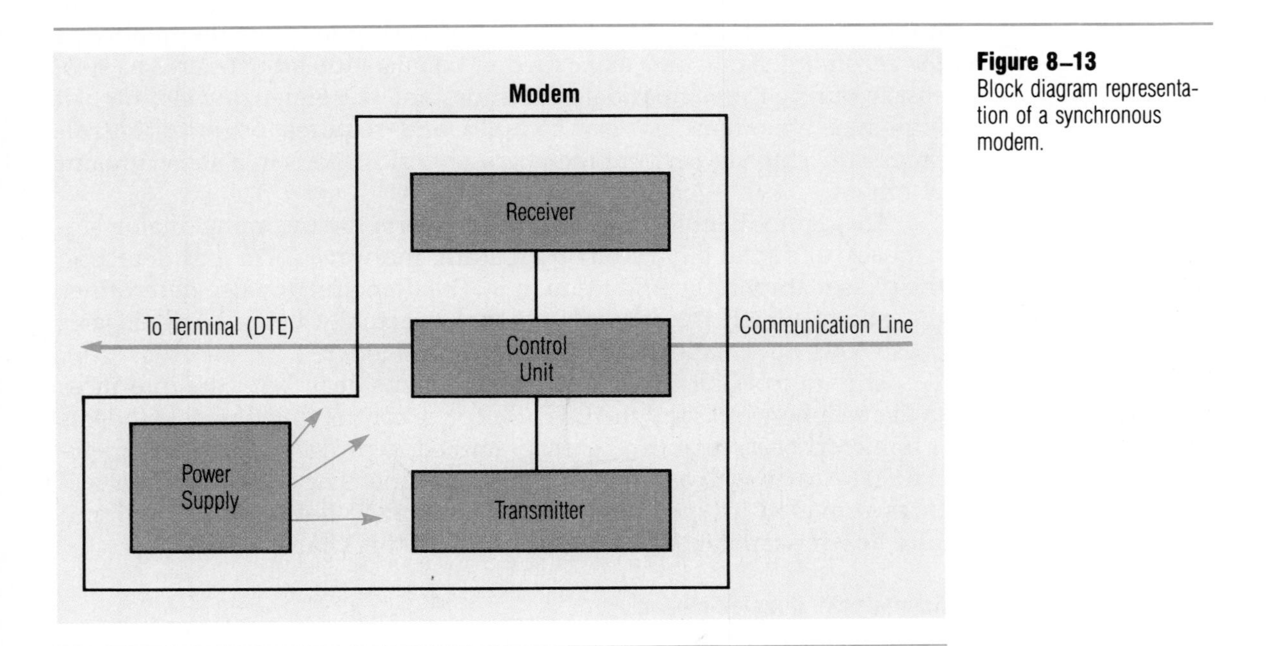

Figure 8–13
Block diagram representation of a synchronous modem.

The Codex 2260 modem is V.32 compatible and transmits at 4,800 bps or 9,600 bps on dial-up telephone lines. (Courtesy of Codex Corporation)

nal of known characteristics is sent from one modem to the other. The received signal is examined, and based on its shape and any errors that have occurred, the receiver's circuitry adjusts itself to the exact characteristics of the incoming wave form. The training signal is then sent the other way, and the other modem adjusts its receiving circuitry appropriately. After the initial training time, which takes from 20 to several hundred milliseconds, the modems constantly monitor the quality of the incoming signal and make further equalization adjustments as necessary during the regular data transmission. The equalizing circuitry in high-speed modems is very complex and requires powerful signal-processing chips to perform the necessary calculations in a short amount of time.

The demodulator in the receiver converts the incoming analog signal back to digital binary bits by reading the wave form and detecting the phase, amplitude, and frequency. The demodulator also determines the rate at which the incoming signal is actually being received and passes this information to clock extractor circuitry.

The control unit of synchronous modems deals with the interfaces to the telephone line and the DTE. Since the control unit of most modems is a microprocessor, it can be programmed to provide a number of special features as well as the standard functions. The most common features provided in even inexpensive modems today are auto dial and auto answer capabilities, discussed later in this chapter.

Acoustically Coupled Modems

An acoustically coupled modem, also known as an acoustic coupler, has two rubber cups into which the user places a telephone handset. This

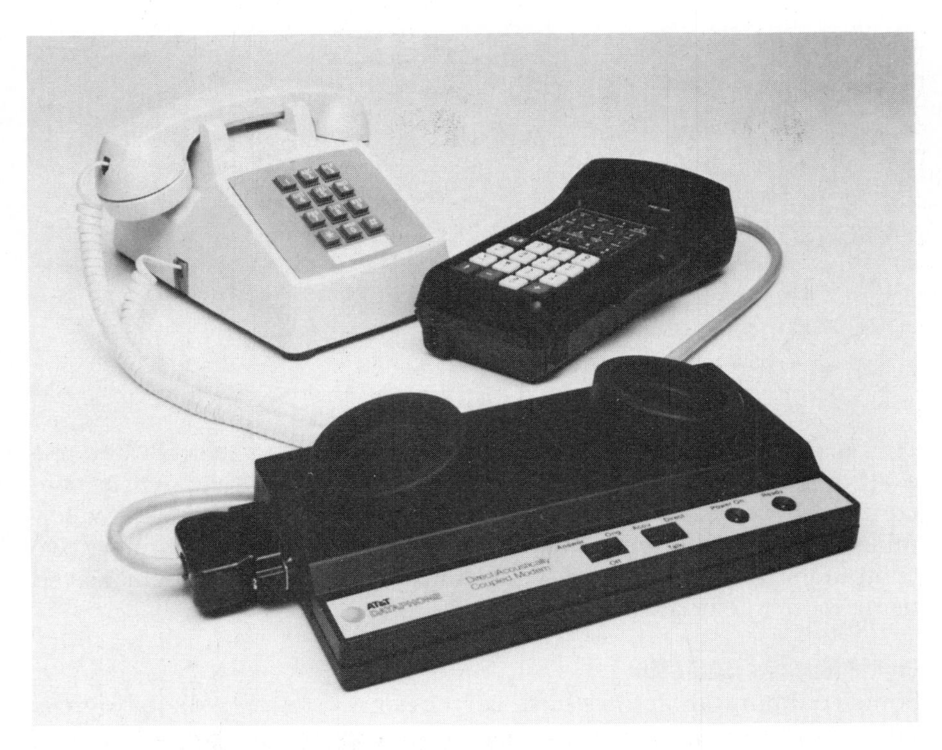

The AT&T Direct/Acoustically Coupled Model can be either directly attached to a telephone line or acoustically coupled via the rubber pads that cradle the telephone handset. (Courtesy of AT&T)

type of connection works adequately with FSK modulation at 300 bps and is appropriate for portable terminals where no direct wired connection is possible or desirable. For higher speeds, a direct electrical connection that is not subject to noise (sound) is necessary.

Limited Distance Modems/Short Haul Modems

Inexpensive modems are available for use when transmission distances are less than approximately 20 miles. There are trade-offs between speed and distance, so that a given modem might operate at 9,600 bps up to 5 miles, 4,800 bps up to 10 miles, and 2,400 bps between 10 and 20 miles. The primary advantage of this type of modem is cost. Savings of 50 percent to 80 percent compared to a conventional modem are possible.

Modem Eliminators/Null Modems

If distances are very short and the DTEs can be connected by a cable, modems may not be needed at all. Simply plugging a cable between the terminal and the computer won't work, however, because both devices deliver data to be transmitted on the same pin of the standard connector used for this type of connection. Conceptually, what is needed is a device with a cable that cross-connects the transmit and receive pins of the connectors so that the transmit pin on one device is connected to the receive pin on the other, and vice versa. A schematic diagram of this type of device is shown in Figure 8–14.

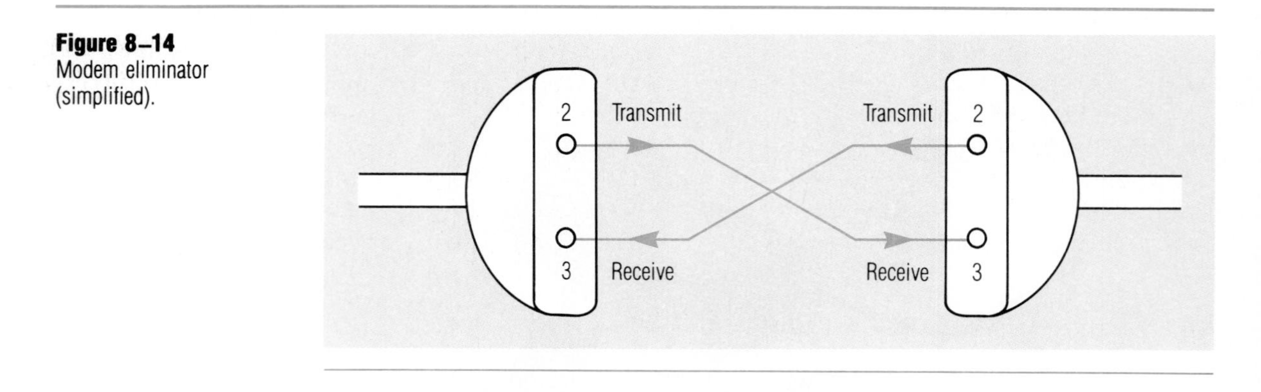

Figure 8–14
Modem eliminator (simplified).

In practice, other pins must be cross-connected as well, and the exact implementation depends on whether the transmission will be synchronous or asynchronous. Modem eliminators are often used for terminals in close proximity to a computer or sometimes to connect two computers. Distances of up to several thousand feet can be achieved using modem eliminators.

Digital Modems (DSU/CSU)

Some communication lines have been designed to carry digital signals rather than analog signals. It would seem that in this case, no modem would be needed because the DTEs emit the digital pulses that the circuit is designed to carry. In reality, a digital modem, usually called a *data service unit/channel service unit (DSU/CSU),* is required. Although the DSU and CSU normally are packaged together in a single unit, it is possible to purchase them separately.

The DSU/CSU ensures that the digital signal that enters the communications line is properly shaped into square pulses and precisely timed. Sometimes it needs to convert the digital pulses coming from a DTE to a form suitable for the line. The DSU/CSU also provides the physical and electrical interface between the DTE and the line. The primary purpose of the CSU portion of the unit is to provide circuitry that can protect the carrier's network from excessive voltage coming from the customer's transmission equipment. DSU/CSUs perform other functions, such as digital loopback and other diagnostic testing. Since they do not perform the analog-to-digital or digital-to-analog function, however, they are much simpler and less expensive than traditional modems.

Using circuits that have been designed for digital transmission, DSU/CSUs can obtain speeds of 56,000 bps and higher. These high-speed transmissions are useful when a great deal of data is being transmitted or when very good response time is desired. Although 9,600 bps is far more commonly used for data transmission today, the wider use of digital circuits will bring with it more extensive use of 56,000 bps and higher speeds in the next few years.

Modems for Fiberoptic Circuits

When the medium carrying the communication circuit is a fiberoptic cable, a special optical modem is used to convert the electrical signals from the data terminal equipment to light pulses. Since light waves have the same characteristics as electromagnetic waves—albeit at a much higher frequency—the same concepts of bandwidth and modulation apply, but the transmission is entirely digital. The conversion that takes place is not from digital to analog but from digital-electrical to digital-optical.

MODEM INTERFACES

A modem has two communication interfaces—one to the telephone line and the other to the data terminal. The specifications of the interfaces include the mechanical characteristics (such as what type of connectors will be used and the number of pin connections), the signaling characteristics (such as how many signals there will be and what they will mean), and the electrical characteristics (such as what voltages and current level will be on each pin). Figure 8–15 shows where these interfaces occur in the communications system.

Interface Between the Modem and the Communications Line

The interface between the modem and the communications line is the simpler of the two. It consists of just two or four signaling wires with physical and electrical specifications that are prescribed by the telephone company. The connection is made with a short two- or four-wire cable that has standard RJ-11 plugs, like those used on telephones, on each end.

Interface Between the Modem and the Data Terminal Equipment (DTE)

The interface between the modem and the DTE is more complicated. Several "standards" have been developed, but none is universally used. The most popular interfaces are discussed here.

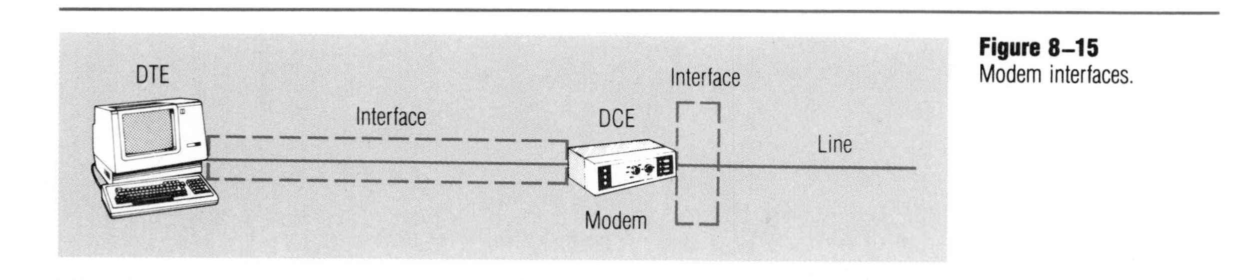

Figure 8–15
Modem interfaces.

RS-232-C Interface Virtually all computer and terminal equipment that uses telecommunications conforms to an interface that was standardized in the United States by the Electrical Industries Association (EIA). The EIA interface is called *RS-232-C*. It has been adapted for international use by the CCITT under the name *V.24*. Although the RS-232-C and V.24 interfaces are not identical, they are functionally equivalent for most applications.

The actual hardware used to connect a DTE and DCE with the RS-232-C interface is a cable with 25 wires and a 25-pin connector at either end, as shown in Figure 8–16. The use of each of the wires or pins is specified in the RS-232-C standards document as well as the levels (voltages) of the signals. The signals sent across the connection are, in all cases, digital and serial. The standard also specifies the mechanical attributes of the interface, specifically that the DCE will have a female connector and the DTE will have a male connector. Another specification is a 50-foot maximum length for the cable connecting the DTE and DCE.

A list of the 25 signals and their functions is given in Figure 8–17. Note that not all of the pins have been assigned a use. This is to provide for future expansion. Note also that pins 2 and 3 are the ones on which the actual data is passed. All of the other pins are used for signaling between the modem and the DTE.

The RS-232-C standard has some limitations that are inconvenient at best and unacceptable in many applications.

- The cable length of 50 feet. At speeds less than 2,400 bps, a longer cable may work, but it is not guaranteed or supported by the equipment manufacturers. In actual practice, cable lengths of up to 100 feet at 9,600 bps are common.

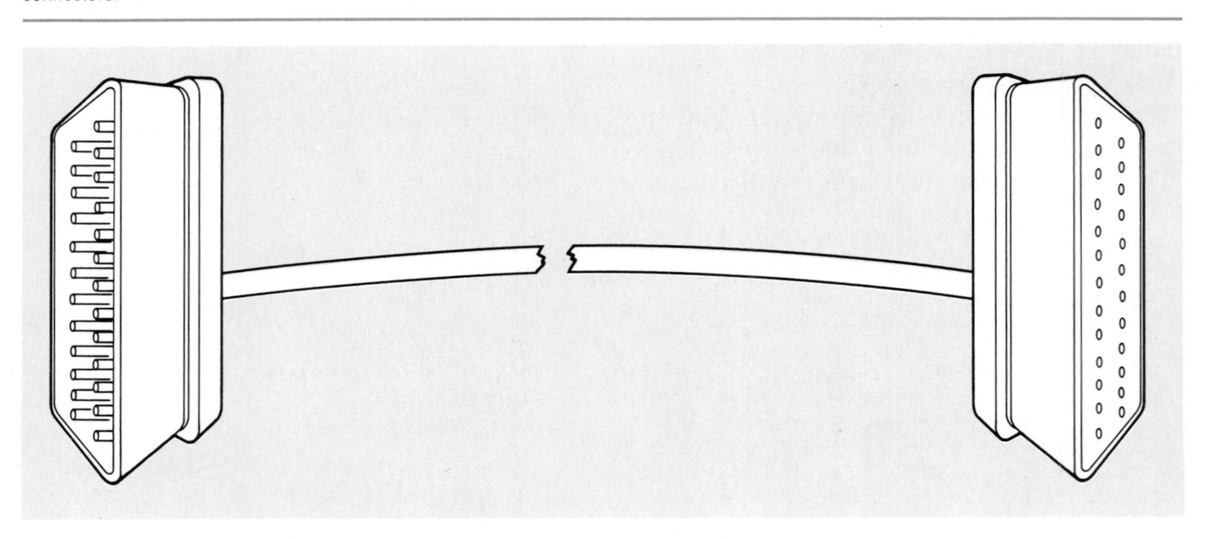

Figure 8–16
An RS-232-C cable with male and female 25-pin connectors.

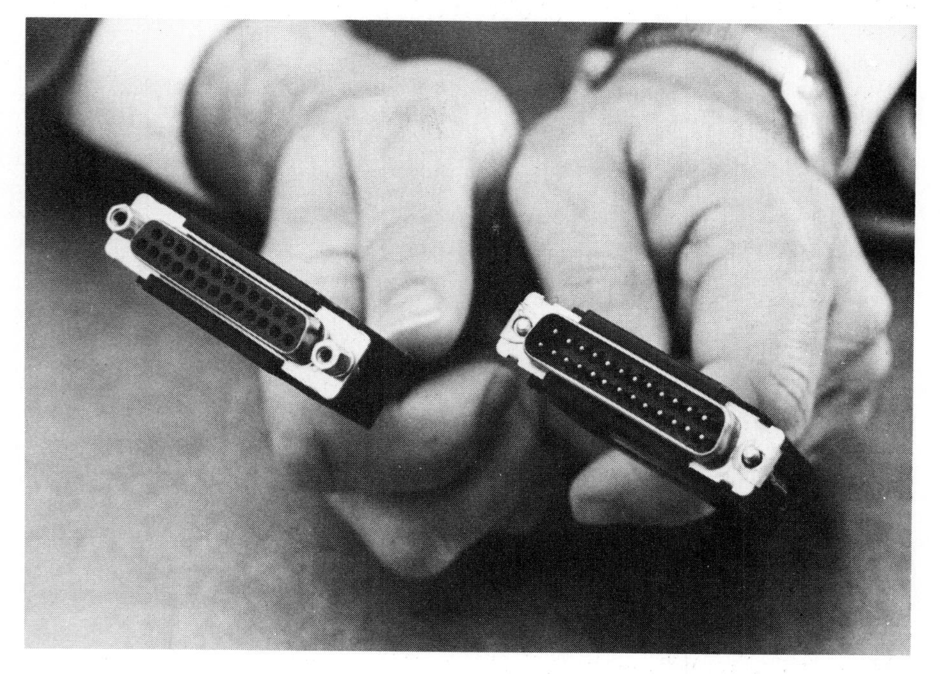

These male and female 25-pin connectors meet the RS-232-C interface standards.

- A maximum speed of 20,000 bps. This is not a problem when transmissions use the public telephone network but is a restriction for private cable within a building or in a campus environment.
- A technical problem with the way electrical grounding is handled that can lead to difficulties in detecting the difference between 0 and 1 bit if the speed is high and the cable is long.

RS-232-D Interface In 1987 the RS-232-C interface standard was revised and renamed RS-232-D. RS-232-D made the following changes to the RS-232-C standard:

- specifications for the 25-pin connector;
- addition of a shield on the cable
- redefinition of the protective ground;
- definition for pin 18 (local loopback), pin 25 (test mode interchange circuit), and redefinition of pin 21 from signal quality detector to remote loopback.

Although these changes solved some of the problems of the RS-232-C standard and made RS-232-D equivalent to the CCITT V.24 standard, the RS-232-D interface has not seen widespread usage.

RS-449 Interface Another interface standard that overcomes some of the problems of RS-232-C is the *RS-449* standard. RS-449 uses 37 signal

Figure 8–17
RS-232-C connector pin assignments.

Pin	Description
1	Protective ground
2	Transmitted data
3	Received data
4	Request to send
5	Clear to send
6	Data set (modem) ready
7	Signal ground
8	Received line signal detector
9	(reserved for modem testing)
10	(reserved for modem testing)
11	unassigned
12	Secondary receive line signal detector
13	Secondary clear to send
14	Secondary transmitted data
15	Transmission signal element timing
16	Secondary received data
17	Receiver signal element timing
18	unassigned
19	Secondary request to send
20	Data terminal ready
21	Signal quality detector
22	Ring indicator
23	Data signal rate selector
24	Transmit signal element timing
25	unassigned

wires, as opposed to 25 for RS-232-C, with the extra ones added primarily for automatic modem testing. The RS-449 standard also overcomes the RS-232-C's length limitation; an RS-449 cable length can run up to 4,000 feet. One of the weaknesses of the RS-449 standard is that it makes no provision for automatic dialing between modems. Although the RS-449 standard was the intended successor to the RS-232-C interface, it has not been well accepted in the marketplace or widely implemented.

RS-336 Interface *RS-336* is an interface standard that does allow for automatic dialing of calls under modem control. However, it does not adequately provide for high-speed data transmission or the use of private circuits. Its primary use has been for applications in which the computer automatically calls numerous remote data terminals for data collection.

X.21 and X.21 BIS Interfaces The *X.21* interface standard was developed by the CCITT and is widely accepted as an international standard. However, it is well ahead of its time. X.21 is based on a digital connection to

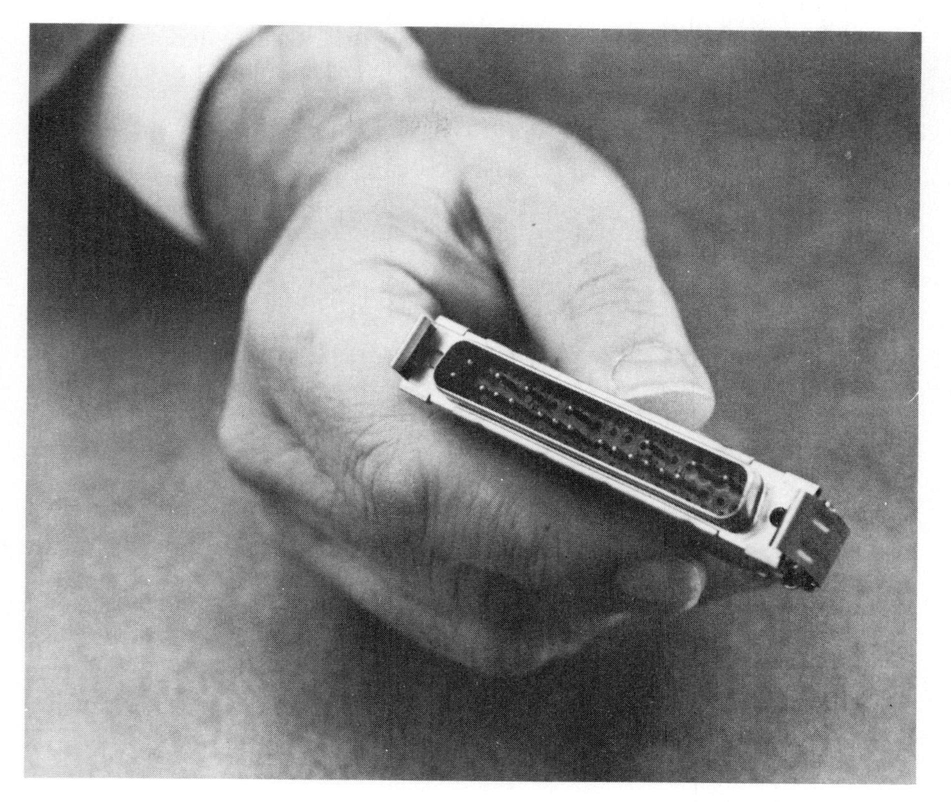

This 37-pin connector is used for the RS-449 interface.

a "digital" public telephone network. Digital public telephone networks do not exist in any widespread form in any country today! X.21 is important as a directional standard, however, because many countries now are installing all-digital circuits. They will come into wider use in the 1990s.

When the X.21 standard is used, data signals are encoded in serial digital form and kept that way throughout the transmission. All other signaling is in digital form, too; for example, digital dial tone is a series of + (plus) signs. The basic rate of a digital channel is 64,000 bps, which not only far exceeds most data transmission speeds today but also can handle digitized voice.

Until digital networks are put in place, the CCITT has defined a temporary standard called *X.21 BIS,* which electrically is virtually identical to RS-232-C and V.24. Its application, however, is for connecting a terminal to a packet switch network via analog lines. Chapter 11 explores packet switching networks.

Current Loop Interface The *current loop* interface, although outdated, is mentioned because it is still in use today. Originally designed for use by teletypewriters, the current loop indicates 1 and 0 bits by the presence or absence of an electrical current. A 1 bit or mark is defined as the flow of either 20 milliamps or 60 milliamps of current, depending on the type of machine. Cables connecting current loop devices can be up to 1,500 feet long and can pass data at up to 9,600 bps. Although the current loop is very popular because it is simple and inexpensive to produce, it is totally nonstandard, which means that one manufacturer's equipment may not be able to connect to another manufacturer's equipment even though both are using a current loop interface.

OTHER MODEM FUNCTIONS AND CAPABILITIES

Reverse Channel

Some slow-speed modems, which do not use all of the available bandwidth of a communications line for primary signaling, use some of the remaining bandwidth to provide a slow-speed *reverse channel*. The modems can use a reverse channel for signaling one another. For example, the receiving modem can send a signal on the reverse channel indicating that it has received blocks of data correctly. The reverse channels typically operate at very slow speeds of 5 or 10 bps. The use of a reverse channel is not considered to be full-duplex operation because of the vast speed difference between the transmission on the primary channel and the reverse channel.

Auto Dial/Auto Answer

The rise in the use of modems with personal computers has led to the development of sophisticated auto dialing capability. *Auto dial* allows the operator to send commands to the modem through the terminal keyboard. With the auto dialer built into the modem, the user sends commands in the form of ASCII data characters from the terminal to the modem control section across the interface. These commands instruct the modem to perform the desired function. The command set has become relatively standardized around the commands developed in the early 1980s by Hayes Microcomputer Products, Inc. The commands tell the computer what number to dial, whether to use dial pulses or tones, how long to wait before answering, and so on. All Hayes commands begin with the letters AT, which stand for *attention*. When telecommunications software is run in a personal computer, the modem commands can be issued automatically. For example, the user could enter one high-level command, "dial stock," and the software would automatically issue all of the necessary detail commands to the modem to dial a stock quotation service, log the user on, request the particular stock prices of interest, log off, and hang up.

The *auto-answer* capability sets up the modem to automatically answer a telephone line when it detects the ringing signal. The incoming call can be handled by a computer or other device. The auto-answer capability is used by personal computer bulletin board systems, commercial communication systems, such as MCI Mail or The Source, and time-sharing services. It is also used by businesses that want to provide their employees with a way to dial in to the company's mainframe computer.

The auto-dial/auto-answer feature also can be used to provide a dial-backup connection when a leased telephone line is normally used for transmission. If the modem detects that the leased line has failed, it automatically dials the receiving modem, establishes a connection through that modem's auto-answer capability, and resumes operation on the public telephone network. Often the backup connection operates at a slower speed than the original transmission. However, in many applications, the slower transmission is preferred to a total outage.

Modem Multiplexing

Another fairly common modem feature, particularly on higher speed modems, is a form of multiplexing called *multiport* or *split stream operation.* As shown in Figure 8–18, a 9,600 bps data stream is subdivided

Figure 8–18
Multiplexing modems.

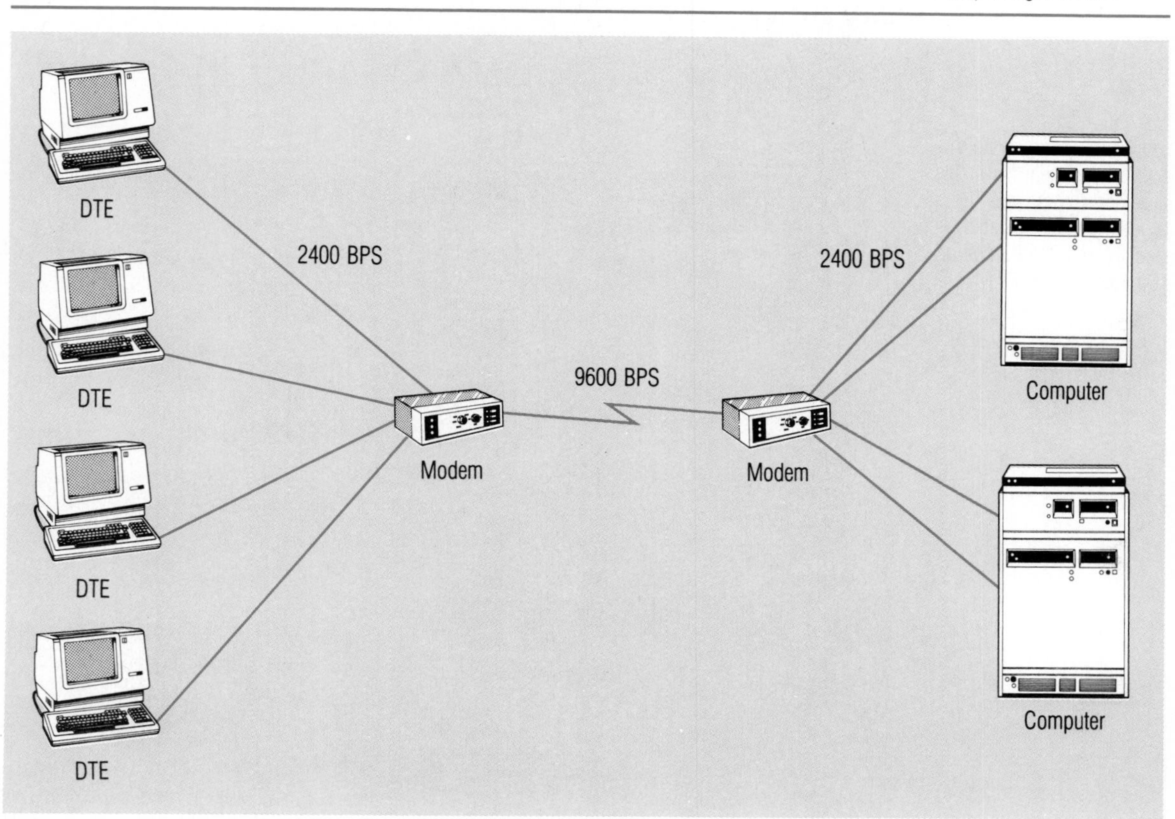

into four 2,400 bps data streams from four terminals. The modem combines the data streams for transmission, and the receiving modem splits them apart. Modems with this capability usually allow different combinations of data stream speeds as long as the maximum data rate of the modem is not exceeded. For example, with a 9,600 bps line speed, one could establish the following four combinations of data streams

- 1 7,200 bps and 1 2,400 bps
- 2 4,800 bps
- 1 4,800 bps and 2 2,400 bps
- 4 2,400 bps

Internal Modems

An *internal modem* is not a feature but a modem contained on a single circuit card that users insert in slots inside a personal computer. The internal modem may have several indicators on a small control panel at the rear of the personal computer, but it usually does not have switches that allow for the options and parameters of the modem to be easily changed. Internal modems cost approximately the same as standalone modems of the same capability. Some are slightly less expensive because they are not packaged in a cabinet.

The AT&T programmable half-card modem fits inside a personal computer. The modem can be operated directly from the keyboard. (Courtesy of AT&T).

Modem Diagnostics

Most modems today contain diagnostic routines that check the circuitry of the modem. Usually, the diagnostics run automatically when the modem is powered on and can be initiated by an operator anytime thereafter. In most modems, the diagnostics cannot run when data is being transmitted because they use the same circuitry. In some cases, diagnostics can be initiated remotely by a control operator or by a remote modem.

Loop Back The *loop back* capability in a modem is a diagnostic capability that allows the transmitting modem to send a special signal to the receiving modem. The receiving modem sends the signal back to the originator, where it can be checked and analyzed for accuracy. If a terminal is having difficulty communicating on a circuit, the loop back test will help determine where the problem is occurring. If the loop back test runs successfully, the problem is not in the modem or circuit but likely to be in the terminal, the computer, or the cables that connect them to the modems.

TYPICAL MODEMS

Over the years, a number of different types of modems have become standardized for transmission at various speeds under differing conditions. Figure 8–19 shows some of the types of modems that have emerged as the most popular. Many have come from the Western Electric manufacturing arm of AT&T, which was an early leader in modem development. These are commonly known as Bell modems. Note that in many cases, the CCITT international standards closely match those of the Bell modems. Typically, the Bell modems were introduced, gained widespread use in the United States, became the de facto standard, and were adopted by the CCITT with only minor modifications. Only recently has the CCITT taken the lead and adopted standards that were not previously implemented in Bell modems.

MODEM SELECTION CRITERIA

Clearly, telecommunications professionals must select modems to match the type of transmission that will be employed. The prospective purchaser must determine

- whether the transmission will be asynchronous or synchronous;
- what line speed will be used;
- what the distance between the modems will be;
- whether the transmission will occur on dial-up or leased lines;
- whether any special type of transmission lines, such as fiberoptics, will be used.
- whether or not it's full duplex or half duplex

Modem Designation	Transmission Type	Speed (BPS)	Signaling Rate (Baud)	Link Type	Transmission Mode	Modulation Type
Bell 103/113	Asynch	300	300	2-wire	FDX	FSK
Bell 202	Asynch	1200	1200	2-wire	HDX	FSK
		1800	1200	4-wire	FDX	FSK
Bell 212A	Asynch	300	300	2-wire	FDX	FSK
	Synch	1200	1200	2-wire	FDX	PSK
CCITT V.22	Either	600	600	2-wire	FDX	PSK
	Either	1200	600	2-wire	FDX	PSK
Bell 201	Synch	2400	1200	2-wire	HDX	DPSK
				4-wire	FDX	DPSK
CCITT V.26	Synch	2400	1200	4-wire	FDX	DPSK
Bell 208	Synch	4800	1600	4-wire	FDX	DPSK
				2-wire	HDX	DPSK
CCITT V.27	Synch	4800	1600	2-wire	HDX	DPSK
				4-wire	FDX	DPSK
Bell 209	Synch	9600	2400	4-wire	FDX	QAM
CCITT V.29	Synch	9600	2400	4-wire	Either	QAM
	Synch	4800	2400	4-wire	Either	PSK
Bell 2024	Synch	2400	1200	2-wire	HDX	DPSK
				4-wire	FDX	DPSK
Bell 2048	Synch	4800	1600	2-wire	HDX	DPSK
				4-wire	FDX	DPSK
Bell 2096	Synch	9600	2400	2-wire	HDX	QAM
				4-wire	FDX	QAM
CCITT V.32	Synch	4800	2400	2-wire	FDX	QAM
		9600	2400	2-wire	FDX	TCM
Bell 2224	Synch	2400	1200	2-wire	FDX	QAM
	Asynch	2400	1200	2-wire	FDX	QAM
	Asynch	1200	1200	2-wire	FDX	PSK
	Asynch	300	300	2-wire	FDX	PSK

Figure 8–19
Characteristics of specific modems and modem types.

In addition, although international standards exist, modem manufacturers sometimes interpret the standards slightly differently. The result is that two modems, whose manufacturers both claim to meet international standards, may not be able to communicate with one another. The safest choice is to use identical modems on both ends of the circuit. Barring that, the purchaser should thoroughly test the modems that will communicate with each other.

SUMMARY

This chapter introduced many technical concepts of data transmission. Transmissions have been categorized several ways. The important distinction between asynchronous and synchronous communications was explained, as well as the pros and cons of each type of transmission. Modems were defined as a specialized signal converter, and several types of modems were examined in detail. The chapter also described contemporary interface standards between terminals and modems. Now you should have a good understanding of data transmission and should be ready to apply this knowledge to the study of various transmission lines, protocols, and networks.

CASE STUDY

Dow Corning's Use of Modems

Dow Corning acquires its modems from a variety of sources. Within the Midland area, some of the data transmission occurs on a large local area network, for which no modems are needed. Other transmission occurs at 56 kbps over limited distances, and on these circuits, limited distance modems by Develcon, Inc., and Avanti Communications Corporation have been purchased. On the nationwide data communications network, the modems are primarily AT&T modems that transmit data at 9,600 bps. Some of the long distance circuits use digital transmission, and on those, AT&T DSU/CSUs have been installed.

Dow Corning's international lines, which run to Europe, Mexico, Canada, Brazil, Japan, Hong Kong, and Australia, have a great diversity of modems. On international circuits, the modems must meet international standards and also must be available in both the United States and the foreign country. Dow Corning's circuits to Europe operate at 56 kbps and use DSU/CSUs provided by AT&T. Paradyne modems operating at 9,600 bps are used on the two circuits to Mexico. The circuit between Midland and Toronto is a digital circuit operating at 9,600 bps; AT&T DSU/CSUs are employed. IBM modems are used on the circuits to Japan, which operates at 14,400 bps. Codex modems are used on the line to Australia, which also operates at 14,400 bps. Obviously, when installing international circuits, there are many unique requirements that dictate different solutions in each country.

simplex

On the circuits to Australia and Japan, data compression devices, made by Simplex, Inc., are employed. These units process all data, using a proprietary algorithm to reduce the number of bits that must be sent. This processing effectively allows more data to be sent on a line of given capacity. Dow Corning had good success in 1988 when Symplex compression units were installed on its 14.4 kbps lines to Europe. European users of the computers in Midland saw much improved response time after the Symplex devices were installed. When the European lines were upgraded to 56 kbps, the data compression devices were moved to the Australian and Japanese circuits.

A variety of dial-in modems have been used, but those from Hayes and AT&T are most common. Dial-in data transmission operates at 1,200 and 2,400 bps. None of the new V.32 modems have yet been installed, although some of the users would like to have the 9,600 bps transmission speed. Dow Corning's telecommunications management believes that the cost of the V.32 modems is too high. Managers are anticipating that the cost will drop in the next year or two.

Review Questions

1. Compare and contrast the meanings of the terms *baud* and *bits per second.*
2. How many unique signal changes do dibits, tribits, and quadbits require in order to be transmitted?
3. Compare and contrast simplex, half-duplex, and full-duplex transmissions.

4. Distinguish between parallel and serial transmission.
5. Describe the purpose of the start and stop bits in asynchronous transmission.
6. Explain the terms *bit synchronization* and *character synchronization.*
7. Explain why synchronous transmission is more efficient than asynchronous transmission when long blocks of data are to be transmitted.
8. Why is synchronous transmission more expensive to implement than asynchronous transmission?
9. Explain the terms *mark* and *space.*
10. The purpose of the clock in synchronous transmission is __________.
11. Distinguish between unipolar, bipolar, NRZ, and bipolar, return-to-zero digital signals.
12. Explain why FSK modulation cannot be used for a 1,200 bps, FDX transmission.
13. Why is phase modulation better suited for high-speed transmission than frequency or amplitude modulation?
14. What are the differences between fixed equalization and adaptive equalization in a modem?
15. Describe how high-speed transmission has been achieved in modems that use the V.32 standard.
16. What are the upper limits to data transmission speed on analog lines?
17. Explain the functions of the DSU/CSU when lines designed for digital transmission are used.
18. Explain the concept of line turnaround in half-duplex transmission.
19. Identify the limitations of the RS-232-C interface.
20. When is the use of an acoustic coupler appropriate? What is its chief limitation?
21. When is the use of a limited distance modem appropriate? Why is it advantageous to use them whenever possible?
22. Describe the purpose of a loop back test.

Problems and Projects

1. A communications system is transmitting 100 character blocks of EBCDIC data at 4,800 bps, HDX. After each block of data is sent, the line must be turned around, and the receiving modem must either acknowledge correct receipt of the data (by sending an ACK character) or signal that the data was received incorrectly (by sending a NAK character). If the data was received correctly, the line is turned around, and the next block of 100 characters is sent. Line turnaround takes 50 milliseconds. Assuming that no transmission errors occur, calculate what percentage of the time the line is used for

transmitting the data and what percentage of the time is spent performing line turnarounds.

2. Do problem 1 assuming the line speed is 9,600 bps instead of 4,800 bps. How does that change affect the results?

3. A personal computer sends blocks of ASCII data to a mainframe that are, on average, 30 characters long. The mainframe returns data blocks that average 500 characters in length. Calculate the transmission efficiency if asynchronous transmission is used, assuming 1 start bit and 1 stop bit per character. Calculate the efficiency for synchronous transmission assuming three synchronizing characters per block. Ignore the effects of line turnaround and assume that no transmission errors occur.

Vocabulary

signaling rate
baud
circuit speed
bits per second (bps)
dibits
tribits
quadbits
simplex transmission
half-duplex transmission (HDX)
full-duplex transmission (FDX)
parallel transmission
serial transmission
asynchronous transmission
start bit
stop bit
start/stop transmission
mark
space
bit synchronization
transmission efficiency
character synchronization
synchronous transmission
digital signal
unipolar
bipolar, nonreturn-to-zero (NRZ)
bipolar, return-to-zero
modem
frequency shift keying (FSK)
phase shift
phase shift keying (PSK)
differential phase shift keying (DPSK)
quadrature amplitude modulation (QAM)
trellis code modulation (TCM)
line turnaround
data service unit (DSU)
channel service unit (CSU)
RS-232-C interface
V.24 interface
RS-232-D interface
RS-449 interface
RS-336 interface
X.21 interface
X.21 BIS interface
current loop
reverse channel
auto dial
auto answer
multiport
split stream operation
internal modem
loop back test

References

Bartee, Thomas C. *Data Communications, Networks, and Systems.* Indianapolis, IN: Howard W. Sams & Co., Inc., 1985.

Black, Uyless. "A User's Guide to the CCITT's V-Series Modem Recommendations," *Data Communications,* June 1989.

Davenport, William P. *Modern Data Communication: Concepts, Language, and Media.* Rochelle Park, NJ: Hayden Book Co., Inc., 1971.

Doll, Dixon R. *Data Communications: Facilities, Networks, and Systems Design.* New York: John Wiley & Sons, 1978.

Edwards, Morris. "Modems Continue to Reach Higher Speeds and to Gain More Intelligence While Getting Smaller," *Communications News,* February 1987, p. 66.

Falk, Howard. *Microcomputer Communications in Business.* Radnor, PA: Chilton Book Company, 1984.

Finneran, Michael F. "Data Comm Focus," *Business Communications Review,* July–August 1985.

Fitzgerald, Jerry. *Business Data Communications: Basic Concepts, Security, and Design,* 2nd ed. New York: John Wiley & Sons, 1984, 1988.

Friend, George E., and John L. Fike, *Understanding Data Communications.* Dallas: Texas Instruments Incorporated, 1984.

Martin, James. *Introduction to Teleprocessing.* Englewood Cliffs, NJ: Prentice-Hall, Inc., 1972.

McNamara, John E. *Technical Aspects of Data Communication.* Bedford, MA: Digital Press, 1982.

Mier, Edwin E. "The Future of Modems: Will It Be Boom or Gloom?" *Data Communications,* September 1986, p. 55.

Rothrock, James. "V.32 Technologies Deliver Leased-Line Speed at Dial-Line Rates," *Telecommunication Products + Technology,* November 1986, p. 58.

Stamper, David A. *Business Data Communications,* 2nd ed. Redwood City, CA: The Benjamin/Cummings Publishing Company, 1986, 1989.

Sterry, Richard E. "Modem Market," *Communications News,* March 1989.

Turner, Steven E. "Modem Makers Aren't Waiting for ISDN," *Business Communications Review,* June 1989.

Von Taube, Eugene. "Proper Modem Selection Ensures Free-Flowing PC Communication," *Data Management,* September 1986, p. 30.

CHAPTER

9

Communications Circuits

OBJECTIVES

After you complete your study of this chapter, you should be able to

- distinguish among a communications line, circuit, and channel;
- discuss various types of communications circuits and their distinguishing attributes;
- describe the characteristics of the various types of media used to carry circuits;
- describe multiplexing and concentrating and tell when they are most productively used;
- describe the major types of errors that occur on communications circuits;
- describe the primary error prevention and detection techniques for communications circuits.

INTRODUCTION

Previous chapters discussed communications lines or circuits without defining them. In this chapter, circuits are defined more precisely, and the many alternatives for the physical media used in a circuit's construction are examined. Information about the ways in which circuits can be configured and used also is presented.

This chapter continues the discussion and description of the first layer of the ISO-OSI model that began in Chapter 8. Communications circuits provide the physical path on which the communications signals flow. The nature of the signals sent on the path is discussed in Chapter 10. The information in this chapter and the next is at the heart of telecommunications. You must understand the characteristics of circuits to learn how telecommunications can be applied to meet the requirements of a particular application. This chapter describes many types of telecommunications circuits, and it gives examples of how they are being used by companies.

DEFINITIONS

Authorities differ somewhat in the details of the definition of the word *circuit*. The commonly accepted definition of a telecommunications circuit, however, is the path over which two-way communications take place. A circuit may exist on many different types of media, such as wire, coaxial cable, fiberoptic cable, microwave radio, or satellite. The word *line* is often used interchangeably with *circuit,* although line gives a stronger implication of a physical wire connection. In fact, many circuits today do not run on wires at all. The longer the distance, the higher the probability that the circuit runs on at least one medium other than wire.

A *link* is a segment of a circuit between two points. For a telephone circuit, one link exists between the residence or business and the local telephone company central office. Other links exist between the central offices. The final link is from the remote central office to the remote residence or business. When the term *data link* is used, it almost always includes the data terminal equipment, modems, and all other equipment necessary to make the complete data connection—software as well as hardware.

Circuits are often subdivided into *channels*, which are one-way paths for communications. Channels may be derived from a circuit by multiplexing, or they may be an independent entity, such as a television channel. A data circuit is sometimes divided into two channels, one of which is high speed for data transmission and the other is low speed for control information. The data channel is called the *forward channel,* and the control channel is the *reverse channel.* As the name implies, the reverse channel carries information in the opposite direction from the data channel. The type of information carried on a reverse channel depends on the rules of communications or protocol used. (Protocols are defined and discussed in Chapter 10.)

A *node* is a functional unit that connects transmission lines. It also can be an end point on a circuit or a junction point of two or more circuits. Typical nodes are telephones, data terminals, cluster controllers, front end processors, and computers.

TYPES OF CIRCUITS

Point-to-Point Circuits

A *point-to-point circuit,* illustrated in Figure 9–1, connects two—and only two—nodes. A typical circuit of this type connects two locations of a company or connects a computer to a terminal or a cluster controller. The standard telephone call between two locations is another example of the use of a point-to-point circuit.

Multipoint Circuits

If there are several nodes connected to the same circuit, as shown in Figure 9–1, it is called a *multipoint* or *multidrop circuit*. With multidrop circuits there is a clear distinction between the circuit and the links. The connections from the host to A, from A to B, and from B to C are links. The overall connection from the host to D is the circuit. Multipoint

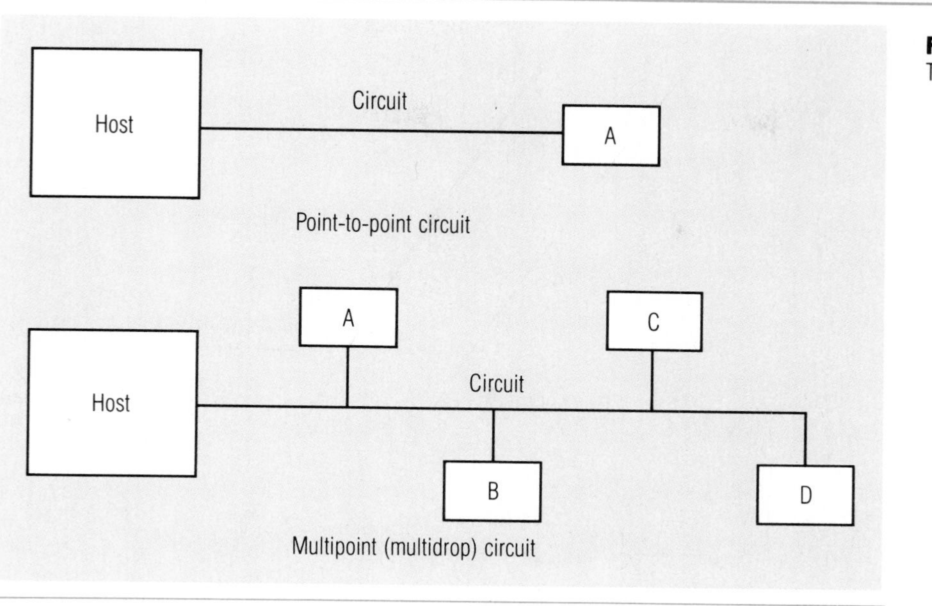

Figure 9–1
Types of circuits.

circuits are used where the volume of traffic between the terminals is sufficiently low that they can share the line and get adequate performance. In most cases, a multipoint circuit is less expensive than four point-to-point circuits, each connecting the host to one of the terminals (A, B, C, and D). Multipoint circuits are used to connect locations that have a relatively low volume of traffic and can share the line without interfering with one another.

Two-Wire and Four-Wire Circuits

Point-to-point and multipoint data circuits can be implemented with either two wires or four wires connecting the points. Normally, two wires are required to carry a communication in one direction. Therefore, a *two-wire circuit* has traditionally been viewed as being a half-duplex circuit and a *four-wire circuit* as a full-duplex circuit. As mentioned in Chapter 8, however, some modems are capable of splitting a circuit into two channels through frequency division multiplexing or other techniques. In that case, it is possible to obtain full-duplex operation on a two-wire circuit. In most cases, however, four-wire circuits are preferable for data communications. With four-wire circuits, two wires provide the forward channel in one direction while the other two wires provide the forward channel in the other direction.

Standard dial-up telephone circuits are two-wire circuits. Four-wire circuits must be ordered from the telephone company and installed on a leased basis. The advantage is that the circuit is then available for full-time use. The disadvantage is that a leased circuit may cost more than

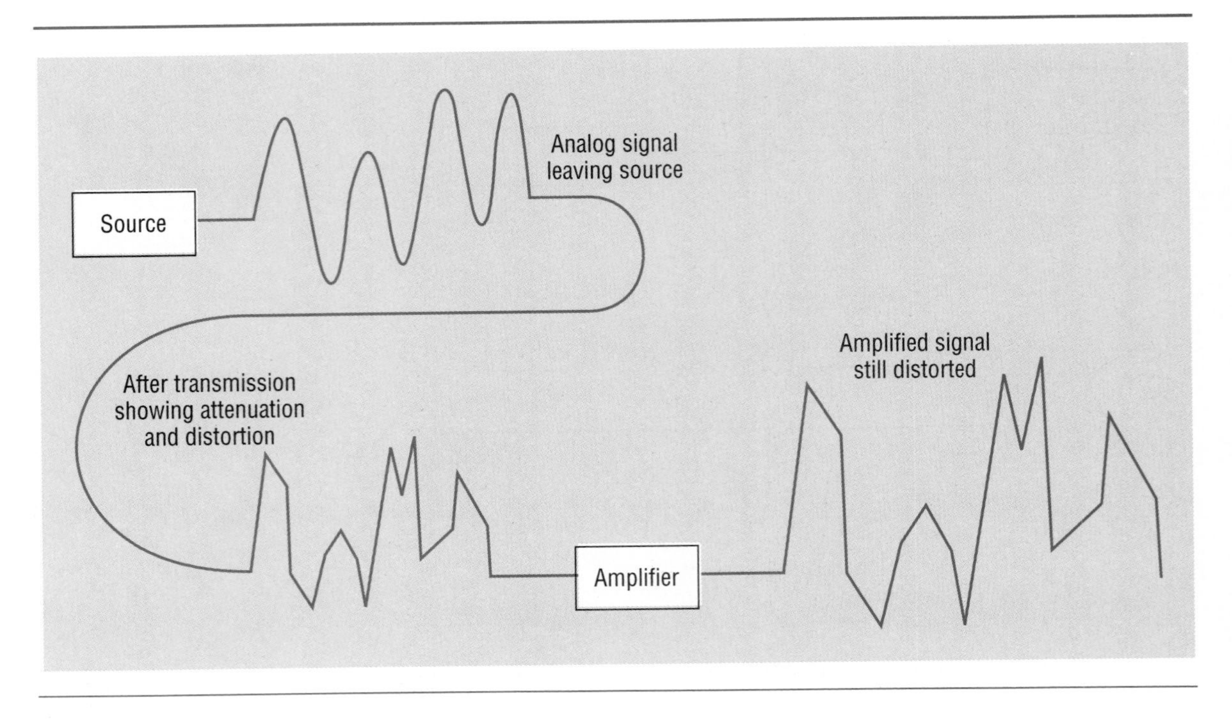

Figure 9–2
Amplification of an analog signal.

a dial-up connection, and it may be uneconomical if relatively little data will be transmitted.

Digital Circuits

Because the public telephone system was designed originally to carry analog voice transmissions, many of the circuits in place today are analog circuits. As was seen in Chapter 8, using analog circuits to carry data requires converting the signal from digital to analog before it can be transmitted and back to digital form at the receiving end. A *digital circuit* is one that has been designed expressly to carry digital signals. The direct digital transmission of data is simple and eliminates the signal conversions at each end. The common carriers have been using digital transmission techniques and digital circuits between central offices for years, but until the late 1970s, digital transmission was not available to end users. Today many new circuits being installed are designed especially for digital transmission.

One of the main advantages of digital transmission is that the distortion of pulses that inevitably occurs along the transmission path is easier to correct than the distortion of an analog transmission. Analog signals are periodically amplified to increase their signal strength, and the distortions in the wave form are amplified as well as the original signal, as shown in Figure 9–2. Digital signals, on the other hand, are made up of simple pulses and are not amplified but regenerated. The

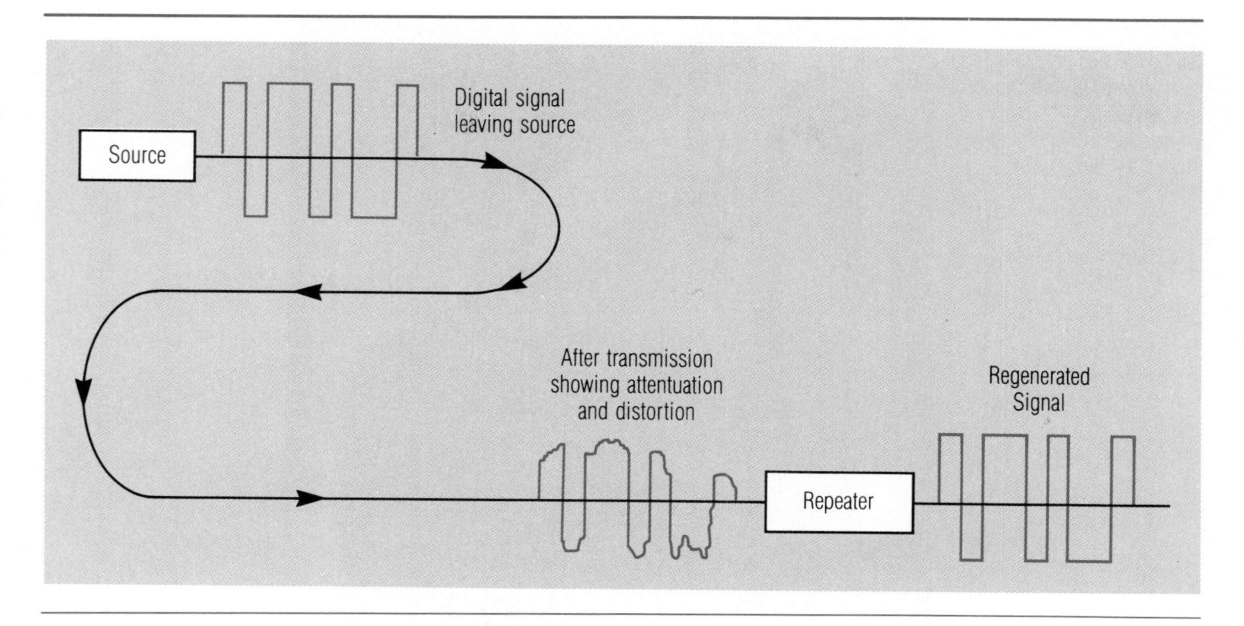

Figure 9–3
Regeneration of a digital signal.

regeneration eliminates any distortion that has occurred, as shown in Figure 9–3. Thus, the signal that arrives at the receiver is cleaner. The result is that digital transmissions have a lower error rate than do their analog counterparts.

Another advantage of using digital transmission is that no analog-to-digital conversion is required. Assuming digital links exist from one end of the circuit to the other, the digital output from the data terminal equipment (DTE) can be simply shaped and timed to conform to the requirements of the digital network and transmitted directly. The device that performs the shaping and timing is called a *data service unit/channel service unit (DSU/CSU)*, which was discussed in Chapter 8. DSU/CSUs are much less complicated and considerably less expensive than modems.

DSU/CSU

Carrier Systems

A family of high-speed digital transmission systems, known as the *T-carrier systems*, has been evolving within the Bell System over the past 30 years. T-carrier systems are designated according to their transmission capacity, as shown in the table below.

Designation	Bit Capacity
T-1	1.544 Mbps
T-2	6.312 Mbps
T-3	44.736 Mbps
T-4	274.176 Mbps

The T-2 and T-4 designations are used primarily by the carriers. T-1 and T-3 circuits are used by both the carriers and their customers.

The Timeplex multiplexer for T-1 circuits comes in several sizes depending on the number of T-1 lines to be handled. (Courtesy of Timeplex Corp.)

A T-1 system uses a standard pair of wires for transmission. Repeaters are spaced about every mile to regenerate the signal and transmit it over the next link of the circuit. A T-1 system can carry 24 circuits of 64,000 bps (24 × 64,000 = 1.536 Mbps; the extra 8,000 bps are used for signaling). Multiplexing equipment is used to combine signals for transmission over the T-1 system and to separate them at the receiving end. A 64,000 bps channel can carry one or two digitized voice signals, depending on whether PCM or ADPCM is used to modulate the signal.

It should be noted that in Europe T-carrier systems are also in use. They are defined slightly differently, however. A European T-1 circuit is made up of 32 to 64 kbps (kilobits per second) channels for a total capacity of 2.048 Mbps. Other T circuits are multiples of the T-1 capacity.

Companies acquire T-1 or T-3 facilities when they have a high volume of voice, data, or video transmissions between two locations. One company found it economical to install a T-1 circuit to carry three 56 kbps data circuits and five voice tie lines. Even though the capacity of the T-1 circuit was not fully used, it was less expensive than if the individual voice and data circuits were leased separately. Generally speaking, the multiplexing equipment is provided by the customer not the carrier. There are several companies in the market, such as Timeplex, Inc., Net, Inc., and General Datacomm, Inc., that specialize in developing and providing T-1 multiplexing equipment.

The major reasons for using T-1 circuits are that they can save large amounts of money, can give flexibility in reconfiguring the T-1 capacity to meet different needs at different times of the day, and can improve the quality of voice and data transmission because the information being transmitted is digitized.

Fractional T-1 A relatively new communications offering, which became available from the long distance carriers (IXCs) in 1989, is called *fractional T-1*. In the past, companies needing digital transmission service had no choice of speed between 56 kbps and T-1's 1.544 Mbps. Fractional T-1 provides companies with other transmission speed choices by subdividing a T-1 circuit into multiples of 64 kbps. Thus, the IXCs now offer leased circuits that operate at any multiple of 64 kbps, although speeds of 64, 128, 256, 512, and 768 kbps are most common. A company can select and pay for only the capacity it needs rather than having to lease a full T-1 circuit. This new capability is particularly interesting to companies with smaller networks that have not been able to justify the cost of a full T-1.

Fractional T-1 was introduced by the IXCs. Potential customers must check to see if their local telephone company provides fractional T-1 service. If not, an alternative must be found to connect from the customer's premises to the IXC office.

CIRCUIT SPEED RANGES

Whereas the modem or DSU usually determines the actual speed at which a circuit operates, carriers offer circuits of three general ranges. *Low-speed circuits,* also known as *subvoice-grade circuits,* are designed for telegraph and teletypewriter usage. These circuits operate at speeds of 45 to 200 bps. They cannot handle a voice transmission. In fact, subvoice-grade circuits are derived by dividing a voice circuit into 12 or 24 low-speed circuits. The public telex network uses circuits of this type.

Voice-grade circuits are those designed for voice transmission. They can transmit data up to 19,200 bps when sophisticated modems are used. These are the type of circuits that have been most commonly described throughout the book. *High-speed circuits,* sometimes called *wideband circuits,* are circuits designed to carry data at speeds greater than voice-grade circuits. They may be digital or analog and are designed to carry thousands of voice conversations or a full-motion color television channel. With multiplexers, the high-speed circuit can be subdivided into channels with speeds suitable for various applications.

CIRCUIT MEDIA

This section discusses the various media used for communications circuits. Although there is a wide overlap in their characteristics and capabilities, each medium has found a particular niche in which it is most commonly used.

Wire

twisted pair

crosstalk

The most common medium used for telecommunications circuits is ordinary wire, usually made of copper. Over the years, the common carriers have laid millions of miles of wire into virtually every home and business in the country, so wire is used for virtually all local loops. Originally, open wire pairs were used. However, they were affected by weather and very susceptible to electrical noise and other interference. Today, the pairs of wires are almost always insulated with a plastic coating and twisted together. This type of wire is called *twisted pair;* it is illustrated in Figure 9–4. Wire emits an electromagnetic field when carrying communications signals, but twisting the pair together has the effect of electrically canceling the signals radiating from each wire. To a large extent, it prevents the signals on one pair of wires from interfering with the signals on an adjacent pair. This type of interference is called *crosstalk.* The wire used for inside applications and local loops normally is 26, 24, or 22 gauge. The smaller the gauge number, the larger the wire.

Where wire enters a building, it is connected to a terminating block, sometimes called a *punchdown block,* with lugs or clips. This terminating block marks the demarcation point between the common carrier and the building owner, who is responsible to provide and maintain all wiring within the building. Before divestiture, the LECs provided and owned all inside wiring. This wiring represented a substantial asset on the accounting books of the LECs. Over time, this large base of previously installed inside wiring is being sold or given to the owners of the buildings where it is installed.

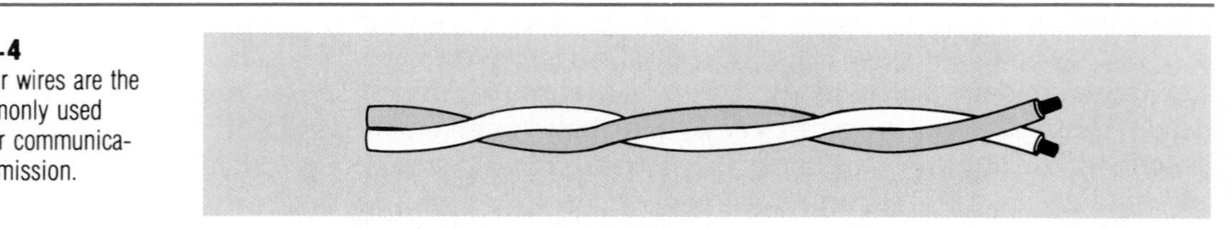

Figure 9–4
Twisted pair wires are the most commonly used medium for communications transmission.

Twisted pair wires are connected together at this punchdown block in the telephone equipment room.

As wires leave a building, they may be directly buried in the ground or suspended from overhead poles, as shown in Figure 9–5. As they approach the central office, they are grouped together in cables that get larger the closer they get. Up to several thousand pairs are grouped together into large cables. They are sometimes surrounded by a wire shielding to provide protection from electrical interference or a heavy metal armor for physical protection.

On local loops, one pair of wires usually is dedicated to one telephone or data circuit. The theoretical maximum data-carrying capacity of a single twisted wire pair is in excess of four million bits per second under controlled conditions. The telephone company frequently uses a twisted wire pair to carry a 1.544 Mbps T-1 transmission, with repeaters every mile. In most cases, however, the actual data rate carried is significantly less.

Shielded Twisted Pair

A variation of twisted pair wiring is *shielded twisted pair*. Twisted pair wire is placed inside a thin metallic shielding, similar to aluminum foil, and then enclosed in an outer plastic casing. The shielding provides further electrical isolation of the signal-carrying pair of wires. Shielded twisted pair wires are less susceptible to electrical interference caused by nearby equipment or wires and, in turn, are less likely to cause interference themselves. Because it is electrically "cleaner," shielded twisted pair wire can carry data at a faster speed than unshielded twisted

Figure 9–5
Wires leaving a building may be buried or hung from poles above ground.

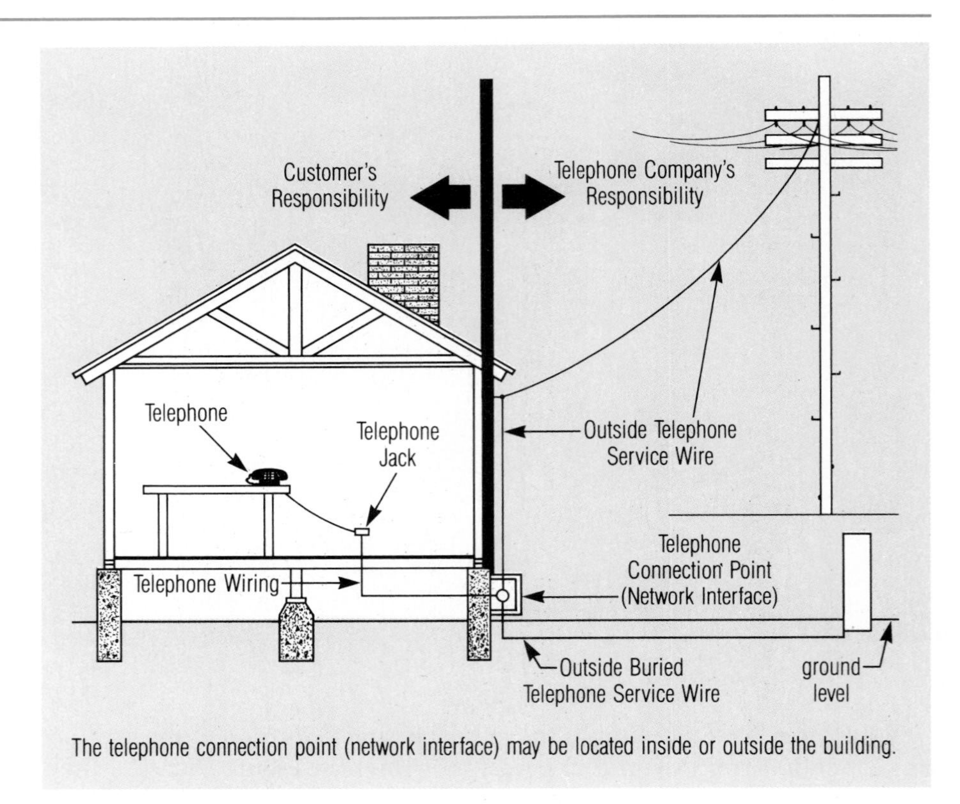

The telephone connection point (network interface) may be located inside or outside the building.

pair wire can. The disadvantage of shielded twisted pair wire is that it is physically larger and more expensive than twisted pair wire, and it is more difficult to connect to a terminating block.

Coaxial Cable

Coaxial cable, as the name implies, is cable made of several layers of material around a central core, as illustrated in Figure 9–6. The central conductor is most often a copper wire, although occasionally aluminum is used. It is surrounded by insulation, most typically made of a type of plastic. Sometimes spacers are put in the cable to keep the center conductor separate from the shielding, and in that case, the insulation material is air or an inert gas. Outside of the insulation is the shielding, which is also a conductor, typically fine, braided copper wire. The shielding is surrounded by the outer insulation, which is almost always a form of plastic that also provides physical protection for the cable.

Coaxial cable has a very large bandwidth, commonly 400 MHz to 600 MHz, and therefore a very high data-carrying capacity. The telephone industry uses pairs of coaxial cable in areas where the population density is high. One coaxial cable can carry up to 10,800 voice conver-

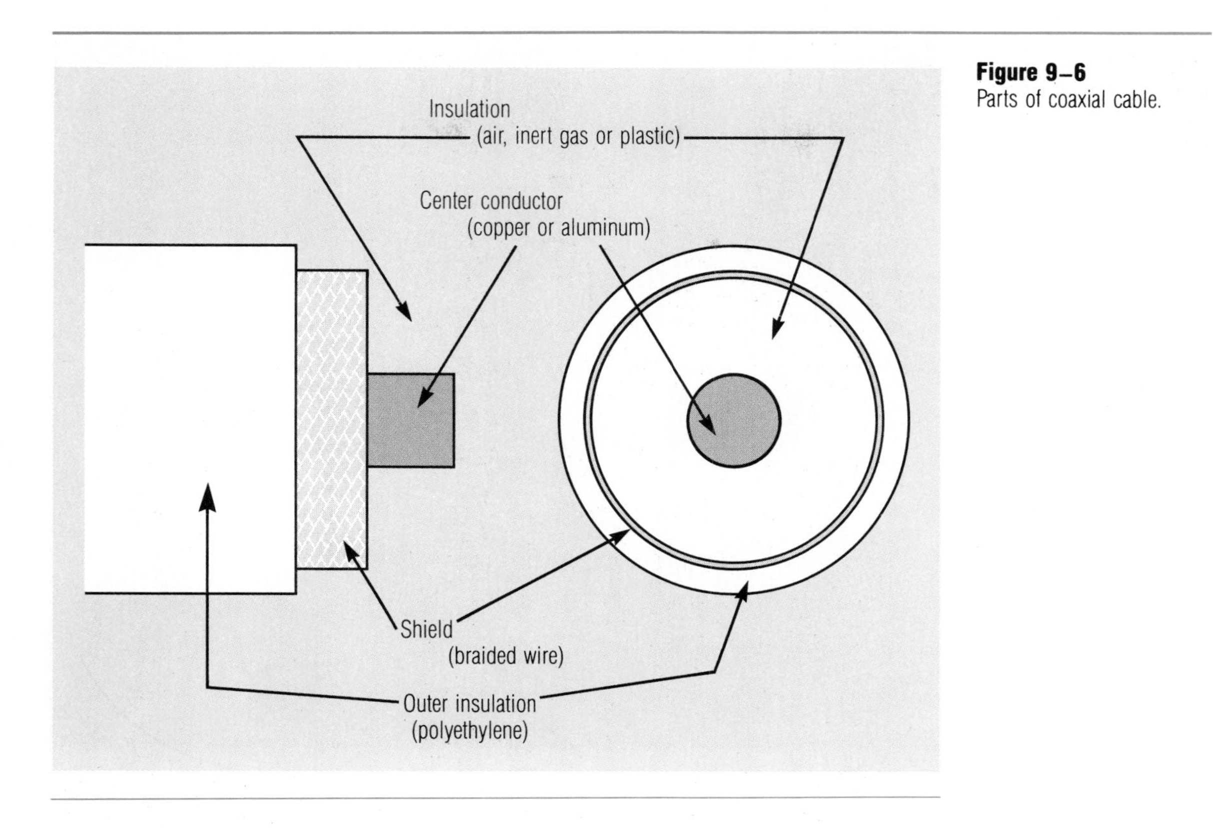

Figure 9–6
Parts of coaxial cable.

sations when amplifiers, spaced about a mile apart, are used to boost the signal. The cable television industry uses coaxial cable extensively for carrying television signals from a central transmitter to individual homes or other subscribers. Over 50 television channels can be carried on a single coaxial cable.

Coaxial cable can be tapped easily. This is an advantage when it is used around an office or factory where many taps are needed but a disadvantage if one is concerned about security and illegal taps. The cable can be bulky and, therefore, difficult to install. Some cable has a rather large bending radius, which must be considered when planning the installation. Because of its shielding, coaxial cable is quite immune to external electrical interference, making it a good candidate for use in electrically noisy environments.

Optical Fiber

Optical fibers are a relatively new technology, although they have been used as a transmission medium by the common carriers for years. The technology is still advancing rapidly, and every few months one reads of the latest developments, many of which have to do with the transmission speed that an optical fiber can handle.

Several coaxial cables are often packed together as shown in this photograph. The 4½-inch cable on the left can carry as many as 40,300 telephone conversations. The optical fiber cable on the right is only 1/2 inch in diameter and can carry more than 46,300 telephone conversations. (Courtesy of AT&T Bell Laboratories)

The optical fiber itself, illustrated in Figure 9–7, is a very thin glass fiber of high purity. The glass *core* at the center provides the transmission-carrying capability. It looks like a very fine fishing line but is in fact a very pure, clear strand of silica glass. The core is surrounded by another type of glass called *cladding*, which is reflective and acts like a mirror to the core. The cladding is covered by a protective covering, usually of plastic. The total diameter of the fiber is less than that of a human hair. Individual fibers often are bundled together in groups around a central metallic wire that provides strength for pulling the fiber cable through conduit when it is laid.

light source

Data is placed on the cable with a light source, either a *light-emitting diode (LED)* or a *laser*. The laser is more powerful and is used when the distance between the transmitter and receiver is greater than 5 to 10 miles. For shorter runs, the less expensive LED can be used. The light stays in the core because the cladding has a low refractive index. When the light beam hits the edge of the core, it is reflected back toward the center by the cladding's mirrorlike surface. The light output of the LED or laser is modulated to provide the variations in the signal that can be interpreted at the receiving end. Without special techniques, a fiber carries a signal in one direction only, so fibers are normally used in pairs.

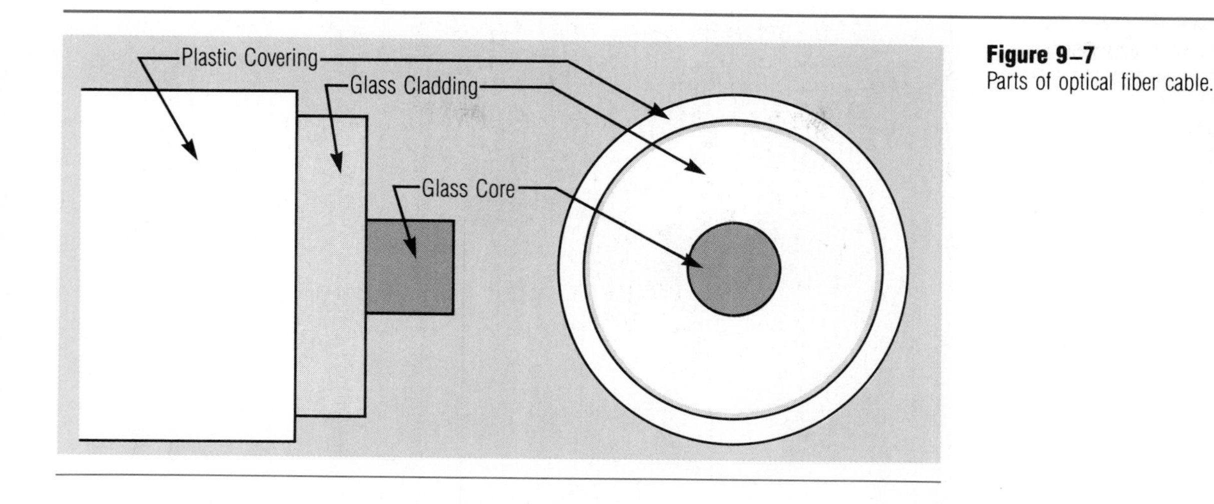

Figure 9–7
Parts of optical fiber cable.

There are two primary types of fiber used. One is called *single mode* and uses a fiber with a glass core approximately 5 microns (.005 millimeters) in diameter. With this very small core size, the light beam travels down the center of the core with little reflection from the cladding. Because the core is so small, however, it requires a very concentrated light source to get the signal into the fiber with adequate strength for long distances.

single mode

The other type of fiber is called *multimode*. It has a core approximately 50 microns (.050 millimeters) in diameter. With the larger core size, it is easier to get the light into the fiber, but there is more reflection from side to side off the cladding. Some of the light rays travel essentially straight down the center of the fiber, whereas others reflect at various angles. Those that travel straight through the fiber arrive at the destination faster than those that reflect, a phenomenon called *dispersion*. The effect of dispersion is that it causes the square pulses of a digital signal to become rounded and effectively limits the signaling rate that can be achieved. The trade-off, then, is that implementing a multimode fiber system costs less than a comparable system built with single-mode fiber. The cable and light source cost less, but the signal-carrying capacity is also less than in a single-mode fiber system.

multimode

One of optical fiber's most notable characteristics is its high bandwidth. Single-mode fibers have carried data at a rate of 20 billion bps over a distance of 50 miles under laboratory conditions. In practical applications, 135 Mbps data rates are achieved routinely for distances of 40 miles or more. For longer distances, repeaters can be inserted in the fiber to regenerate the signal. For shorter distances, much higher rates of 1.7 Gbps and higher can be achieved.

Fiberoptic cables are very difficult to splice, requiring specialized tools and skills. They are best suited for long point-to-point runs in which few or no splices are required. Splices can be detected using a

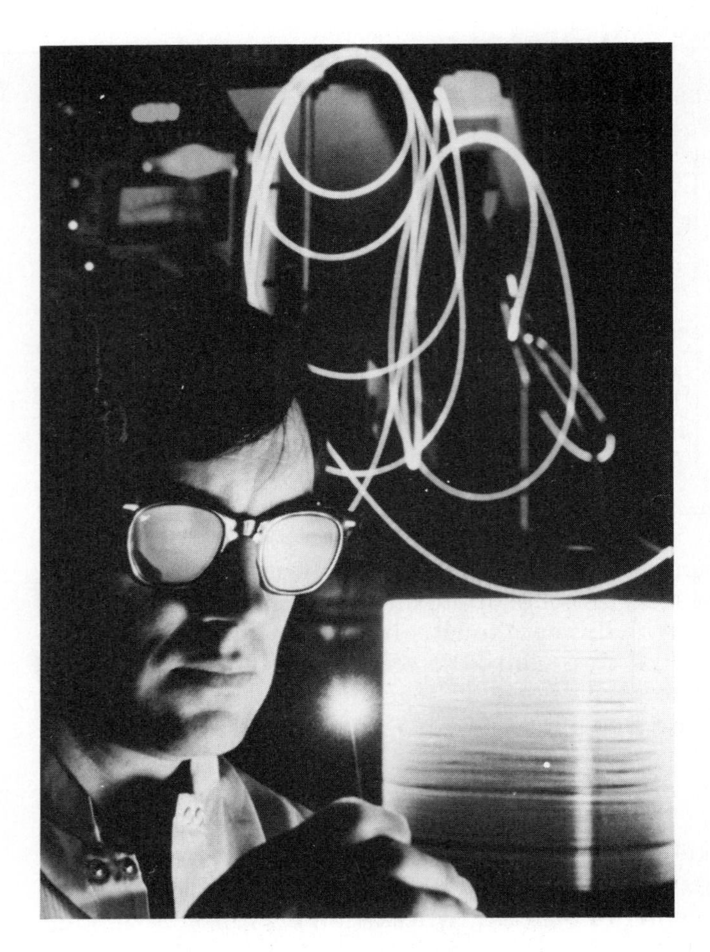

The flexible optical fibers shown in this photograph are approximately the diameter of a human hair. (Courtesy of AT&T Bell Laboratories)

reflectometer, an instrument that sends a light wave down the fiber and measures the reflection that comes back. From a security standpoint, this difficulty in splicing fiberoptic cables and the ability to detect unwanted taps is an advantage. In addition, since the transmission is optical rather than electrical, the cables do not radiate signals, as all electrical devices do. Given these attributes, fiberoptic cables are excellent in situations that require very high security.

Another advantage of fiberoptic cables over wire or coaxial cables is that the fibers are so thin that a fiberoptic cable has a very small diameter and is lightweight. This makes the cables easier to install, since they can be pulled through smaller conduits and bent around corners much more easily than coaxial cable. They can even be installed under carpeting or floor tiles, giving additional flexibility in laying out an office.

Costs of fiberoptic cable and its related components are dropping rapidly and are expected to continue doing so for the foreseeable future. The cost reductions result from an ever-increasing market size and improvements in manufacturing techniques.

A standard for the use of optical fibers in local area networks (LANs) is nearing completion by the American National Standards Institute. This standard, coupled with the cost reductions, virtually ensure that optical fiber will be more widely used for LANs in the future. Because the standard is related to other LAN standards, it is discussed in more detail when LANs are described in Chapter 11.

In late 1988, the first optical fiber undersea cable, TAT-8, was completed between the United States and Europe. TAT-8 effectively doubled the total cable capacity across the Atlantic Ocean and has four times the capacity of the previous trans-Atlantic coaxial cable, which was installed in 1983. The main cable consists of six fibers; two pairs are used for voice and data, and the third pair is for backup. Regenerators are located every 43 miles along the ocean floor. The first undersea cable between the United States and Europe was installed in 1956 and could handle 36 simultaneous calls. TAT-8 can handle 40,000 simultaneous calls, a remarkable advancement in technology.

In summary, the characteristics of optical fiber are

- high bandwidth—very high data-carrying capacity;
- little loss of signal strength—depends on the details of the cable construction, but the overall characteristics are excellent;
- immunity to electrical interference—since it operates in the optical part of the spectrum, electrical noise is not an issue;
- excellent isolation between parallel fibers—crosstalk between fibers does not exist;
- small physical size—lightweight;
- very secure—difficult to tap and splice and does not radiate electrical signals.

One large company in New York recently installed several fiberoptic cables to connect several locations in its community. In one case, the fiber cable was run along a railroad track, and in another situation, the fiber was run through some abandoned underground conduit formerly used for electrical wires. The fibers are primarily carrying voice traffic today to allow a single PBX to serve the company's multiple locations around the city. There are plans to use the fiber to carry data traffic in the future.

Microwave Radio

Microwave radio is the medium most used by the common carriers for long distance communications transmission. Microwave radio transmissions occur in the 4 to 28 gHz frequency range. Specific frequency bands are set aside, and channels are allocated within the bands. Up to 6,000 voice circuits are carried in a 30 mHz-wide radio channel.

At the frequency range in which they operate, microwave radio signals travel in a straight line. Therefore, the transmitter and receiver must be in a direct line of sight with each other. Because of the curvature

This microwave tower is located in New Jersey. Both parabolic and horn antennas can be seen. (Courtesy of AT&T Bell Laboratories)

of the earth, microwave antennas are usually placed on high towers or building roofs to extend the line of sight to the greatest distance possible before another antenna is required. Practically speaking, a range of 20 to 30 miles between towers is common if the terrain is not too hilly. Where the distance to be covered is short, microwave antennas can be placed on the side of a building or even in an office window.

Microwave signals may carry data in either analog or digital form, but analog is more common. Voice, data, and television signals are carried, but each is given its own channel. Depending on the frequency used, some microwave signals are subject to interference by heavy rain. When this occurs, the channel is unusable until the rain subsides. Therefore, when a microwave system is designed, provisions must be made for this possibility. Temporarily stopping the transmission, sending it on an alternate path, and transmitting via a different medium are alternatives to be considered.

Microwave is sometimes installed privately by companies to connect locations that are near but not adjacent to one another. A radio license must be obtained from the Federal Communications Commission (FCC), but no right-of-way permits are necessary. Several companies in New York City have private microwave links connecting offices in different parts of Manhattan. Similar private microwave installations can be found in most major cities.

Microwave systems should be considered when T-1 circuits are not available, when there is a financial advantage or other requirement to have a privately owned transmission system, or when alternative routing is required for certain critical communications links. Sales engineers from the microwave equipment vendors help to obtain the required transmission license, as well as configuring the equipment for the specific terrain, transmission speed, and reliability requirements. Major vendors of microwave equipment are Digital Microwave Corporation, M/A-Com MAC, Inc., Motorola, Inc., Rockwell Communication Systems, and Microwave Networks, Inc.

Satellite

Transmission using an earth satellite is a particular type of microwave radio transmission. In a typical satellite system, a microwave radio signal is transmitted from an antenna on the ground to a satellite in an orbit 22,300 miles high around the earth. At that distance, the circular speed of the satellite exactly matches the speed of rotation of the earth, and the satellite appears to be stationary overhead. This is called a *geosynchronous orbit*. An antenna on the earth can be aimed at the satellite and because the satellite appears stationary, the aim doesn't have to be changed. Although the distance is great, the antennas are definitely in sight of each other. The microwave radio signal is beamed to the satellite on a specific frequency called the *uplink*, where it is received, amplified, and then rebroadcast on a different frequency, called the *downlink*. This is illustrated in Figure 9–8.

Because of its distance from the earth, the satellite can see and be seen from approximately one third of the earth. The signals broadcast from the satellite can, at least theoretically, be picked up by any antenna and receiver in that area. This broadcast attribute is an advantage in some applications and a disadvantage in other situations. It is advantageous for organizations that want to use satellites to reach a mass market. Home Box Office (HBO) uses the broadcast capability to distribute movies to cable TV companies. Financial companies use satellite transmission to distribute stock market information to brokers all over the country.

For companies that want to use the satellite for point-to-point transmission, the broadcast capability presents a security concern. Anyone with the proper equipment can receive the broadcast as it is transmitted down from the satellite. When this concern is serious, encryption devices

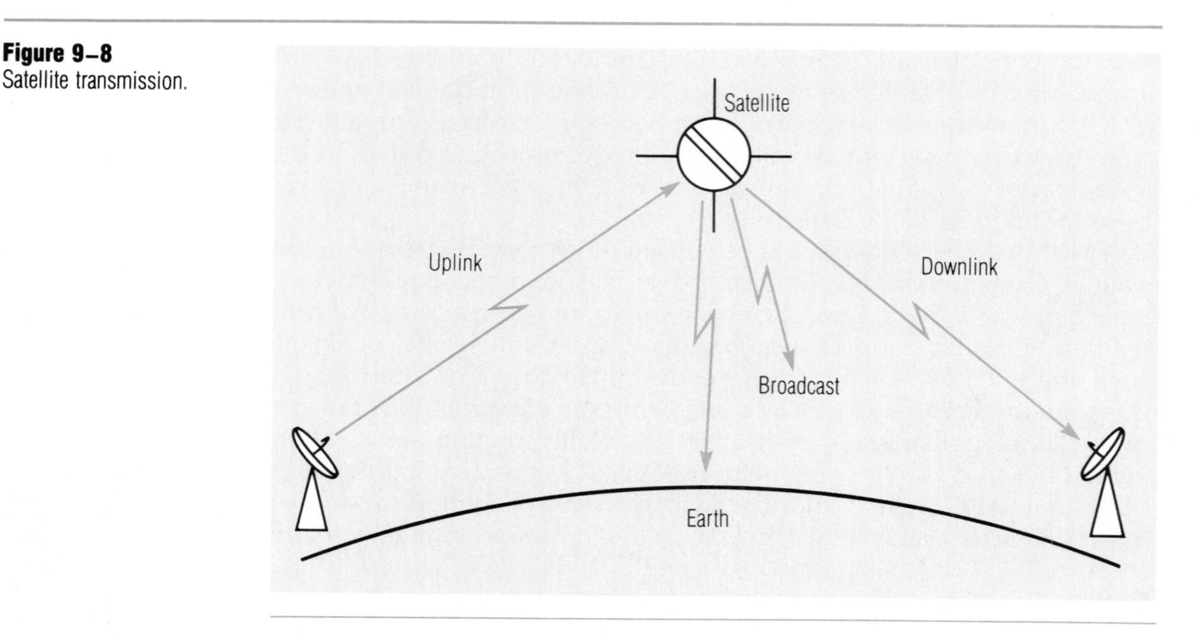

Figure 9–8
Satellite transmission.

can be used to code the data before it is transmitted to the satellite. This makes it difficult to interpret the broadcast on the downlink.

Another characteristic of satellite transmission is that because of the great distance involved—22,300 miles up to the satellite and 22,300 miles back—there is a noticeable delay from the time a signal is sent until it is received. This delay, called *propagation delay,* exists for all communications circuits and radio broadcasts. It is a function of the fact that light signals or radio waves travel at a maximum speed of 186,000 miles per second in a vacuum. A signal traveling 4,000 miles across the United States on a terrestrial circuit will travel somewhat slower than 186,000 miles per second, but an approximation of the propagation delay can be calculated as

4,000 miles/186,000 miles per second = .0215 seconds (21.5 milliseconds)

When the signal goes via satellite, it has to travel 22,300 miles up to the satellite and 22,300 miles back to the receiving earth station. The propagation delay is

44,600 miles/186,000 miles per second = .2398 seconds (239.8 milliseconds)

The satellite signal takes 11 times longer to reach its destination. If a return signal is required, the same amount of delay would be encountered a second time. You may have noticed this satellite delay on some domestic long distance telephone calls a few years ago, and it still can be heard on some international telephone calls today. Most people find

These satellite dish antennas are about 5 meters in diameter. Other dish antennas are as small as 1 meter. (Courtesy of Contel ASC)

the delay annoying, and the common carriers in the United States have switched most voice transmissions back to terrestrial media.

The propagation delay when using a satellite can be extremely significant when data is being sent. Since almost a quarter of a second is added to each transmission in either direction, the transmission time becomes a significant portion of the total response time in an interactive application. We will look at the effect of satellite delay on data transmission in more detail when transmission protocols are discussed in Chapter 10.

Satellite circuits have become very cost effective in recent years due to a glut of satellite capacity and to advancing technology. Smaller antennas are required than in years past, removing one of the major objections to satellite services. Hercules Chemical Company has one of the largest private satellite networks in the world. The Hercules network connects major company locations in the United States and Europe and is used for voice, data and television broadcasts.

The chart in Figure 9–9 summarizes the characteristics of the various media used for telecommunications transmission.

The ASC-1 is one of Contel ASC's fleet of satellites. In geosynchronous orbit, such a satellite can see about one third of the earth's surface. Its antenna is focused on a much smaller area, called the *footprint*, to give maximum signal strength to the areas of interest. (Courtesy of Contel ASC)

CIRCUIT ACQUISITION AND OWNERSHIP

Another way of classifying telecommunications circuits is by the ownership and method of acquisition.

Private Circuits

Private circuits are those installed and maintained by a company other than a common carrier. For example, a company may run coaxial cable between its buildings on a manufacturing site and wire within the buildings to form a data communications network. The typical situation where private circuits are installed is within a building or on a campus environment, where all the property is owned. It becomes more complicated if the circuit must cross property owned by others because permission to cross the property must be obtained. It is even more complicated when public roads or highways must be crossed, but it is possible to obtain permission to cross or bury cable under them. Permission is

	Twisted Pair	Coaxial Cable	Microwave Radio	Satellite	Fiber
Transmission speed	4 Mbps	140 Mbps	275 Mbps	2 Mbps	400 Mbps
Ease of installation	Easy	Moderate	Difficult	Difficult	Difficult
Cost	Least	Moderate	Moderate	High	High
Maintenance difficulty	Low	Moderate	Low	Low	Low
Skill required to install	Low	Moderate	High	High	High
Most common uses	Within buildings	Campus Multi-drop	Point-to-point Short distance	Point-to-point Long distance	Point-to-point
Advantages	Inexpensive, familiar	Carries more information than twisted pair	Speed	Speed, availability	Speed, secure
Disadvantages	Lowest data rate, subject to interference	Bulky	Can be intercepted	Delay, can be intercepted	Difficult to splice, cost
Notes	Shielded twisted pair allows higher speed	Broadband use is more maintenance intensive	Requires radio license from FCC	Private systems not common	Higher speeds coming

Figure 9–9
Comparison of the attributes of the media most commonly used for telecommunications transmission.

obtained from the agency in control of the road; it may be a city, county, state, or the federal government.

In some states, it is possible to get permission to run wire or cable on public utility poles. Usually, the utility company charges a small fee for the use of the pole. Private circuits also have been installed on the shoulders of public roadways, alongside railroad tracks, and on pipeline routes.

When a company installs a private circuit, it is totally responsible for the design, engineering, installation, operation, and maintenance. On the other hand, the circuit is available for the company's full-time, exclusive use, and once the circuit is installed, it is usually very inexpensive to operate. One company, which had many locations in a city, received permission to run a private cable on the poles of the local electrical utility company. The rental charge was $6 per year per pole, an amount the company considered very reasonable. Using the cable, the company was able to connect all of its locations in the city into a common data network.

Leased Circuits

Leased circuits are circuits owned by a common carrier but leased from them by another organization for full-time, exclusive use. Leased facilities are attractive when some or all of these conditions are present:

- it is impossible or undesirable to install a private circuit;
- the cost of the leased circuit is less than the cost of a dial-up connection for the amount of time required;
- four-wire service is required (four-wire service cannot be obtained on dial-up connections);
- high-speed transmission greater than 9,600 bps is required.

The primary advantage of a leased circuit is that it is engineered by the carrier, installed, and left in place so that the same facilities are always used. This means that once the circuit is adjusted and operating correctly, it will continue to operate the same way for long periods of time. For most business data transmission, this consistency and reliability are significant benefits because they go a long way toward helping ensure that the communications service is reliable and trouble-free.

When a leased circuit does fail, the carrier that provides the circuit performs the diagnostic and maintenance work required to restore it to service. Carriers have special testing equipment located at their central offices with which they can examine all of the parameters of a circuit to determine the cause of failure. Furthermore, the technicians at all of the central offices through which the circuit passes can communicate with each other to determine which link in the circuit is experiencing the problem. In most instances, failures are isolated and corrected and the circuit is returned to normal operation within hours. Often, service is restored in minutes.

The price of leased circuits is based on speed and distance. For voice-grade circuits there is a charge for the local channel from the customer premises to the serving central office. The local channel is acquired from the LEC. If the circuit crosses LATA boundaries, it must be carried by an interexchange carrier (IXC). In that case, the local channel and its associated charges extend through the serving CO to the IXC's POP. Added to the local channel charge is the interoffice channel charge for the circuit connecting the POPs. For point-to-point circuits, the carrier computes the shortest airline mileage between the two POPs and bases the monthly charge on that distance. For multipoint circuits, the carrier computes the shortest airline mileage between all the points on the circuit.

A leased four-wire circuit costs about 10 percent more than a leased two-wire circuit. Since additional throughput can be obtained, most users pay the additional cost for four-wire circuits when they want to use half-duplex or full-duplex transmission.

One final note: Leased circuits often are called "private" circuits in common use. The distinction between the two types that has been made in this book frequently is not made by people in the communications industry. It is common to hear someone talk about "our private line" when they really mean "our leased line." For many purposes, the distinction makes no difference, but if in doubt, it is best to request a clarification.

Bypass A topic that has been discussed widely in the telecommunications industry since divestiture is *bypass*. In its simplest form, bypass involves installing private telecommunications circuits to avoid using (to bypass) those of a carrier. The usual reason for considering bypass is to reduce costs, although in some cases, a company may consider bypass in order to obtain a capability the LEC cannot provide.

One application of bypass is to connect a company's facility directly with an IXC, bypassing the LEC. An example where this might be necessary is when the LEC cannot provide fractional T-1 service. A full T-1 circuit or microwave link could carry the voice and data traffic directly from the company's location to the IXC, providing access to the IXC's fractional T-1 service. In another case, bypass circuits might be installed to eliminate the local access charges for circuits that the LEC would normally impose. An economic analysis would have to be performed to determine over what time period the elimination in local access charges would offset the cost of installing the bypass circuit. Another practical consideration may be the company's relationship with the LEC, since the elimination of circuit revenue caused by the bypass will certainly not make the LEC happy.

Switched (Dial-Up) Circuits

Switched or dial-up circuits come from the standard public telephone network, and using them for data is similar to making a normal telephone call. A temporary connection is built between DTEs as though there were direct wires connecting the two. The circuit is set up on demand and discontinued when the transmission is complete. This technique is called *circuit switching*. An obvious advantage of circuit switching is flexibility in network configuration. Transmission speeds of up to 19,200 bps can be obtained on switched circuits, but lower speeds are more common. Charges for switched circuits are based on the duration of the call and the distance, just like a standard telephone call.

Today, when the public telephone network is used for dial-up data transmission, the transmission normally occurs in analog form. AT&T is installing a dial-up digital transmission network, especially designed for data transmission, throughout the United States. The basic unit of speed on the digital network is 56 or 64 kbps.

One factor that must be dealt with when using switched circuits is that when a dial-up connection is made, the actual carrier facilities used in routing the call depend on the facilities that are available at the moment. For this reason, circuits are variable in quality and may be very good on one connection and marginal on another. We have all experienced this phenomenon on long distance telephone calls. Sometimes, it seems as if the person we are talking to is just next door; other times, the volume of the person's voice is low; during other conversations, crosstalk can be heard.

Because of the variability in quality of dial-up connections, the data transmission speed that can be dependably achieved is less than on leased circuits. Some modems are designed to accommodate this situation. These modems start by transmitting at the highest speed of which they are capable. If too many errors are detected, the modems automatically switch to a slower speed until they achieve adequate reliability.

CIRCUIT IDENTIFICATION

Whether a particular facility is a circuit, link, or channel depends to a certain extent on one's point of view. From the user's point of view, a multipoint circuit from Los Angeles to Dallas and Miami has links from Los Angeles to Dallas and from Dallas to Miami. From the carrier's point of view, the user's circuit is a part of a higher-speed facility onto which it is multiplexed. The carrier also sees many more links, in fact, probably thinks of the circuit as being made up of a series of links between each central office through which the circuit passes, plus the links at each end (local loops) connecting the serving central office to the customer premises. As the circuit passes through high-speed, long distance microwave facilities crossing the country, it is likely to be viewed by the carrier as a one-way, two-wire channel within the microwave link. The other channel, running in the other direction, may be on some other microwave link and running totally independently.

The different viewpoints lead to occasional difficulties when communications customers and carriers talk to each other because they may use the same terms to mean different things or different terms to mean the same thing. To partially address this problem, the carrier attaches a circuit number to each circuit for identification purposes. The carrier has blueprints or other documentation that identifies every link making up the circuit and every piece of equipment through which it passes. Customers and carriers use the circuit number when they discuss problems or talk about changes to the circuit.

MULTIPLEXING AND CONCENTRATING

Chapter 5 discussed frequency division multiplexing (FDM). When FDM is used, signals are shifted to different parts of the frequency spectrum, so that a single pair of wires can carry more than one transmission. Data signals in analog form are shifted using FDM techniques.

When transmissions are in digital form, time division multiplexing normally is used.

Time Division Multiplexing (TDM)

Time division multiplexing (TDM) is a technique that divides a circuit's capacity into time slots. Each time slot is used by a different voice or data signal. If, for example, a circuit is capable of a speed of 9,600 bps, four terminals, each transmitting at 2,400 bps, could simultaneously use its capacity. A TDM takes one character from each terminal and groups the four characters together into a *frame* that is transmitted on the circuit. This process is shown in Figure 9–10. At the receiving end, another TDM breaks the frame apart and presents data to the computer on four separate circuits.

Time division multiplexing is totally transparent to the terminal, the computer, and the user. If a terminal has nothing to send at any point in time, its time slot in the frame is transmitted empty. A typical application for TDM is to have multiple slow-speed terminals at one location communicating to a computer at another location. Although four terminals are shown in Figure 9–10, more than four could be multiplexed (however, multiples of four are most common). Eight terminals transmitting at 1,200 bps would work just as well on a 9,600 bps line with the proper TDM equipment.

Figure 9–10
Time division multiplexing.

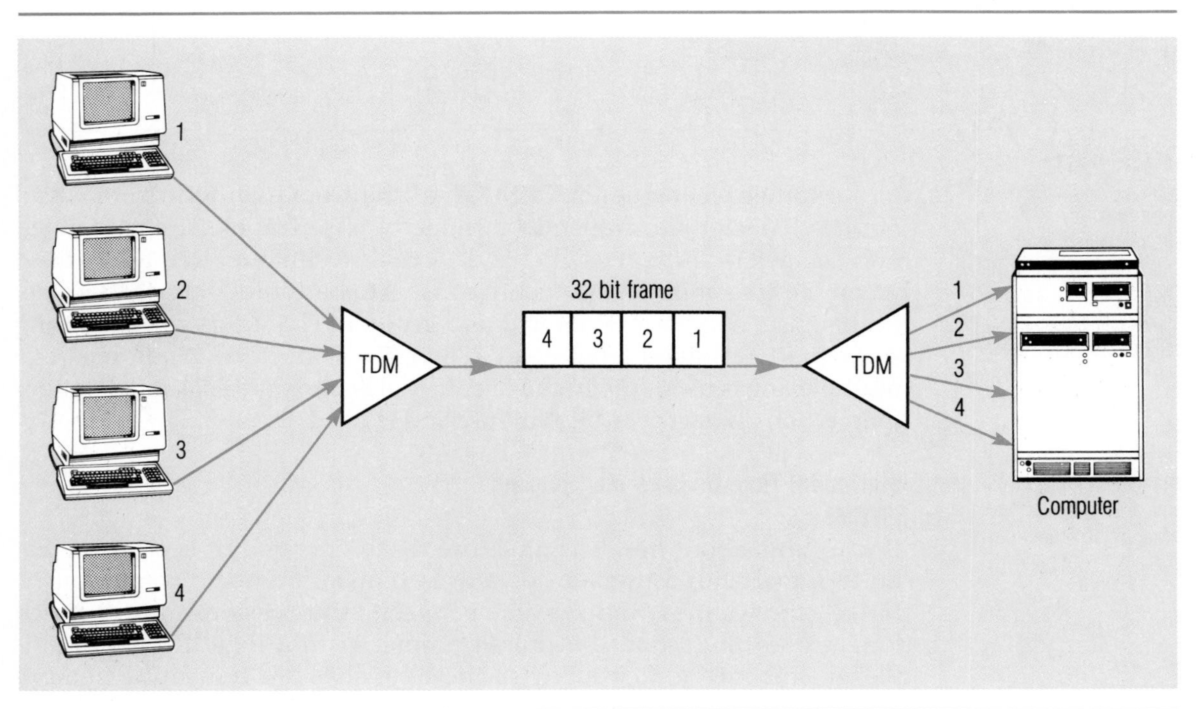

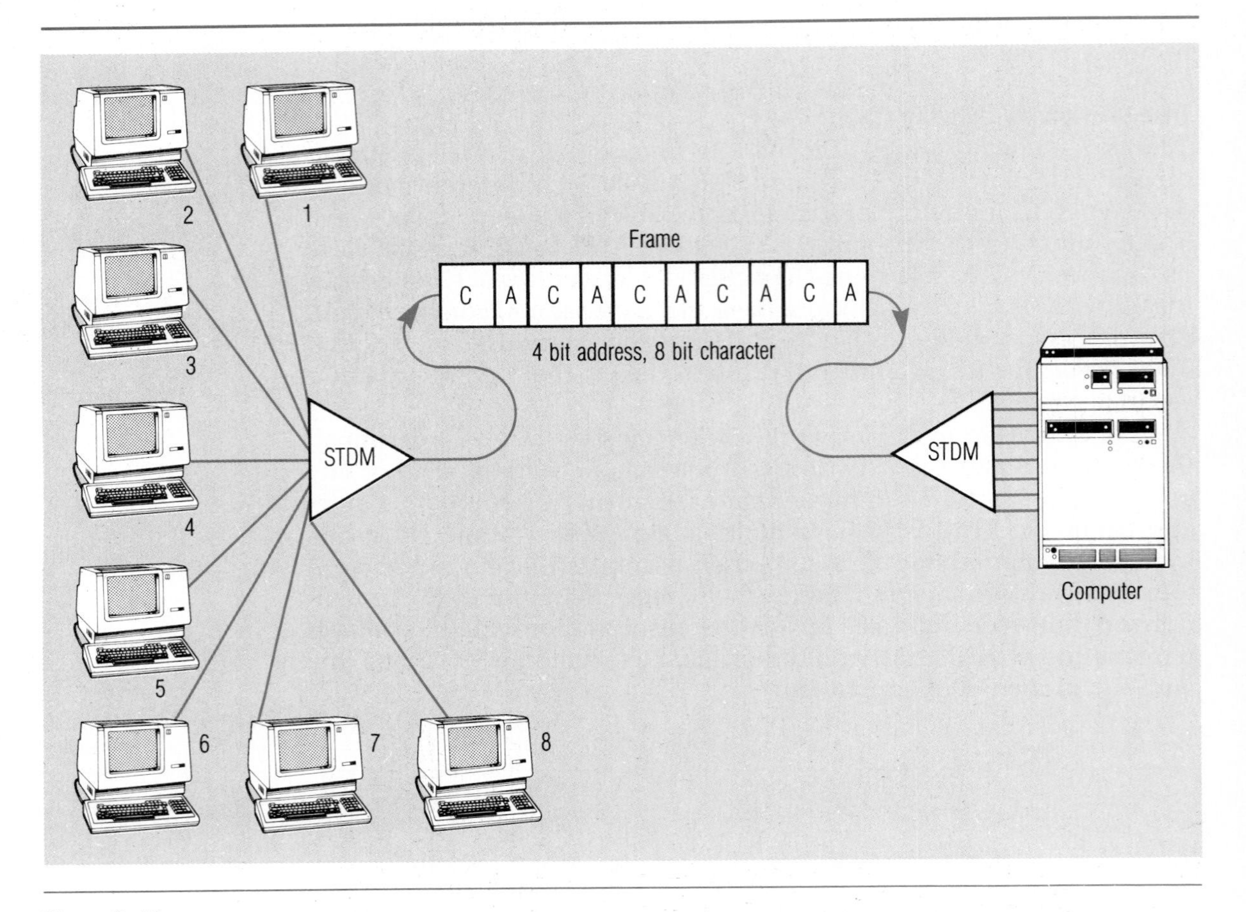

Figure 9–11
The STDM tries to avoid having empty slots in a frame, thereby improving the line use. If a terminal has no data to send in a particular time period, the STDM will see if the next terminal has data that can be included in the time slot. When the STDM at the receiving end breaks the frame apart, it uses the terminal address to route the data to the proper device.

Another technique for TDM takes 1 bit from each terminal instead of one character and transmits a frame of bits. The bits are assembled into characters at the receiving end. A third technique, frequently used when the transmission is synchronous, is to multiplex entire messages. In this case, a *message* is defined as a group of characters not exceeding some predetermined length, say 128 or 256 characters. When message multiplexing is used, the frame of data that is transmitted is much longer than when character or bit multiplexing is used.

Statistical Time Division Multiplexing (STDM)

If you think about how terminals are really operated, it is obvious that no terminal with a human operator is transmitting data continuously. In fact, the wait, or "think," time between transmissions may be much longer than the actual transmission time. With a time division multiplexer, this means that many of the time slots are transmitted empty and the capacity of the circuit is not fully used.

The Codex 6740 Statistical Multiplexer supports a mix of asynchronous, synchronous, and bit-oriented protocols. The 6740 can handle up to eight links ranging in speed from 2,400 bps to 64,000 bps. (Courtesy of Codex Corporation)

A *statistical time division multiplexer (STDM)* is a multiplexer that does not assign specific time slots to each terminal. Instead, the STDM transmits the terminal's address along with each character or message of data, as illustrated in Figure 9–11.

If the address field in an STDM frame is 4 bits long, there are 2^4 combinations and 16 terminals can be handled. With a 5-bit address, 32 terminals can be multiplexed. In any case, extra bits are required for the addresses when STDM is used. For most applications, this additional overhead is a good trade-off. Most of the time, the user will not notice any difference in performance or response time, and the line will be better used.

Occasionally, there are times when most or all of the terminals want to send data simultaneously. During these times, the aggregate data rate

buffer

may be higher than the circuit can handle. For these situations, the STDM contains a storage area or buffer in which data can be saved until the line can accept it. Buffer sizes of 32,000 characters and larger are common for this purpose. The user may experience a slight delay when buffering occurs.

STDM takes advantage of the fact that individual terminals frequently are idle and allows more terminals to share a line of given capacity. Students who are writing and debugging programs on a university timesharing system normally fit this model nicely. They spend some time typing the program into the computer and a great deal of time interpreting error messages and determining how to correct their program's problems. Using an STDM, 12 terminals running at 1,200 bps could be handled by a 9,600 bps line in most cases.

Leading statistical multiplexer manufacturers are Timeplex, Inc., Codex Corporation, and Tellabs, Inc. These vendors and others have a variety of products to handle varying numbers of terminals at diverse line speeds.

Concentration

Concentrators combine several low-speed circuits onto one higher-speed circuit. A concentrator can be thought of as a circuit multiplexer. For example, six 9,600 bps circuits might be concentrated onto one circuit with a 56 kbps capacity, as shown in Figure 9–12. The intelligence and buffering in the concentrator take care of the fact that 6 × 9,600 bps = 57,600 bps, which is greater than the 56,000 bps capacity of the circuit. A primary reason for performing line concentration is economics. In most cases, it is less expensive to lease a 56 kbps circuit between two points than six 9,600 bps circuits.

Figure 9–12
Line concentration.

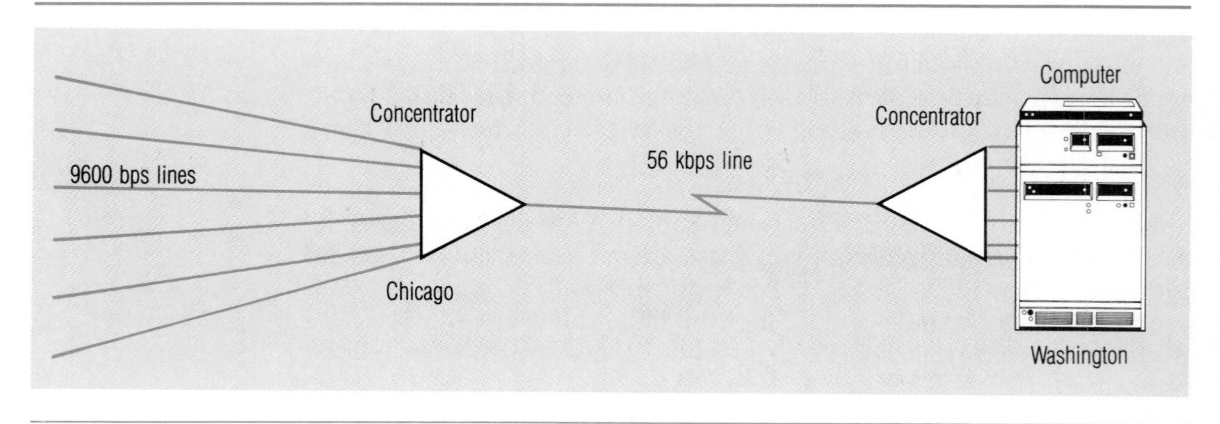

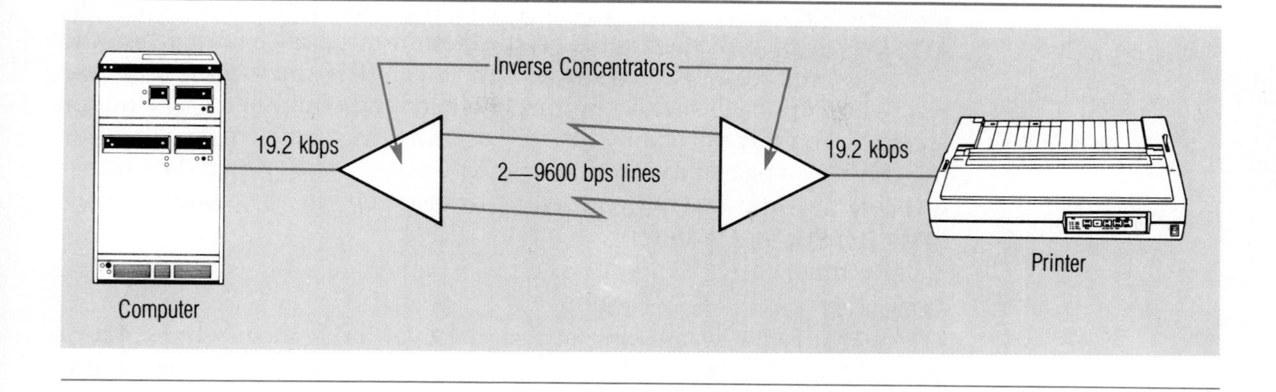

Figure 9–13
Inverse concentration.

Inverse Concentration

In some cases, high-speed circuits are not available between two points, but it is desirable to provide high-speed service. An example is when there is a need to run a remote high-speed line printer. An *inverse concentrator* takes a high-speed data stream from a computer and breaks it apart for transmission over multiple slower-speed circuits, as shown in Figure 9–13. At the remote end, the slow-speed circuits are brought together again, providing a single high-speed data stream to the remote device.

CIRCUIT ERROR CONDITIONS

Communications circuits are subject to many conditions that cause degradation of the transmitted signal so that the receiving end cannot correctly determine what was sent. Some of these conditions arise because of normal characteristics of signal propagation, others are due to faulty circuit design, and still others are due to natural physical phenomena, such as electrical storms. The following discussion involves the most common conditions that can cause errors on a communications circuit.

Background Noise

Background noise, also known as *white noise* or *Gaussian noise*, is a normal phenomenon in all electrical circuitry. It results from the movement of electrons. It is present to some extent on every communications circuit. If the noise is at a high enough level, it can be heard as a hissing sound. It rarely represents a problem for either voice or data transmission because it is a known, predictable phenomenon, and the communications carriers have designed their circuitry and equipment to deal with it.

Impulse Noise

Impulse noise is a sudden spike on the communications circuit when the received amplitude exceeds a certain level. It is caused by transient electrical impulses, such as lightning, switching equipment, or a motor starting. You have probably heard it as annoying static crashes or clicking during a voice conversation. If the noise occurs during a data transmission, the impulse may cause one or more bits to be changed, invalidating the transmission.

Attenuation

Attenuation is the weakening of a signal over a distance, which occurs normally in all communications. Attenuation was illustrated in Figure 5–18. Just as your voice sounds weak at the back of an auditorium or a radio signal fades as you get farther from the transmitter, communications signals traveling through wires fade as the distance increases because of resistance in the medium. When the signal strength gets too low, it is impossible for the receiver to accurately pick out the individual signal changes. Amplifiers or repeaters are inserted in communications circuits often enough so that in normal operation, the signal's strength is boosted before attenuation causes a problem.

Attenuation Distortion

Attenuation distortion occurs because a signal does not attenuate evenly across its frequency range. Without special equipment, some frequencies on a circuit attenuate faster than others. Communications circuits are designed and special equipment is inserted on the circuit so that the signal attenuates evenly across the frequency spectrum. However, attenuation distortion can still occur if equipment is improperly adjusted or there is other maintenance activity.

Envelope Delay Distortion

Envelope delay distortion is an electrical phenomenon that occurs when not all frequencies propagate down a telecommunications circuit at exactly the same speed. The absolute propagation delay is not relevant—only the difference between the delay at different frequencies. Envelope delay distortion can be made worse when the signal passes through filters that are inserted in the circuit to filter out noise. Noise filters tend to delay certain frequencies more than others.

Phase Jitter

Phase jitter is a change in the phase of the signal induced by the carrier signal. It is especially a problem when phase modulation is used because the sudden shift in the phase of the received signal makes it difficult for the receiving modem to sense the legitimate phase changes that the transmitting modem sent.

Echo

Echo is the reversal of a signal, bouncing it back to the sender. Echo occurs on a communications circuit because of the electrical wave bouncing back from an intermediate point or the distant end of a circuit. Echoes are sometimes heard on a voice circuit when the speaker hears his or her voice coming back a fraction of a second after speaking. On a data circuit, echoes cause bit errors.

Carriers install *echo suppressors* on switched circuits to eliminate the problems caused by echoes. Echo suppressors permit transmission in only one direction at a time. Since transmission in the reverse direction is prohibited, the echo cannot bounce back to the source. Although echo suppressors are a great help for voice transmission, they are a problem when data is sent because they take approximately 150 milliseconds to reverse and permit transmission in the opposite direction. Many modems turn signals around in 50 milliseconds or less, so if they send data, the first several hundred bits may be lost because the echo suppressors have not reversed.

The solution to this problem is to have the modem disable the echo suppressors at the beginning of a data transmission. When a connection is made, the modems send a tone that disables the echo suppressors. The suppressors stay deactivated as long as the modem sends a carrier signal on the circuit. When the communications ends and the carrier signal is no longer sent, the echo suppressors automatically reactivate. Leased data circuits do not have echo suppressors.

Crosstalk

As mentioned earlier, *crosstalk* is interference that occurs when the signals from one communications channel interfere with those on another channel. In a voice conversation, you may occasionally hear crosstalk as another conversation lower in amplitude than your own. Crosstalk can be caused by one signal overpowering another, by two pairs of wires that are in close proximity and are improperly shielded, or by frequencies of two or more multiplexed channels that are too close together.

Dropouts

Dropouts occur when a circuit suddenly goes dead for a period of time. Dropouts last from a fraction of a second to a few seconds, and then normal operation resumes. They can be caused by brief transmission problems, switching equipment, and other phenomena.

■ IMPACT OF ERRORS

Many of these errors, especially those caused by noise, occur in short bursts lasting from a few milliseconds to a few seconds. Often a short pause in the transmission or a retransmission of the block of data that is affected will circumvent the problem. An error of a given duration will affect more bits during higher-speed transmission than at lower speeds, as was shown in Chapter 8. One alternative for many types of transmission problems is to reduce the speed at which the data is being transmitted.

The effect of transmission errors is more significant in data transmission than in voice or television. In voice conversations, a crackle of static on the line does not usually cause a big problem. Human speech is very redundant, and the people at both ends of the circuit can interpolate and fill in a missing syllable or word. In television transmission, transmission errors often are seen as white flecks or snow on the screen, but a limited amount does not impair our ability to comprehend or enjoy the picture. In data transmission, on the other hand, an incorrect or missing bit can change the meaning of a message entirely. That is why it is particularly important to identify data transmission errors and take the necessary steps to correct them.

ERROR PREVENTION

Given that errors do occur in data transmissions, it is necessary to take steps to prevent, detect, and correct them. Economics must be considered because many of the error prevention techniques cost money to implement. The manager must judge whether the increased reliability is worth the cost. Certain standard techniques are in common use, however, and they are discussed here.

Line Conditioning

When a leased circuit is acquired from a common carrier, it is possible to request that it be conditioned. A *conditioned line* is one that meets tighter specifications for amplitude and distortion. Signals traveling on a conditioned circuit are less likely to encounter errors than when the circuit is unconditioned. The higher the level of conditioning, the fewer errors that will occur on the circuit. With fewer errors, faster signaling rates and, therefore, faster transmission speeds can be achieved.

Conditioning is accomplished by testing each link of a circuit to ensure that it meets the tighter parameters that conditioning implies. If necessary, special compensating or amplifying equipment is inserted in the circuit at the carrier's central offices to bring the circuit up to conditioned specifications.

Conditioning in the United States is specified as class C or class D. C conditioning adjusts a line's characteristics so that attenuation distortion and envelope delay distortion lie within certain limits. D conditioning deals with the ratio of signal strength to noise strength, called *signal-to-noise ratio*, and distortion.

With the improved sophistication of many newer modems, circuit conditioning is sometimes not required at all. Some modems have enough signal processing capability that they can tolerate the higher error rate of an unconditioned circuit. Modem manufacturers generally specify the type of conditioning that their modems require. In comparison to the total cost of the line, conditioning is relatively inexpensive. Therefore, it is almost always prudent to request it, especially if it is recommended by the modem manufacturer.

Conditioning cannot be obtained on switched circuits because the physical facilities used to make the connection vary from one call to the next. The variance is wide enough that conditioning facilities cannot realistically or cost effectively be provided.

Shielding

Shielding of a communications circuit is best understood by looking at Figure 9–6, which shows a coaxial cable. A metallic sheath surrounds the center conductor. Frequently, this shielding is electrically grounded at one end. Shielding prevents stray electrical signals from reaching the primary conductor. In certain situations, shielding on critical parts of a communications circuit may reduce noise or crosstalk. Electrically noisy environments, such as in factories or where circuits run near fluorescent light fixtures or elevator motors, are situations in which communications circuits can benefit from shielding.

Electronic Versus Mechanical Equipment

Though not entirely a user option, the replacement of mechanical equipment in the common carrier's central office or the company's equipment rooms with modern electronic equipment can lead to improved circuit quality. Not only is mechanical equipment more prone to failure, but it is also electrically noisier and more likely to induce impulse noise on circuits that pass through it. If a central office containing electromechanical switches is upgraded to an all-electronic configuration, it would be natural for the circuits running through that office to be of better quality and more trouble-free.

ERROR DETECTION

Even if circuits are well designed and good preventive measures, such as circuit conditioning, are implemented, errors will still occur. Therefore, it is necessary to have methods in place to detect these errors so that something can be done to correct them and data integrity can be maintained. *Error detection* normally involves some type of transmission redundancy. In the simplest case, all data could be transmitted two or more times and then compared at the receiving end. This would be fairly expensive in terms of the time consumed for the duplicate transmission. With the error rates experienced today, a full duplication of each transmission is not necessary in most situations.

A more sophisticated type of checking is to calculate some kind of a check digit or check character and transmit it with the data. At the receiving end, the calculation is made again and the result compared to the check character that was calculated before transmission. If the two check characters agree, the data has been received correctly.

Echo Checking

One of the simplest ways to check for transmission errors is called *echo checking*, in which each character is echoed from the receiver back to the transmitter. This is done in some timesharing systems in which a terminal operator can verify immediately that what he or she typed is what appears back on the screen as echoed from the computer. Of course, the original data transmission could be correct, and an error might occur on the transmission of the echo message back to the sender. In either case, the operator can rekey the character.

Vertical Redundancy Checking (VRC) or Parity Checking

The next most sophisticated error detection technique is called *vertical redundancy checking (VRC)* or *parity checking*. This technique was introduced in Chapter 7 and is illustrated in Figure 7–3.

Unfortunately, noise on the communications line frequently changes more than 1 bit. If two 0 bits are changed to 1s, the VRC will not detect the error because the number of 1 bits will still be an even number. Therefore, additional checking techniques are employed in most data transmission systems.

Longitudinal Redundancy Checking (LRC)

Horizontal parity checking is called *longitudinal redundancy checking (LRC)*. When LRC is employed, a parity character is added to the end of each block of data by the DTE before the block is transmitted. This character, also called a *block check character (BCC)*, is made up of parity bits. Bit 1 in the BCC is the parity bit for all the 1 bits in the block, bit 2 is the parity bit for all of the 2 bits, and so on. For example, assuming even parity and the use of the ASCII code, the BCC for the word "parity" is shown below.

	p	a	r	i	t	y	BCC
Bit 1	1	1	1	1	1	1	0
Bit 2	0	0	0	0	0	0	0
Bit 3	1	0	1	0	1	1	0
Bit 4	0	0	0	1	0	1	0
Bit 5	0	0	0	0	1	0	1
Bit 6	0	0	1	0	0	0	1
Bit 7	0	1	0	1	0	1	1
VRC	0	0	1	1	1	0	1

A VRC check will catch errors in which 1, 3, or 5 bits in a character have been changed, but if 2, 4, or 6 bits are changed, they will go undetected. Thus, VRC by itself will catch about half of the transmission errors that occur. When combined with LRC, the probability of detecting an error is increased. The exact probability depends on the length of the data block for which the block check character is calculated. In any case, even when used together, VRC and LRC will not catch all errors.

Cyclic Redundancy Checking (CRC)

Cyclic redundancy checking (CRC) is a particular implementation of a more general class of error detection techniques called *polynomial error checking*. The polynomial techniques are more sophisticated ways for calculating a BCC than an LRC provides. All of the bits of a block of data are processed by a mathematical algorithm by the DTE at the transmitting end. One or more block check characters are generated and transmitted with the data. At the receiving end, the DTE performs the calculation again and the check characters are compared. If differences are found, an error has occurred. If the check characters are the same, the probability is very high that the data is error free. With the proper selection of polynomials used in the calculation of the BCC, the number of undetected errors may be as low as 1 in 10^9 characters. To put this in more familiar terms: on a 9,600 bps circuit transmitting 8-bit characters 24 hours per day, one would expect no more than 1 undetected error every 231+ hours, or 9.6 days.

Several standard CRC calculations exist; they are known as CRC-12, CRC-16, and CRC-CCITT. The standard specifies the degree of the generating polynomial and the generating polynomial itself. CRC-12 specifies a polynomial of degree 12; CRC-16 and CRC-CCITT specify a polynomial of degree 16. For example, the polynomial for CRC-CCITT is $x^{16} + x^{12} + x^5 + 1$, where x is the bit being processed. CRC-16 and CRC-CCITT generate a 16-bit block check character that can

- detect all single bit and double bit errors;
- detect all errors in cases where an odd number of bits are incorrect;
- detect two pairs of adjacent errors;
- detect all burst errors of 16 bits or fewer;
- detect over 99.998 percent of all burst errors greater than 16 bits.

Cyclic redundancy checking has become the standard method of error detection for block data transmission because of its high reliability in detecting transmission errors.

ERROR CORRECTION

In most applications, data validity and integrity are of prime importance, so once an error is detected, some technique must be employed to correct the data.

Retransmission

The most frequently used and usually the most economical form of error correction is the retransmission of the data in error. Although there are many variations, the basic technique used is that when the receiving DTE detects an error, it signals the transmitting DTE to resend the data. This is called an *automatic repeat request (ARQ)* technique, and it is a part of the line protocol. In order for ARQ to work, the transmitting station must hold the data in a buffer until an acknowledgment comes from the receiver that the block of data was received correctly. Another requirement is that there must be a reverse channel for signaling from the receiver to the transmitter.

Stop and Wait ARQ In the *stop and wait ARQ* technique, a block of data is sent and the receiver sends either an acknowledgment (ACK) if the data was received correctly or a negative acknowledgment (NAK) if an error was detected. If an ACK is received, the transmitter sends the next block of data. If a NAK is received, the data block that was received in error and is still stored in the transmitter's buffer is retransmitted. No data is transmitted while the receiver decodes the incoming data and checks it for errors. If a reverse channel is not available, the line must be turned around for the transmission of the ACK or NAK and then turned around again for the transmission (or retransmission) of the data block. Stop and wait ARQ is most effective where the data blocks are long, error rates are low, and a reverse channel is available.

Continuous ARQ Using the *continuous ARQ* technique, data blocks are continuously sent over the forward channel while ACKs and NAKs are sent over the reverse channel. When a NAK arrives at the transmitter, the usual strategy is to retransmit beginning with the data block that the receiver indicated was in error. The transmitting station's buffer must be large enough to hold several data blocks. The receiver throws away all data received after the block in error for which the NAK was sent because it will receive that data again.

An alternate strategy is for the transmitter to retransmit only the block in error. In this case, the receiver must be more sophisticated because it must insert the retransmitted data block into the correct sequence among all of the data received. Of the two approaches to continuous ARQ, the first strategy, sometimes called "go back N blocks," is more commonly used.

Continuous ARQ is far more efficient than stop and wait ARQ when the propagation times are long, as they are in satellite transmission.

Forward Error Correction (FEC)

VRC, LRC, and CRC checking methods are effective in detecting errors in data transmission. However, they contain no method for automatically correcting the data at the receiving end. By using special transmission codes and adding additional redundant bits, it is possible to include enough redundancy in a transmission to allow the receiving station to automatically correct a large portion of any data received in error, thus avoiding retransmission. This technique is called *forward error correction (FEC)*.

Research into FEC techniques has been conducted by organizations such as Bell Laboratories and the military. The military's interest lies in being able to make one-way transmissions to submarines or aircraft, knowing that messages will arrive with a predetermined but very low probability of error. Three well-known error correcting codes are the Bose-Chaudhuri code, the Hagelbarger code, and the Hamming code. The Bose-Chaudhuri code, in its original version, uses 10 check bits for every 21 data bits and is capable of correcting all double bit errors and detecting up to 4 consecutive bit errors. The Hagelbarger code will correct up to 6 consecutive bit errors if the group of bits in error is followed by at least 19 good data bits. The Hamming code, in its 7-bit form, allows single bit errors in each character to be corrected. However, only 16 unique characters are allowed in the character set. Other modifications to the Hamming code allow a larger character set, with a corresponding increase in the number of checking bits.

FEC codes have a high cost in terms of the number of redundant bits required to allow error correcting at the receiving end. In certain applications—particularly where only one-way simplex transmission is allowed or possible—the cost is well justified. Since the FEC techniques are sophisticated, a specially programmed microcomputer in the DTE or DCE normally is used to calculate the FEC codes and perform the error correction.

WIRING AND CABLING

Though not strictly a circuit issue, the subject of wiring and cabling within a building or facility is closely related to circuit installation and operation. This is particularly true since divestiture, when the ownership and responsibility for inside wiring began to be transferred from the common carriers to their customers. Most buildings today contain several pairs of twisted wires, originally installed by the telephone company, running to each office. One pair is for the telephone, and the others are spares. If the office contains a computer terminal, it is, in all likelihood, connected to the computer or cluster control unit via a totally separate wire or cable that was installed by the data processing department. Where private television exists, a third cabling system—perhaps "owned" by the audiovisual department—also may be found.

A commonly stated requirement for the future is that all of these telecommunications wiring systems need to be merged and consolidated so that a single communications outlet is installed in each office.

This outlet would provide jacks for connecting all of the communications equipment in the office. The wire behind the outlet would run to a local wiring distribution center or equipment room on the floor of the office building, as shown in Figure 9–14. The distribution centers would be connected via high-capacity cable, perhaps one of optical fiber, to the wiring center for the building or site. The communications wiring of the future should be similar to the electrical wiring of today, where, for most offices, a single type and size of wire runs to one or more standard outlets conveniently located.

To achieve this ideal standardized communications wiring plan, a long history of nonstandard communications wiring must be overcome. Each vendor of communications equipment traditionally has set individual standards for data communications wiring and for the media that connect terminals to its computers. Of course, these vendors' standards bear little or no relationship to the standards for telephone wiring that the telephone companies have used for years. Even within a single vendor's product line, many standards exist. IBM has used a 92 ohm coaxial cable for its 3270 series of terminals, twinaxial cable for connecting terminals on smaller computers, and other types of cable in other circumstances.

Within the last few years, there has been some movement toward using standard twisted pair telephone wire for connecting data terminals to computers. The two largest vendors in the communications industry, AT&T and IBM, both have developed wiring architectures or wiring plans that purport to be the universal answer to all wiring problems now and in the future. Neither plan has been widely accepted by communications users, however.

The major problem that most companies face, even if they agree that either of the wiring plans was a panacea, is the effort to replace all existing wiring with wire that meets the requirements of the new "standard." In most cases, the only practical, affordable alternative is to convert existing wiring when such opportunities as building renovations arise and to install the new type wiring in all newly constructed buildings.

■ SUMMARY

This chapter looked at the characteristics of telecommunications circuits, various ways of classifying them, and the media with which they can be implemented. The telecommunications network designer is faced with a wide variety of choices for circuits. Only by understanding the characteristics and trade-offs between various circuit types and media can the designer reach a reasonable solution for a given network or application.

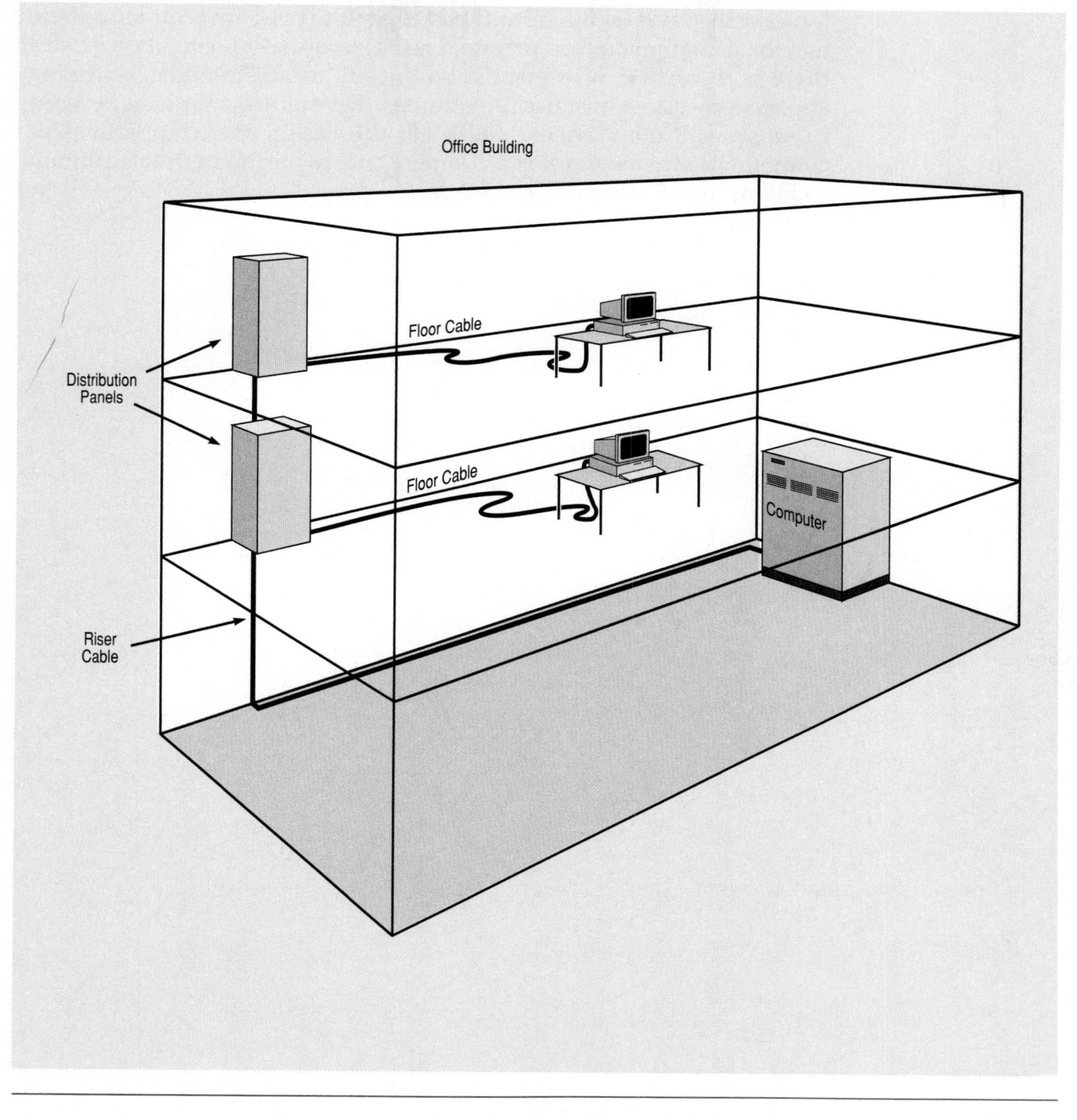

Figure 9–14
Simplified wiring diagram of an office building.

More often than not, a company's diverse communications needs indicate that several different types of circuit types are required. There may be a combination of private, leased, and dial-up circuits. Usually, there is a mixture of transmission speeds to accommodate differing applications and transmission volumes. Different media may be used to satisfy still other needs. All in all, the design of a comprehensive communications network is a complex task requiring both telecommunications and application knowledge.

CASE STUDY

Dow Corning's Data Communications Circuits

One of Dow Corning's philosophies about its data communications system is that it must deliver excellent response time to its users. For that reason, the circuits that make up the system operate at the highest speeds that the company believes it can justify economically.

At the corporate headquarters, a large local area network was installed in 1988 and 1989. Terminals and personal computers attached to the local area network are able to communicate with the mainframe computers at a speed of 4 Mbps. Generally, the users of these terminals experience response times that are less than 1 second.

At other locations in the Midland area, 56 kbps digital circuits are installed. They are all point-to-point circuits because of a restriction in the IBM 3174 cluster control unit; multipoint 56 kbps circuits are not supported by the CCU hardware. Users on these circuits also generally experience good response time, partially because the volume of data coming from the VDTs and personal computers does not tax the data-carrying capability of the circuits. As a result, the 56 kbps circuits are not extremely busy.

Outside of Midland, a combination of circuit type and speeds are used. Leased multipoint circuits operating at 9.6 kbps connect the company's sales offices and warehouses. These locations do not generate a lot of data, so the multipoint configuration works well. Connecting Midland with the larger plant locations are 56 kbps digital circuits. In one case, a major plant is located in an out-of-the-way location in Kentucky, and the local telephone company does not provide 56 kbps service. Dow Corning's telecommunications staff studied various ways to provide 56 kbps service to the site and selected a satellite circuit. Service on the satellite circuit has been adequate, but it has not been as realiable as Dow Corning's terrestrial lines.

In 1986, Dow Corning installed its first private fiberoptic circuit. The fiber cable connects the corporate headquarters building with a building located 1.1 miles away. The fiber was justified because it provides the capacity to carry television channels for security monitoring. Because the site can be monitored from a remote location, it was not necessary to hire security guards at the building.

GTE won the bid to install the cable and buried it in the ground on Dow Corning property. Installation was fairly simple because all of the property between the two buildings is owned by Dow Corning, and only one road had to be crossed. Permission was obtained from the county to bury cable under the road. The fiberoptic circuit is being used to carry data communications between buildings as well as several television channels used for security monitoring.

Dow Corning also has several T-1 circuits installed. One T-1 circuit connects the corporate center with the Midland plant 5 miles away, and another runs to the plant in Hemlock, Michigan, 15 miles away. A third T-1 circuit connects the corporate center with an office building located in Midland, and a fourth T-1 runs from the corporate center to the Michigan Bell central office and is used for voice circuits. All of the T-1 circuits are leased from Michigan Bell Telephone, which did the installation. The circuits are less expensive than the multiple voice and data circuits that they replaced.

Multiplexing equipment was installed by Dow Corning on both the fiberoptic and T-1 circuits to divide the capacity into video, voice, and data circuits. Even though the company has maintenance contracts on the multiplexing equipment, it was thought that a Dow Corning technician should be trained to understand the multiplexers so that he or she could reconfigure them and diagnose problems when they occur.

The company has installed a broadband coaxial cable transmission system in the Midland plant. This system carries data from instruments to process control computers throughout the laboratories and plant. A part of the cable also is used for data transmission between the IBM terminals on the site and a concentrator. The concentrator forwards the data on high-speed circuits to the mainframe computers at the corporate headquarters. A third use of the cable is for a television channel providing news and other information to plant employees. The broadband cable system was installed by Midland plant personnel in the early 1980s in response to some unique requirements for television and data transmission. The cable has been upgraded to a higher quality since the original installation and is operated and maintained entirely by plant people.

Internationally, Dow Corning's data communications circuits tie headquarters in Midland to area headquarters in Brussels, Belgium, Hong Kong, and Tokyo, Japan. In addition, several circuits run from Midland to major operating centers within foreign areas, such as Barry, Wales, Wiesbaden, Germany, Sydney, Australia, Toronto, Canada, Sao Paulo, Brazil, and Mexico City, Mexico. The lines to Hong Kong and Sao Paulo are shared with The Dow Chemical Company. The international circuits operate at various speeds and primarily carry traffic from IBM terminals or computers located in the foreign locations. The international circuits are leased from AT&T.

Review Questions

1. Explain how a line, a circuit, a link, and a channel differ.
2. Describe a multipoint circuit.
3. Why is a four-wire circuit preferable to a two-wire circuit for data transmission?
4. Compare and contrast the functions of a modem and a DSU/CSU.
5. What is the data-carrying capacity of a T-1 circuit? Can a T-1 circuit be used to carry analog voice signals?
6. What is the normal maximum data transmission speed on a voice-grade circuit?
7. Identify five different media used for carrying communications signals and discuss under what circumstances each is most appropriately used.
8. What are the advantages of shielded twisted pair wire compared to ordinary twisted pair?
9. Under what circumstances would it be most appropriate to use a broadband coaxial cable instead of a fiberoptic cable?
10. List the characteristic of optical fiber.
11. What are some potential disadvantages of using microwave radio for data transmission?
12. Explain the term *propagation delay* and why it is important in satellite transmission.

13. Under what circumstances would a company consider installing a private fiberoptic link connecting two of its locations?
14. Under what circumstances would a company consider bypassing its local telephone company?
15. What are the advantages of STDM over traditional TDM?
16. Discuss the circuit attributes of attenuation, envelope delay distortion, phase jitter, and crosstalk.
17. Why are communications line errors more significant when data is being transmitted than when voice is being transmitted? If voice is transmitted digitally, do line errors become more significant?
18. What is the purpose of communications line conditioning?
19. What is the difference between LRC and CRC checking?
20. Compare and contrast the stop and wait ARQ and continuous ARQ techniques.
21. Why is it desirable for a company to manage its communications wiring?

Problems and Projects

1. Using the VRC and LRC parity checking techniques and the ASCII code, calculate the parity bit and block check character for your last name.
2. A company has a leased satellite circuit between its New York and San Francisco locations. The circuit is routed on a satellite. When 500-character blocks of data are sent from New York to the San Francisco office using a stop and wait ARQ technique, what percentage of the line time is used for actual data transmission and what percentage is spent waiting for acknowledgments and line turnarounds, assuming

   - line turnaround takes 50 milliseconds;
   - no errors occur during the transmission;
   - line speed is 9,600 bps;
   - an acknowledgment message is 5 characters long.

   How does the percentage change if 5,000 character blocks are transmitted?
3. Identify the trade-offs a company would have to consider when deciding whether to implement dial-up data or leased communications circuits. The company has locations throughout the United States. The applications are primarily basic business transactions, such as customer order entry, shipping, purchasing, accounts receivable, and accounts payable.
4. Visit a company that has installed fiberoptic links. Why did they install optical fiber? What difficulties did they have when installing the fiber? What error rate are they experiencing for data transmission on the fiber? How often has it failed? Overall, how

has the fiber operated compared to the previous communications technology that was installed?

5. Investigate whether it is possible to obtain digital circuits in your area. Also find out whether 56 kbps or 64 kbps circuits can be obtained. Is fractional T-1 service available? If these services are available, must they be leased, or is switched service available?

Vocabulary

circuit
line
link
data link
channel
forward channel
reverse channel
node
point-to-point circuit
multipoint circuit
multidrop circuit
two-wire circuit
four-wire circuit
digital circuit
data service unit (DSU)
channel service unit (CSU)
T-carrier system
fractional T-1
low-speed circuit
subvoice-grade circuit
voice-grade circuit
high-speed circuit
wideband circuit
twisted pair
crosstalk
punchdown block
shielded twisted pair
coaxial cable
optical fiber
core
cladding
light emitting diode (LED)
laser
single mode
multimode
dispersion
microwave radio
geosynchronous orbit
uplink
downlink
propagation delay
private circuit
leased circuit
bypass
circuit switching
time division multiplexing (TDM)
frame
statistical time division multiplexing (STDM)
concentrator
inverse concentrator
impulse noise
background noise
white noise
Gaussian noise
attenuation
attenuation distortion
envelope delay distortion
phase jitter
echo
echo suppressor
dropouts
conditioned line
signal-to noise ratio
shielding
error detection
echo checking
vertical redundancy checking (VRC)
parity checking
longitudinal redundancy checking (LRC)
block check character (BCC)
cyclic redundancy checking (CRC)
polynomial error checking
error correction
automatic repeat request (ARQ)
stop and wait ARQ
continuous ARQ
forward error correction (FEC)

References

Axner, David H. "Stat Muxes Are Alive and Well," *Telecommunications Products + Technology*, June 1988.

Bowen, Terry. "Wiring the Open Office for Fiber," *Telecommunications*, April 1988.

Doll, Dixon R. *Data Communications: Facilities, Networks, and Systems Design*. New York: John Wiley & Sons, 1978.

Effron, Joel. *Data Communications Techniques and Technologies*. Belmont, CA: Wadsworth, Inc., 1984.

Fitzgerald, Jerry. *Business Data Communications: Basic Concepts, Security, and Design*. 2nd ed. New York: John Wiley & Sons, 1984, 1988.

Hornsby, Thomas. "Fractional T1—Can Users Get More for Less?" *Communications News*, October 1989.

Keller, John J., Pete Engardio, Kenneth Dreyfack, and Russell Mitchell. "The Rewiring of America: Making Phone Lines the Transportation System for the Information Age," *Business Week*, September 15, 1986, p. 188.

Kirvan, Paul. "T-1 Communications: Technology, New Developments and Planning Guidelines," *ICA Communique* 40, No. 3 (October 1986), p.14.

Martin, James. *Future Developments in Telecommunications*. 2nd ed. Englewood Cliffs, NJ: Prentice-Hall, Inc., 1977.

Martin, James. *Telecommunications and The Computer*, 2nd ed. Englewood Cliffs, NJ: Prentice-Hall, Inc., 1976.

Meir, Edwin E. "Coming Soon to an Outlet Near You: Premises Fiber," *Data Communications*, February 1989.

Pruitt, James. "Microwave Systems Reach Out to Fill Gaps in Private Networks," *Networking Management*, October 1989.

Robins, Marc. "It's T-1 Time," *Teleconnect*, February 1988.

———. *Telecommunications and You*. Order number GE20–0790. Armonk, NY: IBM Corporation, 1986.

CHAPTER

10

Data Link Control Protocols

OBJECTIVES

After studying the material in this chapter, you should be able to

- explain what a data link protocol is and the function it performs;
- describe desirable attributes of protocols;
- distinguish between contention and polling protocols;
- explain the distinguishing characteristics of character-oriented, byte-count oriented and bit-oriented protocols;
- explain in general terms how several protocols work;
- describe how LAN protocols work;
- explain protocol conversion and situations in which it is likely to be useful.

INTRODUCTION

Chapter 9 looked at telecommunications circuits in detail. The circuits are like the roads in a highway system; they provide a mechanism over which communications can occur. In themselves, however, circuits have no "rules of the road." Just as there must be rules for the use of the highways to ensure efficient and safe transportation, there must be rules for the use of the circuits to ensure that data is transported efficiently and accurately. The rules for using data communications circuits are called *data link protocols*. This chapter deals with the protocols of the data link control layer, layer 2, of the ISO-OSI reference model. The data link control layer defines how data is structured into blocks or frames and surrounded by control characters before being transmitted on the communications circuit.

DEFINITION AND THE NEED FOR PROTOCOLS

In general terms, a *protocol* is a set of rules or guidelines that govern the interaction between people, between people and machines, or between machines. Protocols exist for all types of social situations. When two people meet on the street and one says, "Hello, Jim," to which Jim responds, "Hello, Bill, How are you?" they are following a simple protocol. This simple exchange gets the conversation going, and after the initial words of greeting, the two men launch into any of a variety of topics. Similarly, when a telephone rings, the convention or protocol is that the answering party says "Hello" and the calling party identifies herself by saying something like, "Hello, Mary? This is Jane Smith." How the conversation proceeds from there depends on the relationship between the individuals and the reason for the communications.

In data communications, protocols are the rules for communicating between data terminal equipment, which for our purposes can be between terminals and computers, between computers, or between terminals. In order for communications between equipment to occur, there are several types of protocols that need to be in place. This chapter only concerns the data link protocol—the rules for operating the circuit and sending messages over it.

DATA LINK PROTOCOL FUNCTIONS

Data link protocols must have rules to address the following situations that occur during a data communications.

- Communications startup—There must be rules to specify how the communications will be initiated: whether there is automatic startup, whether any station can initiate the communications, or whether only a station designated as the "master" station may initiate communications.
- Character identification and framing—There must be a way for the data terminal equipment to separate the string of bits coming down the communications line into characters. Furthermore, there must be ways of distinguishing between control characters, which are a part of the protocol, and data characters, which convey the message being communicated.
- Message identification—The data terminal equipment must separate the characters on the communications line into messages.
- Line control—Rules must exist to specify how the receiving terminal signals the sending terminal whether it has received data correctly, how and under what circumstances the line will be turned around, and whether the receiving terminal can accept more data.
- Error control—The protocol must contain rules specifying what happens when an error is detected, what to do if communications suddenly and unexplainably cease, and how communications are reestablished after they are broken.
- Termination—Rules must exist for ending the communications under normal and abnormal circumstances.

In normal conversation between two people, there are rules or protocols that specify how all of these situations are handled. Because of the human intelligence at both ends of the communications, the rules need not be precisely specified, and there is room for considerable variation. Whether we end a telephone conversation by saying "Goodbye, Mr. Smith," "See 'ya later," or "Bye" matters little. The intention is clear and understood by both parties. Data communications between two pieces of equipment, however, must be somewhat more precise and rigid in the application of the rules because even the growing intelligence of the terminals and computers is still no match for the human intellect.

DESIRABLE ATTRIBUTES OF DATA LINK PROTOCOLS

In addition to the functional requirements specified earlier, there are several other desirable attributes for a data link protocol to have.

- Transparency—It is very desirable for a DTE to be able to transmit and receive any bit pattern as data. The complication occurs because certain bit patterns are assigned to represent control characters that have a specific meaning within the protocol.
- Code independence—It is desirable for the protocol to allow the transmission of data from any data coding system, such as ASCII, EBCDIC, Baudot, or any other. (Some protocols require a certain code to be used.)
- Efficiency—The protocol should use as few characters as possible to control the data transmission so that most of the line capacity can be used for actual data transmission.

PROTOCOL IMPLEMENTATION

A data link protocol is implemented by transmitting certain characters, or sequences of characters, on the communications line. The specific characters are determined by the code and protocol being used. If you look at the table of ASCII characters shown in Figure 7–5, you will see that there is a whole set of characters on the left side of the table that begin with the bits 000 and 001. These are control characters in the ASCII code. Some of them are used to implement the data link protocol.

control characters

By way of contrast to the other characters in the table, which are sometimes called the graphic characters, control characters usually are not printable or displayable by a terminal or printer. Not all of these characters are used in the data link protocol. Some of them are used for other control purposes, such as skipping to a new line on a terminal, tabbing, and printer control.

PREDETERMINED COMMUNICATIONS PARAMETERS

Many communications parameters used to establish compatibility between terminals or between a terminal and a computer are set manually by switches or are specified as parameters via software. The data code (ASCII, EBCDIC, or some other) usually is predetermined. The code determines how many bits make up a character, such as seven in the case of ASCII. Whether there will be a parity bit and whether odd or even parity will be used is often determined by a switch setting in the data terminal equipment (DTE). The transmission speed usually is determined by a switch setting on the modem, although some newer modems are capable of determining the speed automatically from the first one or two characters that are transmitted.

Preestablishing these parameters is necessary so that the control characters of the protocol may be correctly received and interpreted, allowing the communications to be established and maintained.

PROTOCOL CONCEPTS—A GENERAL MODEL

This section explores protocols in a generic sense. The elements that make up a protocol will be introduced, and they will serve as a general model that can be used for comparison when specific protocols are examined later in the chapter.

Line Access

Before a DTE can begin communicating, it must gain control of (access to) the circuit. There are three primary ways in which access can be obtained: contention, polling, and token passing.

Contention Contention systems are quite simple. A DTE with a message to send listens to the circuit. If the circuit is not busy, the DTE begins sending its message. If the circuit is busy, the DTE waits and keeps checking periodically until the line is free. Once a DTE gains access to the circuit, it can use the circuit for as long as necessary. In simple contention systems, there are no rules to limit how long a terminal may tie up the circuit.

The contention approach can work very well on a point-to-point circuit in which either DTE is capable of sending. Usually on a point-to-point circuit, the DTE that sends a message first requests permission to transmit by sending a specific control character to the other DTE. The receiving station indicates whether it is ready to receive by returning a different control character or sequence. If the receiving station is ready, transmission begins. Contention systems do not work as well on a multipoint circuit or where the circuit is heavily loaded with traffic or of relatively slow speed. In those cases, the DTE with a message to send would frequently find the circuit busy.

CSMA/CD *CSMA/CD* stands for *character sense multiple access with collision detection*. CSMA/CD is the primary form of contention access to a circuit that is used in local area networks. It was originally developed by Xerox Corporation and has been refined by Digital Equipment Corporation and Intel. It is the contention scheme used on the Ethernet LAN, which is described in Chapter 11.

CSMA/CD is a broadcast protocol. There is no master station on a network that uses CSMA/CD; all stations are equal. When a terminal has a message to send, it examines the carrier signal on the network to determine whether or not a message is already being transferred. If the network is free, the station begins transmitting, indicating with an address the destination terminal that is to receive the message. All connected terminals monitor the network at all times but act on messages only if they see their address characters in the message.

At times, two stations on the network decide to transmit simultaneously, causing a data collision that garbles the transmission. When the collision is detected, the stations that caused it wait a random period of time, determined by circuitry in their communications interface, and then try transmitting again. The length of the random delay is critical because if it is too short, repeated collisions usually will occur. On the other hand, if the delay interval is too long, the channel remains idle.

CSMA/CD works quite well when the traffic is light. Its biggest weakness is the nondeterministic nature of its performance when the network is heavily loaded. As the number of terminals or the amount of traffic grows, the number of collisions increases, and the network delays become unpredictable. Thus, CSMA/CD is not a good protocol to use when response time must be consistent, such as in a manufacturing plant where control signals need to be sent to a machine. CSMA/CD can work well, however, in an office application where inconsistencies in response time (as long as they aren't too great) can be tolerated.

Polling

Roll Call Polling *Roll call polling* is the most common implementation of a polling system. One station on a line is designated as the master, and the others are slaves. The master station sends out special polling characters that ask each slave station in turn whether it has a message to send. If the slave has no messages, it sends a control character indicating so. If the slave does have a message, it sends the message. Each station on the line responds only to its unique polling characters, which usually correspond to its address. In the simplest case, the master station polls each terminal in turn. In most systems there is a sequenced list, called the *polling list*, which specifies the sequence in which terminals are polled. Terminals that normally have more data to send may be polled—and given the opportunity to use the circuit—more frequently than other terminals with less traffic.

Hub Polling Another type of polling is called *hub polling*. It is most easily visualized if one thinks of a multipoint circuit. One station begins by polling the next station on the line. That station responds with a message if it has one. If it does not have a message, it passes the polling message on to the next station on the line, and so on. Hub polling reduces the amount of polling character traffic on the line. It requires more complicated circuitry in each terminal because each is performing a part of the overall line control (a function that is concentrated in the master terminal when the roll call polling technique is used).

Token Passing Token passing is similar in concept to hub polling, and the access technique is more complicated than the CSMA/CD technique. A single *token,* which is a particular sequence of bits, is passed from node to node. The node that has the token at any point in time is allowed to transmit, and a node with a message to send waits until a free token passes. The node changes a bit in the token, thereby changing the token's status from "free" to "busy" and attaches its message to the "busy" token.

As with CSMA/CD, all of the stations on a token passing system are considered equal, but at least one of the stations must have logic to ensure that there is always one, but only one, token circulating on the circuit. The advantage of a token passing technique is that the response time of the network is deterministic or predictable. Token passing is discussed again later in this chapter and is referred to again in Chapter 11.

Message Format

Once the line has been seized, whether by a contention, polling, or token passing, and the DTEs have both indicated they are ready, the communications begins.

A data message consists of three parts: the *header*, the *text* of the message, and the *trailer*, as shown in Figure 10–1. The header contains information about the message, such as the destination node's address, a sequence number, and perhaps a date and time. The text is the main part of the message—it's what the communication is all about. In most protocols, the trailer is quite short and contains only checking characters.

In some protocols, each part of the message begins and ends with special control characters, such as the *start of header (SOH)* and *start of text (STX)* (which also marks the end of the header) characters. In other protocols, the header may be of a fixed length, in which case a special end of header character is unnecessary. The end of the text is either marked by a special character, such as the *end of text (ETX)* character, or its location can be calculated from a "text length" field in the header. In

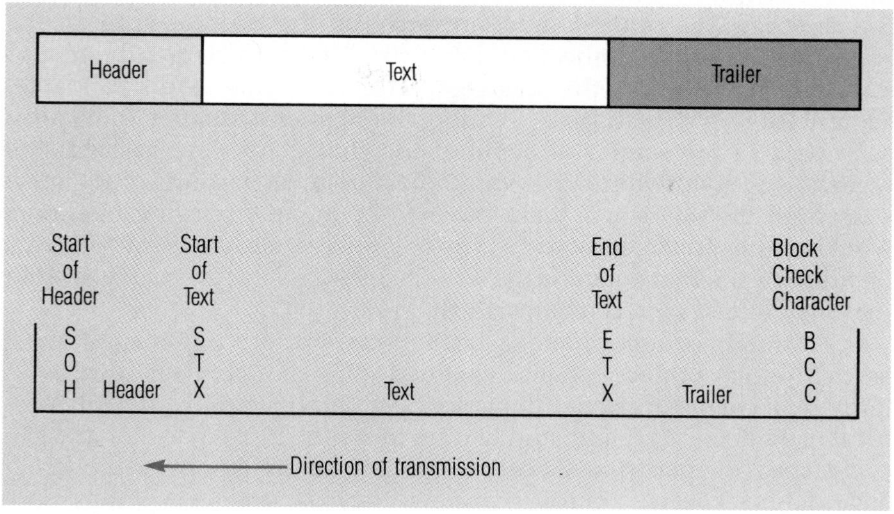

Header Text Trailer

Start of Header | Start of Text | End of Text | Block Check Character

SOH Header STX Text ETX Trailer BCC

Direction of transmission

Figure 10–1
Message format.

many protocols, the trailer consists of only a *block check character (BCC)*, which is generated by the error checking circuitry or software.

During the transmission of a message or a group of messages, error checking and synchronization occur at points in time determined by the specific protocol in use. Error checking may occur at the end of each block or message, at which time the line is usually turned around and the receiving terminal sends an ACK or NAK to the sending terminal, indicating whether the data block that was just received was correct. If a continuous ARQ technique is being used, the line turnaround may only occur periodically, and blocks of data can be sent essentially continuously until an error occurs.

Synchronization characters (SYN) are inserted in the data stream from time to time by the transmitting station in order to ensure that the receiver is maintaining character synchronization and properly grouping the bits into characters. The synchronization characters are removed by the receiver and do not end up in the received message.

ASYNCHRONOUS DATA LINK CONTROL PROTOCOLS

Asynchronous data transmission is the oldest form of data communications. It evolved from the early days of teletypewriter operation, where operators sat at the teletypewriters at both ends of a point-to-point circuit. With operators at both ends and with most transmissions occurring manually, little protocol was needed. Usually, it was as simple as the operator with a message to send looking to see if his or her teletypewriter was printing a message. If it wasn't, the operator could assume that the circuit was free. He or she would then type the message, and the message was printed simultaneously on the teletypewriter at the receiving end of the connection. Error checking consisted of the receiving operator reading the message, and if it was understandable, acknowledging it, perhaps by sending the sequence number of the message back to the sender. If some of the characters were garbled, the receiving operator would ask for a repeat of all or portions of the message until the characters were correct. Frequently, sending operators would automatically repeat the words or numbers, such as dollar amounts or addresses, when the message was originally sent as a way of ensuring that they would be received correctly.

Although communications have become more automated, most asynchronous protocols have remained quite simple. They deal with individual characters rather than blocks of data. Typically, the only protocol used is the start and stop bits surrounding each character and the parity checking performed if certain codes are used. Although the ASCII code defines special control characters, they are used only when more elaborate protocols are employed. The typical asynchronous transmission today is still character oriented.

As terminals came to have more intelligence and communications between terminals and computers rose in popularity in the 1960s and early 1970s, synchronous communications actually grew at a faster rate than asynchronous transmission because it is a more efficient way to transmit data. With the development of the personal computer in the mid-1970s and the desire to add communications capability to it, asynchronous communications began regaining favor. Asynchronous communication enjoyed increasing popularity primarily because it provided adequate capability for most personal computer communications, and requires less sophisticated—and therefore less expensive—equipment. As the desire for more automatic verification of messages and file transfers between personal computers occurred, it became necessary to define and implement more elaborate protocols. The more elaborate asynchronous protocols allow the transmission of data in blocks. They perform additional checks on the blocks of data to ensure that transmission is correct. Two of the better known block-oriented asynchronous protocols are Xmodem and Kermit.

The Xmodem Protocol

One of the most widely used asynchronous protocols is the *Xmodem protocol* developed for use between microcomputers, especially for transfers of data files between them. Using this protocol, one microcomputer is designated by its operator as the sender, and the other is designated by its operator as the receiver. The receiver indicates that it is ready to receive by sending an ASCII NAK character every 10 seconds. When the transmitting system receives a NAK, it begins sending blocks of 128 data characters surrounded by a header and trailer. The header consists of a start of header (SOH) character, followed by a block number character, followed by the same block number with each bit inverted. The trailer is a checksum character, which is the sum of the ASCII values of all of the 128 data characters added together and divided by 255.

checksum character

At the receiving end, the message is checked to ensure that the first character was an SOH, that the block number was exactly 1 more than the last block received, that exactly 128 characters of data were received, and that the checksum computed at the receiving end is identical to the last character received in the block. If all of these conditions are true, the receiver sends an acknowledge (ACK) back to the transmitter, and the transmitter sends the next block. If the data was not received correctly, a NAK is sent, and the transmitter resends the block of data that was in error.

The entire message or data file is sent in this way, block by block. At the end, the transmitter sends an end of text character that is acknowledged by the receiver with an ACK, and the transmission is complete.

Although the Xmodem protocol is quite simple, its error checking is not very sophisticated. Therefore, the reliability of the received data is not as good as with other protocols. Xmodem is considered to be a half-duplex protocol because the sender waits for an acknowledgment (stop and wait ARQ) of each block of data before sending the next block. Thus, if a full-duplex transmission facility is used, Xmodem can only use it inefficiently.

As microcomputers continue to grow in power and sophistication, it is reasonable to assume that they will be capable of processing more sophisticated protocols with significantly better error checking incorporated in them. Such protocols are beginning to emerge.

SYNCHRONOUS DATA LINK PROTOCOLS

Classification

Synchronous data link protocols typically deal with blocks of data, not individual characters. Synchronous protocols may be divided into three types according to the way the start and end of message are determined. The three types are character-oriented protocols, byte-count-oriented protocols, and bit-oriented protocols.

Character-Oriented Protocols A *character-oriented protocol* uses special characters to indicate the beginning and end of messages. For example, the SOH character is used to indicate the beginning of a message and the ETX character to indicate the end. The best known character-oriented protocol is the Binary Synchronous Communications protocol, also known as BSC or BISYNC.

Byte-Count-Oriented Protocols *Byte-count-oriented protocols* have a special character to mark the beginning of the header, followed by a count field that indicates how many characters are in the data portion of the message. The header may contain other information as well. It is followed by the data portion of the message. The data portion of the message is followed by a block check character or characters. The best known byte-count-oriented protocol is Digital Equipment Corporation's Digital Data Communications Message Protocol (DDCMP).

Bit-Oriented Protocols A *bit-oriented protocol* uses only one special character, called the *flag character*, which marks the beginning and end of a message. Within the message, the header and the fields within it are of a predefined length, and the header is followed by the data field with no intervening control character. No special control character is used to mark the beginning of the trailer segment of the message, if one exists. The flag character also marks the end of the message. The receiving terminal knows that the bits preceding the flag are the check characters for the message. The best known bit-oriented protocol is IBM's Synchronous Data Link Control (SDLC), which is a proper subset of the International Standards Organization's High-Level Data Link Control (HDLC). Other bit-oriented protocols include the CSMA/CD and token protocols that are used on local area networks.

We will now look at three of the most common synchronous data link protocols—BSC, DDCMP, and SDLC—in more detail.

Binary Synchronous Communications (BSC, BISYNC)

The *Binary Synchronous Communications (BISYNC)* protocol was introduced by IBM in 1967. It has been implemented by many manufacturers for a wide variety of equipment. BISYNC supports only three data codes: the 6-bit transcode (SBT), which is rarely used anymore, ASCII, and EBCDIC. Certain bit patterns in each code have been set aside for the required control characters: SOH, STX, ETB, ITB, ETX, EOT, NAK, DLE, and ENQ. Some additional control characters are really two-character sequences: ACK0, ACK1, WACK, RVI, and TTD. All of these control characters and sequences are defined in Figure 10–2.

BISYNC operates in either nontransparent or transparent mode. *Transparent mode* means that any bit sequence is permissible in the data field, even if it is the same as a character used for control. Transparency is needed when transmitting binary data, such as computer programs and some data files. In transparent mode, the data link escape (DLE) character is inserted before any control characters. DLE STX initiates transparent mode, and DLE ETX or DLE ETB terminates the block. Since the combination DLE ETX could also occur in the middle of the data portion of the message and inadvertently terminate the transmission, the transmitter scans the text portion of the message. Whenever it finds

	Description	Purpose
SOH	Start of header	Marks the beginning of the header of a transmission.
STX	Start of text	Marks the end of the header and the beginning of the data portion of the message.
ITB	End of intermediate	Marks the end of a data block, but does not reverse the line or require the receiver to acknowledge receipt. A block check character follows the ITB for checking purposes.
ETB	End of text block	Marks the end of a data block. Requires the receiver to acknowledge receipt.
ETX	End of text	End of data block and no more data blocks to be sent.
EOT	End of transmission	Marks the end of transmission that may have contained several blocks or messages.
ACK0 ACK1	Positive acknowledgment	Previous block was received correctly. ACK0 is used for even-numbered blocks; ACK1 for odd-numbered ones.
NAK	Negative acknowledgment	Previous block was received in error. Usually requires a retransmission.
WACK	Wait before transmit	Same as ACK, but receiver is not ready to receive another block.
SYN	Synchronization	Sent at beginning of transmission to ensure characters will be received correctly.
ENQ	Enquiry	Requests use of the line in point-to-point communication.
DLE	Data link escape	In transparent communication, creates two-character versions of ACK, WACK, and RVI.
RVI	Reverse interrupt	Positive acknowledgment and asks transmitter to stop as soon as possible as receiver has a high priority message to send.
TTD	Temporary text delay	Transmitter uses this to retain control of the line when it is not ready to send data.

Figure 10–2
Binary Synchronous Communications control characters.

a DLE character, it inserts an additional DLE. On the receiving side, the data stream also is scanned. Whenever two DLEs are found together, one is discarded. The receiver also knows that when it sees the sequence DLE DLE ETX in the data portion of the message, the sequence is not to be interpreted as a set of control characters indicating message termination, but only as data.

Synchronization of Characters and Messages Character synchronization is accomplished in BISYNC by sending a special synchronization character or characters (SYN) at the beginning and periodically in the middle of each transmission. The exact number of SYN characters is somewhat dependent on the hardware. Generally, at least two are transmitted. The hardware scans the line searching for the SYN character (e.g., 00010110 in ASCII). Once the SYN character is found, character synchronization is established, and the following characters can be interpreted correctly.

Message synchronization in BISYNC is accomplished with the SOH, STX, ETB, and EOT characters. These control characters mark the beginning and end of each message being sent.

Block Checking BISYNC uses various techniques for error detection and an ARQ approach for error correction. When ASCII code is being used, a parity check (VRC) is performed on each character and an LRC (horizontal parity) is performed on the whole message. When EBCDIC or 6-bit transcode is being used, no parity check is made, but a CRC is calculated for the entire message.

In all cases, if the block check character or characters transmitted with the data do not agree with the characters calculated by the receiver, a NAK is sent back to the transmitter, which is then responsible for resending the block in error. When the block check characters agree, a positive acknowledgment is sent, ACK0 for an even numbered block and ACK1 for an odd numbered block. The alternation between ACK0 and ACK1 provides an additional check against totally missing or duplicated blocks of data. If two ACK0 acknowledgments are received in succession, the transmitting station knows that a block of data has been lost.

Transmission Sequences BISYNC has well-defined rules that govern the sequence of transmission, acknowledgment, and the placement of control characters in various situations. Three common sequences are described here.

Point-to-Point Operation In point-to-point operation, a station with a message to send first requests permission to use the line from the other station. Once permission is granted, the transmission proceeds as shown in Figure 10–3. After the positive acknowledgment of the last

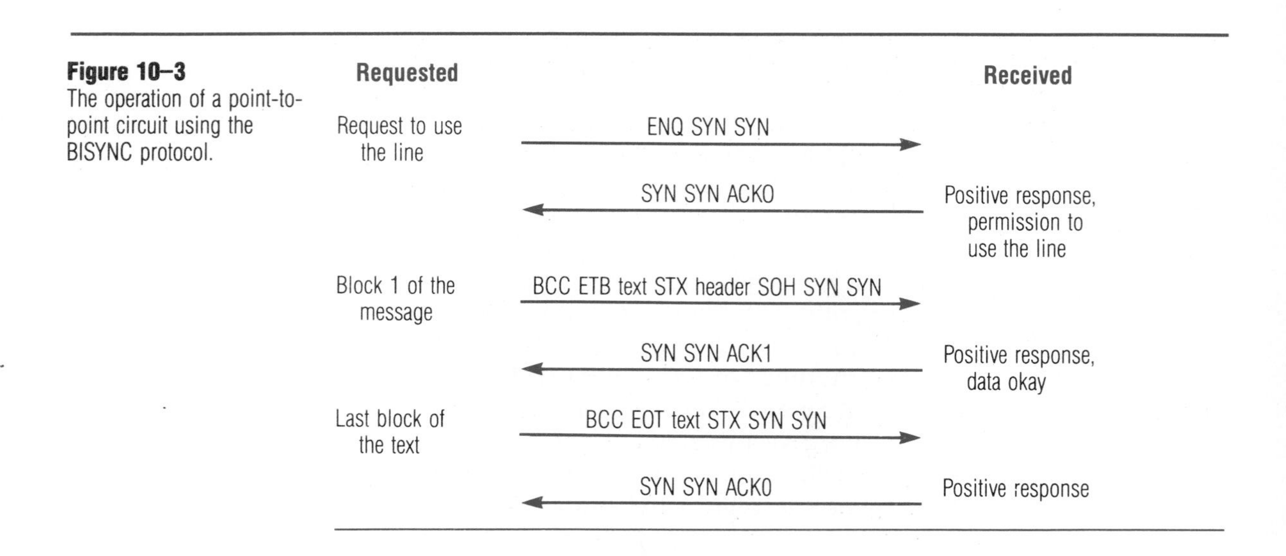

Figure 10–3
The operation of a point-to-point circuit using the BISYNC protocol.

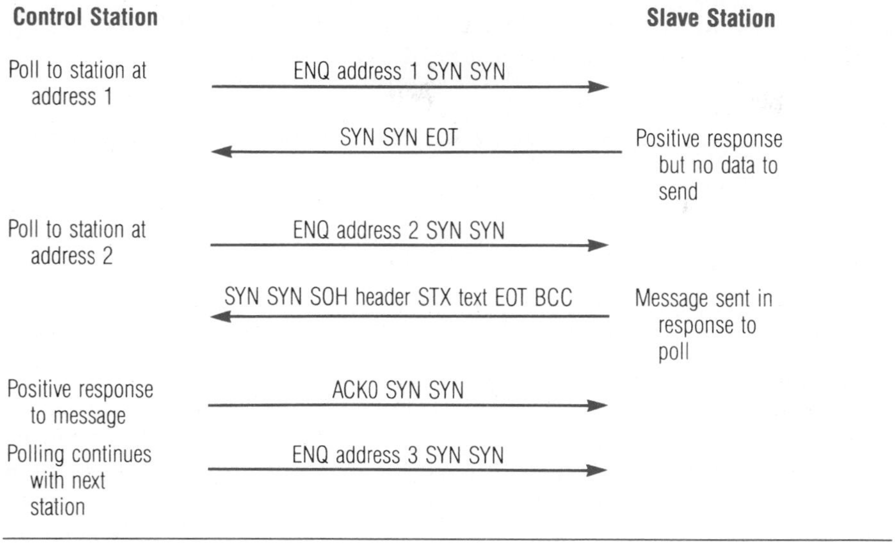

Control Station		Slave Station
Poll to station at address 1	ENQ address 1 SYN SYN →	
	← SYN SYN EOT	Positive response but no data to send
Poll to station at address 2	ENQ address 2 SYN SYN →	
	← SYN SYN SOH header STX text EOT BCC	Message sent in response to poll
Positive response to message	ACK0 SYN SYN →	
Polling continues with next station	ENQ address 3 SYN SYN →	

Figure 10–4
The polling operation on a multipoint circuit using the BISYNC protocol.

block of data, the line is again available, and either station may request its use with the SYN SYN ENQ sequence.

Multipoint Operation In multipoint operation, polling solicits input from the stations on the circuit. This is illustrated in Figure 10–4. When the control station has a message to send, it uses the addressing sequence shown in Figure 10–5. After the last acknowledgment, the line is again free.

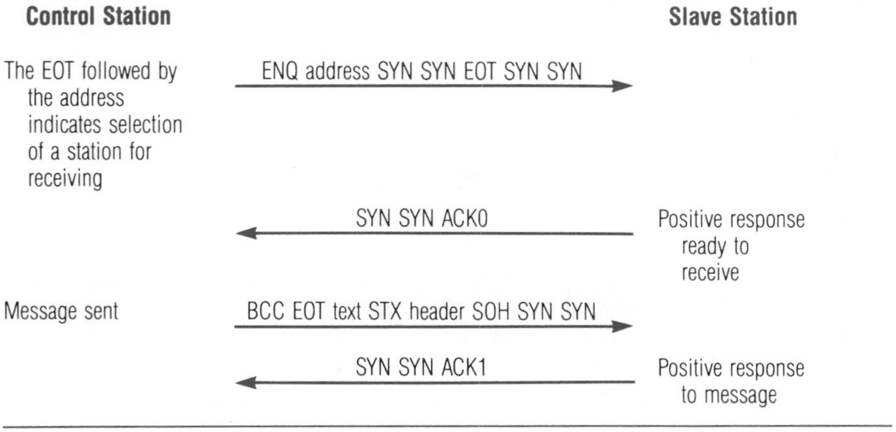

Control Station		Slave Station
The EOT followed by the address indicates selection of a station for receiving	ENQ address SYN SYN EOT SYN SYN →	
	← SYN SYN ACK0	Positive response ready to receive
Message sent	BCC EOT text STX header SOH SYN SYN →	
	← SYN SYN ACK1	Positive response to message

Figure 10–5
The addressing operation on a multipoint circuit using the BISYNC protocol.

Overall, BISYNC is a relatively efficient protocol that is easy to understand and implement. This has led to its wide popularity among computer and terminal vendors. Its primary drawbacks are

- it is not code independent;
- it is a half-duplex protocol and cannot take advantage of full-duplex circuits;
- its implementation of transparency is cumbersome.

Despite these drawbacks, BISYNC is likely to be in widespread use for years to come.

Digital Data Communications Message Protocol (DDCMP)

Digital Data Communications Message Protocol (DDCMP) is a byte-count-oriented protocol. The general format for byte-count-oriented protocol messages is shown in Figure 10–6. The header of the message is a fixed length, preceded by at least two SYN characters and a special SOH character. The SOH character is the only unique character required in the protocol. Therefore, the implementation of transparency is relatively easy. One of the fields in the header indicates the number of characters that the message contains, and the data can be a variable length. Both the header and the data portions of the message are checked with a block check character.

DDCMP has two message types. The data message is similar in format to the general format for byte-count-oriented protocols shown in Figure 10–6. Additional fields in the DDCMP header, shown in Figure 10–7, allow the data message sent in one direction to also acknowledge the receipt of a data message sent in the other direction. (More on this later.) The other DDCMP message type is the control message. Control messages are of fixed length, the same length as the header of the data message. Control messages are used to initiate communications between two stations, to send an ACK or NAK about a previously received message, and to request an acknowledgment of previously sent messages.

Figure 10–6
General format of byte-count protocol data message.

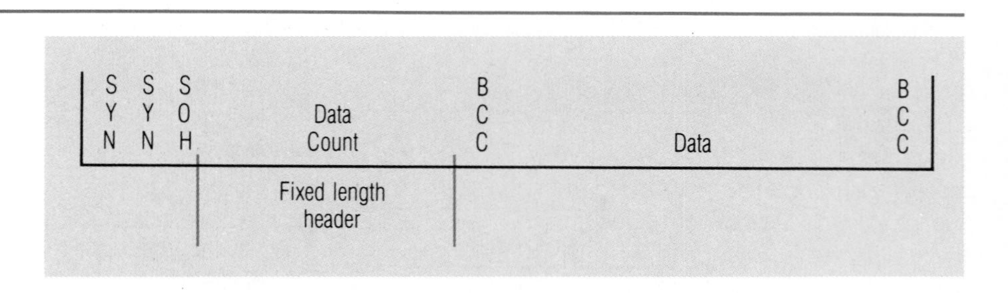

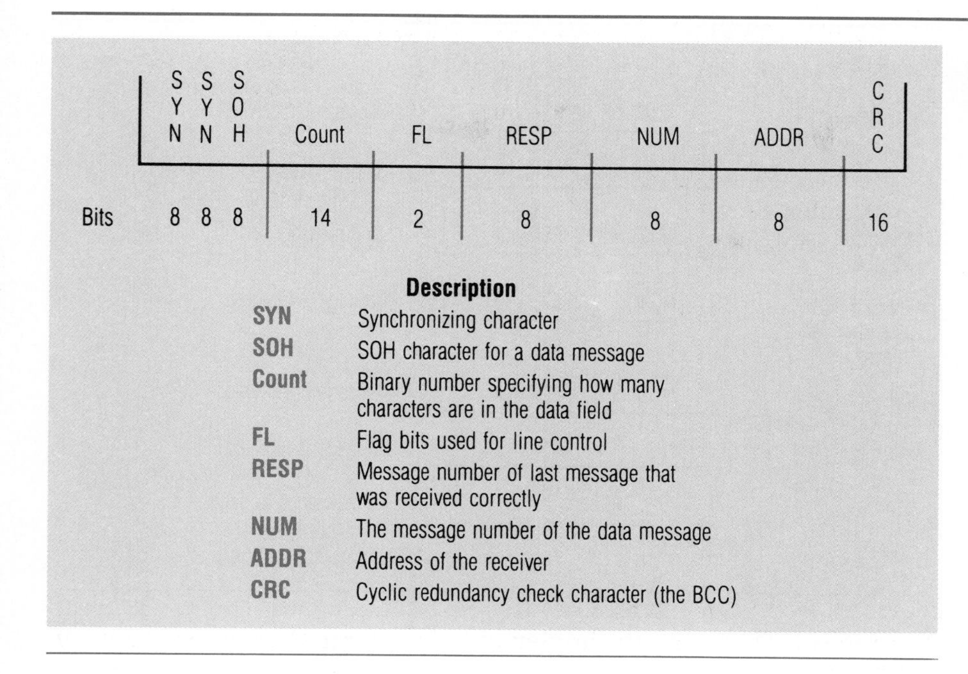

Figure 10–7
The fields in the header of a DDCMP data message.

DDCMP operates in point-to-point or multipoint configurations on either half-duplex or full-duplex lines. In full-duplex operation, both stations can transmit simultaneously. As previously mentioned, the data message from one station can contain an ACK or NAK for a message received from the other station. A station can send several messages in sequence without receiving an acknowledgment, but if the receiver cannot handle them, they may have to be retransmitted. However, there can never be more than 255 outstanding messages before an acknowledgment is made.

Message Sequence Numbers One significant difference between DDCMP and BISYNC is the implementation of a message sequence number. With DDCMP, each transmitted message is assigned a unique and increasing sequence number. When the receiver provides an acknowledgment, it only needs to indicate the sequence number of the last message received correctly. This tells the sender that all messages up through that message have been received and checked. The implication is that not all messages need to be specifically acknowledged. This adds to the efficiency of the DDCMP protocol.

Compared to BISYNC, DDCMP and other byte-count protocols implement transparency in a much more efficient manner. Furthermore, with message sequencing, full-duplex operation is easily implemented. An example of the DDCMP protocol in a half-duplex, point-to-point environment is shown in Figure 10–8.

Figure 10–8
DDCMP protocol: half duplex, point-to-point.

DTE 1		DTE 2
Send message 1	NUM = 1 RESP = 0 →	Message 1 received
	← NUM = 1 RESP = 1	Send message 1 from DTE 2, acknowledge msg 1
Messages 1 and 2 received	← NUM = 2 RESP = 1	Send message 2
Send message 2, acknowledge messages 1, 2	NUM = 2 RESP = 2 →	Message 2 received
Send message 3	NUM = 3 RESP = 2 →	Message 3 received
Send message 4	NUM = 4 RESP = 2 →	Message 4 received
	← RESP = 4	No messages to send Acknowledge message 2, 3, 4

Synchronous Data Link Control (SDLC)

Synchronous Data Link Control (SDLC) was introduced by IBM in 1972 and was the first bit-oriented protocol to be defined. Several years later, minor additions were made to the protocol, and it was standardized by the International Standards Organization (ISO) as *High-Level Data Link Control (HDLC)*, which is now internationally recognized. Today, however, SDLC is used more widely than HDLC, so it is discussed here.

SDLC and all bit-oriented protocols use a special character to mark the beginning and ending of messages. This special character is called the *flag*, and its unique pattern is 01111110. Because of the number of consecutive 1 bits in the flag character, it also serves as a synchronization character, so no SYNs are required. To ensure its uniqueness, no other sequence of six consecutive 1 bits can be allowed in the data stream. To accomplish this, the bit stream of all transmissions is scanned by the hardware. Using a technique called *bit stuffing*, a 0 bit is inserted after all strings of five consecutive 1 bits in the header and data portion of the message. At the receiving end, the extra 0 bit is removed by the hardware. Bit stuffing is illustrated in Figure 10–9.

bit stuffing

SDLC operates on full-duplex or half-duplex lines in either a point-to-point or multipoint configuration. Switched lines are supported only in half-duplex point-to-point arrangements. SDLC is a positional protocol; each field except the data field has a specific length and location. The block check calculation in SDLC is a special form of a CRC check that provides a high probability of detecting bit errors.

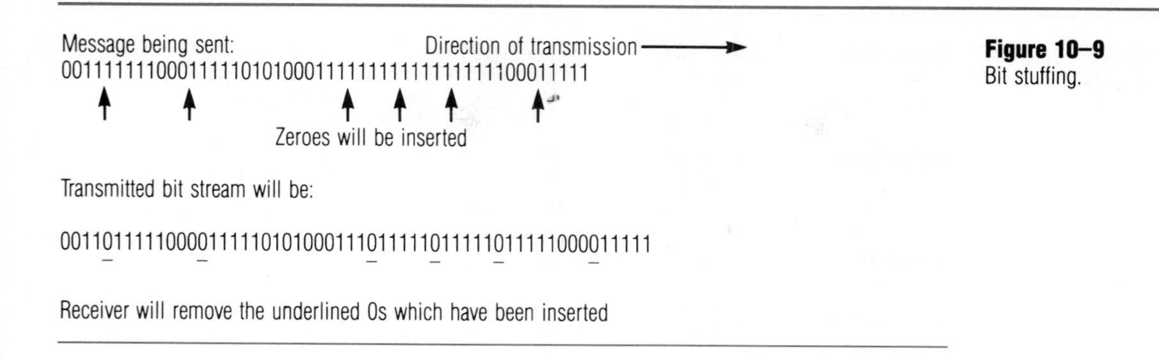

Figure 10–9
Bit stuffing.

Frames The basic operational unit for SDLC is a *frame*, as shown in Figure 10–10. Frames have three different formats:

- the supervisory format;
- the nonsequenced format;
- the information format.

Figure 10–11 shows how the format of the frame is determined by the 8 bits in the control field. The supervisory format is used to initiate and control the information transfer. The commands of the supervisory format are used to give a positive acknowledgment that frames up to a specified sequence number have been received correctly and to indicate whether the receiver is ready to receive more data.

The nonsequence format is used for setting operating modes and initializing transmissions.

The information format is used for data transmission. Like the byte-count-oriented protocols, information frames are assigned sequence numbers that are transmitted in the headers. Thus, an additional check is provided to ensure that no blocks are missing or duplicated. Two

Fields	Beginning Flag 01111110	Address	Control	Data	Frame Check Character	Ending Flag 01111110
Bits	8	24	8	$8 \times n$ Any multiple of 8 bits	16	8

Figure 10–10
SDLC frame format.

Figure 10–11
SDLC control field format.

Frame Type	Command	Bit Configuration (bit 0 sent first) 01	23	4	567
Supervisory	RR	10	00	P/F	NR
	RNR	10	10	P/F	NR
	REJ	10	01	P/F	NR
Information	I	00	NS	P/F	NR
Nonsequenced	NSI	11	00	P/F	000
	RQI	11	01	F	000
	SIM	11	10	P	000
	SNRM	11	00	P	001
	ROL	11	11	F	000
	DISC	11	00	P	010
	NSA	11	00	F	110
	CMDR	11	10	F	001
	ORP	11	00	1	100

NR—the number of the last message received correctly
NS—the number of the last message sent
P/F—the poll/final bit used to poll a terminal or indicate the final block of a transmission

sequence counts are maintained in the control field of the information frame: the number sent (or Ns) count and the number received (or Nr) count. Transmitting stations increment the Ns count, and receiving stations increment the Nr count.

The frame's address field is 24 bits long and is used to direct a message to the proper destination. The data field in the frame can be any multiple of 8 bits. This does not imply, however, that the data coding scheme must be an 8-bit code. If seven 7-bit characters are sent, the 49 data bits are padded with an additional 7 bits to bring the total number of bits to 56, a multiple of 8.

Message Flows Message flows in SDLC can become quite complicated. Like many other protocols, SDLC operates with master and slave stations, with the master station performing the polling and addressing. Figure 10–12 shows half-duplex message traffic in SDLC and illustrates the use of the polling and final bits.

SDLC uses a continuous ARQ technique. The protocol can have up to 128 transmissions outstanding before an acknowledgment is received. This capability is of great benefit on circuits for which the propagation delay is long, such as satellite circuits. It allows blocks of data to be sent continuously until a transmission error occurs, rather than the line having to be turned around and an acknowledgment received for each block, as is true with the BISYNC protocol. SDLC provides considerably bet-

DTE 1		DTE 2
Poll DTE 2	FLAG CRC RR NR=0 P=1 ADDR=2 FLAG →	
	← FLAG ADDR=1 NR=0 NS=1 F=0 CRC FLAG	DTE 2 responds with message 1, block 1
	← FLAG ADDR=1 NR=0 NS=2 F=1 CRC FLAG	Send block 2, last block (F=1)
Acknowledge block 1, 2 and send own block 1	FLAG CRC F=1 NS=1 NR=2 ADDR=2 FLAG →	
	← FLAG ADDR=1 RR NR=1 F=1 CRC FLAG	Acknowledge block 1 indicates ready to receive (RR) and nothing to send (F=1)

Figure 10–12
SDLC protocol: half-duplex, point-to-point.

ter throughput on satellite circuits than protocols that use a stop and wait ARQ.

LOCAL AREA NETWORK (LAN) PROTOCOLS

Local area networks, which are described in detail in Chapter 11, use bit-oriented protocols that are really a combination of a circuit access technique and a protocol as we have studied them thus far.

Logical Link Control

At the heart of LAN protocols is *logical link control (LLC),* a data link control protocol defined for use on LANs. LLC, or IEEE 802.2 as it is better known, is a bit-oriented protocol that is similar but not identical to HDLC. LLC's frame is called a *protocol data unit (PDU),* and its format is shown in Figure 10–13. The destination address identifies the node to which the information field is to be delivered, and the source address identifies the node that sent the message. The control field contains the commands, responses, and sequence numbers necessary to control the data link. The information field can contain any multiple of 8 bits, and any combination of bits (transparency) is acceptable.

CSMA/CD Protocol

As its name implies, the CSMA/CD protocol uses the contention access technique to allow a node to gain control of a circuit. The basic data format is a frame. The header, called the *preamble,* serves to synchronize

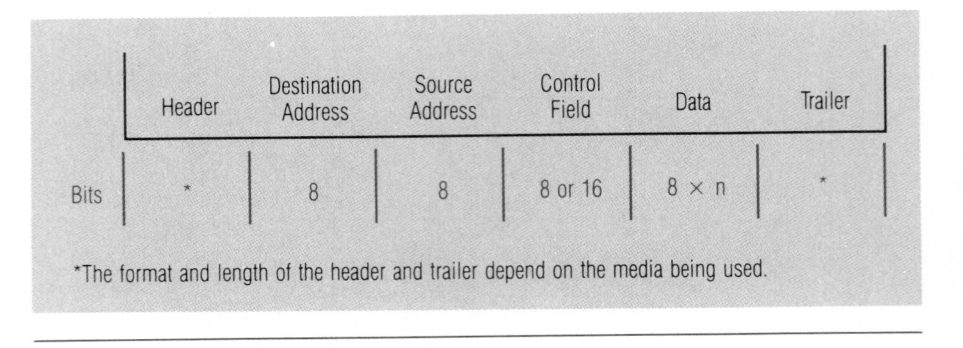

Figure 10–13 LLC protocol data unit (PDU) format.

the transmitter and receiver, thereby eliminating the need for any SYN characters. The control field can take on various meanings but normally is used to indicate the type of data being transmitted. A 32-bit CRC field is used, which provides extremely good error detection capability.

Messages are broadcast on the medium. All nodes receive the message and look at the address field to see if the message is for them. Only the station that is addressed by the message takes action, such as printing the message on a printer or storing it on a disk.

Token Passing Protocol

As was described earlier, a token circulates on the circuit until a station with a message to send acquires it, changes the token's status from "free" to "busy," and attaches a message. The token and message move from station to station, and each station examines the address of the message to determine whether it is the intended receiver. When the message arrives at the intended destination, the station copies it. If the CRC check is okay, the receiving station sends an aknowledgment to the sender. The token and acknowledgment return to the originating station, which removes the acknowledgment from the circuit and changes the token's status from "busy" to "free." The token then continues circulating, giving other stations an opportunity to use the circuit.

The performance of a circuit using the token passing protocol is predictable. However, if the number of stations on the circuit is large, it may take a long time for the token to get around to a station that has a message to send. Therefore, although response time is predictable, it may be longer than on a CSMA/CD circuit.

PROTOCOL CONVERSION

With the proliferation of data link protocols, it is often desirable to convert a data transmission from one protocol to another. This is the function of the *protocol converter*. This equipment takes a transmission in one

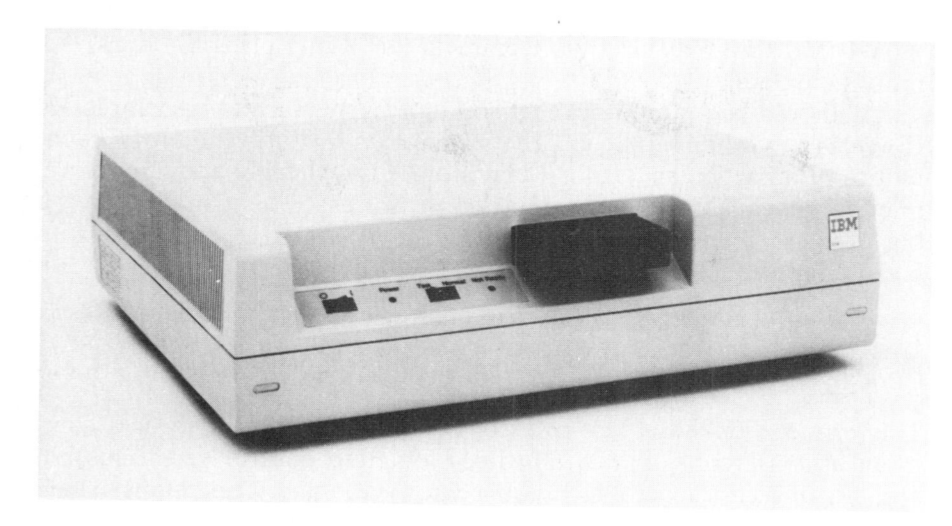

The IBM 3708 protocol converter handles up to eight dial-in lines simultaneously. Many conversion options are available. (Courtesy of IBM Corporation)

protocol and changes it as required to a different protocol. The most common protocol conversion being performed today is from a character-oriented asynchronous protocol to a block-oriented synchronous protocol, such as SDLC. This particular conversion is popular because it allows inexpensive, dumb asynchronous terminals to be used to communicate with an IBM mainframe computer that communicates synchronously and uses the SDLC protocol.

Protocol conversion involves, at a minimum, changing the control bits or characters that surround the data. Because there is not always a one-for-one correspondence of control characters in different protocols, the process is more complex than a simple character translation. Sometimes blocks of data must be reformatted or changed to a different length. The protocol converter may have to receive an entire message, made up of several blocks, before it can do the reformatting. Protocol converters also frequently perform error checking on the incoming data stream and request retransmission if an error is detected.

Another common function of protocol converters is code translation from one transmission code to another. In the common example mentioned before, the protocol converter must also convert from ASCII to EBCDIC and back because inexpensive terminals almost always use the ASCII code and IBM mainframe computers use EBCDIC.

Protocol converters can be located in various places in the network. One of the most common locations is at the site of the host computer. Usually, multiple leased or dial-up circuits can be connected because most protocol converters can handle several lines at a time. Another location for the converter is at the terminal. In fact, some terminals have protocol conversion capability built into them so that they can communicate in several protocols. In this way, they can emulate or appear to the network as one of several different types of terminal. The DEC

terminal emulation

VT-100 and IBM 3101 are two terminals that many other vendors' terminals emulate.

It should be pointed out that some protocols cannot be completely converted to other protocols. The implication is that some of the capabilities of a particular terminal may not be translatable and, therefore, may be unavailable after the translation process. This usually is not a big problem. However, it is something to be aware of and taken into consideration.

SUMMARY

This chapter looked at the rules or protocols under which data communications circuits operate. The need for circuit control was discussed, and circuit access control techniques were examined. The details of several specific protocols, all of which are in widespread use throughout industry, were explored. Protocols unique to LANs also were described. Data link protocol standardization is paving the way toward greater interconnection of terminals and computers from all manufacturers.

CASE STUDY

Protocols Used in Dow Corning's Data Network

Dow Corning's data network is for the most part an IBM SNA network and as such, it uses the SDLC protocol. The details of the protocol are not something that the data communications people concern themselves with unless there is a problem. In that case, Dow Corning's technical communications specialists get IBM involved and may use an instrument called a *protocol analyzer* or a trace of all of the activity on the line to try to identify the problem.

In the manufacturing plants, computers from Digital Equipment Corporation (DEC) are being installed. These computers use various types of communications, but the CSMA/CD protocol is the most common.

Protocol conversion is required between the DEC and IBM communications systems. A specialized protocol converter, specifically designed for DEC-to-IBM protocol conversion, is employed. This converter allows IBM terminals to use the DEC computers and DEC terminals to use the IBM computers. Because the protocols cannot be converted precisely, the terminals have some restrictions in their capabilities. However, for most applications, the conversion is adequate.

Dow Corning also supports the connection of a variety of asynchronous ASCII terminals through to its network. Protocol conversion is necessary, and the company has IBM 3174 and 3708 protocol converters that make a variety of devices emulate IBM 3270 VDTs so they work with the company's mainframe computers. This type of protocol conversion is used primarily on dial-in links with personal computers or non-IBM VDTs.

For Dow Corning, the decision about which communications hardware and software to use has been made at a high level with a concern for meeting the requirements of the people using the system. Selecting a protocol has never been a key decision point. The protocol used has been a by-product of other decisions that were made.

Review Questions

1. Define the term *protocol.*
2. Why are data link protocols required?
3. Identify the six functions of a data link protocol.
4. Explain the term *transparency* as it relates to data link protocols.
5. What is meant by "predetermined communications parameters?"
6. Compare and contrast contention techniques for accessing a circuit with polling techniques.
7. Explain how the CSMA/CD access technique operates.
8. In a list of terminals to be polled, why might a terminal's address be listed more than once?

9. What is the function of the header of a message?
10. Why is the Xmodem protocol not a particularly reliable means of transmitting large blocks of data?
11. Can the BISYNC protocol transmit Baudot code? Why or why not?
12. How is character synchronization established in BISYNC?
13. Describe BISYNC's approach to error correction.
14. Give some reasons why DDCMP is more efficient than BISYNC.
15. What is the purpose of the flag in SDLC?
16. What is bit stuffing?
17. Explain how the token passing protocol operates.
18. A token passing access technique is said to be deterministic. Explain this concept and why a CSMA technique is nondeterministic.
19. What is the purpose of a protocol converter? Why do protocols need to be converted?

Problems and Projects

1. Make a chart of the way the Xmodem protocol operates similar to the charts shown in the text for BSC and DDCMP.
2. Which data access technique provides the fastest access to a circuit? Are there any limitations to the speed? In what situations or applications is a predictable, known rate of access more important than having the fastest access?

Vocabulary

protocol
carrier sense multiple access with collision detection (CSMA/CD)
roll call polling
polling list
hub polling
token
header
text
trailer
start of header (SOH)
start of text (STX)
end of text (ETX)
block check character (BCC)
synchronization character (SYN)
Xmodem protocol
character-oriented protocol
byte-count-oriented protocol
bit-oriented protocol
Binary Synchronous Communications protocol (BSC, BISYNC)
Digital Data Communications Protocol (DDCMP)
Synchronous Data Link Control protocol (SDLC)
High-Level Data Link Control protocol (HDLC)
flag
bit stuffing
frame
logical link control (LLC)
protocol data unit (PDU)
protocol converter

References

Bartee, Thomas C. *Data Communications, Networks, and Systems.* Indianapolis, IN: Howard W. Sams & Company, Inc., 1985.

Doll, Dixon R. *Data Communications: Facilities, Networks, and Systems Design.* New York: John Wiley & Sons, 1978.

Falk, Howard. *Microcomputer Communications in Business.* Radnor, PA: Chilton Book Company, 1984.

Lane, Malcom G. *Data Communications Software Design.* San Francisco: Boyd & Fraser Publishing Company, 1985.

Martin, James. *Introduction to Teleprocessing.* Englewood Cliffs, NJ: Prentice-Hall, Inc., 1972.

McNamara, John E. *Technical Aspects of Data Communications.* Maynard, MA: Digital Equipment Corporation, 1982.

Stamper, David A. *Business Data Communications,* 2nd ed. Menlo Park, CA: The Benjamin/Cummings Publishing Company, 1986, 1989.

CHAPTER

11

Communications Networks

OBJECTIVES

When you complete your study of this chapter, you will be able to

- explain how communications circuits are connected together to form networks;
- explain the major types of communications networks as characterized by their
  - topology
  - ownership
  - purpose or type of transmission
  - geography;
- distinguish between LANs, MANs, and WANs;
- discuss several ways that networks can be connected to one another;
- describe several ways that messages are routed within a network;
- describe the IEEE standards that apply to local area networks;
- describe the Ethernet and token ring LANs and their method of operation;
- explain the necessity of and some of the problems of managing LANs.

INTRODUCTION

In Chapter 9, you studied the characteristics and attributes of single telecommunications circuits of various types. Now you will expand your horizon and look at how several circuits and other comunications equipment can be connected together and arranged into a network. The material in this chapter describes and explains layers 3 and 4 of the ISO-OSI model. These layers, the network control and transport control layers, describe how a transmission is addressed and routed through the network from its source to its destination.

It should be noted that throughout this chapter there will be references to the term *protocol*. You should recognize that throughout a network there are many protocols. Data link protcols, which you studied in Chapter 10, are only one of the protocols employed in a network. You will be introduced to other, higher-level protocols in this chapter.

NETWORK DEFINITION AND CLASSIFICATION

A *network* is one or more communications circuits and associated equipment that establishes connections between nodes. Usually when we think about a communications network in a business sense, we think about the way in which users or locations in the company are connected together. Most companies have several networks. They may have a voice

Figure 11–1
Several ways to classify communications networks.

By Topology
- Star
- Hierarchical
- Mesh
- Bus
- Ring
- Hybrid

By Ownership
- Private
- Public

By Purpose or Type of Transmission
- Telex/Electronic Mail Network
- Value Added Networks (VAN)
- Packet Data Networks (PDN)
- Integrated Services Digital Network (ISDN)

By Geography
- Wide Area Networks (WAN)
- Metropolitan Area Networks (MAN)
- Local Area Networks (LAN)

network that is separate from their data network, or the two may share certain facilities, such as T-1 or higher speed lines. There may also be a facsimile network that consists of the facsimile machines connected together when needed by dial-up telephone calls. With the rapid growth in the use of communications terminals in the last few years, most businesses are finding that they have multiple communications circuits and nodes to arrange into one or more networks.

Networks may be classified in several different ways, as shown in Figure 11–1. In most companies, these network types overlap. That is, one company's network may have a combination of WAN and LAN geographies. The WAN portion may be arranged into a basic star, but with some mesh connections. You'll see how this is possible in the next pages.

NETWORKS CLASSIFIED BY TOPOLOGY

The way that circuits are connected together is called the network *topology*. If a map is drawn of all of the circuits and nodes, showing how they are connected to one another but without regard for the geography of where they are located, the topology of the network can be seen. Network topologies fall into six major categories.

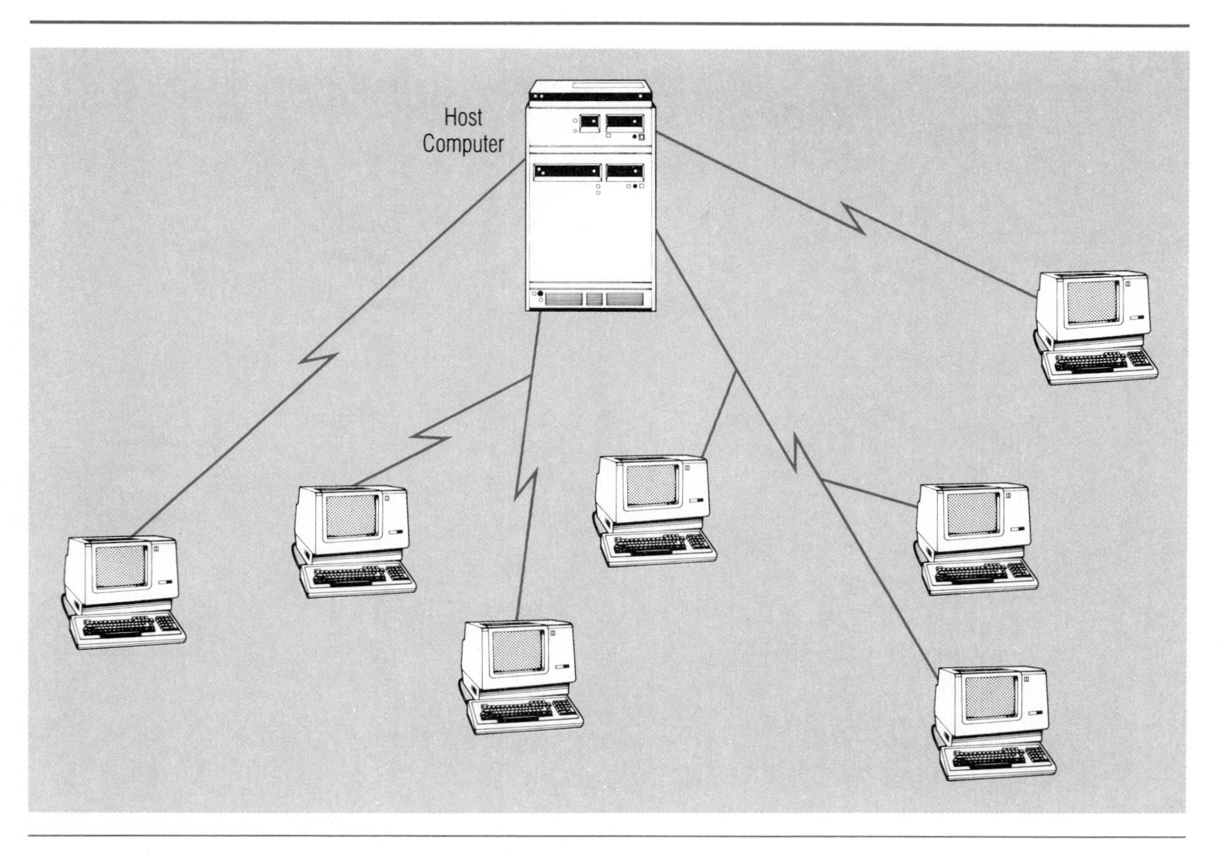

Figure 11–2
Star network.

Star Network

A *star network* is illustrated in Figure 11–2. All circuits radiate from a central node, often a host computer. The circuits can be point-to-point or multipoint or a combination of both. The star network puts the central point in contact with every other location, which makes it easier to manage and control the network than in some other configurations. There is no limit to the number of arms that can be added to the star or the length of each of the arms. Thus, it is relatively easy to expand a star network by adding more nodes. On the other hand, since all transmissions are controlled by the central point and they typically all flow through it, the central node is a single point of failure. If it is down, the entire network may be out of service. Another potential problem with star networks is that in times of peak traffic, the central node may become overloaded and unable to keep up with all of the messages that the outlying stations want to transmit.

Many business data communications networks have the star topology. Where there is a central computer that supports many terminals, the star configuration is easy to implement. Since the central computer

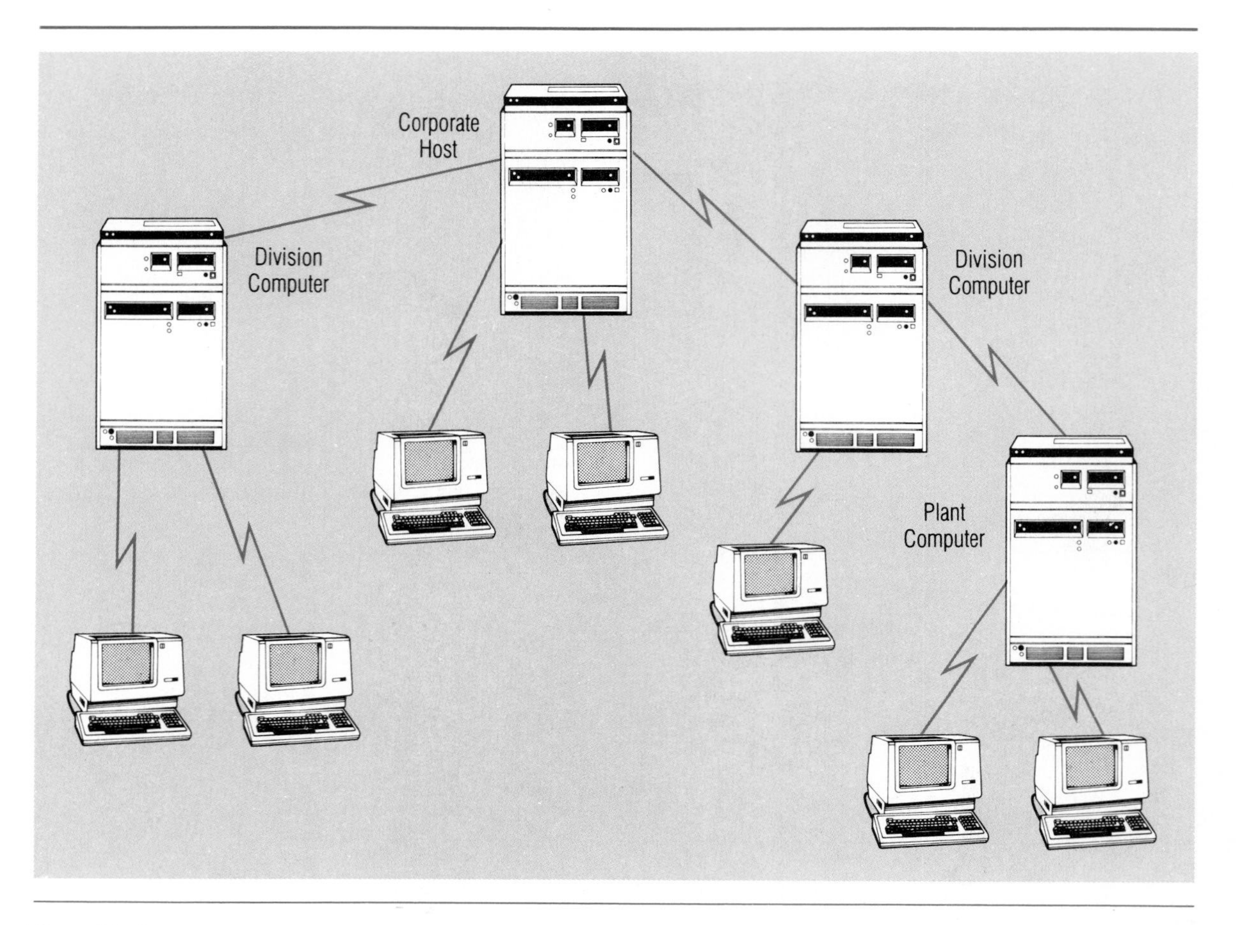

Figure 11–3
Hierarchical network.

is the master node and controls the network, the rules for network operation are relatively simple. The star topology also applies to PBXs to which all of the telephones connect.

Hierarchical Network

A *hierarchical network* is illustrated in Figure 11–3. It is sometimes also called a tree structure, and the top node in the structure is called the root node. You may notice that the hierarchical structure mirrors a typical corporate organization chart, and in fact it is in this setting that a hierarchical configuration is most likely to be found. Compare if you will the root node to a corporate headquarters on an imaginary organization chart and the nodes immediately underneath in the second level to a divisional level. Under each division are nodes corresponding to plants in the case of manufacturing divisions and district offices in the case of sales and marketing divisions. Under the district offices, one might find nodes in local sales offices.

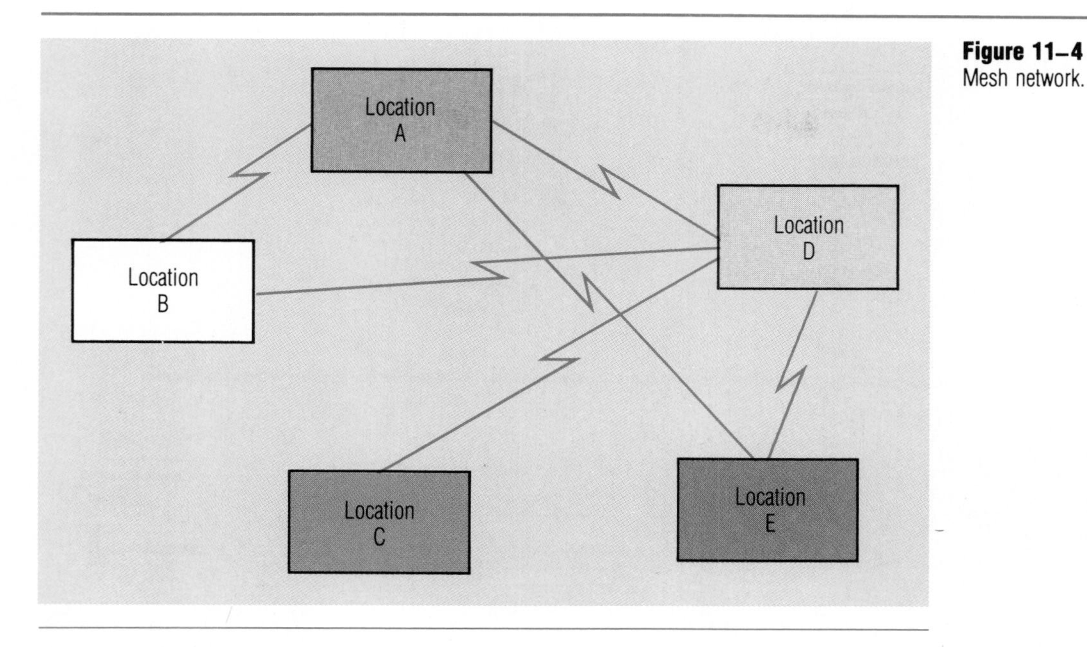

Figure 11–4
Mesh network.

This type of network would most likely be implemented where the lower level nodes at the second or third level are in themselves computers. One can envision a district office computer being connected to the local sales office computers and collecting data from them. The district data would be consolidated at corporate headquarters. In that configuration, there is no single point of failure in the network. If one division's computer or network fails, the other divisions are not affected. Even if the root node at corporate headquarters fails, the divisions can go on doing their daily processing and can send and receive data from lower levels in the hierarchy. The results of divisional communications and processing can be transmitted to the higher level in the network when it is restored to service.

Mesh Network

A *mesh network,* illustrated in Figure 11–4, is similar to a hierarchical network except that there are more interconnections between nodes at different levels. In fact, the levels may not exist at all. In a fully interconnected mesh, each location is connected to every other location, but for cost reasons, this is seldom implemented. Usually, the major nodes are connected together, whereas minor nodes are connected to one or more locations depending on their need and criticality. The public telephone network is an example of a mesh network that because of good design and a high level of redundant connection, provides many alternate paths between nodes. The heavy interconnection makes the telephone network virtually fail-safe.

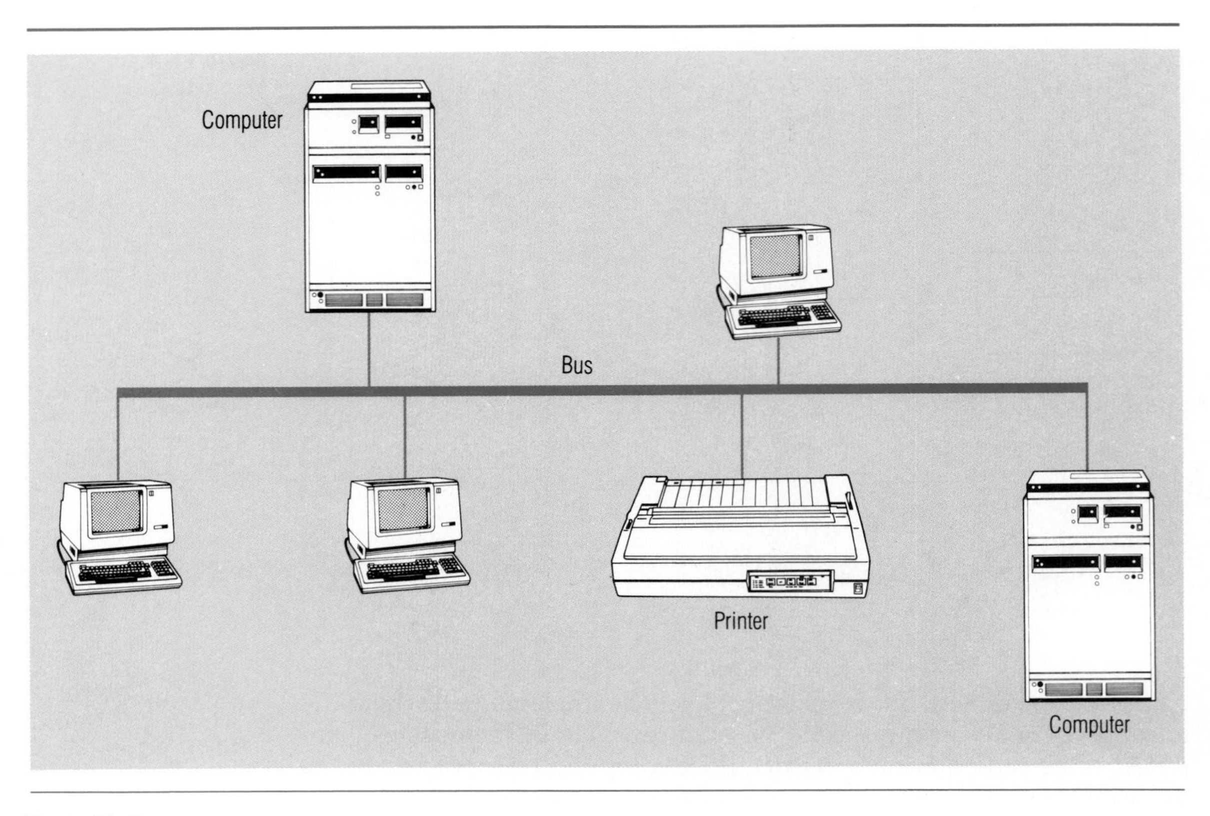

Figure 11–5
Bus network.

Bus Network

A *bus network* is shown in Figure 11–5. Conceptually, a bus is a telecommunications medium to which multiple nodes are attached. The term *bus* implies very high-speed transmission, and bus networks are usually implemented in situations where the distance between all of the nodes is limited, such as within a department or building.

Devices are attached to the bus by a tap connection that breaks into the bus cable. Each tap causes a certain amount of signal loss on the cable, and, therefore, bus networks have a limit as to the number of devices that can be attached to them. Another problem with bus networks is that when problems do occur, faults can be difficult to locate. Because all of the devices are connected serially on the bus, each device may have to be checked in sequence to locate the problem.

Although the bus may look similar to a star or hierarchical network, the major difference is that on a bus, all stations are independent of one another. There is no single point of control or failure as in a star network. The loss of a single node on a bus has no impact on the other nodes, and, in fact, unless the bus itself fails, the reliability of a bus network is excellent. On the other hand, because of their high-speed operation, bus networks usually are limited in the distance they can traverse, which is why the topology is normally found only in LANs.

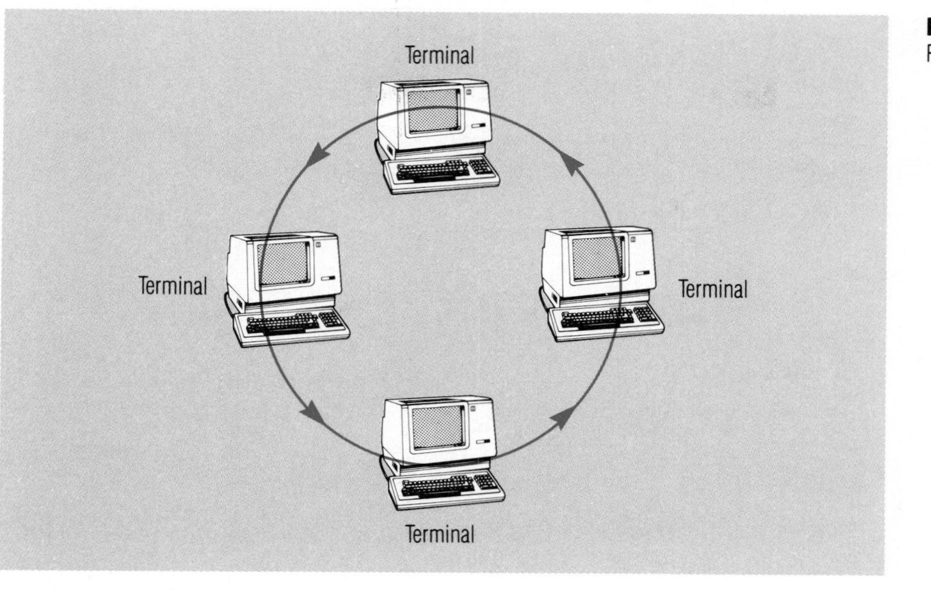

Figure 11–6
Ring network.

Ring Network

A *ring network* is illustrated in Figure 11–6. Ring networks are also usually associated with networks where the nodes are relatively close together. Each device is connected to the ring with a tap similar to the type found on a bus network. As communications signals pass around the ring, a receiver/driver unit in each device checks the address of the incoming signal and either routes it to the device or regenerates the signal and passes it on to the next device on the ring. This regeneration is an advantage compared to bus networks; the signal is less subject to attenuation. Furthermore, each device can check the signal for errors, which allows for more sophisticated error control and network management.

All of the stations on the ring are essentially equal and must be active participants to make the ring work. If one of the nodes fails, the potential exists that the entire ring will be out of service because messages cannot be passed through the failing node. Figure 11–7 illustrates a ring with two channels that transmit the data in opposite directions. This allows transmission to continue in case one of the links or nodes fails. All stations can be reached from either direction by transmitting on one channel or the other.

Hybrid Networks

The various network configurations discussed so far can be combined into *hybrid networks*. For example, a star network might have a ring on it, as shown in Figure 11–8, or a hierarchical network could have a star radiating from one of the nodes. As long as all of the networks use the same protocol and basic operating technology, connecting them together

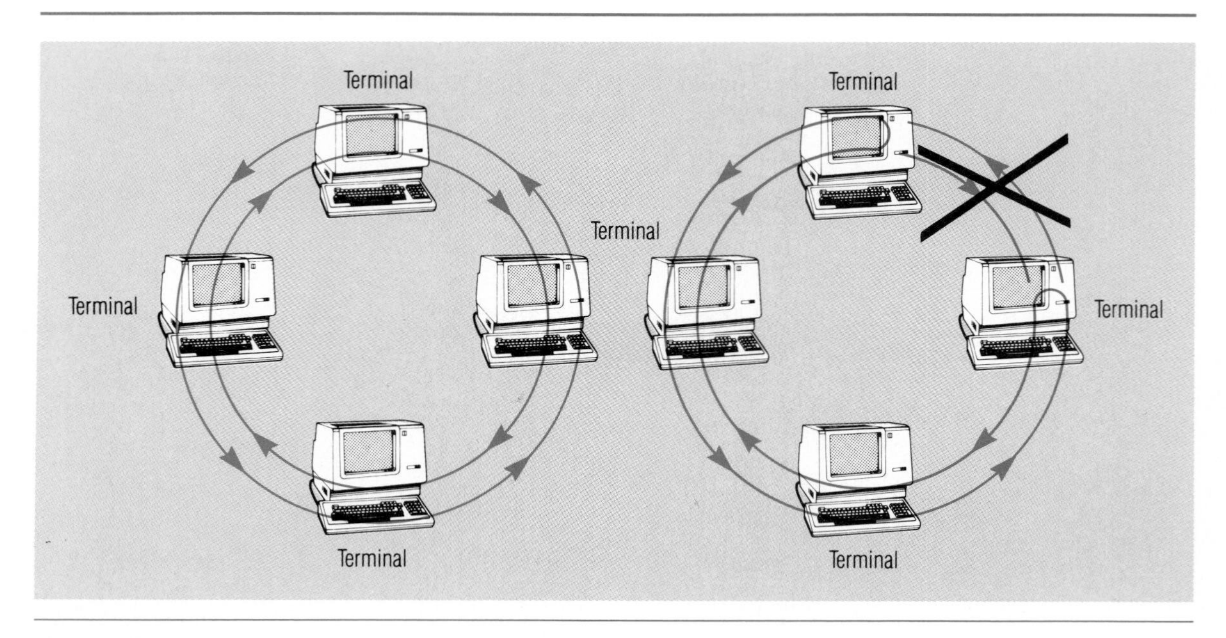

Figure 11–7
Ring network with two channels.

is easy. If they use different protocols, conversion devices are needed, and they are discussed later in this chapter.

In practice, most networks are hybrid networks of one type or another. Only very small networks are purely one topology. The different topologies, combined with other network attributes to be discussed, meet diverse telecommunications needs, and most companies find that they need a combination of capabilities.

NETWORKS CLASSIFIED BY OWNERSHIP

Another way of classifying networks is by their ownership. Two broad categories exist, private networks and public networks. We will look at each category and its variations.

Private Networks

A *private network* is built by a company, normally for its exclusive use. The network is built from circuits available from a variety of sources and may include a combination of privately installed and operated circuits and leased or switched facilities. There also may be connections to one or more of the public networks. One of the advantages of a private network is that it can be designed to address specifically the voice or data communications requirements of the company. Since it is built around particular traffic patterns or communications flows, it can potentially make better use of circuits than could be achieved if a public network

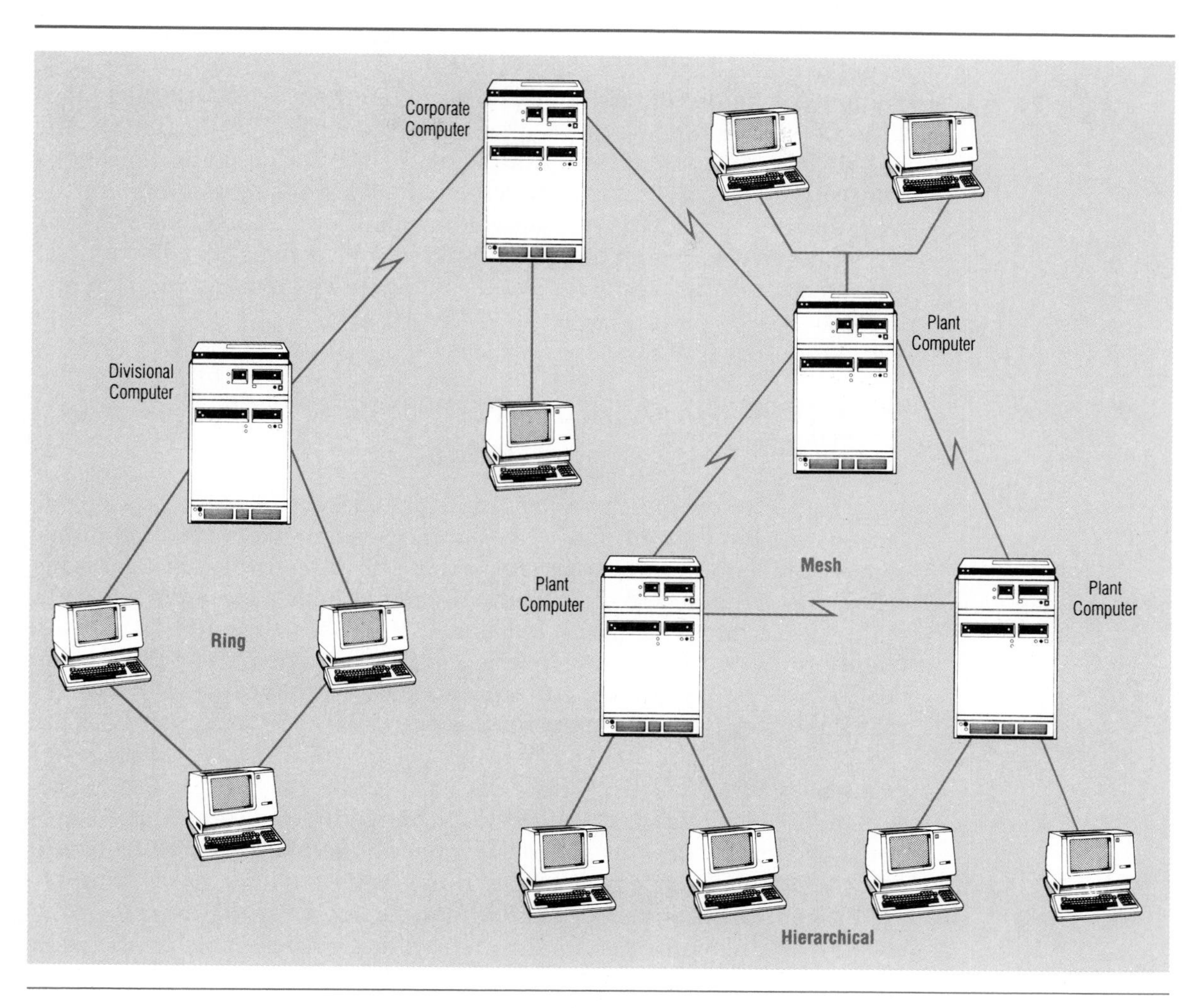

Figure 11–8
Hybrid network.

were used. Another advantage is that it gives the company full control of the network's operation and potentially better security. Communications using a private network may, however, be more expensive than comparable communications on public networks. Because of the flexibility to tailor the private network to a company's exact requirements, however, most major companies have one or more private networks connecting their locations.

Public Networks

A *public network* is a network built and owned by a communications company or a common carrier for use by its customers. The most familiar example of a public network is the one we use to make telephone calls, the PSTN. Another familiar, widely used network is the public *telex network* used for exchanging messages between subscribers.

The advantage of a public network is that it provides services or access to locations that a company might not otherwise be able to afford. Companies frequently use public data networks to exchange messages with small locations such as a single-person sales office, where the installation of a private line cannot be justified. The communications company or common carrier may have many similar customers in the area and can spread the implementation and operational costs of providing service across all of its customers. The carrier can achieve good utilization of its network while at the same time providing high-quality service.

NETWORKS CLASSIFIED BY PURPOSE OR TYPE OF TRANSMISSION

Networks also can be classified according to their purpose or the type of transmission they employ. For example, we have discussed the public switched telephone network that we use to place ordinary telephone calls. We've mentioned that a company might have a facsimile network, made up of all its facsimile machines, that uses the public switched telephone network to connect with one another when they have facsimiles to exchange. Now we will look at several other types of networks that serve a unique purpose or use a particular type of transmission.

Public Telex Network

The public telex network is used to exchange written (typed) messages between subscribers. Like the public telephone network, the telex network is global in scope because the individual networks of each country are all tied together and operate under common standards and protocols. An individual or company that wants to participate in the telex network pays a network access fee and purchases or rents a terminal that is compatible with the telex network standards. The subscriber is given a telex number, much like a telephone number, that can be given to correspondents. There are many public and private directory services throughout the world that list telex numbers, much like a telephone book lists telephone numbers.

Messages are sent by dialing the number of the correspondent from the telex terminal, receiving a brief acknowledgment that the number has been reached, and then typing the message directly. Alternately, the message may be prepared ahead of time and stored on paper tape or a magnetic medium. Then when the connection is made, the message is sent as fast as the communications circuit can accept it. Telex systems are run by computers so that even if the correspondent teminal is busy or out of service, the computer usually will accept the message and deliver it later when the recipient is back online.

Telex was the original electronic mail system. It became almost ubiquitous throughout the world. Every company and institution of any size had at least one telex machine because the service was inexpensive. Originally, the error checking was not very good because a 5-bit code was used, but that problem was largely solved by the introduction of equipment that used 7- and 8-bit codes and computers to convert messages from machines that used the older 5-bit code. In the industrialized countries, telex is being replaced by newer public systems that operate at higher speeds, such as the Teletex system, which is gaining popularity in Europe.

Value Added Networks (VAN)

A *value added network (VAN)* is a particular type of public data network that in addition to offering transmission facilities, contains intelligence that makes the basic facilities better suited for satisfying the communications needs of a particular type of user. The intelligence might provide code or speed translation, or it could store messages and deliver them at a later time (store and forward). This intelligence provides the "added value" from which the generic name of this type of network is derived.

The intelligence in a VAN is provided by computers located at network nodes that may assist in the routing of messages or perform other communications-related processing, such as code translation or speed conversion.

Because computers are employed for code conversion and storing messages in the public telex network, it has become a VAN. Other examples of industry-oriented VANS include the SWIFT network that connects international banks, the SITA network that connects the airlines' networks, the IVANS network connecting many U.S. insurance companies, and, in the U.K., the Tradanet service that connects major retailers and their suppliers. Each of these networks contains intelligence to perform certain processing for its users.

Packet Data Networks (PDN)

One type of VAN is a *packet data network (PDN)*. Packet data networks are normally based on *packet switching* technology. When packet switching is used, messages are segmented into blocks of a predetermined size, called *packets*, before they are transmitted. Packets are sent through the network somewhat autonomously. Packets from the same message may be sent on different routes and may arrive at the receiving node out of sequence. Computers at each network node route the packets through the network toward the destination node. The computer at the receiving node is responsible for reassembling the packets into the correct sequence before passing them on to the user.

In addition to the original piece of data, an address is attached to each packet to tell the destination of the packet. Other checking or control fields also may be added to the packet to ensure data integrity.

Packet switching networks are designed so that there are at least two and usually several alternative high-speed paths from one node to another, as shown in Figure 11–9. A message from Detroit to San Fran-

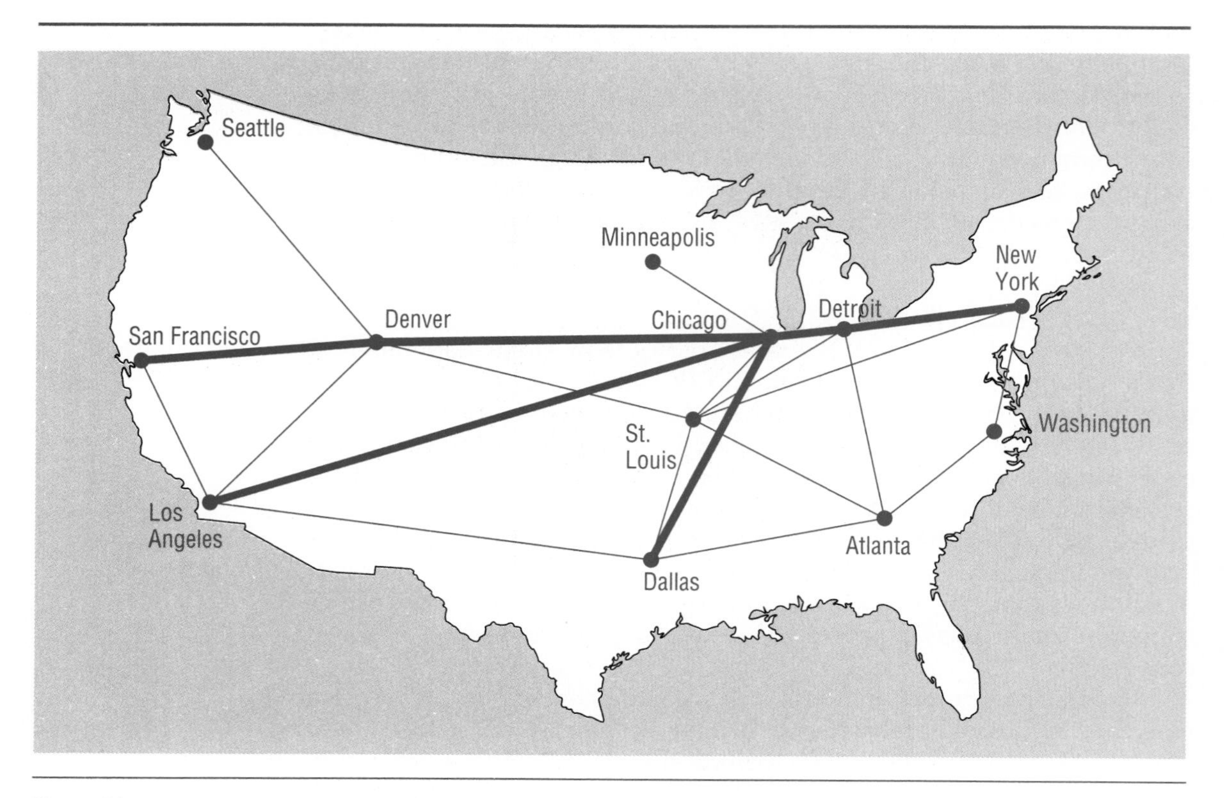

Figure 11–9
A hypothetical packet switching network (heavier lines indicate multiple circuits).

cisco might normally be routed through Chicago. However, if the Chicago node or the link from Detroit to Chicago is down or extremely busy, the computer node in Detroit would know, using internal tables, to route the message via Saint Louis instead. This provides a redundant, fail-safe capability and implies that the route a specific message takes is in part a function of the condition and traffic on the network when the message is sent.

Packets from different messages—indeed from different users—are interspersed on the network at any point in time. A packet switching network does not establish an actual physical link between two nodes that wish to communicate. Packets are routed, much like telephone calls are, with the distinction that telephone connections are set up and maintained for the duration of a conversation, whereas with packet switching, each packet is individually routed. The only direct connection is between the user's terminal and the nearest node of the packet network at each end of the connection.

Since a direct connection is not established between the sender and receiver on a packet switching network, there can be delays in communications. Therefore, the use of packet switching networks for interactive access to computers must be studied carefully to be sure that the

The Tymnet Compact XL is an intermediate size node in a Tymnet packet switched data network. (Courtesy of BT Tymnet)

response time that the packet network can deliver will be adequate. If the primary application is message switching to relay messages that do not require instantaneous delivery, the use of a packet switching network may be ideal.

Packet switching has been operational since 1969, having been first developed by the Advanced Research Projects Agency (ARPA) of the U.S. Department of Defense. The first packet switching network was known as ARPANET, and it was used to connect ARPA research centers. In 1975, the first commercial packet switching network, known as Telenet, became available in the United States.

Many public packet switching networks have been established in the world. In the United States, Telenet, Tymnet, General Electric Information Services (GEIS), and Uninet are the four principal networks. In Canada, the packet switching network is called Datapac, in France, it is Transpac, and in Japan, the network is called Venus P. All of these networks provide access in major cities in their country, and virtually all of the networks are interconnected so that it is possible to send messages to subscribers in other countries.

The packet networks are compatible thanks to the standard transmission rules (protocol) that they all use. Its name is X.25. X.25 became an international standard in the late 1970s, and standard networks using that protocol began to be implemented soon thereafter.

The control room of a nationwide packet switching communications network. VDTs are used to watch the status of the network, keep track of outstanding problems, and enter commands that affect the network's operation. (Courtesy of BT Tymnet)

PDNs provide another alternative for a company's message or data communication. If, for example, a company wanted to provide its customers with access to its computer or information, it could install a leased circuit from its computer to the nearest PDN node. Customers all over the world would contact their local PDN node, usually via dial-up telephone call, and with proper authorization, would be able to sign on and retrieve information. Similarly, salespeople or other employees with terminals could access the company computer while travelling simply by making a telephone call to the nearest PDN node and then logging on.

Yet another alternative for a company is to install a private network using PDN technology. Several vendors sell node computers and software allowing a company to build its own network that uses packet switching. Leased or dial-up lines may be employed, depending on the traffic volumes between specific locations. A few companies have chosen this method for handling their internal message or data traffic, particularly where rapid response time is not required.

Continental Grain, a multibillion dollar exporter and processor of agricultural commodities, has used GE Information Systems as the backbone of its international packet network. From its New York headquarters, Continental ties together its 175 domestic locations with 45 international locations in South America, the Far East, and Europe.

Public packet networks provide many capabilities that make them attractive for a variety of applications:

- reliable service—because of multiple connections and alternative routing between nodes;
- nationwide service—packet switching networks may be accessed from most locations in the United States and many locations throughout the world with a local telephone call. For high-volume users, leased lines may be connected directly to the packet switching node;
- low error rates—within the packet switching networks, the data is fully checked and retransmitted until correct. The actual error rate is primarily dependent on the local loops at the ends of the connection;
- variety of transmission speeds—although originally designed for 300 bps or 1,200 bps asynchronous communications, packet switching networks today are capable of transmitting data at up to 64,000 bps using both asynchronous and synchronous techniques;
- cost effectiveness—the user of a packet switching network pays a connection charge for each fraction of an hour of connection to the network and a packet charge for each packet transmitted. Other charges exist for leased line connection to the network and for dedicated dial-up ports. In many cases, however, the total cost of using a packet switching network is less than the cost of other alternatives.

Integrated Services Digital Network (ISDN)

The *Integrated Services Digital Network (ISDN)* is, in 1990, less of a network and more a set of standards that the name implies. The ISDN standards are being developed by the CCITT as a vision for the direction that the world's public telecommunications systems should take. Although there are still some standards to be hammered out, enough have been agreed upon that ISDN is moving from vision to reality at an accelerating pace.

ISDN can best be visualized as digital channels of two types, as shown in Figure 11–10. One type, the "B" (bearer) channel, carries 64 kbps of digital data. The other type, the "D" (delta) channel, carries 16 kbps of data and is used for signaling. These two types of channels are packaged, according to ISDN standards, into two types of access services. The "basic access," also known as 2B + D, provides two 64 kbps B channels and one 16 kbps D channel. Basic access ultimately will be provided on the line side of a PBX or central office, thereby making it available in homes and offices on standard twisted pair wiring.

basic access

The other type of access is known as "primary access," or 23B + D. Primary access provides 23 64-kbps B channels for carrying data and 1

primary access

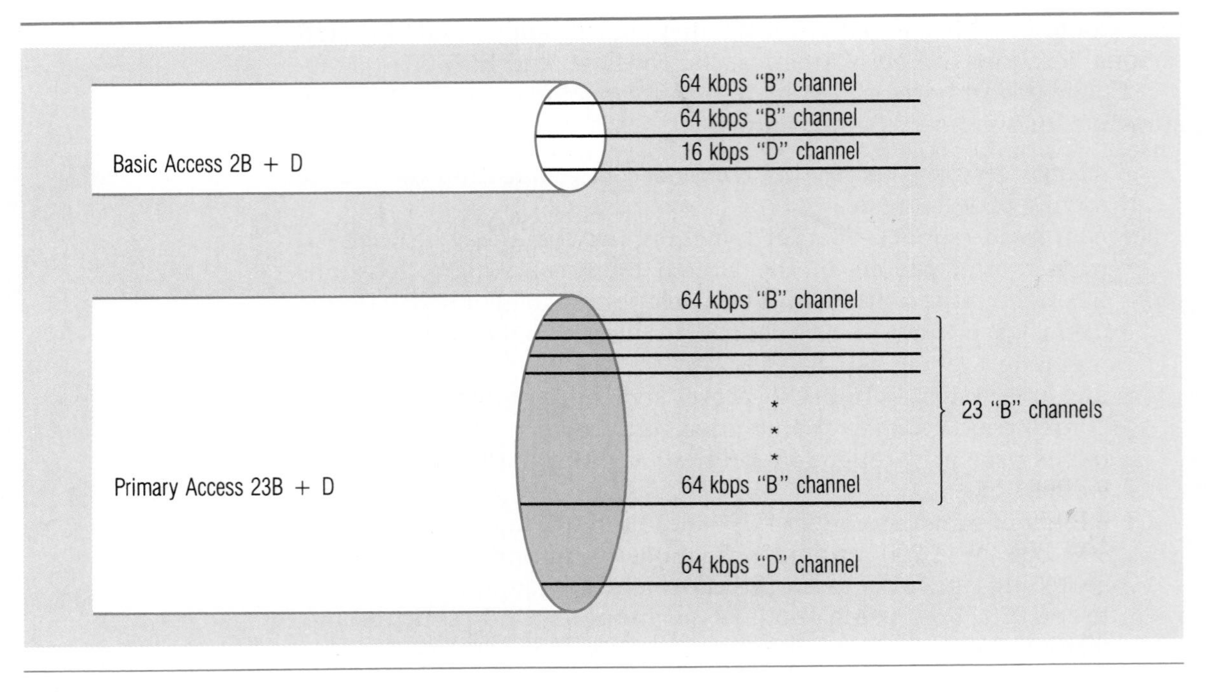

Figure 11–10
ISDN basic and primary access arrangements.

16 kbps D channel for signaling. The total capacity of 24 64-kbps channels happens to equal the carrying capacity of a T-1 circuit. In Europe, primary access is defined as 30B + D, which matches the capacity of a European T-1 circuit. Figure 11–10 shows the two types of access schematically.

The large bandwidth provided by ISDN circuits will be able to be used for digitized voice and data. With basic access in a home, a person could be having a telephone conversation and simultaneously transmitting data at 64 kbps from a personal computer. When fully implemented, ISDN will be able to display the telephone number of the calling party on the ringing telephone. A person could decide whether or not to answer the telephone based on who is calling! This capability is also becoming available on some non-ISDN telephone systems and is raising a storm of controversy because some people claim that it infringes on the privacy of the calling party.

A business with a primary ISDN access group could subdivide the bandwidth in any way necessary to meet the needs of its application set. Half of the capacity, or 772 kbps, might be used for a television transmission while the other 772 kbps is further subdivided to support 12 voice conversations at 32 kbps each and 6 64-kbps data channels. At another time, the bandwidth could be configured differently, all under the control of the customer. In fact, these configuration changes could be programmed into a computer so that the ISDN capacity is automatically reconfigured at certain times of the day. There could be a daytime configuration and a nighttime configuration or any other combination to meet the business needs of the company.

Many companies have been using early versions of ISDN systems. Virtually all of the major telephone companies in the United States have been conducting ISDN trials with major customers for the past several years. Many of these early trials have focused on the voice aspects of ISDN, but there also have been tests of the data handling capability. Most users have been quite pleased with the ISDN performance and the new services that ISDN delivered that were not possible before.

The benefits of ISDN are that

- it provides efficient multiplexed access to the public network;
- it has the capability to support integrated voice and data;
- it has a robust signaling channel, which is important for network management;
- it provides an open system interface that is internationally defined. This will go a long way toward making multivendor telecommunications systems a reality.

Although it will take years to fully implement ISDN capabilities throughout the public networks, steady progress toward the full ISDN can be expected in the next few years.

NETWORKS CLASSIFIED BY GEOGRAPHY

Yet another way of classifying networks is by the geographic expanse they cover. The most common designations are wide area networks (WAN) and local area networks (LAN). A third category now receiving some discussion in trade journals and research papers is the metropolitan area network (MAN).

Wide Area Networks (WAN)

Wide area networks (WAN) are generally understood to be those that cover a large geographic area, require the crossing of public right-of-ways, and use circuits provided by a common carrier. Wide area networks may be made up of a combination of switched and leased, terrestrial and satellite, and private microwave circuits. Since carrier facilities are used, the transmission speeds are most commonly 9,600 bps and below. However, 56,000 bps circuits are available in most parts of the country. The usual topology of WANs is star, hierarchical, or mesh, with star networks being the most common.

The WAN for a large multinational company may be global, whereas the WAN for a small concern may cover only several cities or counties. General Electric Corp. recently announced plans for a private global network to transmit voice, data, and video signals to GE offices in 25 countries. A large bank has built a private WAN to link its offices in Massachusetts, Hong Kong, London, Munich, and Sydney. Circuits operating at speeds of 56/64 kbps provide 3-second response time anywhere in the world, the bank reports. These are but two examples of large, international WANs that companies use.

Metropolitan Area Networks (MAN)

The Institute of Electrical and Electronic Engineers (IEEE) has an active project to define *metropolitan area networks (MAN)*. The work grew out of the committees that define LANs and an unresolved concern for how LANs could be connected across distances of greater than a few kilometers. Another aspect of the IEEE work is to provide support for other types of communications besides data. (LAN standards strictly relate to data transmission, as you will see in the next section.) The work of the IEEE committee is focused on support for data, voice, and television communications at distances of 5 to 50 kilometers. The group is concentrating on coaxial cable and optical fiber as the primary media on which a MAN would be implemented.

In the past few years, a number of small companies have begun businesses by installing fiberoptical cable in metropolitan business areas. Rights-of-way are obtained in many ways—on telephone poles, in subway tunnels, under city streets, and so on. These companies primarily sell high-speed digital bandwidth and typically offer no value added or other services. Their primary customers are large companies that need to communicate within the metropolitan area. The customer is usually responsible for providing equipment, such as multiplexers and DSUs.

The primary market is the customer that needs a lot of high-speed digital service, such as multiple T-1 or T-3 circuits. The MAN providers typically offer lower prices than the telephone companies and offer diverse routing, providing backup in emergency situations. They also claim to offer quicker installation and better service than the telephone companies.

Local Area Networks (LAN)

Local area networks (LAN) have joined the networking scene in the last few years. Because of some of the capabilities they offer, they are rapidly being installed in companies of all sizes. A discussion of LANs must deal with many topics beyond those usually associated with networks. For this reason, a special section later in this chapter is dedicated to a complete coverage of LANs.

CONNECTING NETWORKS

Companies are finding that they have several physical networks that they would like to connect together into a single logical network. From a user's perspective, it's ideal if all other users are, or appear to be, on the same network and can be easily communicated with. Without a single standard, however, many types of networks have evolved, and an organization frequently finds that it has a WAN that it wants to connect to one or more LANs. Often LANs need to be connected together, and sometimes different WANs need to be connected. The United States government and universities have been leaders in requiring network interconnection and have done pioneering work to develop ways in which dissimilar networks can be linked. As the connection of networks has evolved, the term *internet* has come to be used to refer to an interconnected set of networks.

Transmission Control Protocol/Internet Protocol (TCP/IP)

Transmission Control Protocol/Internet Protocol (TCP/IP) was developed to connect incompatible processors used by military suppliers and researchers. TCP/IP is a set of protocols that operate at the OSI layer 3, the network layer, and layer 4, the transport layer. TCP/IP is used primarily to connect WANs with one another.

The Internet Protocol (IP) is implemented in each computer and each gateway to another network. Blocks of data are sent out through the network and through as many computers or gateways as needed until they reach their intended destination. The IP software must have information that allows it to know where to route a block of data.

The TCP ensures that data is delivered error free, in sequence, and with no loss or duplication. Whereas the IP only handles blocks of data and is not concerned with errors or sequences, the TCP has the responsibility, on the user's behalf, to achieve reliable data transfer. In addition, the TCP allows users to indicate the urgency or priority of their transmission and to assign a security classification to it.

One of the major uses of TCP/IP is to transfer files from one computer to another. TCP/IP provides an extensive file transfer capability that allows the users to make an online request for a batch file transfer. Data compression and transparency can be invoked if required.

Using TCP/IP, huge "networks of networks" have been built. The Internet is the name for the largest operational superset of research-oriented internets. Its three major subnetworks are ARPANET (Advanced Research Projects Agency Network), CSNET (Computer Science Network), and NSFNET (National Science Foundation Network), each of which is a network of computer networks. A very large fraction of the engineering and computer science research community interact with one another on The Internet. No one knows how many computers and terminals are connected to this network!

Over time, TCP/IP will be replaced by software that fully implements the OSI model. TCP/IP is compatible with OSI but does not implement all of its capabilities. Because of the compatibility, organizations that use TCP/IP today should have a fairly easy time migrating to OSI software when it becomes available.

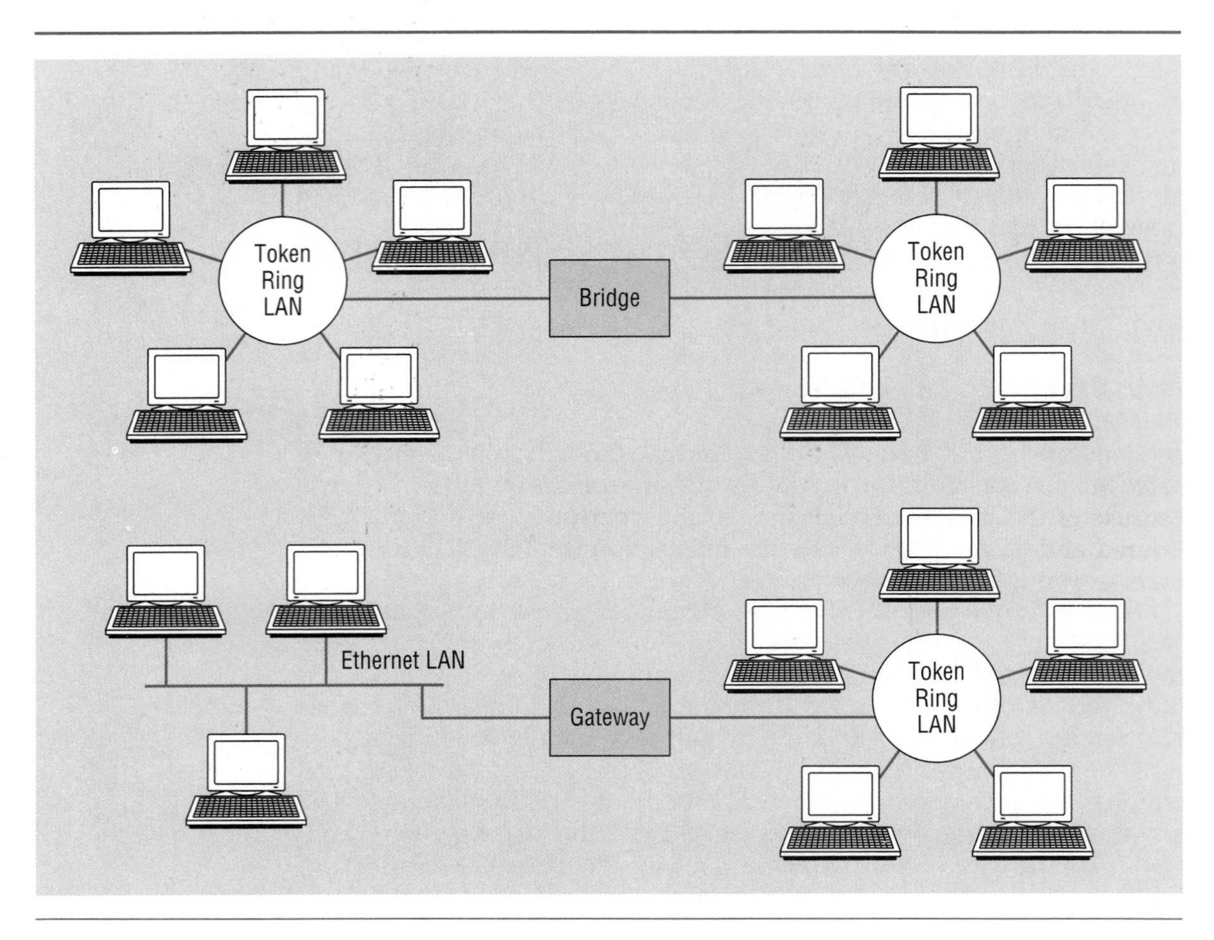

Figure 11–11
Bridges connect networks that use the same protocols. Gateways connect networks that use dissimilar protocols.

Bridges and Gateways

Local area networks are connected together using *bridges* or *gateways*, as shown in Figure 11–11. Bridges connect networks that operate according to the same rules or protocol. A bridge operates at layer 2 of the OSI model and allows data to be sent from one network to another so that terminals on both networks can communicate as though a single network existed. Bridges "listen" to all of the traffic on one network and pass on data intended for the other network. Because they have no translation to perform, bridges are simpler and operate at higher speeds than gateways.

Bridges have a variety of applications, such as extending a LAN to larger distances and greater numbers of ports. One of the more important uses is to divide a busy LAN into two smaller LANs to improve the throughput and reduce the congestion.

If two networks operate according to different protocols, a gateway is used to connect them. Gateways operate at OSI layers 4 and up and

	Static	Dynamic
Centralized	Relatively simple Least flexible Single point of failure Potential performance bottleneck	More flexible Reacts to changing traffic conditions
Distributed	No single point of failure Routing table updates may be a burden on the network	Most flexible Most complicated to implement
Broadcast	Simplest Adequate with small network	Ineffective with moderate to heavy traffic loads

Figure 11–12
Advantages and disadvantages of various network routing techniques.

basically translate the protocols in order to allow terminals on the two networks to communicate. Gateways can suffer from slow performance because of the protocol translation, so their performance must be considered and tested when a gateway installation is contemplated. Gateways perform an important role in allowing an organization to interconnect different types of LANs so that, to the user, the network appears as a single entity.

ROUTING TRAFFIC IN THE NETWORK

Routing messages from their source to their destination is one of the important responsibilities of the network layer. There are many ways in which messages can be routed, ranging from fairly simple techniques, suitable for small networks, to elaborate, adaptive techniques that only large networks need and can afford. The chart in Figure 11–12 shows the attributes of the various routing techniques.

Broadcast Routing

One very simple technique, *broadcast routing*, used by the CSMA/CD protocol, is to broadcast all messages to all stations on the network. The station for which the message is intended copies it. All other stations ignore it.

Centralized Routing

As networks grow beyond a few stations, however, broadcast routing becomes impractical. The next level of sophistication is *centralized routing*. A central or control computer keeps a table of all of the terminals on the network and the paths or routes to each. All traffic flows to the central computer, which then uses the table to determine the best route for the message to take to its destination.

Centralized routing is used in star or hierarchical networks, where a central computer controls all communications flows. IBM-SNA networks frequently use this approach. Centralized routing is relatively simple, but it is normally static. Routes are established when the software table is created and do not normally change based on network traffic loads or communications line failures. Furthermore, the central computer can become a performance bottleneck or, worse, a point of failure. Despite these disadvantages, centralized routing is used in many networks and can be very successful. Planning is required, however, to ensure that the proper routes are built into the software table.

Distributed Routing

Distributed routing places the responsibility for building and maintaining routing tables on at least some of the nodes in the network. Each node that performs routing is responsible for knowing which paths or links are attached to it and their status. The node also must advise other nodes in the network when the status of paths changes so that they can update their tables. This transmission of path status information can put a communications burden on the network.

Distributed routing avoids the single-point-of-failure problem associated with centralized routing. It is more complicated to implement, especially when the practical problems of updating routing software in each of the nodes are considered.

Static and Dynamic Routing

Static routing means that the route between two nodes on the network is fixed. Messages must always be sent on the predefined route. If it is down, messages must be held. Some static routing schemes allow alternate paths to be statically defined so that some type of backup exists.

Dynamic routing is considerably more flexible in that each node chooses the best path for routing a message to its destination. Consecutive messages to the same destination can be routed by different paths, if necessary, and the node can adapt to changing network traffic volumes, error rates, or other conditions. As would be expected, dynamic routing is more difficult to implement, particularly when it is combined with distributed routing, but the benefits may make the implementation cost worthwhile.

Hardware Routers

Hardware *routers* perform the function of routing messages from one network to another and translating the destination address to the format required by the network receiving the message. In simple terms, a router can be thought of as a sophisticated gateway designed to connect networks that use different technologies. To the traditional gateway functions, routers add the capability to determine the best route for a message, usually on a dynamic basis. If several routers are installed in an internet, they communicate with one another to keep their path tables up-to-date.

Router technology is new and is developing rapidly. As more companies find a need to connect their diverse networks or to connect to networks of other organizations, router technology will play an increasingly important role.

 LOCAL AREA NETWORKS (LAN)

Introduction and Characteristics

Local area networks (LAN) are limited distance networks usually existing within a building or several buildings in close proximity to one another. The media used for LANs are almost always privately owned and installed. Therefore, a regulatory agency is not involved with LANs. LANs transmit data at rates much higher than can be achieved over switched or leased circuits obtained from a common carrier. Also, because the distances are short, the probability of errors occurring is low.

Another characteristic of LANs is that in most cases, each terminal or computer connected to the network has an equal ability to access all other devices. There is no master–slave relationship. Networks with star and hierarchical topologies do not technically qualify as LANs because the terminals and computers do not have equal connectivity. Bus and ring topologies are most common in LANs, and bandwidths are high.

Personal computers normally are attached to a LAN via a cable connected to a LAN interface circuit card inside the PC. When an organization installs a LAN, it must purchase these LAN attachment cards for all PCs that will be connected to the LAN. Dumb terminals can also connect to LANs, but the connection is usually made through a cluster control unit or other hardware that is shared by several dumb terminals.

On many LANs, one or more special computers, called *servers,* are attached. Servers provide unique capabilities that can be shared by all participants on the LAN. The most typical servers are a *file server* that provides one or more large capacity disk drives, a *print server* that provides one or more printers, frequently of high speed or high quality, and a *communications server* that provides access to other LANs or a host computer. Servers are often personal computers with extra circuit cards added. Sometimes they are used as a workstation as well as providing server capability to others on the LAN, but most frequently, the servers are dedicated machines. Dedicated servers are able to provide better performance to all users. Server computers have special software that will be discussed later in the section on LAN software.

The typical university campus, hospital, corporate headquarters, large manufacturing site, and research center are good candidates for LAN installation. LANs are also being installed within departments so that users of personal computers can share expensive hardware, such as laser printers or large disk storage units. Other LANs are installed simply to provide a fast data path and good response time from terminals or personal computers to a large central computer.

Transmission Techniques

baseband transmission

There are two transmission techniques used for operating a LAN. One technique, called *baseband transmission*, uses a digital signal. When baseband transmission is used, the medium is directly pulsed, and the entire bandwidth is used for a single signal. Baseband transmission typically occurs at speeds of 1 Mbps and higher. Baseband transmission is used when data is being transmitted but is only suitable for voice or television if the signals are digitized.

broadband transmission

The other transmission technique is called *broadband transmission*, and the signal is transmitted in analog form. The capacity of the cable is subdivided, using frequency division multiplexing, into whatever circuits or channels are required for the particular applications. The difference between broadband and baseband transmission is in how the bandwidth of the circuit is used and not necessarily in the capacity or the medium. A chief advantage of broadband transmission is that many communications can be going on simultaneously. The broadband system can be used for multiple purposes, such as data, voice, and television, without having to install separate lines for each type of traffic.

The majority of today's LANs use baseband transmission. LANs often are used to connect personal computers together so that they can communicate with one another or so they can share a large disk or a printer. Since only one transmission can be carried at a time, baseband techniques rely on handling each transmission very quickly using the high speed at which baseband operates. A transmitting node gets control of the medium and transmits its message. Since the transmission speed is high, the transmission is completed quickly, and the medium becomes free for another node to use.

Broadband transmission is typically found in an environment where there are diverse requirements for many types of transmissions. The broadband system can be divided as if it had several baseband channels inside it. In addition, broadband systems typically carry transmissions that do not meet the technical qualifications of a LAN. In a manufacturing plant, for example, a broadband system might be used to connect

- robots to a minicomputer servicing a production line;
- VDT terminals to a host computer;
- laboratory instruments to a central computer used for quality testing;
- a series of personal computers connected to each other in LAN fashion (on one channel of the broadband system);
- a plant television studio to television sets located around the plant

for broadcasting plant news, notice of job availabilities, and other information of interest to the employees.

All of these transmissions could occur simultaneously on a broadband system, and each would operate at a speed appropriate for the type of transmission. Thus, a broadband system has a greater ability to handle a wider range of signals than a baseband system.

Broadband and baseband systems should be seen as complementary not competitive. Both have distinct advantages. Baseband systems are simpler; new terminals can be attached by simply tapping into the cable. Broadband systems require modems to modulate the signal to the proper frequency range. Thus, they are more expensive to implement. Broadband cable is normally much larger than baseband cable, hence it is more expensive and difficult to install. On the other hand, broadband systems can handle more diverse communications requirements.

Topology and Media

LANs are normally implemented using either a bus or a ring topology, as illustrated in Figures 11–5, 11–6, and 11–7. The medium may be twisted pair or shielded twisted pair wire, coaxial cable, or optical fiber cable. Because of the speeds at which LANs operate, there are distance limitations that must be observed. Equipment manufacturers specify the distances over which their equipment will operate, but the limitations have a physical basis related to the propagation delay of the signal on the circuit and the distance the signal can travel without being amplified or regenerated.

LANs with a ring topology are most frequently installed as a physical star, connected into a logical ring, as shown in Figure 11–13. In most office buildings, there is an equipment room on each floor where the telephone and other network wiring and equipment are located. Wire is run from the equipment room to each office in star fashion. In the equipment room, the wire is connected together to form the ring. The advantage of this type of installation becomes evident when problems occur. Rather than having to trace a ring all over the floor of the building, the technician can work with the ring in the equipment room. Once he or she determines which part of the ring is causing the problem, that leg, running to an office, can be followed or traced. This type of installation has similar advantages in a plant or factory.

Manufacturers design their LAN equipment and software to use specific media according to standards that are discussed in the next section. The characteristics of twisted pair, shielded twisted pair, and coaxial cable were discussed in Chapter 9. For LAN installations, twisted pair wire is very popular because it is inexpensive and, in many cases, wiring that is already installed can be used. Optical fiber cable is not often used for LANs today because of its cost. However, the cost of the

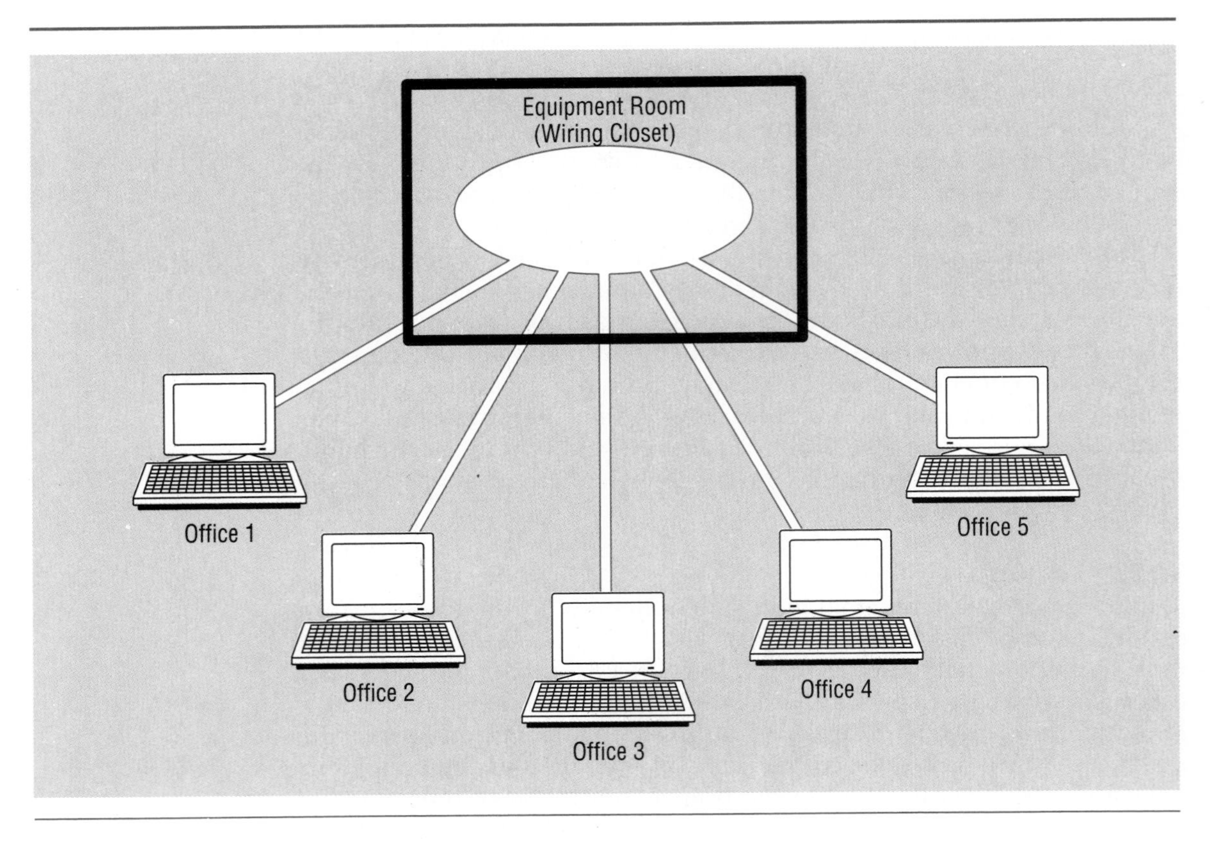

Figure 11–13
A ring LAN installed as a physical star.

fiber and its installation is continually dropping, and it is almost assured that optical fiber will be extensively used for LANs in the near future.

LAN Standards

The IEEE has done a great deal of work to standardize the definitions of LANs and how they operate. The IEEE 802 Committee has published four standards, numbered 802.2, 802.3, 802.4, and 802.5, for LAN networks and protocols. Manufacturers have designed their equipment and software to conform to these standards.

It should be noted that the IEEE standards define the physical and data link layers, layers 1 and 2, of the ISO-OSI model. Although these layers were covered basically in earlier chapters, it is appropriate to keep the discussion of LANs together with other network topics.

IEEE 802.2: LLC Protocol Standard 802.2 describes the LLC protocol for use on LANs. This protocol was explained in Chapter 10. The other three IEEE LAN standards specify certain combinations of access techniques and media that, when combined with the LLC protocol, provide a complete definition for a LAN implementation.

IEEE 802.3: CSMA/CD Baseband Bus This type of LAN uses a coaxial cable or twisted pair wire as the transmission medium in a bus topology. At the physical level, the standard defines the electrical characteristics of the media and the type of connectors that will be used. The CSMA/CD access technique is used, and the standard defines a specific algorithm to randomize the time at which colliding stations will again attempt to transmit. This type of LAN is designed to operate at a rate of up to 10 Mbps over a maximum distance of 2,500 meters. When coaxial cable is used, there can be a maximum of 100 stations connected to the cable.

The 802.3 standard is based on research originally done by the Xerox Corporation. Xerox called this type of LAN *Ethernet*, a name that is still used today. Ethernet LANs are the most popular and widely implemented.

IEEE 802.4: Token Passing Bus The IEEE 802.4 standard defines a token passing access method operating on a coaxial cable arranged in a bus topology. As in the CSMA/CD protocol, all of the stations on a LAN that uses a token passing protocol are considered equal. Since the bus topology does not provide a natural sequence of stations, each node must be assigned a sequence number, and the token is passed from one station to another according to the sequence number assigned.

The 802.4 standard specifies analog broadband signaling on a 75 ohm TV-type coaxial cable. Within that specification, several techniques are available for using the cable's capacity. Data rates of 5 to 10 Mbps are possible.

The advantage of a token passing technique is that the response time of the network is predictable. Therefore, the 802.4 technique is particularly suited for manufacturing environments where processes or machines must be guaranteed a certain response time by the network.

IEEE 802.5: Token Passing Ring The IEEE 802.5 standard defines a token passing access method for use in a ring topology. Several types of cable can be used for token rings, but twisted pair wiring is the most common. Using twisted pair wire, data rates of up to 4 Mbps can be achieved. Using shielded twisted pair wiring, data rates of 16 Mbps are possible. Vendors are working to improve these rates, and it is expected that 16 Mbps will be possible on twisted pair wiring in the near future.

The 802.5 token ring standard is based on research done by IBM Corporation. Like the 802.4 standard, the performance of the token passing ring is predictable. However, if the number of stations on the loop is large, it may take a long time for the token to get around to a station that has a message to send. Therefore, although response time is predictable, it may be longer than in a CSMA/CD network.

Figure 11–14 summarizes the LAN standards discussion in a chart that shows the four major attributes that describe a LAN. Figure 11–14 identifies the alternatives for each of the attributes and shows combinations used by some real-world LANs.

Figure 11–14
LAN attributes and options, and the combinations used by certain commercially available LANs.

LAN Attributes	Options	Ethernet	PC–Network	Token Ring	Starlan
Transmission Technique	Baseband	X		X	X
	Broadband		X		
Access Control	CSMA/CD	X	X		X
	Token			X	
Topology	Bus	X	X		X
	Star				
	Tree				
	Ring			X	
Medium	Twisted Pair			X	X
	Coaxial Cable	X	X		X
Speed		10 Mbps	2 Mbps	4 Mbps 16 Mbps	1 Mbps

The Fiber Distributed Data Interface (FDDI) Standard A newly emerging standard is the *Fiber Distributed Data Interface (FDDI)*. This standard is being developed by a subcommittee of the American National Standards Institute (ANSI) and was completed in 1990. As LANs based on the IEEE 802 standards reach capacity, optical fiber LANs, based on the FDDI standard, are likely to become the preferred growth path. In their first implementations, FDDI LANs will provide high-speed *backbone* connections between other LANs.

The FDDI standard defines a LAN based on two counterrotating 100 Mbps token rings. Data moves on the rings in opposite directions so that, should a node or link fail, network operation can be maintained by sending the data on the second ring. In this way, the failing device or path can be bypassed. This type of implementation is shown in Figure 11–7. The token passing technique defined by FDDI is similar to the one defined by IEEE 802.5 but with significant enhancements in the areas of fault tolerance and topology.

The FDDI standard allows up to 500 stations to be connected to the ring in a priority system. High priority stations can access the ring for longer periods of time. The maximum length of the ring can be up to 200 kilometers, and stations must be located no more than 2 kilometers apart.

It is easy to see that FDDI greatly expands the speed and operating distances of LANs. Now it will be possible to connect a large campus or corporate complex into a single LAN or to easily connect smaller LANs in buildings or departments where that is desirable. Eventually, there will be a need to have the FDDI speed to many desktops to support applications, such as full motion, color video, high resolution graphics and photographs, traditional data, and voice—all concurrently!

Commercially Available LANs

Ethernet The original work on LANs was done by Xerox Corporation in the 1970s. In 1980, Xerox teamed with Digital Equipment Corporation and Intel and jointly announced a LAN product called Ethernet. Ethernet is an 802.3 LAN and, as such, operates at 10 Mbps using baseband transmission. Traditionally, Ethernets have been implemented using coaxial cable, but recently, hardware that supports the use of twisted pair wires has become available. Other characteristics of Ethernet have been discussed in the sections describing the CSMA/CD protocol and the 802.3 standard.

Ethernet frequently is used to connect personal computers together. It works particularly well when the data from any given terminal occurs in short bursts, with relatively long pauses in between. Digital Equipment Corporation uses Ethernet as its primary communications technique for connecting terminals to each other or to its larger computers.

Other 802.3-Based LANs AT&T has a LAN product called Starlan that operates at 1 Mbps on two twisted pairs of wires. Starlan is also a baseband network that follows the 802.3 standard.

IBM's PC Network product also follows the 802.3 standard and transmits data at 2 Mbps on coaxial cable using broadband transmission. The PC Network only uses a part of the capacity of the cable. The rest is available for other uses.

Hardware and software for building 802.3-based LANs is widely available from a variety of companies. Vendors have found that the market for these products is large, since the LANs are relatively easy to install and maintain, at least when they are small.

Token Ring IBM's premier LAN offering is the token passing ring. IBM's basic research for this product served as the foundation for the IEEE 802.5 standard. IBM's implementation passes data at 4 Mbps on twisted pair wires or 16 Mbps on shielded twisted pair. Token ring hardware and software are available also from several other companies, such as Novell, Inc., and 3-Com, Inc.

Among all of the LANs, a dual standardization is emerging. Recognizing that Ethernet and token ring have different applications suitability, users are standardizing on both of them, allowing the application to dictate which is the most suitable. Strong product support from multiple vendors in each case reinforces the dual standardization and places the vendors of proprietary standards at a disadvantage.

MAP and TOP

General Motors Corporation and the Boeing Company have been especially aggressive in attempting to get standards defined. Both companies sensed a strong need to have the ability to interconnect products made by hundreds of different vendors. In the late 1970s, both GM and Boeing, working independently, began vigorous efforts to further define the layers of the OSI model. GM's work has become known as the *Manufacturing Automation Protocol (MAP)*, and Boeing's work is known as the *Technical Office Protocol (TOP)*. Products using MAP are the primary users of the 802.4 standard. MAP networks use broadband coaxial cable at 1, 5, and 10 Mbps.

Both MAP and TOP define all seven layers of the OSI model, and, in that context, their use of the word "protocol" is really the definition of a complete set of rules for the entire communications process, far more than just a data link protocol. MAP is based on the IEEE 802.4 token-bus standard, whereas TOP is based on 802.3, CMSA/CD. At higher layers, MAP and TOP are identical until the application layer, where they diverge.

GM and Boeing have had large numbers of vendors and other potentially interested customers working with them to define and implement the MAP and TOP standards. MAP's primary focus is the communications necessary between machines on a factory floor, including robots. TOP concentrates on electronic mail, document content, and graphics. Compatibility between the networks has been publicly demonstrated. Specialized gateway hardware performs the translations necessary for messages from a user on one network to be sent to a user on the other.

It is clear that both sets of standards will be important in the future, and a company will do well to monitor MAP and TOP developments as they unfold.

LAN Software

At the heart of any LAN, such as the one shown in Figure 11–15, is a variety of software that provides much of the intelligence that tells the hardware how to operate. Each user workstation connected to the LAN must have software that tells it how to communicate with the servers. Servers require software to perform their function. Applications software for a LAN may be different from the software for a single personal computer because the probability exists that the software, and perhaps data files, will be accessed by more than one user at a time. Finally, there is a need for management software to provide commands and status information to the individual who manages the LAN. Each of these types of software are discussed in more detail.

Software for the Workstation For personal computers, the basic communications software for a LAN can be viewed as an extension of the oper-

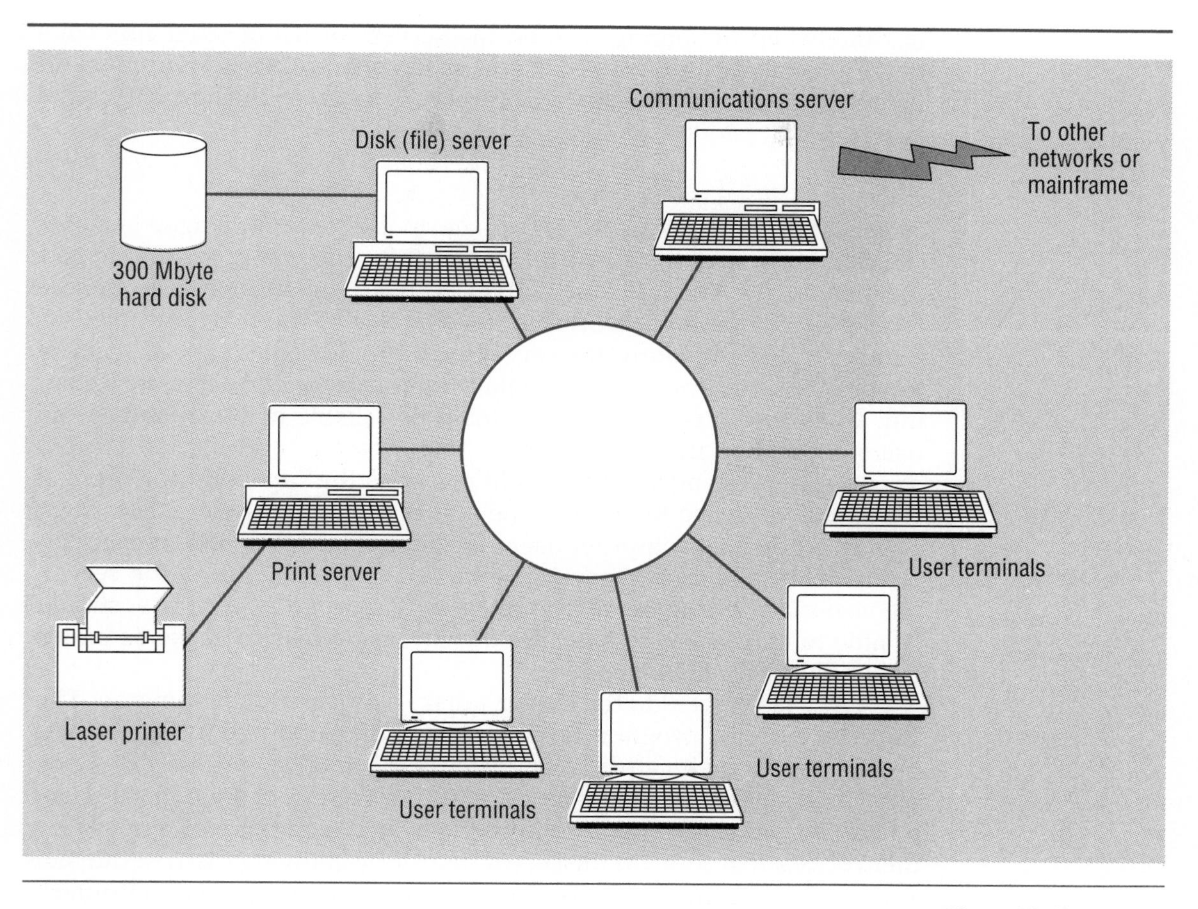

Figure 11–15
A ring LAN with servers attached.

ating system. In the IBM world, this extension is called *NetBIOS*, which stands for *Network Basic Input Output System*. NetBIOS operates at OSI layers 2 and 5, the Data Link layer and the Session Layer. At the data link layer, it provides for application programs to send and receive short messages, but without acknowledgment. At the session layer, it supports two-way, reliable communications between application programs, with acknowledgments that messages have been received. At a higher level, a LAN workstation program uses NetBIOS and provides the capability for application programs in the PC to use servers and otherwise share network resources.

Other vendors, such as Novell, Inc., 3-Com, Inc., and Banyan Systems, Inc., provide LAN workstation and NetBIOS-like programs that have similar capabilities, often with proprietary architecture. These vendors' products frequently have their own special features that give them a competitive edge in the marketplace.

Dumb terminals need only basic communications software that lets them interface with the LAN. Since they cannot run programs, they are not interested in sharing files on file servers or using other functions that personal computers need. Thus, their software is much simpler and usually is provided by a cluster control unit to which the dumb terminal is attached.

Server Software Software for servers normally comes packaged in a special operating system or control program for the computer that is performing the server function. Server software must be able to handle requests from multiple workstations attached to the LAN and process them rapidly. Frequently, the server must provide some type of security to permit users to protect their data from unauthorized access or updating. In all cases, server software must be reliable because many LAN users depend on it.

Disk server software normally provides the capability for users at workstations to use the disk of the server as though it were their own. That is, a personal computer user may have a hard disk of 30 megabytes on his or her PC, but when it is attached to a LAN with a file server, the user may have access to a 600 megabyte server disk. The user should be able to store programs or files on the server disk as easily as on his or her own PC's disk.

With easy access to the server's disk, new possibilities emerge. The user can access a program on the server's disk and run it. Files stored on the server disk can be designated as shareable, thereby allowing other users to look at the data or, if authorized, to change it. Files can be transferred easily to other individuals or programs, either by giving them access to a copy on the server disk or by allowing them to transfer a copy of the file from the server's disk to their own personal computer's disk.

File sharing leads to some new data integrity problems, however. Since two or more users could be given access to update a file on the server's disk, some form of protection must be instituted to ensure that two users don't update the same data at the same time. If user A and user B both look at the same record in a file simultaneously and both make changes to it, the last one who updates the record will wipe out the changes made by the first person. Good file server software provides a mechanism that prevents two or more users with update authority from accessing the same record in a file at the same time. If one user has the record and a second tries to access it, he or she receives a message that the record is unavailable.

Print server software allows users of workstations attached to the LAN to use printers attached to the server. Frequently, the server has printers that provide high-quality output or that operate at high speeds. Often, these types of printers are too expensive for an individual to purchase, but the print server allows several users to share the printer and the cost.

Since more than one user may try to use a printer at the same time, some type of queuing capability must be provided by the print server software. This queuing capability is called *spooling*, and it allows data that is to be printed to be temporarily stored on disk if the printer is busy printing another user's output. When the printer finishes printing the first user's output, the print server software takes the second user's output from the disk and prints it.

spooling

Spooling software is transparent to the user or application program that requests the print operation. Data is accepted by the print server regardless of whether the printer is busy or not. Good spooling software provides data integrity by taking care of operational problems, such as the printer running out of paper. The software ensures that all data is printed properly.

Communications server software provides the capability to communicate with other networks or computers. The software establishes the connection with the other network and surrounds the data with the appropriate protocol for transmission. Communications server software most often works in conjunction with network interface hardware, called routers, bridges, and gateways. This hardware is discussed later in the chapter.

Application Software Application software, such as word processing, spreadsheet, and graphics programs must be written so that they can take advantage of the features that a LAN, with its disk, print, and communications servers, offers. If the application has been programmed in such a way that it cannot use the server's disk or printers, the LAN is of little use. Probably the biggest problem that most applications have is the data integrity of shared files. Most applications were not written with the idea that multiple people might be simultaneously accessing or updating the data.

For these reasons, many application software packages have separate versions that are designed for LAN use. When written for a LAN, the software is often licensed to the user and priced so that only one copy need be purchased for multiple users. The software is stored on the disk of the file server where it can be accessed by all authorized LAN users. The software often contains additional program code to provide data integrity and to supplement the capability provided by the server's software.

In many companies, individual departments have found installing a LAN with shared software to be of great benefit. In one company, the tax department, with the help of the computer department, set up a LAN so that several tax accountants could use an application package designed to help in the preparation of the company's tax returns. Several tax accountants were able to work on different tax schedules at the same time, all sharing a common data file of basic corporate accounting data on which the return was based.

LAN Management Software It is very desirable for LAN software, on the workstations and the servers, to provide commands and information that assist in the daily operation of the LAN. Users may not be in close proximity to the LAN's servers, or the servers may be located in a locked room for security reasons. Therefore, it is helpful if the users can inquire as to the status of the servers. Knowing how many jobs are in the print queue may give some indication of when one's output will be printed. Knowing how many users are logged on to the LAN may give an indication of the response time and other performance that can be expected. Having the ability to send a message to all users on the LAN may be useful for the LAN manager to inform people that the LAN will be taken down for maintenance.

In addition to these daily operational types of commands, the LAN software also should collect statistics that show usage over periods of time, such as a day, week, or month. This helps the LAN manager monitor growth and anticipate the need for additional disks, faster printers, or even a faster LAN.

LAN Management

When companies or departments consider installing a LAN, one of the most underestimated functions is LAN management. Often LANs are proposed and installed by people who have very good technical skills but who lack an understanding of the necessity to manage the LAN on an ongoing basis and to provide appropriate policies and procedures. Many times, the need to physically protect the LAN hardware, to provide backup for the data stored on the file servers, and to establish procedures for authorizing access to the data are ignored. Sometimes the need only becomes apparent after a problem has occurred. LANs have a way of growing and changing that mandates that some controls be in place.

Any time a LAN is established, the following management items should be considered.

- Organization and management: A LAN administrator or manager should be formally designated. Policies and procedures should be established and communicated to all LAN users.
- Physical safeguards: LAN hardware, such as the cable and servers should be physically protected from unauthorized tampering. Keeping the servers in a locked room is a preferred technique.
- Documentation: The LAN must be documented. Documentation should include the network topology, types and locations of attached workstations, names and version numbers of installed software, authorized users and their capabilities, and so on.
- Hardware and Software control features: Hardware and software control features should be acquired or developed. All features should be used and documented.

- Change controls: Procedures should be established for making changes to the network or its software. These procedures should include notification to users, testing, and fallback plans.
- Hardware and software backup: Effective backup provisions and contingency plans for both the hardware and software must be made. Users will depend on the LAN to do their job.
- Access to network facilities: Adequate security mechanisms should be provided to restrict access to network hardware, software, and data.
- Network application standards: There should be adequate controls and training regarding applications to ensure compatibility, integrity, and effective application usage.
- Network performance monitoring: Network performance monitoring mechanisms should be established to ensure effective network throughput, load leveling, and overall performance reporting.

From the preceding list, it is evident that establishing and operating a LAN is a task that needs good planning and control. In many ways, operating a LAN is like operating a small computer center. The LAN manager must consider and develop plans for most of the same items as the company's data processing organization. For that reason, in many organizations, the installation and control of LANs are the responsibility of a central group. Although this would seem to take away some of the departmental or workgroup autonomy often associated with LANs, it does help to ensure that the LANs are managed properly and are protected from unforeseen problems.

SUMMARY

Communications networks come in many shapes and sizes. Networks can be classified several different ways: by their topology, their ownership, their purpose or type of transmission, and their geography. The OSI layers that deal with networks must perform many functions, and there often are differences as to how the functions are performed in different types of networks.

Local area networks are an important, fast-growing subset to the total network picture. The technology is developing rapidly, and as quickly as new standards are finalized, others are under development. LANs based on fiberoptics will be an important part of the scene in the near future.

The telecommunications network designer is faced with a wide variety of choices for building a network. Only by understanding the characteristics and trade-offs between various types of networks can the designer reach a reasonable solution for a particular application. More often than not, a company's diverse communications requirements indicate that a hybrid network is required. There may be a combination of WANs and LANs, each with a different topology. All in all, the design of a comprehensive communications network is a complex task requiring both telecommunications and applications knowledge.

CASE STUDY

Dow Corning's Data Communications Network

Dow Corning's data communications network, as shown in Figure 11–16, is a classic star configuration with the center being in Midland, Michigan. In all but a few cases, the data circuits are leased, primarily from AT&T. Because it is a star network, it is quite easy to add additional circuits to support new locations or growing traffic volumes. The company has found that a major overhaul or redesign of the network has been necessary only every 8 to 10 years.

Smaller networks, also with a star topology, are also in place in Europe and in Japan. Although the single most important node of the network is in Midland, there also are major nodes in Brussels and Tokyo. These nodes are the focal points for the network in Europe and Japan.

Dow Corning's lines to Hong Kong and Brazil are shared with the Dow Chemical Company, one of Dow Corning's parent companies. Leased lines run from Dow Corning's headquarters in Midland to Dow Chemical's headquarters, also in Midland. The Dow Corning circuit is multiplexed onto a high-speed circuit that runs to the foreign location. At the foreign site, the high-

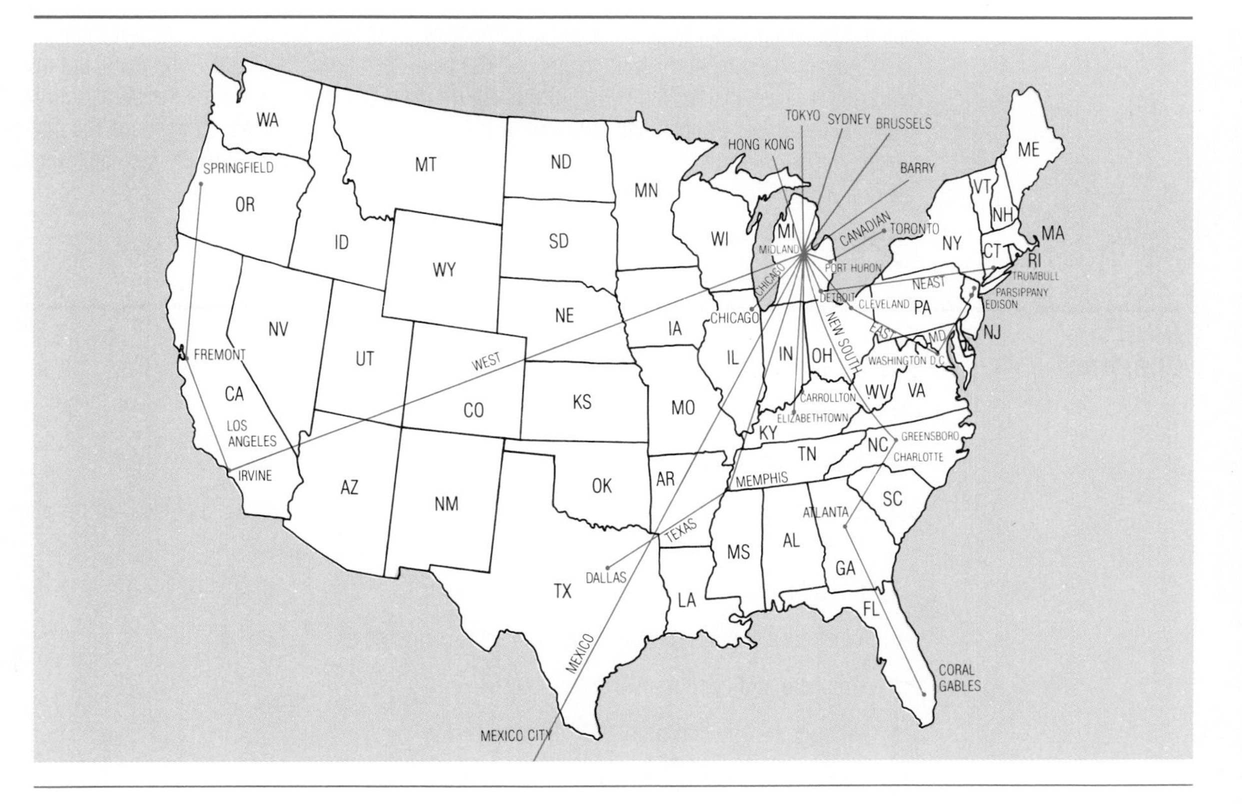

Figure 11–16
Dow Corning's U.S. data communications network has a classic star topology, with the center at the company's headquarters in Midland, Michigan.

speed circuit is de-multiplexed, and the Dow Corning circuit is run to the Dow Corning office. The primary reason for this arrangement is to split the cost of the expensive international circuits between the companies.

An interesting backup arrangement has been designed for the Hong Kong and Australian locations, which also takes advantage of the Dow Chemical network. Dow Chemical has a leased circuit between Hong Kong and Sydney. The communications software at Dow Corning has been defined to use the Hong Kong–Sydney circuit as a backup. Therefore, if Dow Corning's circuit from Midland to Sydney fails, data for Australia is routed to Hong Kong and then, automatically, to Sydney. This backup arrangement was actually used during the aftermath of the October 1989 San Francisco earthquake. The Midland–Sydney circuit was down for about 2 days, during which time data was routed on the backup circuit via Hong Kong. Users experienced slower response time than usual but were still able to conduct business.

A primary goal of Dow Corning's network is to allow users at any Dow Corning location, anywhere in the world, to log on to any computer for which they are authorized. This goal has been achieved, and employees routinely make intercontinental connections to computers in other countries as a part of their job.

Dow Corning's use of LANs is growing rapidly. Several LANs have been installed at the corporate center location for the primary purpose of providing faster access to the mainframe computers. These LANs also provide traditional LAN capabilities, such as file sharing or printer sharing among personal computers. The LANs at the corporate center use IBM's token ring technology and operate at 4 Mbps. In the manufacturing and research locations, Ethernet is used to connect DEC terminals and computers. Other dedicated LANs, specifically designed to share data and printers between personal computers, have been installed in several locations.

Across the company, several types of LANs have been tested in the past few years, but the company believes that for the future, IBM token ring and DEC Ethernet will be the primary LAN technologies used. Bridges and gateways will be installed where needed to continue the philosophy of allowing users anywhere in the company to log on to any computer for which they are authorized.

Review Questions

1. Define the word *network*.
2. Describe four ways that networks can be classified.
3. Explain the characteristics of a star network. A ring network.
4. Describe some situations in which public networks may be more appropriate than private networks.
5. Describe how a packet switching network operates.
6. What are the purported advantages of an ISDN network?
7. Compare and contrast WANs and LANs.
8. What is the function of the TCP/IP protocol?
9. Compare and contrast bridges, gateways, and routers.

10. What are the advantages and disadvantages of a centralized static routing system?
11. What are the advantages and disadvantages of a distributed dynamic routing system?
12. What are the main characteristics of a LAN?
13. Describe the advantages and disadvantages of broadband and baseband transmission.
14. Why do LANs have distance limitations?
15. Distinguish between the IEEE 802.3, 802.4, and 802.5 standards.
16. What are the main characteristics of the FDDI standard?
17. Compare and contrast Ethernet with a token ring LAN.
18. What are the functions of a print server on a LAN? A file server?
19. Why must personal computer software often be changed in order to operate on a LAN?
20. Why is it important for a company to place emphasis on LAN management?
21. What are some of the security considerations that LAN administrators must deal with?

Problems and Projects

1. Calculate the theoretical amount of time it would take to transmit a 1 megabyte file from one location to another on networks that operate at the following speeds: 9,600 bps, 64 kbps, 4 Mbps, 100 Mbps. Why will the actual transmission times be longer than the ones you calculated?
2. Draw diagrams and describe the operation of LANs implemented using the 802.3, 802.4, and 802.5 standards.
3. Think of applications where the token passing bus might not provide fast enough response time. In what applications might a CSMA/CD LAN not provide consistent enough response time?
4. Visit a company that has a LAN installed for the purpose of studying the LAN management techniques they have in place. Report your findings to your instructor or the class.

Vocabulary

network
topology
star network
hierarchical network
mesh network
bus network
bus
ring network
hybrid network
private network
public network
telex network
value added network (VAN)
packet data network (PDN)
packet switching
packet

None

- Integrated Services Digital Network (ISDN)
- wide area network (WAN)
- metropolitan area network (MAN)
- local area network (LAN)
- internet
- Transmission Control Protocol/Internet Protocol (TCP/IP)
- bridge
- gateway
- broadcast routing
- centralized routing
- distributed routing
- static routing
- dynamic routing
- router
- server
- file server
- print server
- communications server
- baseband transmission
- broadband transmission
- Ethernet
- Fiber Distributed Data Interface (FDDI)
- backbone
- Manufacturing Automation Protocol (MAP)
- Technical Office Protocol (TOP)
- Network Basic Input Output System (NetBIOS)
- spooling

References

Bairstow, Jeffrey, "GM's Automation Protocol: Helping Machines Communicate," *High Technology,* October 1986, p. 38.

Bartee, Thomas C. *Data Communications, Networks, and Systems.* Indianapolis, IN: Howard W. Sams & Company, Inc., 1985.

Boule, Richard, and John Moy. "Inside Routers: A Technology Guide for Network Builders," *Data Communications,* September 21, 1989, p. 53.

Dalrymple, Rick, "LAN Standards Efforts Begin to Pay Off," *Mini-Micro Systems,* March 1986, p. 93.

Doll, Dixon R. *Data Communications: Facilities, Networks, and Systems Design.* New York: John Wiley & Sons, 1978.

Glass, Brett, "Understanding NetBIOS," *Byte,* January 1989, p. 301.

Gregory, Rosemary. "Local Area Networks: An Overview," *Infosystems,* October 1985, p. 80.

Kolodziej, Stan. "Users Move Toward MAP," *Computerworld Focus,* March 19, 1986, pp. 17 and 20.

Lefkon, Richard G. *Selecting a Local Area Network.* New York: American Management Association, 1986.

Martin, James. *Introduction to Teleprocessing.* Englewood Cliffs, NJ: Prentice-Hall, Inc., 1972.

McNamara, John E. *Local Area Networks: An Introduction to the Technology.* Bedford, MA: Digital Press, 1985.

McQuillan, John M. "Routers as Building Blocks for Robust Internetworks," *Data Communications,* September 21, 1989, p. 28.

Messmer, Ellen. "Global Networks: Putting the Pieces Together," *Information Week,* July 24, 1989, p. 24.

Mier, Edwin E. "LAN Gateways: Paths to Corporate Connectivity," *Data Communications,* August 1989, p. 72.

Milligan, Gene E. "FDDI Emerges as a High-Speed Fiber LAN," *Telecommunications Products + Technology,* April, 1988, p. 43.

Moskowitz, Robert A. "TCP/IP: Stairway to OSI," *Computer Decisions,* April 22, 1986, p. 50.

Mulqueen, John T. "The Rise of the MAN Handlers," *Data Communications,* October 1989, p. 67.

Retz, David. "TCP/IP: DOD Suite Marches Into the Business World," *Data Communications,* November 1987, p. 209.

Schultz, Brad. "The Evolution of ARPANET," *Datamation,* August 1, 1988, p. 71.

Spiewak, Eric. "MAP and TOP: What, Why and How," *Manufacturing Systems,* March 1986, p. 42.

Stallings, William. "Tuning Into TCP/IP," *Telecommunications,* September 1988, p. 23.

Swastek, Mary Rose, David J. Vereeke, and Darrell R. Scherbarth. "Migrating to FDDI on Your Next Big LAN Installation," *Data Communications,* June 21, 1989, p. 35.

Thurber, Kenneth J. "Getting a Handle on FDDI," *Data Communications,* June 21, 1989, p. 28.

CHAPTER

Connecting the Circuit to the Computer

OBJECTIVES

After studying the material in this chapter, you should be able to

- discuss alternative methods of attaching a circuit to a computer;
- describe the functions of a front-end processor;
- describe the role of a host computer in a communications network;
- explain the functions of the various pieces of software that help to operate a communications network.

INTRODUCTION

Chapters 9, 10, and 11 presented the characteristics of data communications circuits, the data link protocols or rules under which the circuits operate, and networks—the way the circuits are connected together. In this chapter, we will look at the ways in which a circuit or network is connected to the host computer and at alternative ways the required functions can be provided.

The hardware and software capabilities described in this chapter are necessary for the operation of a computer-controlled telecommunications network. Depending on the complexity of the network, the functions may be performed to a greater or lesser degree by the computer or another piece of hardware called the front-end processor.

CIRCUIT TERMINATION ALTERNATIVES

A telecommunications circuit may be connected to a host computer in two primary ways. The first is a direct connection between the circuit and the computer so that as bits arrive, they are stored directly in main memory. This is the approach used in many microcomputers. It works well when only a single circuit is connected.

The advantage of this approach is low cost. The problem is that although there is usually specific circuitry dedicated to assembling groups cycle stealing

of bits into characters, the CPU must be interrupted to store each character into main memory. These interruptions, called *cycle stealing*, can put a significant drain on the resources of the computer and have an impact on other work it is doing.

With the second type of connection, when several circuits terminate at the same host computer they may be connected to a *front-end processor (FEP)*, also known as a *telecommunications control unit* or *transmission control unit*. This is usually a separate piece of hardware that, at minimum, assembles the bits into characters and the characters into blocks before feeding them to the main computer. The computer is interrupted fewer times. For outbound transmission, the transmission control unit accepts blocks of characters from the computer and splits them into bits for transmission. Although these two alternatives have been described as if they were discrete alternatives, in fact there is a range of possibilities between the two.

The FEP may go one step farther and assemble blocks of incoming data into complete messages. This requires the FEP to be a programmable computer itself. If the FEP is programmable, it also can be programmed to do additional editing of incoming messages before passing them to the computer. This approach further minimizes the number of interruptions to the main computer but, of course, requires more elaborate hardware and software in the FEP.

Which of these line connection approaches is chosen depends to a large extent on the size of the host computer, the techniques the computer uses, and even the philosophy of the computer manufacturer. Microcomputers most often attach circuits through an interface card directly to the processor. Minicomputers often have special circuitry in the CPU dedicated to the FEP function. Large mainframe computers usually have a separate piece of hardware that provides the FEP function.

Another set of alternatives that must be considered by the computer designer is whether the front-end function will be provided in hardware, software, or a combination of the two. The advantage of implementing it in software is flexibility and adaptability. Software can be modified or corrected much more easily than hardware can. New protocols or terminal types can be supported by adding program code. On the other hand, the advantage of implementing the function in hardware is speed of operation. Once a function has been designed and thoroughly checked out, it can be put into a circuit chip for very little cost and will operate much faster than a software implementation.

FRONT-END PROCESSORS (FEP)

Regardless of which implementation alternatives are chosen, certain capabilities must be provided to interface the communications line to the computer. This discussion assumes that the FEP has been imple-

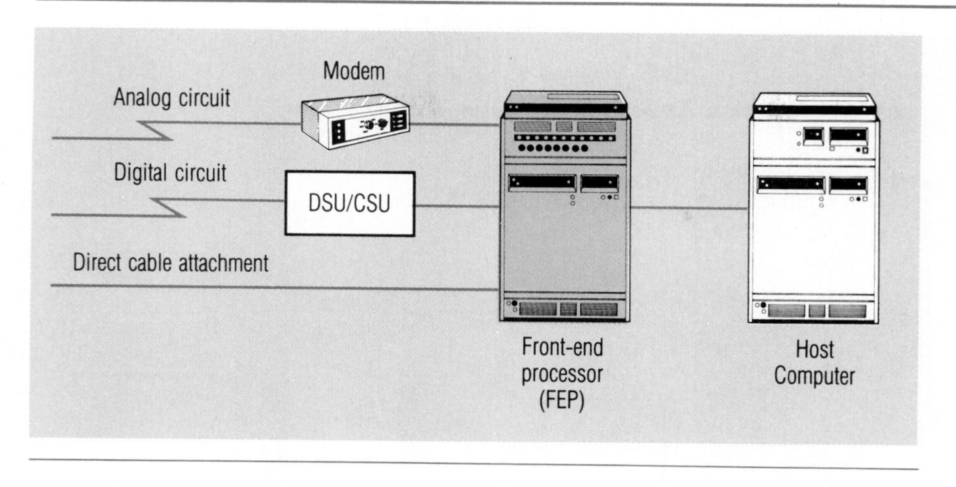

Figure 12–1
Location of the front-end processor in the network.

mented as a separate piece of hardware because it is easier to visualize. It is important to remember, however, that each vendor decides how it will implement the communications line interface.

A full-capability, standalone FEP is a specialized computer specifically designed to control a data communications network. It links the host computer with the network as shown in Figure 12–1. FEPs are most often provided by computer manufacturers, although there are third party vendors in the market. Clearly, the computer manufacturer is in the best position to understand what is required to make the mainframe computer and FEP operate compatibly.

Figure 12–2 shows the basic internal structure of a typical front-end communications processor. The heart of the device is a computer, the communications processor CPU. The CPU requires memory in order to store the software programs and data used during communications processing. The console may take several forms. It simply may be some lights and switches on the control panel of the FEP, or it may be a separate VDT or printing terminal through which an operator can check status and enter commands. The channel interface supports the connection(s) to the host computer.

port

On the network side, the FEP provides interfaces to the telecommunications circuits. Each interface in the FEP is called a *port*. Standalone FEPs range in capacity from 10 to 500 or more ports. Usually, the port capacity depends on the speed of the circuits that will be connected, because the FEP has an inherent limit to the number of bits per second it can process. In other words, the slower the speed of the connected circuits, the more that can be handled. With higher-speed circuits, fewer can be connected.

driving the FEP

Normally, the FEP manufacturer also provides the software to drive it. This programming, commonly called the *network control program (NCP)*,

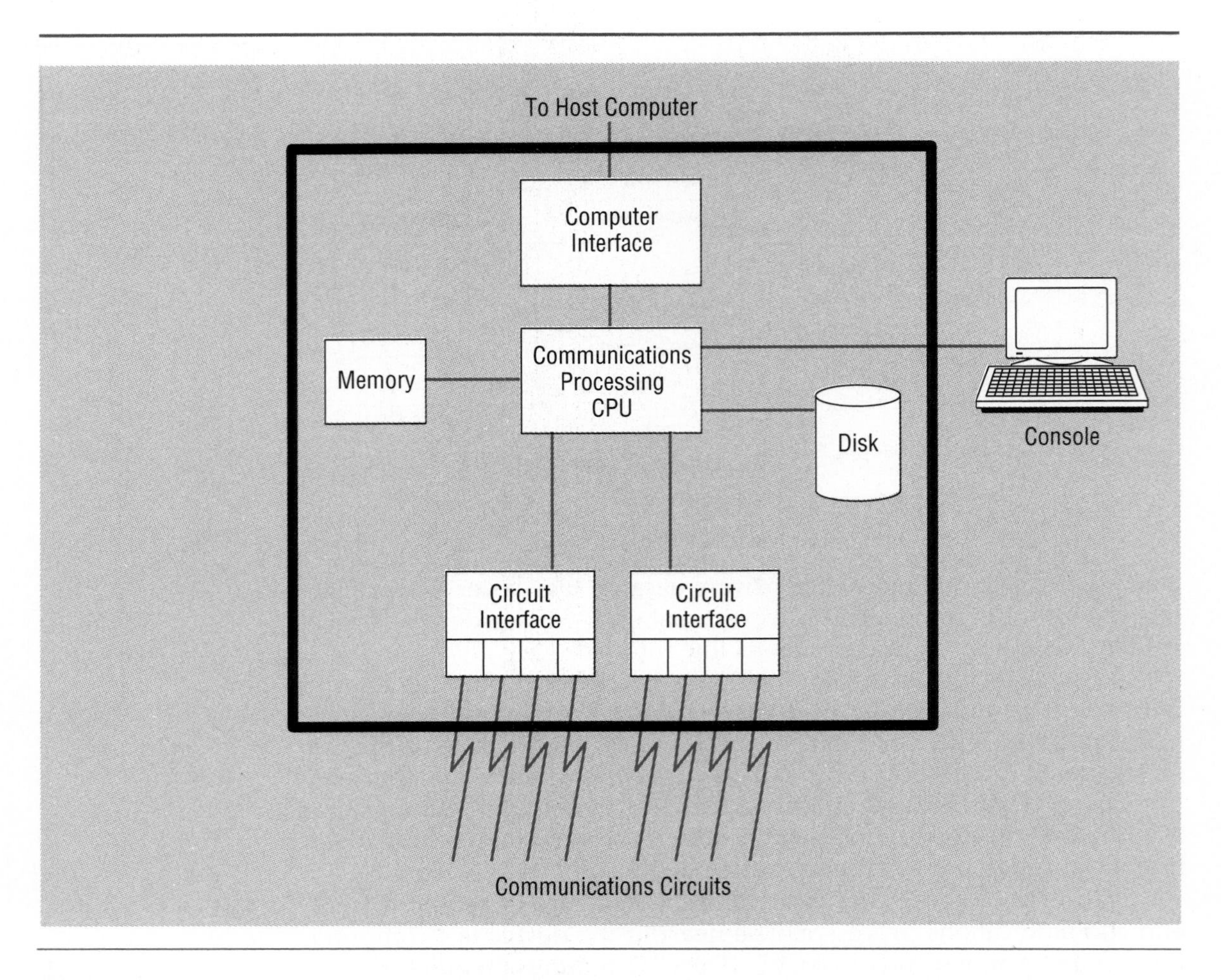

Figure 12–2
The internal structure of a front-end processor.

is very complicated. Just as users seldom write computer operating systems, they seldom write NCPs, although they may modify the software provided.

Front-End Processor Functions

The functions an FEP performs are

1. Circuit control
   (a) polling and addressing terminals
   (b) addition of protocol to outgoing messages; removal of protocol from incoming messages
   (c) answering dial-in calls; automatic dialing for outgoing calls
   (d) protocol conversion—such as an asynchronous protocol to SDLC
   (e) code conversion—for example, ASCII to EBCDIC

The NCR Comten 3600 series of front-end communication processors come in a wide range of sizes supporting up to several hundred communications lines. (Courtesy of NCR Corporation)

   (f) character recognition, including the distinction between control characters and data characters
   (g) accommodating circuit speed differences
   (h) multiplexing

2. Assembly of characters into blocks or messages
   (a) assembling of incoming bits into characters
   (b) assembling of characters into blocks of data or complete messages, depending on the design of the interface with the host computer

3. Message queuing or storing
   (a) holding outgoing messages when they arrive from the computer faster than the communications lines can accept them
   (b) holding incoming messages from the communications line when the computer is down or otherwise unable to accept them
   (c) adjusting the sequence in which messages are passed to the computer or the line so that higher-priority messages are handled first

4. Error control
   (a) VRC, LRC, and CRC checking
   (b) requesting retransmission of blocks that contain errors
   (c) converting computer codes to transmission codes and vice versa

5. Administrative functions
   (a) noting when terminals or circuits are not operating correctly and sending a message to a network control operator
   (b) logging all messages to a tape or disk
   (c) keeping statistical records of line use, terminal use, and response time
   (d) keeping records of the various type of transmission errors
   (e) performing online diagnostics of circuits or remote terminals
   (f) switching all circuits to a backup computer if the primary computer fails

The functions that are implemented in a particular FEP can be determined by reading the manufacturer's literature. Some FEPs have terminals connected directly to them that act as consoles. These terminals can change the operating parameters or conditions of the FEP or its software. Sometimes a disk or magnetic tape drive is also connected to the FEP. These tapes or disks are used for

- storing a copy of the FEP software so that it can be reloaded automatically in case of failure;
- logging copies of all or selected messages that pass through the FEP;
- recording statistical information about the message traffic, circuit utilization, and errors;
- storing a configuration of the network so that the operating software knows which terminals are on which lines, their addresses, names, and other characteristics.

On the host side of the FEP there are two alternatives for connecting to the computer. The first is to have multiple connections between the FEP and host. With this arrangement, each telecommunications circuit effectively passes through the FEP and directly into the host on a separate physical connection. The other alternative is to have a single high-speed interface between the FEP and the host, such as a computer channel. Using this alternative, the FEP multiplexes the data from all of the telecommunications circuits onto the channel. Special software residing in the host demultiplexes it and passes it to the proper application program.

It is worth noting the wide difference in speed between communications lines and computers that FEPs must manage. A communications line operating at 9.6 kbps is capable of passing

9,600 bits ÷ 7 bits per character = 1,371 ASCII characters per second

Disks and tapes typically pass data to a computer at 1 to 3 million characters per second. Even if the communications line is speeded up to 56 kbps, it can only pass ASCII data at 8,000 characters per second. Clearly, to a computer, a communications line operates at a very slow speed. The FEP (being a computer) can easily manage many communications lines and have time left over for other network management functions.

FEP computer models

Some typical FEPs are the IBM 3720 and 3745, the Amdahl 4745, and the NCR-Comten 5675. These units come in a large number of models that support networks of greatly varying sizes. Prices range from $26,000 to over $1.6 million, depending on the number of lines supported and various other FEP features.

THE ROLE OF THE HOST COMPUTER

The functions performed by the host computer are complementary to those performed by the FEP. To some extent, what the FEP does not do, the host computer must do, and vice versa. However, many functions are performed exclusively in the host, and virtually all of them are controlled by software.

Types of Host Software

Many types of host computer software are involved in a network's operation. We will examine them in this section; they are illustrated in Figure 12–3.

Operating System The *operating system* is the central control program that governs the computer hardware operation. The operating system uses the hardware capabilities to provide interrupt handling, multiprogramming, and at least high-level management of all input/output devices, including telecommunications lines.

Access Method An *access method* is computer software and is normally a part of the operating system. The access method controls the passage of data to and from an input/output device. There are access methods for magnetic tape, disk, communications lines, and all other input/output devices. For telecommunications circuits, the access method handles the data passed between the FEP and the host computer.

One of the objectives of an access method is to mask the specific characteristics of the hardware from other software and application programs so that the programs do not have to adapt to them. The access method provides the ability to read or write a data record or message. It performs the transformations needed to actually place the data on the hardware device or transmit the data on the telecommunications line. The access method that controls the communications lines is called the *communications access method (CAM)* or the *telecommunications access method (TCAM)*. IBM's primary telecommunications access method is called the *virtual telecommunications access method (VTAM)*.

Figure 12–3
Telecommunications software.

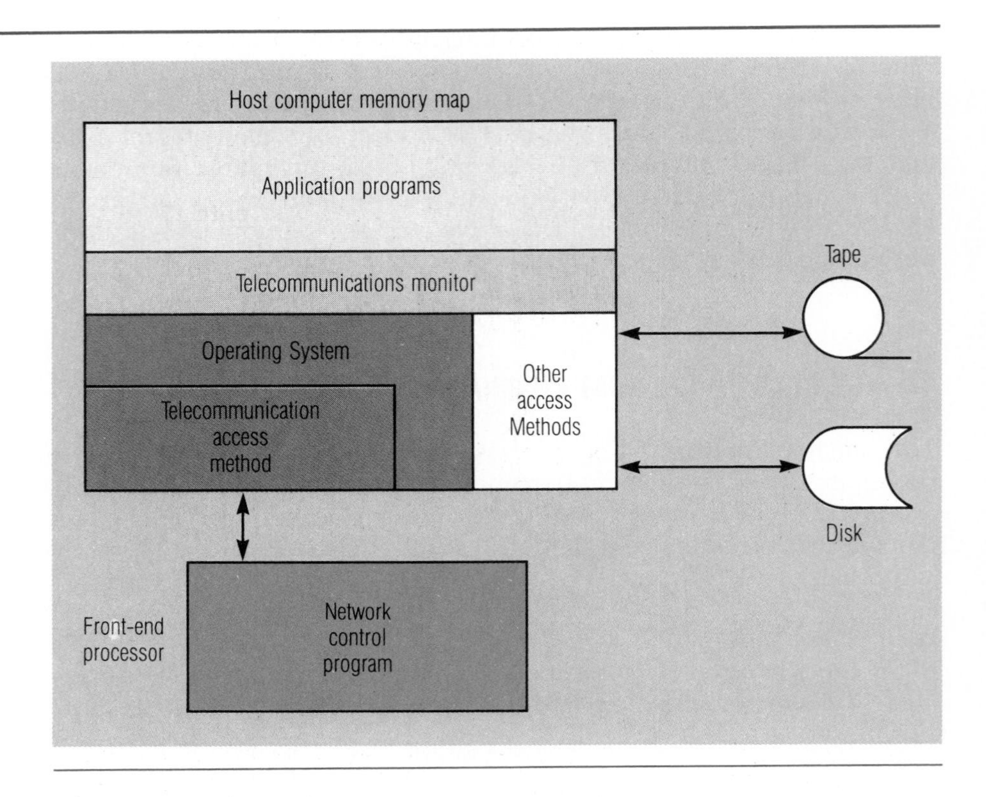

Another function of the access method is to provide the logical connection between a terminal and an application program. In the past, the access method functions were built into each application program. Each telecommunications line and terminal was dedicated to an application. If a company sales office wanted to use both the order processing application and the accounts receivable/credit application on the host computer, it had to have two separate terminals, each connected via a separate telecommunications line to the host computer. Naturally, as terminal usage grew and networks expanded, this redundancy became unacceptably expensive.

A modern access method allows a terminal to sign onto an application, perform the required operation, sign off, and then sign onto a different application. This permits terminals and lines to be shared and eliminates the redundancy and cost inherent in previous generations of network software.

Telecommunications Monitors (TCM) The functions of *telecommunications monitors (TCM)* overlap those of access methods in some operating systems. The functions most frequently attributed to a telecommunications monitor are described here.

Security The telecommunications monitor, operating in conjunction with the operating system and the network control program in the FEP, provides security to ensure that only authorized users or terminals are able to use application programs and access data. Techniques used to ensure security include

- validating user identification codes and passwords entered by operators at terminals;
- checking the user's authority against tables or files stored in the computer to determine whether the user can use the specific program he or she is requesting;
- checking the same tables to determine whether the terminal operator has the authority to read, write, update, or delete data files or can do only some or one of these functions;
- accepting the user ID and password from an operator on a dial-up circuit; then terminating the connection and calling the operator back at a predefined telephone number.

Communications security is discussed in more detail in Chapter 14.

Buffer Management Incoming messages are stored in portions of the computer memory called *buffers*. Buffers are almost invariably of a fixed length, whereas messages are almost always of variable length. Consequently, managing the buffers is a significant job for the TCM. Incoming messages must be assigned to buffers as they arrive and before they are passed to the application program for processing. Response messages from the application are stored in buffers until they are passed to the FEP for transmission. Once the messages are sent, the buffers are freed and added to a queue of available buffers. Efficient buffer management can help ensure good message handling performance and can reduce the storage requirements of the TCM.

Routing a Message to a Program Messages that come into the telecommunications monitor may be destined for one of several application programs. The TCM can examine the contents of the incoming message and determine which application program it is destined for.

Queuing Messages When messages arrive faster than they can be processed by the application program, they are held in a queue or sequential waiting line (which may be in the computer's main memory or on disk) until the application program can handle them. In some systems, incoming messages are held in a queue until a specified number have arrived. The application program then processes all of the messages in the queue at one time. Queuing also may occur if an application program only operates during certain hours of the day, whereas messages for it arrive anytime. Messages are held in a queue until the application program is ready to accept them.

Scheduling Application Programs The TCM may tell the operating system when to bring an application program into memory for execution. Its scheduling may be based on elapsed time since the program was last scheduled, or it may be dependent on the arrival of incoming messages. In other words, when the queue of waiting messages reaches a certain level, it may trigger the telecommunications monitor to schedule the application program to process those messages.

Providing Continuity Between Parts of a Transaction Some transactions arrive in several parts. For example, a customer order processing transaction may require the terminal operator to first enter the customer number and order number. This data is transmitted to the computer, where the TCM schedules the application program. The application program examines the customer number and looks up the customer's name and address in a file. The name and address are then transmitted back to the terminal operator for verification. The operator then enters the identifying number of the product the customer wants to order. When the number arrives at the computer, the TCM must determine that it is associated with the first part of the order that was processed previously. The way that this is achieved is to save the parts of the transactions in memory or on disk and effectively build an order from the data in the transactions as they come in.

Another type of continuity may be required if an operator is in the middle of processing a complicated transaction but stops for an extended period of time to answer the telephone, go to lunch, or take a break. Undoubtedly, he or she will want to resume processing the transaction after the break without having to reenter previously submitted data. The telecommunications monitor can provide continuity in this situation.

Message Formatting Another function of the TCM is to convert the format of messages from that required or generated by the application program to or from the format required by the specific terminal. This message formatting is also called *presentation services*. Presentation services are discussed in more detail in Chapter 13.

virtual terminal

The objective of the message formatting function is to provide a *virtual terminal* interface to the application and its programmer. The idea is to allow the application program to pass messages to the telecommunications monitor in a standard format and without regard to the specific type of terminal for which it is ultimately destined. Thus, if new terminals become available and are added to the network, only the TCM software must be changed to format messages for the new terminal; the application programs do not have to be modified.

The virtual terminal concept works in reverse, too. Incoming messages from different terminal types may arrive at the TCM in different formats. The responsibility of the TCM is to put them into a standard format before presenting them to the application program for processing.

Checkpoint/Restart The TCM has a responsibility, often in conjunction with the database management software or the operating system, for providing the ability to restart processing after a failure of the FEP, the host computer, or the software has occurred. Whereas with traditional batch computer processing, the usual approach is to simply rerun the jobs that were processing when the failure occurred, this technique is not adequate or even possible in communications-based message processing systems. One cannot tell terminal operators at 4:00 on a Friday afternoon, "The computer has just failed and we have restarted it, but you are going to have to reprocess the entire day's work. Please resubmit all of the transactions you have sent since 8:00 this morning."

The concept behind checkpoint/restart processing is to periodically record on disk or tape the status of the computer processing, including the contents of memory, the programs that were active, the messages that were being processed, and the database or file records that had been accessed. This status information is called the *checkpoint record*. After a failure occurs and the computer is restarted, the checkpoint record is loaded into memory and used to reset the computer to the conditions that existed when the checkpoint record was taken. The only processing that has to be redone is any that occurred since the last checkpoint record was written.

checkpoint record

Determining the frequency with which to take checkpoint records requires knowledge of the telecommunications network and the applications. The determination normally is made by a group of telecommunications and data processing people working together. If checkpoints are taken too frequently, checkpoint processing can place a significant load on the host computer. Furthermore, because in many systems all other processing must stop while the checkpoint record is being written, terminal operators may see unusually long response times or other unexplainable delays. On the other hand, if checkpoints are not taken frequently enough, many transactions may have to be reprocessed after a failure occurs, causing frustration and redundant work for many people.

Preventing Messages from Being Lost or Duplicated The two biggest problems that can occur in communications-driven business systems are missing messages and repeated messages. If a message containing a customer order for two tank cars of an expensive chemical is lost or otherwise not processed, not only will the customer be angry, but also the supplier will have lost the opportunity to sell some product and enjoy the resulting profit. On the other hand, processing the message twice in an automated order processing system and sending four tank cars of the chemical to the customer is equally undesirable. Duplicate shipping costs, the cost to reship the material to another customer, and the potential lost opportunity to sell it during the erroneous shipment eat into the product's profitability. In addition, the supplier looks somewhat foolish in the eyes of the customer.

Telecommunications monitors and software written to handle business applications must contain adequate controls using techniques, such as checkpoint/restart and sequence numbers, to ensure that all incoming messages are processed once, but only once!

Application Programs Application programs perform basic business functions, such as customer order processing, inventory control, and accounts receivable. If the TCM, access method, operating system, and network control program are properly designed and well-written, writing an application program that performs transaction processing from terminals connected via communications lines should be no different from writing any other program. To the application programmer, messages to be processed should be obtained with simple GET or READ commands, much like obtaining data from a disk or tape file. After processing, the response message should be able to be sent with a PUT or WRITE command. With proper implementation of the virtual terminal concept, the application program should be shielded from the unique attributes, characteristics, and idiosyncrasies of the telecommunications terminals. An application programmer should not need to be any more aware of the operation of a telecommunications circuit than he or she is of the operation of a channel on a computer that connects a disk drive to the CPU.

Software for Network Management Another type of telecommunications software that exists, at least partially, in the host computer is programs for network management. Like most other communications software, there are alternatives for where these programs are located. They may reside entirely in the host computer, or they could be in the FEP. Alternatively, they may be shared between the two. If the network management software is in the host, it may be a part of the access method, the telecommunications monitor, or the operating system. It also could be a separate program that can work closely enough with the other software to direct the network's operation.

Network management software performs several functions. One is to monitor the status of the network and display data pertinent to the network's operation. Some networks have large status maps with lights controlled by the network monitoring software. When a line or control unit is operating normally, its corresponding light is green. If a problem exists, the light is changed by the software to yellow or red, depending on the severity of the problem. Other systems display the network status on a VDT screen that may be dynamically updated as the status changes. Circuits or nodes that are fully operational may be shown in green. Those with some problems might be displayed in yellow. Units that are out of service might be shown in red.

log file

Another function of the software is to log pertinent network operational data to a data file stored on magnetic disk or tape. The log file may contain statistical records about the network's operation. These records can be analyzed to calculate performance measures, such as network availability and response time. The log file also may contain error records. These records help to pinpoint problems, especially those that occur infrequently but over a long time span. For example, an intermittent line problem might cause transmission errors two or three times a day and might not be noticed by users of the network, especially if the transmission errors are automatically corrected by the hardware. An analysis of the error records would show the pattern of errors and provide data for service personnel.

A third function of network management software is to provide the network control operator with commands to control and monitor the status of the communications network. Using these commands, the network operator can manually control certain aspects of the network's operation.

Network Control Commands Some of the types of commands required by the network control operator are

- start line—this command tells the software to start polling, addressing, or otherwise using a communications line. It could be used when new lines are installed but is used more frequently to resume operation after a circuit has been out of service;
- stop line—this command provides the inverse capability of start line and instructs the software to stop using one of the circuits. Another use for this type of command might be to shut down operations on a circuit at the close of a business day. If a host computer located in Seattle has a network that stretches to the east coast, the network operator may want to stop the east coast circuits at 5 P.M. or 6 P.M. EST because the terminal operators will have gone home. Since it is only 2 P.M. or 3 P.M. in Seattle, the local terminals would continue to operate for several more hours;
- start terminal/stop terminal—these commands are similar to the start and stop line commands, but they only affect a specific terminal on a circuit. They might be used to stop a terminal that is failing and to restart it after the problem has been corrected;
- check status—the check status command may have many variations that allow the network control operator to check any part of the communication network operation. The operator could use one variation to check the status of a particular terminal. Another variation checks the status of a line, and still other variations could allow the links or contents of message queues to be examined;
- start network/stop network—these commands are used to start normal network operations at the beginning of a business day and to stop operations at night. All of the stop commands usually examine the status of terminals and circuits and allow any traffic in progress to be completed normally. Since this process can take some time with long messages, there are sometimes variations of the stop commands, such as stop normal, just described, and stop immediate, which instantly interrupts all transmissions and brings the network to a crashing halt.

user-written commands

Other commands are available in most networks. Some software allows *user-written commands* to be defined by the network operators. Using this capability, a network user can write commands unique to the particular requirements of each company's network.

SUMMARY

This chapter looked at the ways in which a communications circuit is interfaced to a host computer. The computer manufacturers make design decisions about which functions will be implemented in the host computer, the FEP, hardware, and software. Communications software is relatively complex and is most often written and provided by the computer and FEP manufacturer. If the communications software is properly designed and implemented, the work of the application programmer is simplified so that reading and writing messages from a telecommunications line are similar to reading and writing disk records.

Proper controls must be built into both the communications and applications software to ensure that all messages are processed but that none are processed more than once. Although the vendor-written communications software can simplify this task, the ultimate responsibility is with the business for which the network and processing exist.

Having capable network management software is necessary to successful network operations. Operators must be informed about the status of the network through status displays, and must be able to control the network by issuing commands.

CASE STUDY

Dow Corning's Use of Front-End Processors

Dow Corning's primary FEPs are IBM 3745 communications control units that are attached to both of the company's mainframe computers and shared between them, as illustrated in Figure 12–4. A single communications controller provided adequate capacity for many years, albeit with no redundancy. Two years ago, when additional capacity was required, telecommunications management decided to install a second communications controller to provide redundancy. This

Figure 12–4
Dow Corning has two IBM 3745 front-end processors to provide backup capability in case one fails.

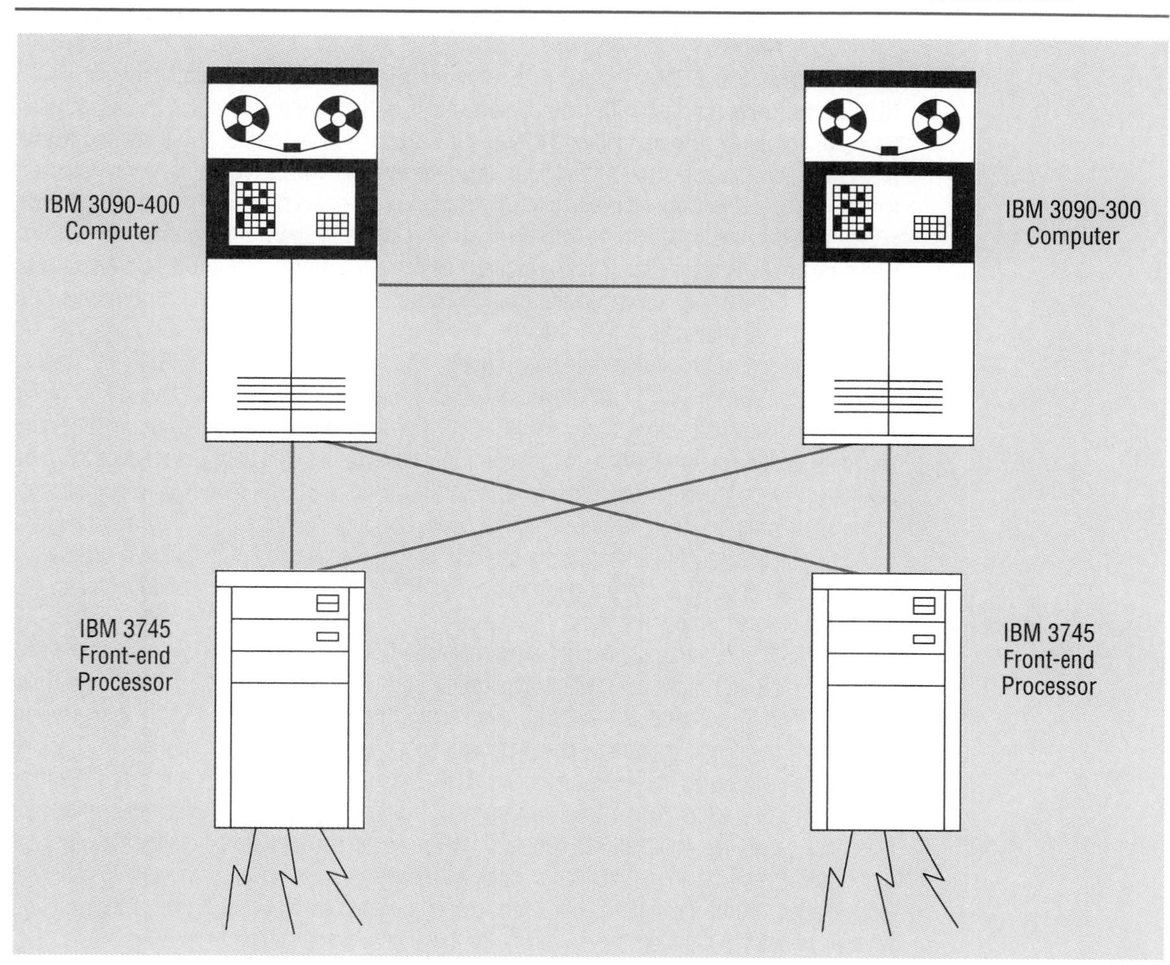

decision was based on the company's increasing reliance on the data communications network for conducting business and the concern for the potential impact on business operations if the network were out of service for a significant length of time. The second IBM 3745 controller was added at that time.

In 1986, a smaller FEP, the IBM 3720, became available. The smaller capacity, coupled with the growing desire for redundancy in the network, made it very attractive for Dow Corning to consider. The company installed 3720s in all of the major plants in the United States to serve as data communications concentrators. One or more 56 kbps circuits were run between the 3745 at the computer center and the remote 3720s at the plants. From the plant 3720s, 56 kbps circuits are run to IBM 3174 cluster control units that serve the terminals. This provides 56 kbps service to users in the plants without having to run a circuit for each cluster controller, an option for which the cost would be prohibitive.

A 3745 controller is also installed at the computer center in Brussels. It controls the operation of the European portion of the network and also is connected to the 3745s in Midland via a 56 kps circuit. In Mexico and Japan, 3720s are installed to manage the networks in those countries.

The 3745 and 3720 FEPs use IBM software called the Network Control Program (NCP). NCP works in conjunction with the access method of the mainframe, which is called virtual telecommunications access method (VTAM). NCP and VTAM allow the 3745s serving several host computers to be connected together in an arrangement known as the multisystem networking facility (MSNF). Dow Corning is using MSNF to connect its data centers in Midland and Brussels. MSNF provides the capability for any terminal user on the network to log onto any CPU for which he or she is authorized. Thus, a properly authorized terminal user at Dow Corning's plant in Munich, Germany, or Tokyo, Japan, can log onto the company's computer in Brussels or to the computer in Midland.

The telecommunications monitor program that Dow Corning uses is IBM's Information Management System (IMS). IMS also serves as a database manager. All checkpoint restart activities are provided automatically by IMS. Dow Corning has a strong policy of not modifying vendor-supplied software. SIM management firmly believes that this policy is productive for the company in that it avoids the problems of both the vendor and Dow Corning making changes to the same program code. At the same time, Dow Corning recognizes that they may be giving up some efficiency by not changing the software to fit their specific needs, but they believe that the potential benefits do not outweigh the costs. Therefore, they use the software "as is," the way it is delivered from the vendor.

In 1983, an internal study concluded that Dow Corning should provide a comprehensive electronic mail and office automation application on the mainframe. IBM's Professional Office System (PROFS) software was selected after competing products from several vendors were evaluated. The second mainframe computer was added to handle the new application. The new application area, called end user computing (EUC), has grown rapidly. By the end of 1989 there were over 5,000 users. One of the strengths of the EUC system is its electronic notes and messages capability, which allows people anywhere in the company to communicate with each other quickly and reliably. Other tools, such as a word processing program and spreadsheet, are provided on the mainframe, but users are encouraged to make their own decision about whether to use the tools on the mainframe or those on a personal computer.

Review Questions

1. Why is it unlikely that the cycle stealing technique would be used on a mainframe computer that controls several communications circuits?
2. What are some of the factors a computer/FEP manufacturer would consider in deciding whether to implement the line interface function in a separate piece of hardware or in the CPU?
3. Describe the functions of a telecommunications access method.
4. Describe the functions of a telecommunications monitor.
5. Describe the functions of checkpoint/restart processing.
6. Describe some situations when queuing would occur in a telecommunications system.
7. Explain several security techniques that might be employed to ensure that users of a telecommunications network are properly authorized.
8. Explain the concept of a virtual terminal.
9. Why is there so much concern about ensuring that messages are processed once, but only once?
10. Why is it important to have network management software?

Problems and Projects

1. Using your imagination, "invent" several commands other than those listed in the text that would be useful when the network control operator operates the network or diagnoses problems.
2. Can you think of some situations in which it would not make any difference if a message was processed two or three times? What if it was processed 500 times?
3. Visit a computer installation that has a telecommunications network. Find what security techniques are being used to ensure that users of the system are properly authorized.
4. What statistics about a network's operation would it be useful for the network operations and management people to have?

Vocabulary

cycle stealing
front-end processor (FEP)
telecommunications control unit
transmission control unit
port
network control program (NCP)
operating system
access method
communications access method (CAM)
telecommunications access method (TCAM)
virtual telecommunications access method (VTAM)
telecommunications monitor (TCM)
virtual terminal
checkpoint/restart

References

Axner, David H. "What's New in Communications Processors?" *Telecommunications Products + Technology,* September 1986, p. 65.

Fitzgerald, Jerry. *Business Data Communications: Basic Concepts, Security, and Design,* 2nd ed. New York: John Wiley & Sons, 1984, 1988.

Martin, James. *Introduction to Teleprocessing.* Englewood Cliffs, NJ: Prentice-Hall Inc., 1972.

Reynolds, George W. *Introduction to Business Telecommunications.* Columbus, OH: Charles E. Merrill Publishing Company, 1984.

Spanier, Steve. "FEPs Ease Migration to New LAN Protocols," *Mini-Micro Systems,* September 1986.

Stamper, David A. *Business Data Communications,* 2nd ed. Menlo Park, CA: The Benjamin/Cummings Publishing Company, Inc., 1986, 1989.

CHAPTER

Telecommunications Architectures and Standards

OBJECTIVES

After studying the material in this chapter, you should be able to

- explain the difference between communications architectures and communications standards;
- explain the need for a communications architecture;
- describe the seven layers of the ISO-OSI model architecture;
- describe a vendor's communications architecture, IBM's SNA;
- discuss the advantages and disadvantages of layered architectures.

INTRODUCTION

This chapter examines telecommunications architectures and standards. The first telecommunications architectures were developed in the 1970s to tie together many of the individual pieces of telecommunications into a unified whole. In Chapter 4, the International Standards Organization's OSI reference model was introduced. Reference to the OSI model was made frequently as you looked at the technical details of telecommunications in Chapters 8 through 12. Now that you have learned how telecommunications systems are put together, you are in a better position to understand more details about the OSI model.

We will first discuss the difference between architectures and standards and then look at the OSI layers in more detail than before. As we study its layers, we will relate them to the subject matter we have studied in the preceding chapters. Then we will look at two specific architectures developed by computer manufacturers, Systems Network Architecture (SNA) developed by IBM and Digital Network Architecture (DNA) developed by the Digital Equipment Corporation.

DEFINITION OF ARCHITECTURE

In general terms, an *architecture* is a plan or direction that is oriented toward the needs of the user. It describes "what" will be built but does not deal with "how." An architect must consult with the eventual occupants of a home to ensure that it matches the family's lifestyle or special requirements. If a family is active in sports, the architect may design special closets or racks to store sports equipment. There might also be shelves in the family room to hold sports trophies.

network architecture

In the same way, a telecommunications system architect must be aware of the needs of communications users in order to design an architecture to meet their needs. A *network architecture* is a set of design principles used as the basis for the design and implementation of a communications network. It includes the organization of functions and the description of data formats and procedures. An architecture may or may not conform to standards.

Since most architectures (plans) are made for the long term, the architect must consider not only today's requirements. He or she must also ensure that the architecture is flexible enough to meet new requirements and support new capabilities that will arise in the future. This is particularly challenging in a field such as communications, where technology is advancing so rapidly and visions become reality in only a few years.

COMMUNICATIONS STANDARDS

Communications standards are established to ensure compatibility among similar communications services. Communications standards are the flesh on the architectural skeleton. They specify "how" a particular communications service or interface will operate.

telex and TWX

In many cases, the United States has developed standards for domestic use ahead of the rest of the world. More recently, when the need arose in other countries, they improved on the U.S. standards, and those improvements became the international standards. It is not uncommon to find one communications standard in use in the United States and another in the rest of the world. For example, the telex system for international message communications is an international standard. It uses the 5-bit Baudot code for individual characters, and the transmission speed is 66 words per minute. In the United States, AT&T developed a similar system, the teletypewriter exchange system (TWX), which uses an 8-bit code and transmits at 100 words per minute. The two systems operated according to different sets of standards.

There are a number of organizations in the world that establish communications standards. Many of these organizations are shown in Figure 13–1. Because of the recognized need for common international communications standards, there has been a good deal of cooperation among these organizations, especially in recent years.

Standards Organization	Main Telecommunications Focus
International:	
Consultative Committee on International Telephones and Telegraph (CCITT)	telephone and data communications
Consultative Committee on Radio	radio frequencies
International Standards Organization (ISO)	communications standards of all types (coordinates with the CCITT)
United States:	
American National Standards Institute (ANSI)	data communications in general
Electrical Industries Association (EIA)	interfaces, connectors, media, facsimile
Institute of Electrical and Electronic Engineers (IEEE)	802 LAN standards
National Bureau of Standards (NBS)	standards of all types
User/Vendor Forums: (these groups feed information to the standards-setting bodies listed above)	
European Computer Manufacturers Association (ECMA)	computer and data communications standards (feeds input to ISO)
Corporation for Open Systems (COS)	computers and data communications
Manufacturing Automation Protocol (MAP), Technical Office Protocol (TOP)	standards for manufacturing and office communication

Figure 13–1
Some of the organizations involved in setting telecommunications standards or in passing input to the standards-setting bodies.

GENERAL NEED FOR ARCHITECTURES AND STANDARDS

As early communications systems were conceived and developed, the need for an architecture was not evident. The designers of the first communications networks and applications could not envision how rapidly online terminal-based systems would grow. Early communications systems were built for specific applications that had substantial justification, for example, a high return on investment, or payback.

The most famous example is the SABRE airline reservation system, which was built jointly by American Airlines and IBM in the late 1950s. SABRE, like most early telecommunications systems, was a customized design. It was built for the specific application of airline reservations, and it used nonstandard hardware and specially written software designed to optimize the capability and performance of the system. In the late 1950s there were no standard rules for data communications and no standard "off the shelf" products that the network designer could use.

Special terminals had to be designed, and all software had to be written from scratch.

With the success of early telecommunications systems and with the reduction of communications costs based on improving technology, other online applications became justifiable. Because there had not been an overall plan or architecture for telecommunications or a vision of future telecommunications applications, the existing networks were not flexible enough to support new requirements. As a result, separate networks had to be built for each application. This meant that one location of a company might have had multiple communications lines running to it, each attached to a separate set of terminals. Lines and terminals could not be shared because of the use of different protocols for transmitting data on the lines and also because the communications software often was built into the application programs on the host computer. Each application contained specialized communications programming designed to operate the terminals for that application. There were no access methods and no telecommunications monitor programs. Little consideration was given by application designers to sharing programs, terminals, or circuits with other applications. As each new application was justified and approved, its designers and programmers started over. They made design decisions to meet what they perceived to be the unique requirements of their application, with little regard for what had been done before or what might follow.

USER NEED FOR ARCHITECTURES AND STANDARDS

As telecommunications networks and systems evolved, it became obvious to network designers and users that there had to be a better way. It didn't make good business sense to continually start over and build new, unique communications networks for each new application. There was a growing recognition of the need for an overall plan or architecture to guide future network and application developments. What gave rise to this realization?

First, users recognized that systems were becoming too complex. Because of the lack of an overall plan and the desire to optimize communications systems to particular applications, a proliferation of transmission protocols, programs, and communications services had evolved. Most of them were incompatible with one another. It became very difficult for a company to manage the diversity of hardware and software in their communications networks. Making changes to a network was difficult and risky because of the complexity and the possibility that the implications of a change would not be fully realized and ensuing problems would crash the system.

Second, users had a desire to be isolated from the complexities of the network. Communications networks do change. There is a need to add lines and terminals almost constantly and a strong economic incentive to add new applications as they are justified and developed. Terminal users should be isolated from these changes. They should not have to worry about modifications to the network infrastructure. A person using the network should not see changes in how the network reacts from day to day. Consistency, from the user's perspective, is a virtue! On the other hand, the network designer must have the ability to make changes in the topology of the network, the services it provides, or other characteristics without affecting current users. As user requirements change, traffic volume increases or decreases, and new terminals or other products come to the market, the designer should be free to incorporate them into the network.

Also, network users wanted to connect different types of devices to the network. Networks had to be able to service different types of terminals. Both interactive devices, such as VDTs, and batch terminals that contain high-speed line printers were needed by most companies. Furthermore, it is desirable for a company to be able to acquire these terminals from different vendors and to take advantage of new technology improvements as they become available.

Fourth, distributed processing became a network issue. With the rapid increase in the number of minicomputers and personal computers, the ability to do data processing on a distributed basis exists and is often desirable. However, these distributed processors do not exist in a vacuum. There is a need to communicate between them in order for them to share and refer to common data.

Finally, users developed a need for integrated network management. It is very desirable to have an integrated set of tools with which to manage the communications network. Diagnostic and performance measurement capability can help ensure that the communications network is delivering the service for which it is designed efficiently and effectively.

Clearly, to be most effective and to meet user requirements for network capabilities today, a telecommunications architecture is required to mask the physical configuration and capabilities of the network from the logical requirements of the user. Discipline is required to get the full benefits of networking and distributed processing. The chaos of multiple terminals on multiple lines using multiple protocols can be avoided only if an overall plan (architecture) is in place.

Fortunately, several vendors and the *International Standards Organization (ISO)*, a group made up of representatives from the standards organization in each member country, recognized the need for a telecommunications architecture and for standards to support it. Working with the CCITT, ISO developed the *open systems interconnection (OSI) reference model* in 1978. The OSI model is gaining rapidly growing acceptance throughout the world.

THE ISO-OSI MODEL

Objectives

The architects of the OSI model had as their primary objective to provide a basis for interconnecting dissimilar systems for the purpose of information exchange. For their purposes, a system is viewed as consisting of one or more computers, associated software, and terminals capable of performing information processing. The intent was to define communications rules that, if followed, would allow otherwise incompatible systems made by different manufacturers to communicate with each other.

As we have seen, the OSI model uses a layered approach, each layer representing a component of the total process of communicating. A diagram of the seven OSI layers is shown in Figure 13–2. In reality, each end of the communication (a computer or a terminal) must have an implementation of the seven-layer architecture because each layer in one end system communicates with its peer on the other end system. Layer 7 in a remote terminal talks to layer 7 in a host computer, layer 6 to layer 6, and so forth.

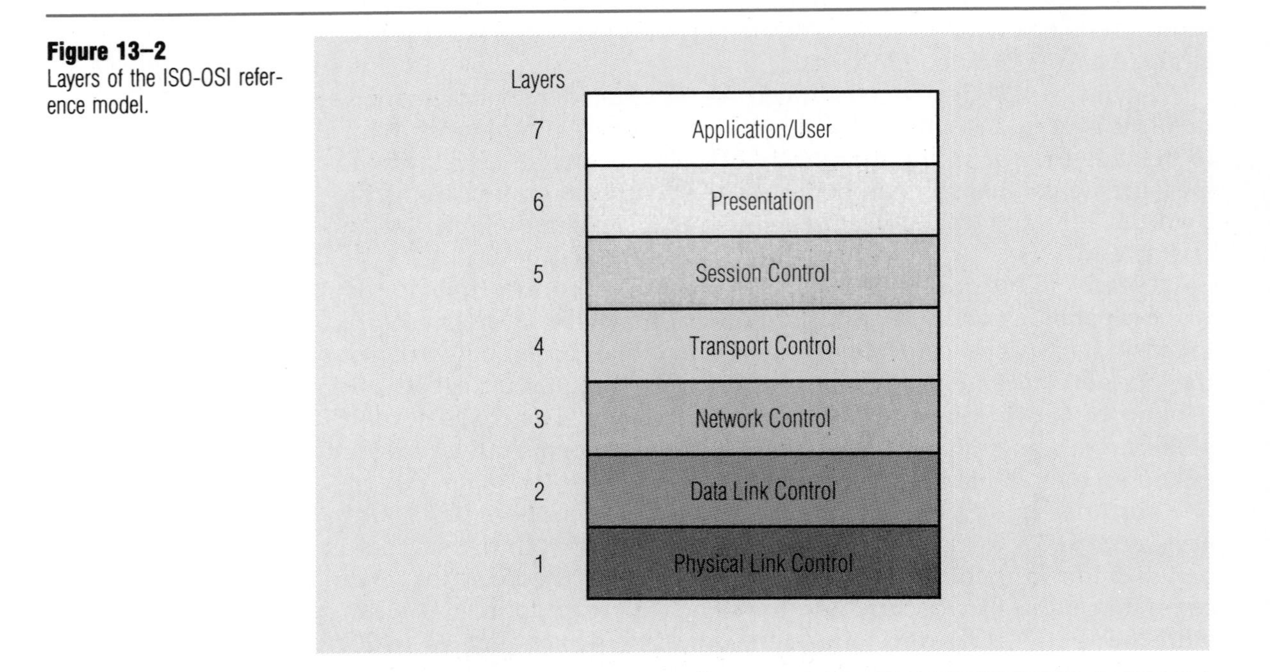

Figure 13–2
Layers of the ISO-OSI reference model.

A helpful way to visualize the layers is to picture an onion and think of the way a computer system is layered. At the heart is the computer hardware. The hardware is surrounded by operating system software. Outside the operating system is other specialized software, perhaps a communications program or a database system. Finally, the outermost layer is the application program. The application program is shielded from the complexities of the hardware by the operating system and other software. At the same time, the operating system and other specialized software provides useful services to the application program.

Another way to view the interaction of the layers is that each one requests services from the layer below it. The lower layer honors the request using its own capability or, if necessary, requesting services from layers below it. For example, the network control layer, layer 3, assumes that the data link control layer, layer 2, has passed good data to it, but it is not concerned with the details of how that data was transmitted or how any errors were corrected. In the OSI model, the highest layer, layer 7, is the end user or a program that makes the original request for service. The lowest layer, layer 1, communicates across the transmission line to the lowest layer in the other system.

In the OSI model, the first three layers are well-defined. Standards have been written and agreed to, and these layers have become more widely known and understood than the other layers. The combination of these three layers is the *X.25 standard for data transmission* used in packet switching networks.

X.25 standard

The Seven Layers of the OSI Model

Layer 1: Physical Link Control Layer Layer 1 defines the electrical standards and signaling required to make and break a connection on the physical link and to allow bit streams from the DTE to flow onto the network. It specifies the modem interface between the data terminal equipment and the line. For analog circuits, the most common interface is the V.24 or RS-232-C standard. The standard for the interface to digital circuits is X.21. As we saw in Chapter 8, the physical layer is concerned with voltage levels, current flows, and whether the transmission is simplex, half-duplex, or full-duplex. Besides the interface standards, the physical link control layer defines how connections will be established and terminated. Other functions defined in this layer include monitoring channel signals, clocking on the channel, handling interrupts in the hardware, and informing the data link layer when a transmission has been completed.

Layer 1 is the only level of the architecture in which actual transmissions take place. Although we talk about layers 2 through 7 "communicating" with each other, it is important to remember that they must pass the information they wish to communicate down through the layers to layer 1 for transmission to the other correspondent in the communication.

Layer 2: Data Link Control Layer Layer 2 defines standards for structuring data into frames and sending the frames across the network. It is concerned with such questions as

- How does a machine know where a frame of data starts and ends?
- How are transmission errors detected and corrected?
- How are polling and addressing handled?
- How are machines addressed?

The answers to these questions make up the data link protocol. The OSI protocol is High-Level Data Link Control (HDLC), which was discussed in Chapter 10.

For LANs, the data link layer is divided into two sublayers: the media access control layer (MAC) to define access to the network (the two alternatives are CSMA and token techniques, discussed in Chapter 10) and the logical link control (LLC), which defines the protocol, a subset of HDLC, for use on LANs.

It is important to know that the IEEE, which is establishing the 802 standards for LANs, fully subscribes to the OSI architecture.

Layer 2 is normally located in the front-end processor (if one exists) at the host end of the circuit and in a remote cluster controller or intelligent terminal. Layer 2 must work closely with the modem to accomplish its work. Whereas modems ensure that bits are accurately sensed from the communications line at the receiving end, layer 2 groups the bits into characters for further processing.

Layer 2 is also responsible for checking the VRC, LRC, and CRC codes transmitted with the data, for acknowledging the receipt of good frames or blocks, and for requesting the retransmission of those that are in error. Another type of error that layer 2 handles can occur when data is received from the circuit faster than the receiver can handle it. When this happens, layer 2 signals the transmitter to either slow down or stop transmitting until further notice. This is called *pacing*.

Layer 3: Network Control Layer The primary functions of layer 3 are network addressing and routing. Layer 3 also generates acknowledgments that an entire message has been received correctly. When a message is received from layer 4 for transmission, layer 3 is responsible for breaking it into blocks or packets of a suitable size for transmission. In a packet switching network, for example, the packet size might be 128 characters. Layer 3 assigns the correct destination address to the packet and determines how it should be routed through the network. Specifically, it decides which communications circuit to send a packet or block on.

At the receiving end, layer 3 reassembles the packets or blocks into messages before passing them up to layer 4. Layer 3 might receive a message from another computer that is not destined for its location. In that case, it turns the message around and sends it back down to level 2 for forwarding through the network to the ultimate destination. This situation occurs in multinode networks where some nodes relay data for others.

Over the past 30 years, countries and companies have established their own standards for layers 1 through 3 (without thinking of them as layers). Many networks have been implemented using these proprietary standards. Therefore, the changeover process to international standards is difficult and time-consuming even if a country or company wants to make the change.

Layer 4: Transport Control Layer Layer 4 selects the route the transmission will take between two DTEs. A favorite analogy for layer 4 is to compare it to the postal system. A user mails a letter but does not know (or care) exactly how the letter is transported to its destination, as long as the service is reliable. In a telecommunications network, layer 4 selects the network service to be used to transport the message, when options exist. For example, if a computer has both leased lines and a packet switching network connected to it, layer 4 decides which of the two services should be used to transport a particular message. To a terminal or computer, layers 4, 3, 2, and 1 together provide the transportation service for the user's message.

Layer 4 also contains the capability to handle user addressing. At the transmitting station, this means that network addresses that are meaningful to the user, such as location codes, terminal names, or other mnemonic codes, are converted to addresses that are meaningful to the network software and hardware. These network addresses usually are binary numbers. On the receiving end, network addresses must be converted back to user addresses.

Another function of layer 4 is to control the flow of messages so that a fast computer cannot overrun a slow terminal. This flow control works in conjunction with the flow control at layer 2, but layer 4 is concerned with controlling entire messages, whereas layer 2 is concerned with controlling the flow of frames. Layer 4 could allow a message to be sent, but the individual frames that make up the message could be delayed by layer 2 because of slowdown signals that it had received from layer 2 at the receiving end.

Layer 4 also prevents the loss or duplication of entire messages. Whereas more detailed checks occur at lower levels to ensure that frames and blocks of data are received correctly, layer 4 must implement controls to ensure that entire messages are not lost and, if necessary, to request their retransmission.

Other functions of layer 4 include multiplexing several streams of messages from higher levels onto one physical circuit and adding appropriate headers to messages to be broadcast to many recipients.

The transport control layer is sometimes implemented in the host computer and sometimes in the front-end processor, when one exists. The implementation depends on the particular computer manufacturer's design.

Layer 5: Session Control Layer A *session* is a temporary connection between machines or programs for an exchange of messages according to rules that have been agreed on for that exchange. The session is the first part of the communications process that is directly visible to the user. Users directly request the establishment of sessions between their terminals and computers when they begin the sign-on or logon process.

Before a session can begin, the machines or programs must agree to the terms and conditions of the session, such as who transmits first, for how long, and so on. Clearly, there will be differences in these rules between interactive sessions and batch sessions and between terminals of different types. Given an appropriate terminal, there is no inherent reason why a user cannot have multiple sessions in progress simultaneously.

Layer 5 establishes, maintains, and breaks a session between two systems or users. If a session is unintentionally broken, session control must reestablish it. Session control also provides the ability for the user to abort a session. For example, the BREAK key or ESC key on the terminal may be used for this purpose.

The implementation of priorities for expediting some messages or traffic occurs in this layer, as do certain accounting functions concerning session duration. These are used to charge the user for network time.

The session layer is usually implemented in the host computer access method software.

Layer 6: Presentation Layer The presentation layer deals with the way data is formatted and presented to the user at the terminal at the receiving end of a connection. It also performs similar formatting at the transmitting end so that the data from a terminal is presented to the lower layers in a constant format for transmission. An application programmer on the host computer writes programs to talk to a standard virtual terminal, as was discussed in Chapter 12, and the layer 6 software performs a transformation to meet the specifications of the real terminal in use. Screen formatting, such as matching the message to the number of characters per line and the number of lines per screen, would also be done.

Other functions that occur in layer 6 are code conversion, data compaction, and data encryption. With the exception of data encryption, which is often implemented in hardware, the rest of layer 6 is almost always performed by software in the host computer.

Layer 7: Application or User Layer Layer 7 is the application program or user who is doing the communicating. This is the layer at which data editing, file updating, or user thinking occurs. This layer is the source or ultimate receiver of data transmitted through the network.

A great deal of effort is being expended to define and standardize some common elements of applications that operate in layer 7. The activity has six major thrusts.

- Common application service elements (CASE)—this work is aimed at defining standards for such things as logon and password identification, as well as checkpoint, restart, and backup processes;
- Job transfer and manipulation (JTM)—this defines standards for the transfer of batch jobs from one computer to another;
- File transfer, access, and management (FTAM)—this work is aimed at defining standards for the transfer of files between systems and for providing record-level access to a file on another computer;
- Message oriented interchange systems (MOTIS)—concerned with defining standards for interconnecting the world's many message exchange systems, this work is also known as the CCITT X.400 standard;
- Office document architecture/office document interchange facility (ODA/ODIF)—this work is aimed at providing standards to allow the transfer, edit, and return of documents across systems from multiple vendors;
- Virtual terminal services (VTS)—this work is concerned with precise definition of the virtual terminal concept, including character, graphics, and image terminals, as well as hardcopy devices.

Obviously, many standards are required to address all of these areas, and the work will go on for years. It is encouraging to see that there is a recognized need for standardizing the communications and computing process and that progress is being made toward their definition.

The CCITT X.400 STANDARD

The CCITT *X.400* standard for electronic mail deserves special mention because it is more completely defined than some of the other standards, and systems based on X.400 are coming into widespread use. The X.400 standard operates at layers 6 and 7 of the OSI model and was developed to allow users on incompatible public electronic mail systems to communicate with each other. Since adoption of X.400, its support has been announced by every major computer and communications vendor. X.400-based electronic mail systems are provided by AT&T, MCI, Telenet, and many other carriers worldwide. What makes X.400 so attractive is the ability to connect public and proprietary electronic mail systems that operate according to different protocols and standards. The connection is transparent to the users—they may not even know that their correspondent is on a different network.

A CAVEAT ABOUT TELECOMMUNICATIONS STANDARDS

It is important to understand that writing the specifications for a telecommunications architecture and the related standards is difficult work where precision is needed but difficult to achieve. The architecture and standards are always subject to interpretation. Therefore, it is entirely possible for two companies to implement networks that they believe correspond to a set of standards, but the resulting networks cannot communicate with each other. True compatibility in telecommunications is difficult to achieve and must be confirmed by extensive testing of the components thought to be compatible.

To this end, several testing laboratories have been established to test hardware and software to ensure that it meets OSI standards. The independent verification these laboratories provide will go a long way toward achieving consistent interpretation of the standards as they are developed.

MANUFACTURERS' ARCHITECTURES

Even before the International Standards Organization began its work developing the OSI model, the major computer manufacturers, such as IBM, Digital Equipment Corporation, Burroughs, and Univac, were working to develop architectures on which to base their communications products. In the early 1970s, many of these companies found themselves supporting a wide variety of communications terminals and protocols. Most companies felt they could not afford to continue proliferating incompatible communications hardware and software, so they began working to develop telecommunications architectures on which to standardize their future products. In 1974, IBM announced the Systems Network Architecture (SNA). Soon after, Digital Equipment Corporation announced Digital Network Architecture (DNA), Burroughs announced Burroughs Network Architecture (BNA), and so on.

Note, however, that the focus of the manufacturers was different from that of the International Standards Organization. The manufacturers were primarily interested in developing proprietary architectures on which to base their future products. In most cases, their work was very practical, oriented toward solving immediate problems, and driven by economics. ISO's primary goal was to connect dissimilar networks and systems. The OSI model is intended to define standards for connecting the proprietary architectures of the manufacturers, as well as to connect to public packet switching networks.

We will look now at several manufacturers' architectures and compare them with the OSI model.

IBM Corporation's System Network Architecture (SNA)

Historical Basis IBM's *System Network Architecture (SNA)* was announced in 1974. Product development followed, and by 1978 when the OSI model was announced, IBM had already implemented a number of SNA hardware and software products. IBM's largest customers were beginning to jump on the SNA bandwagon, and IBM saw little reason at that time for changing SNA to meet the OSI model and the international standards.

Certain countries in Europe developed public packet switching networks based on the X.25 standard in the late 1970s. Some of IBM's large European customers wanted to use these networks to connect their IBM equipment, and they put pressure on IBM to support X.25 as a part of SNA. In 1980, IBM announced support for X.25, available to European customers only. A few years later, the X.25 support for IBM hardware and software became available from IBM in the United States also.

SNA Concepts SNA is conceptually similar to the OSI model but not directly compatible with it. Like OSI, it can be viewed as a seven-layer architecture, but some of the layers are defined differently, as shown in Figure 13–3. Rather than making an exact layer-by-layer comparison between the OSI model and the SNA layers, we will examine some concepts and terminology that are basic to SNA and important to its understanding.

Physical Units SNA views hardware as being of four specific types called *physical units (PU)*. The four types of physical units are numbered 1, 2, 4, and 5 with no type 3 currently defined. The four types are

Physical Unit Type	Type of Hardware
1	terminals
2	cluster controllers
4	front-end processors
5	host computers

Figure 13–4 shows how this hardware might be connected together in an SNA network and identifies some of the IBM model numbers associated with the different types of equipment.

Logical Units SNA users can be either people at terminals or application programs. Users are represented in the system by entities called *logical units (LUs)*, which are implemented in software. The communications between two system users is really a communications between LUs and is called a *session*.

Sessions LUs can request several different types of sessions, specifically terminal-to-program, terminal-to-terminal, or program-to-pro-

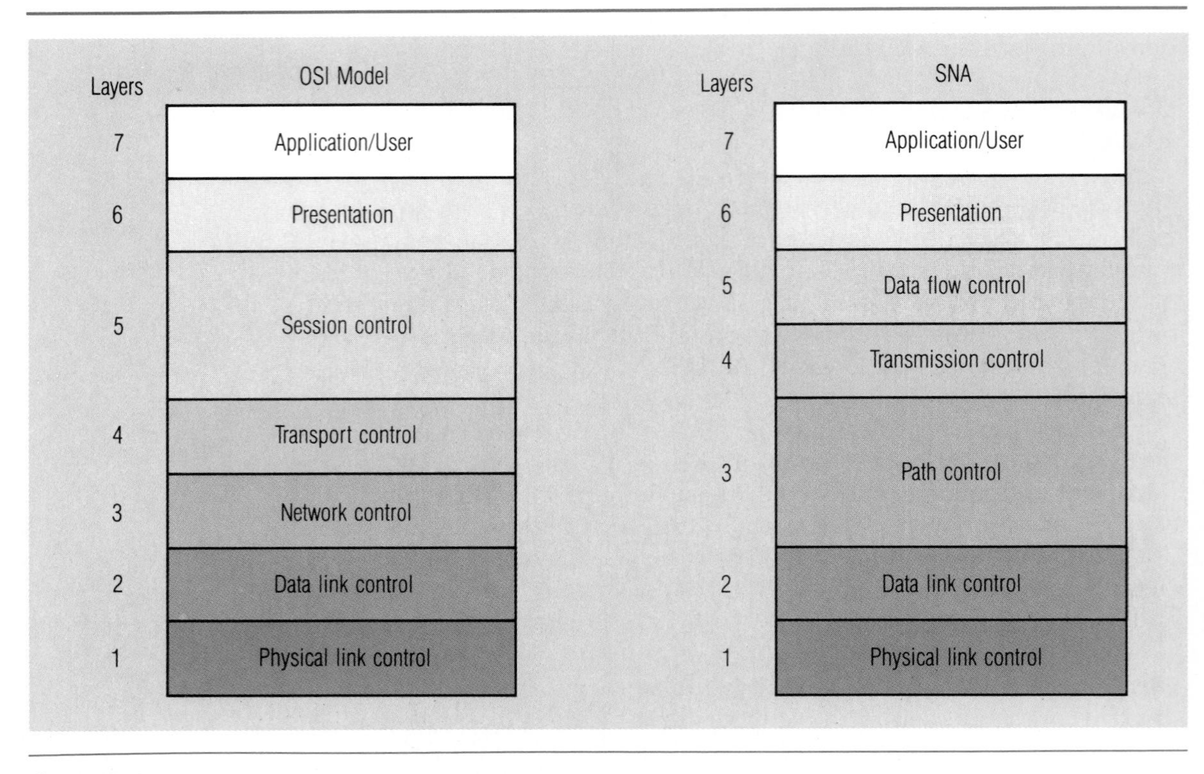

Figure 13–3
ISO-OSI layers compared to IBM's SNA layers.

gram. Terminal-to-program sessions are requested when a user wants to use an application program on a host computer. Terminal-to-terminal sessions are established when one user wants to communicate directly with another. Program-to-program sessions are established when one program needs to pass data to another.

Sessions also are classified as interactive, batch, or printer sessions. In addition, a user can have multiple simultaneous sessions in progress, each with its own logical unit. This gives the user at a terminal the ability to communicate with two or more computers or with two or more programs on the same computer simultaneously.

Network Addressing In SNA, all LUs and PUs are called *network addressable units (NAU),* and each has its own unique network address. SNA addresses are 24 bits in length, which allows for a very large number of nodes to be connected.

Data Link Protocols The primary data link protocol used within SNA is Synchronous Data Link Control (SDLC), developed at the same time as the SNA architecture. In fact, the terms SNA and SDLC often are confused and incorrectly used interchangeably. However, SNA can also operate with the BISYNC and X.25 protocols. SDLC is a proper subset of OSI's HDLC.

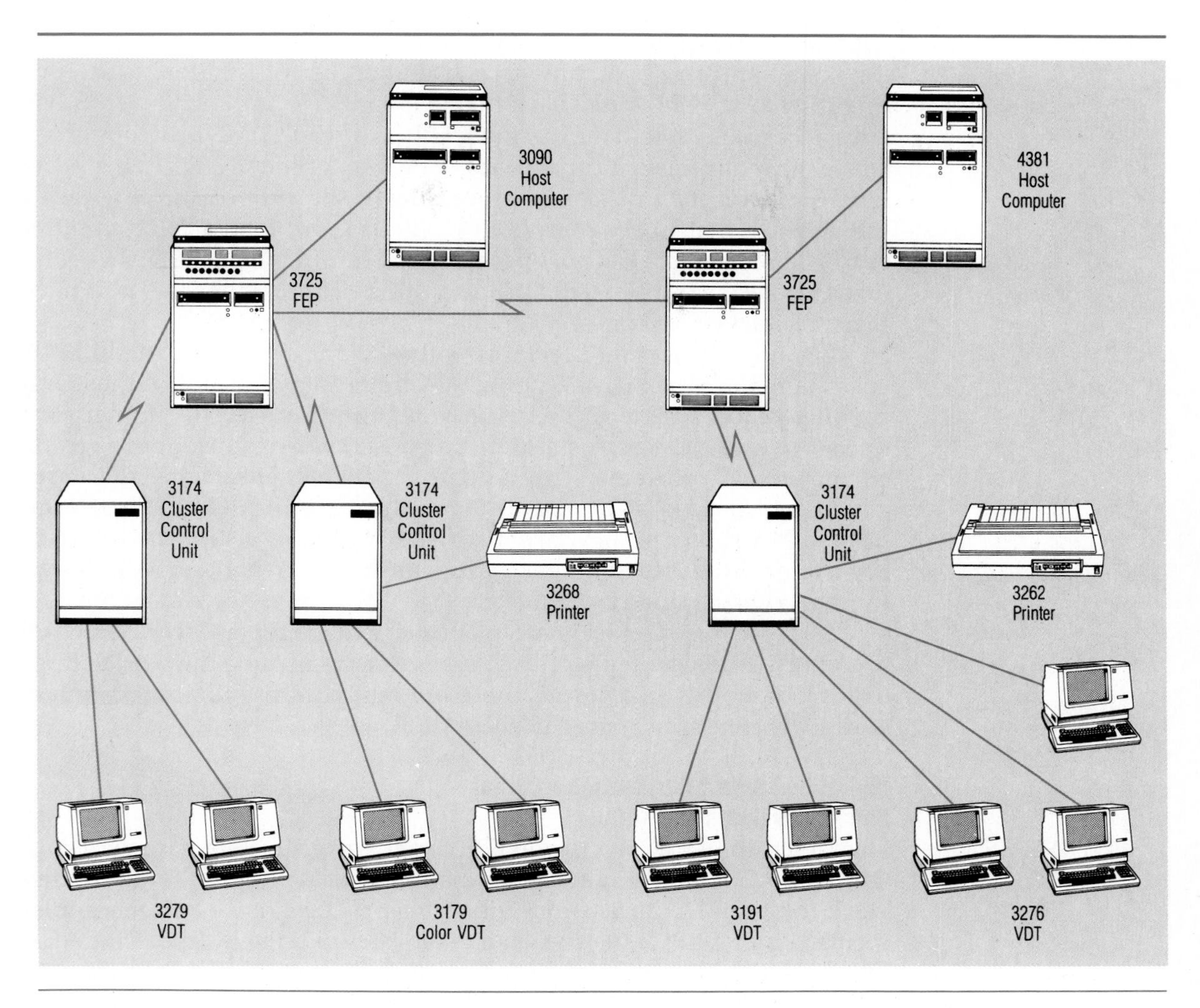

Figure 13–4
Typical IBM SNA network.

SNA Software Although the physical link control layer and part of the data link control layer are implemented in hardware in the IBM front-end processors, most of the SNA layers are implemented in software. The network control program software in the front-end processor is called the network control program, or NCP. The other major software component is the telecommunications access method, which resides in the host computer. It is called VTAM, which stands for virtual telecommunications access method.

IBM has several pieces of software that perform the functions of the telecommunications monitor. These are the Customer Information Control System (CICS), the Information Management System (IMS), and the Transaction Processing Facility (TPF). These telecommunications monitors are not directly a part of the SNA implementation, however.

CICS is the most widely used of the three and can provide good processing speed and response time to a large number of terminals. The primary focus of IMS is database management. First priority in IMS has always been given to database integrity. As a result, IMS provides excellent checkpoint/restart capabilities. TPF was originally written for the very high-volume transaction processing of the airlines and was called the Airline Control Program (ACP). TPF is more than a telecommunications monitor; it is also an operating system. In the past several years, it has started to be used by banks and other organizations that also have high transaction volumes.

Several supporting pieces of software are available to help SNA customers in managing the network. The most significant of these is a product called Netview, which provides the interface to the network operator's console. Netview also keeps statistics about transmission errors, circuit errors, problems with modems, and response time, measures circuit use, and assists in diagnosing problems with either the hardware or the SNA software. IBM has made the specifications for the interface to Netview available to other vendors in the hope that they will design their products to interface with it.

SNA has received wide acceptance. There are reportedly over 36,000 SNA networks installed worldwide. SNA users include the world's largest corporations. As a result, there are many thousands of computers and users communicating via SNA today.

Digital Equipment Corporation's Digital Network Architecture (DNA)

Digital Equipment Corporation (DEC) defined an architecture very similar to the OSI model. *Digital Network Architecture (DNA)* is the framework for all of Digital's communications products and has been implemented in a family of hardware and software collectively called *DECNET*. DECNET establishes a peer relationship between all nodes in the network; no central controlling node is required. Three data link protocols are supported by DECNET: Digital Data Communications Message Protocol (DDCMP), X.25, and Ethernet for use on local area networks. DEC also supplies an interface to IBM's SNA architecture in the form of a combination of hardware and software called the SNA Gateway.

DECNET contains five layers, as shown in Figure 13–5. The physical link control layer, data link control layer, transport layer, and network services layer correspond almost exactly to the lowest four layers of the OSI model. DECNET's application layer is a combination of the OSI presentation and application layers. DECNET does not have a separate session layer.

The data link control layer can use any of the previously mentioned protocols interchangeably. As a result, DEC networks have a very well-developed capability to interconnect DEC hardware, other vendors' hardware that subscribes to the X.25 standard, and equipment con-

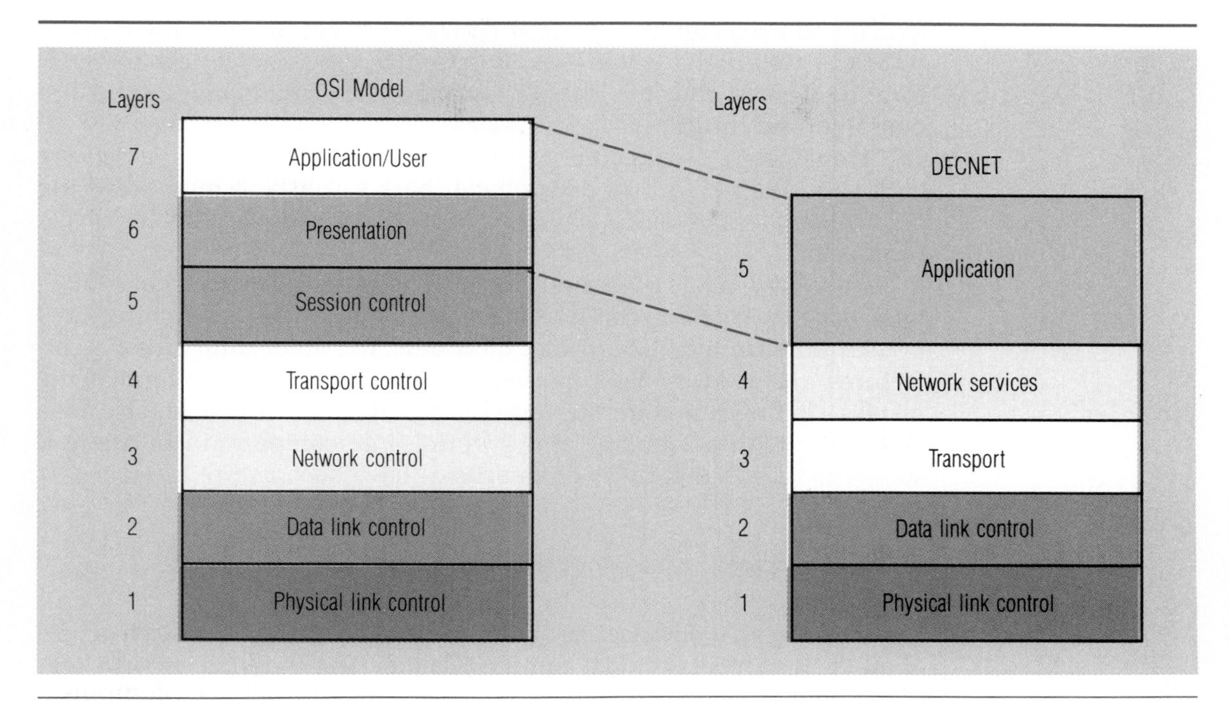

Figure 13–5
ISO-OSI layers compared to Digital Equipment Corporation's DECNET layers.

nected to local area networks that communicate using Ethernet. This capability is unmatched in any other vendor's products at the present time. Thousands of DECNET networks are in place that interconnect terminals and computers of all sizes.

Undoubtedly, the largest DECNET in place is Digital Equipment Corporation's own network called Easynet. Designed as an internal network to support the company's operations, Easynet connects over 27,000 computers in twenty-six countries. The network is used by employees at all levels. For example, engineers use the network to communicate with their peers during product development. Marketing, sales, and manufacturing people also can get details about the intended product. Whereas many companies restrict communications along formal organization lines, at DEC anyone can send messages to anyone else, including messages to executive management. DEC management believes that Easynet has changed the way the corporation operates because of the ubiquitous, easy-to-use communications capability it provides.

Other Computer Manufacturers' Architectures

Burroughs developed a data communications architecture called *Burroughs' Network Architecture (BNA)*. At the data link level, BNA used either the X.25 protocol or Burroughs' data link control (BDLC). Sperry Univac's architecture was called *Distributed Communications Architecture*

(DCA) and used either the X.25 or Univac's Universal Data Link Control (UDLC) protocol. Now that Burroughs and Sperry Univac have merged, forming Unisys, they are leaning toward an implementation of the OSI model for the future.

Honeywell's architecture is called *Distributed Systems Environment (DSE)* and uses HDLC as its protocol. NCR's architecture is called the *Communications Network Architecture (CNA)* and uses the X.25 or SDLC protocols.

In reality, IBM's SNA and Digital Equipment Corporation's DECNET have emerged as the two main telecommunications architectures in use in the United States today. Of course, the other companies' architectures are used by their customers, but because 80 percent of the installed mainframe computers are IBMs or IBM compatible and because DEC has a high percentage of the world's minicomputer installation, it stands to reason that these architectures have become predominant.

ADVANTAGES OF LAYERED ARCHITECTURES

Layering forces a modularization of function that simplifies the structure of all of the aspects of data communication. The thinking process that is required to define the layers in the first place forces clarification of ideas and resolution of troublesome areas. The output of the layering definition process is standards that help ensure that implementers who subscribe to the standards will produce products that can communicate with one another. The combination of the layering and the standards is an aid to understanding the entire telecommunications process.

To the extent that the interfaces between layers are clearly defined, the implementation of one layer can be changed or modified without affecting the other layers as long as the interface specifications are met. This means, for example, that certain layers of the architecture could be implemented in software or hardware, depending on the cost and the requirements for performance and flexibility. Furthermore, different implementations could be substituted transparently. For example, by changing layer 3, leased lines could be substituted for packet switching lines, or vice versa. This again provides a way to meet changing requirements in a modular, nondisruptive manner.

DISADVANTAGES OF LAYERED ARCHITECTURES

The implementation of a layered architecture requires reasonably sophisticated intelligence at each end of the connection. This intelligence is required to establish the connection at the beginning by agreeing to the rules for the transmission and to manage the connection during its duration. In a typical terminal-to-computer connection, the intelligence at the computer end is provided by the computer or a front-end processor. At the terminal end, the intelligence must be provided by the terminal or a cluster control unit to which it is attached. Alternatively, many of the transmission parameters must be predetermined.

economics

Economics play a big part in determining the amount of intelligence and where it is placed. Many applications cannot justify a terminal with the intelligence required to support a layered architecture. Attaching several dumb terminals to an intelligent cluster control unit allows the cost of the intelligence to be spread over the terminals and may be an economic solution for some applications. Still other applications can be economically justified only when the simplest of terminals using the simplest transmission protocols are used.

Personal computers are another way that intelligence can be provided at the terminal. The increasing capability and decreasing cost of personal computers makes them suitable for many applications. Software and circuit cards are widely available to allow most personal computers to communicate with IBM's SNA networks, taking advantage of all its capabilities.

Another disadvantage of layered architectures is that the communications equipment and protocols are likely to be more complex than if a nonlayered approach is used. The complexity is related to the intelligence that is required at each end of the connection and also to the additional capability that architected communications can provide. This additional complexity must be managed more carefully than more rudimentary systems.

For most applications, the advantages of using a layered telecommunications architecture greatly outweigh the disadvantages. Most companies are implementing layered architecture telecommunications for their major communications applications.

SUMMARY

Telecommunications is moving from being relatively unstandardized to a position where a high degree of standardization exists. Worldwide standards organizations are cooperating to develop standards that work together.

Telecommunications architectures provide a framework for the design and implementation of communications systems. When supplemented with standards, layered architectures provide a modularity of implementation that allows flexibility to meet current and future requirements. The International Standards Organization has developed a model for open systems interconnection (OSI) that is partially standardized. Work to complete the standards is progressing rapidly. Most major computer manufacturers have defined and implemented their own architectures that more or less conform to the OSI model.

Computers and terminals that communicate using a layered architecture must have intelligence in order to make connections and negotiate the rules for specific transmissions. The ability to provide the intelligence required at the terminal end of a connection is not a technological issue but an economic one. The economics are dictated by the application for which the terminal will be used.

In general, telecommunications architectures have wide acceptance throughout the telecommunications community. The benefits of a standardized, architected approach to telecommunications far outweigh the costs of implementing and operating such a system.

CASE STUDY

Dow Corning's SNA Decision

Dow Corning was an early user of IBM's SNA. The company made the decision to switch from older communications technology to SNA in 1975 for one primary reason: SNA allowed the consolidation of two separate communications networks that were serving many of the same locations. One of the old networks was providing RJE capability, and the other was serving the VDTs. The consolidation of the two networks eliminated several redundant telecommunications lines, and the company realized significant savings. It was also simpler to manage one network than two.

Dow Corning found the SNA software to be more complicated than the older software it had been using. It wasn't until the company hired a skilled telecommunications professional that it really began to take advantage of the additional flexibility and new capabilities SNA provided.

The company is very interested in OSI developments because of its international operations. Despite the fact that it is a heavy user of IBM computer and communications equipment, Dow Corning is interested in maintaining its flexibility. The company is very interested in having the ability to use competitive equipment when it provides a new capability or is more cost effective. Furthermore, Dow Corning is connecting computers from various manufacturers into a single network but has established the requirement that, to users, the network must continue to appear as a single entity. That is, regardless of where a user is located or what kind of equipment is used, he or she must be able to access any computer or data for which he or she has a need and is authorized. This requirement presents a challenge to the telecommuncations department, since even with the growing number of international communications standards, it is difficult to provide universal connections that are easy to use and economically effective.

To assist in defining network requirements and parameters, a committee consisting of users and telecommunications department members has been formed. This group focuses on both short-range and long-range requirements and at the moment is especially interested in LANs and communications in the manufacturing plants. The intent of the group is to ensure that Dow Corning's telecommunications architecture meets todays needs and provides a foundation for the network that will be required in the future.

Review Questions

1. Explain what a telecommunications architecture is.
2. Why were telecommunications architectures not devised when data communications systems first began to be used?
3. What were some of the problems that companies experienced when they installed early data communications networks? How did an architecture help to solve those problems?
4. What role does the ISO perform for communications architectures?

5. The __________ layer defines the electrical standards and signaling required to make and break a connection on a physical circuit.

6. The OSI data-link protocol is called __________. It is defined in the __________ layer of the OSI architecture.

7. Describe X.25 in terms of the layers of the OSI model.

8. Layers 4 through 1 together provide the __________ for the user's message.

9. Explain the term *pacing*.

10. Explain the term *session*.

11. When data is received, the presentation layer is responsible for __________ before presenting it to the user.

12. Why might two telecommunications systems, both implemented according to the OSI model, not be able to communicate with each other?

13. ISO and vendors had different reasons for developing communications architectures. What were they?

14. Are SNA and SDLC different names for the same thing? Explain your answer.

15. Explain the function of an SNA logical unit.

16. Discuss the relative advantages and disadvantages of using a layered architecture.

17. Why does it take so long to develop an internat'l standard.

Problems and Projects

1. SNA and DECNET assume that a fundamentally different network structure will be used. Explain what characteristics each architecture assumes the network will have.

2. Describe the functions that a gateway between DECNET and SNA must perform.

3. Explain why the OSI model has been more widely and rapidly accepted in Europe than in the United States.

4. Thinking about all of the material covered in the book thus far, make a list of the important trends that are happening in communications. Your list should include both technical and nontechnical items.

5. The integration of voice and data is a topic that is discussed frequently by telecommunications professionals. Using the knowledge you have gained from the text to this point, describe several ways in which voice–data integration could be accomplished in a communications system. (*Hint:* Identify where the voice and data signals could be brought together to share a circuit.)

Vocabulary

architecture
network architecture
communications standards
International Standards Organization (ISO)
open systems interconnection (OSI) reference model
X.25 standard for data transmission
pacing
session
X.400 standard for electronic mail
System Network Architecture (SNA)
physical unit (PU)
logical unit (LU)
network addressable unit (NAU)
Digital Network Architecture (DNA)
DECNET

References

Aschenbrenner, John R. "Open Systems Interconnection," *IBM Systems Journal,* Vol. 25, p. 369, 1986.

Backman, Frank F. *Advanced Function for Communications—IBM's Implementation of Systems Network Architecture.* Armonk, NY: IBM Corporation, 1976.

Cypser, R.J. *Communications Architectures for Distributed Systems.* Reading, MA: Addison Wesley, 1978.

Czubek, Donald. "What are the Differences Between SNA and DECNET?" *Communications Week,* February 9, 1987, p. 54.

Digital's Networks: An Architecture with a Future. Maynard, MA: Digital Equipment Corporation.

Doll, Dixon R. *Data Communications: Facilities, Networks, and Systems Design.* New York: John Wiley & Sons, 1978.

Eagleson, Dick. "Here is a Map to MAP", *Systems User,* August 1986, p. 5.

Fitzgerald, Jerry. *Business Data Communications: Basic Concepts, Security, and Design,* 2nd ed. New York: John Wiley & Sons, 1984, 1988.

Helm, Leslie. "How the Leader in Networking Practices What It Preaches," *Business Week,* May 16, 1988, p. 96.

Meijer, Anton, and Paul Peeters. *Computer Network Architectures.* Rockville, MD: Computer Science Press, 1982.

Randesi, Stephen J. "IBM Communications Are a Lot More Than Just SNA," *Information Week,* January 19, 1987, p. 17.

Rutledge, J. H. *OSI and SNA: A Perspective.* Armonk, NY: IBM Corporation, 1981.

Spiewak, Eric. "MAP and TOP—What, Why and How," *Manufacturing Systems,* March 1986, p. 42.

Stamper, David A. *Business Data Communications,* 2nd ed. Menlo Park, CA: Benjamin/Cummings Publishing Company, Inc., 1986, 1989.

Systems Network Architecture—General Information. Armonk, NY: IBM Corporation, 1975.

Telecommunications and You. Order number GE20-0790. Armonk, NY: IBM Corporation, 1986.

"What is X.400 and Who Supports It?" *Telecommunications,* August 1989, p. 32.

PART THREE

Managing and Operating the Telecommunications Department

The Purpose of Part Three is to describe how the telecommunications department and network are managed. Often, telecommunications management is viewed narrowly as simply the operation of the communications network. In fact, the total picture is much broader. The telecommunications department can be viewed as a company within a company. It has all of the same functions and responsibilities as the enterprise of which it is a part—marketing, product development, finance, and manufacturing. It is important to understand this broader view of telecommunications, for without it one's knowledge of the subject is incomplete.

Chapter 14 explains the broad scope of telecommunications department management. Alternative ways of organizing the department are discussed, management responsibilities are described, and several management issues unique to the telecommunications department are explored.

Chapter 15 describes the process of communications network analysis and design. A phased approach to the activity is used, with emphasis on understanding and defining the requirements for the network before any design work is done.

Chapter 16 delves into the network operations and technical support organizations. This is the group responsible for running the communications network on a day-to-day basis and solving problems as they arise. The necessity of taking a broad approach to the responsibilities of this group is explained, and six specific responsibilities are discussed.

Chapter 17 looks at the future of telecommunications. Technological, regulatory, and standards directions are explored. The chapter also looks at some of the future telecommunications applications that will soon be available to us.

After studying Part Three, you will be knowledgeable about the way the telecommunications department fits into the organization and how it is managed. You will also understand what the department's responsibilities are and how they are carried out. Finally, you will gain some perspective on the future uses for telecommunications technology.

CHAPTER

14

Telecommunications Management

OBJECTIVES

After studying the material in this chapter, you should be able to

- explain why it is desirable to manage voice and data communications together;
- describe where the telecommunications department fits within the organization;
- describe the functions of the telecommunications department;
- describe the administrative activities of the telecommunications department;
- describe in detail the telecommunications management responsibilities;
- discuss several other issues that telecommunications management faces and with which it must be familiar.

INTRODUCTION

This chapter looks at the management of the telecommunications department, the organizational entity that is responsible for planning, designing, and operating telecommunications facilities for a business. We will look at how the department is organized and where it fits into the overall structure of the company. The primary functions of the department are covered in detail in subsequent chapters, so they are introduced only briefly here. The basic management responsibilities of staffing, planning, directing, and controlling and how they are applied to the telecommunications department are examined in detail. Other issues important to the telecommunications manager and staff also are discussed in this chapter.

THE NEED FOR MANAGEMENT

In many organizations, telecommunications management has received little attention. Only the largest companies have devoted any significant time to managing the telecommunications resource. In the last 10 years, however, the telecommunications technology and regulatory environments have been changing so rapidly that most companies are now taking steps to ensure that they are properly managing their communications facilities and people. You should understand some of the factors that have contributed to the increasing interest in communications management. First, deregulation of the telecommunications industry has given businesses more choices and opportunities for communications services, but deregulation has brought with it the need for more analysis, judgment, and decision making. Before deregulation, companies had only one choice for their telephone and data communications equipment circuits and services—the telephone company. Since deregulation, there are many competitive communications products.

deregulation

A second factor is the technology that is making the integration of voice and data communications feasible. In the past, there was little overlap of the two and few possibilities for sharing services or facilities, such as circuits. Therefore, there was little incentive to manage the voice and data together. Today, telephones are being used as computer terminals, and voice is being digitized so that electronically it looks like data. Therefore, for many organizations, it is more feasible and desirable than ever before to group all communications activity together under one management.

voice and data technology

Third, the increases in the number of online, realtime computer systems and in the number of computer terminals using those systems are causing many companies to take a fresh look at what is being managed. In the past, most companies viewed their computer operations department as having the responsibility to run a computer with terminals attached via communications lines, as shown in Figure 14–1. With the rapid increase in the number of terminals and personal computers, many of which are interconnected, companies are beginning to find that the operations department needs to focus primarily on running a communications network, as seen in Figure 14–2. This view is further supported when one thinks of the modern digital PBX as being a specialized computer accessed by specialized terminals called telephones. This change in emphasis from computer center management to communications network management is subtle but significant. It is adding new importance to the responsibilities of the communications managers within the firm.

managing the communications network

The fourth factor is the broader recognition that communications can be a strategic asset to the business. In certain industries, such as those that do heavy marketing by telephone or the transportation industry, this recognition came early. Companies in those industries have always been on the leading edge of communications development and usage. Now companies in other industries are recognizing that a good communications system adds value to the products and services they offer. Customers are being given access to their vendors' databases, and documentation is being exchanged electronically, all aimed at providing better service and making it easier to do business. From the seller's

value of good communications

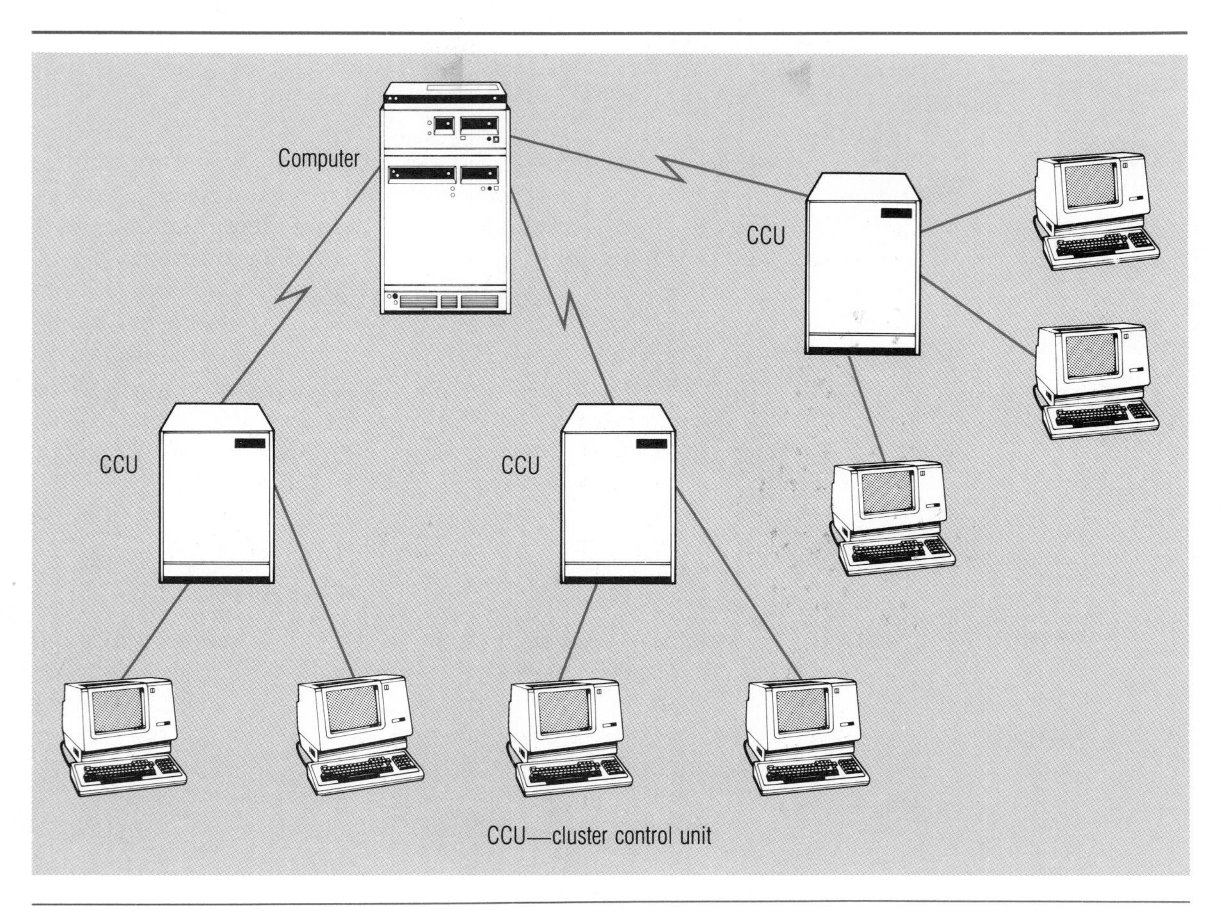

Figure 14–1
Computer with a network of terminals attached.

perspective, the idea is to give customers more reasons to do business with them rather than with the competition.

costs of communications systems

Finally, the amount of money being spent on communications is becoming a significant and noticeable part of the overall cost structure of many companies. Network expenses for all companies were about $5 billion in 1986, and this amount is expected to grow to $9 billion by 1991. Previously, communications expenses were small enough that they could be, and often were, ignored. As more money is spent, however, the cost becomes more visible to senior managers who want to be reassured that the money is being well spent and that the growth is under control.

It is important to distinguish between managing the communications department and managing the communications network. Many times when people speak of communications management, they are thinking of network management or network operations. Although operating the communications network is an important part of the tele-

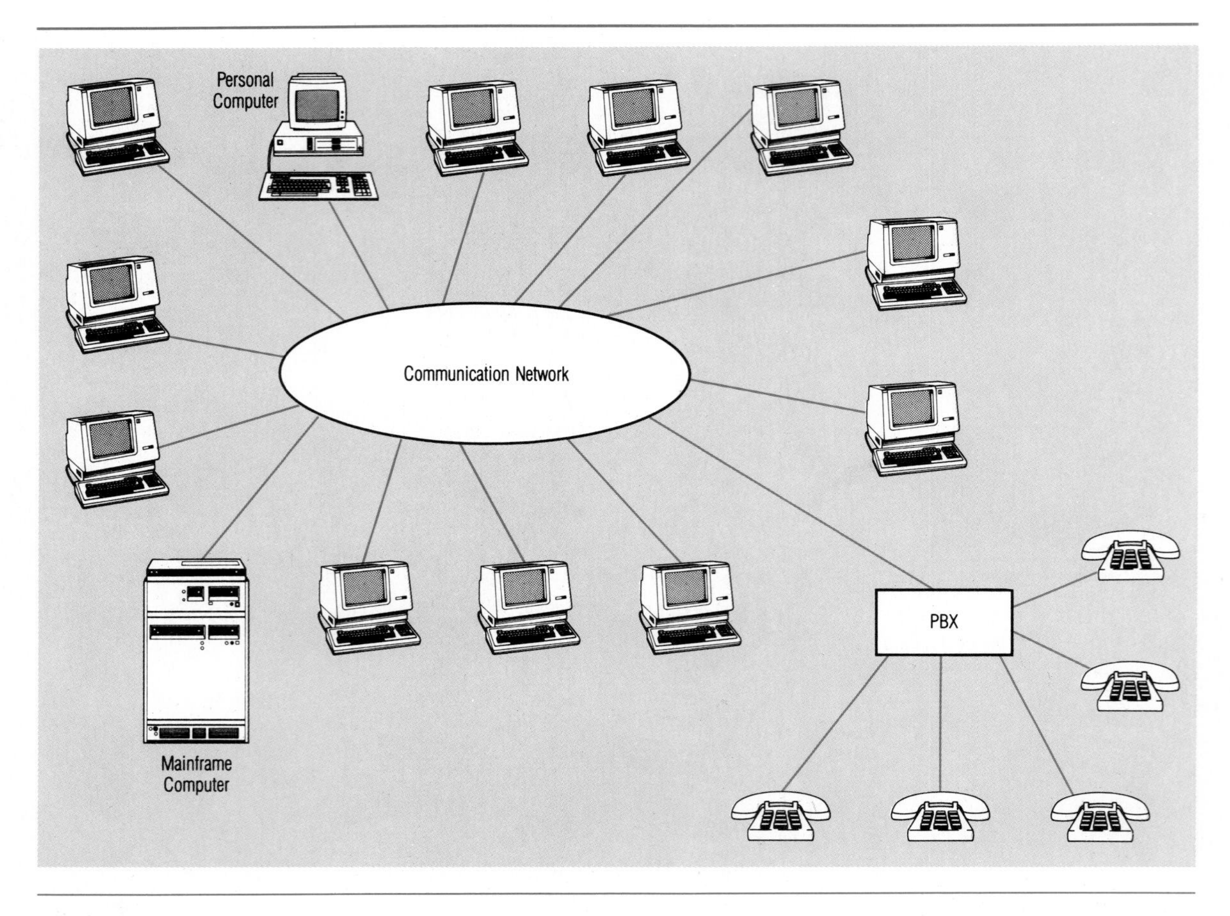

Figure 14–2
Communication network with terminals and computers attached.

communications department's responsibility, it is only part of the total communications management job. Planning and design of the communications facilities, budgeting and controlling costs, and the ongoing interaction with senior management and users are other facets of communications management that must be included to make the picture complete.

WHERE THE TELECOMMUNICATIONS ORGANIZATION FITS WITHIN THE COMPANY

voice

Historically, the voice and data communications facilities of a company have been managed in two different parts of the organization. Voice communications management usually has been handled by the administration function or, in some cases, the engineering department. Often the same person who handled the telephone system was also responsible for other office services, such as copy machines, printing, facilities design, office furniture, and company cars. With so many responsibilities, the telephone system got scant attention unless there was a problem. Traditionally, there has been a great reliance on the salesperson from the telephone company to understand the business well enough to specify the right solution for a company's voice communications problems. It is fair to say that in the predivestiture days when the telephone company was a monopoly, voice communications in most companies was not really managed, only administered.

data

Data communications has grown up within the data processing department. It is a much newer discipline because computer-controlled data communications capabilities have only existed for about 25 years. As contrasted with the relatively slow, steady growth of telephones, data communications has experienced explosive growth as the use of terminals has blossomed. The management of data communications is generally regarded as being more complex than voice communications management. Often the technical capability of data communications people has been higher than that of the voice communications people. Data communications management people have operated in a less regulated environment than their voice management counterparts because they have been able to acquire their terminals from unregulated companies. They have grown accustomed to working with both unregulated and regulated suppliers.

combining voice and data functions

The voice and data communications management in most companies have spent the last 25 years operating independently of one another. They have communicated with each other infrequently if at all, installed separate networks of lines, and, in some cases, offered competing capabilities or services to other departments in the company. Clearly, there is synergism to be achieved by combining the management of the voice and data communications facilities. Although many companies have not yet made this organizational change, there is a definite trend toward bringing the two together. The primary reasons cited by companies that have combined the two activities are efficiency, productivity, and cost advantage. With today's technology, communications circuits can be shared between voice and data, computerized PBXs require physical facilities like those that already exist in the mainframe computer room, redundant wiring in offices can be avoided, and the combining of staffs, equipment, and budget gives a communications manager more flexibility and opportunities for making trade-offs.

The reason that voice and data management have not been brought together in many companies is largely political. There is redundancy in the staffs of the two departments from the manager on down to the technician, and neither group wants to relinquish its responsibilities. However, as in any merger, there almost always is some pain and personnel dislocation when the two departments are brought together. The decision to merge the voice and data communications groups is a tough one, but it must be faced if the company is going to build an efficient, effective communications department and network for the future.

Another issue is where the combined communications organization will report. There are four basic choices:

- to administration;
- to data processing;
- to some other function;
- as a separate function reporting directly to a member of executive management.

Although there is no right answer, the trend throughout U.S. industry is clearly one of merging the voice communications group into the data communications group and having the combined organization report within the information systems department. There are several reasons why this is occurring:

- The information systems department has a high profile and a history of coping with technological issues and problems;
- It is easier for data communications people to learn voice communications than for voice communications people to learn about data communications;
- The physical facilities needed for a PBX already exist in the computer room;
- Communications is viewed as being too technical to be set up as a separate function except in companies whose entire product line or service is communications based.

On yet a broader basis, many companies are looking at a merger of activities in addition to communications and information systems. Office automation is another discipline that has become computer-based in the past several years. Office automation had its roots in word processing, which often got started in the marketing or administration department. Modern office automation began with memory typewriters in the early 1970s and then moved through a phase of dedicated office machines for word processing in the late 1970s and early 1980s. Now, word processing is being done on personal computers or at least computerlike devices, many of which can communicate with other computers or terminals that are a part of the network. Communicating word processors perform data communications, which tends to push them into the information systems department's domain.

Figure 14–3 shows how a company might have been organized when data communications, voice communications, and office automation or word processing were separate activities managed in separate parts of the organization. Figure 14–4 depicts how the organization would look after voice and data communications had been merged. Figure 14–5 depicts an organization in which the office automation activ-

information resource management

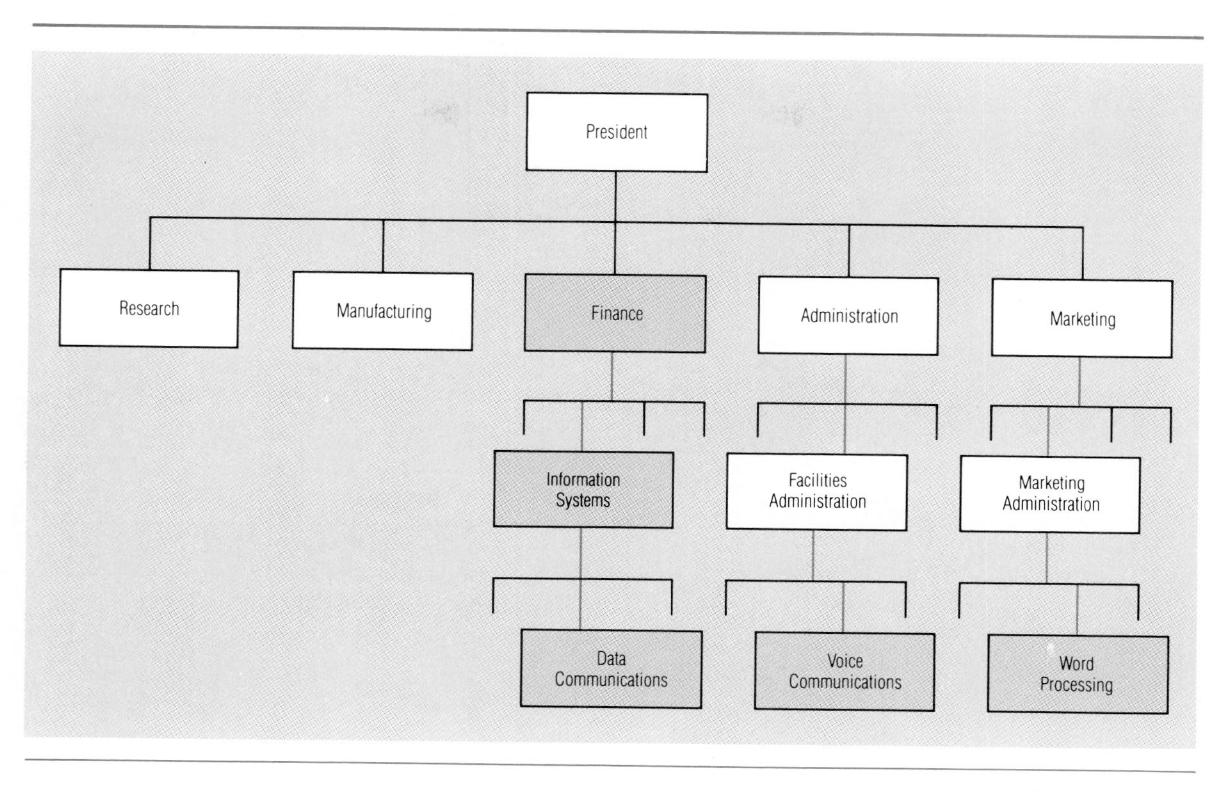

Figure 14–3
Traditional placement of communications and word processing departments within the organization.

ities have been combined into a department called *information resource management (IRM)*, which has responsibility for all data processing, communications, and office automation in the company.

In a vast majority of companies, the data processing department has historically reported within the finance function, typically to the vice president of finance or the controller. The new information resource management department may continue to report to finance, as shown in Figure 14–5, or it could be set up as a separate function itself, as shown in Figure 14–6. Setting IRM up as a separate function usually raises the IRM manager to a vice presidential level. In some companies, he or she is given the title of *chief information officer (CIO)*. The chief information officer oversees all of the company's information handling technology, including data processing, office automation, and telecommunications. This person concentrates on long-term strategy and planning, leaving day-to-day operations to subordinates. Usually he or she reports directly to a high-ranking executive, such as the chief executive officer or chairman of the board. The CIO often is assigned the complicated job of figuring out how to integrate a confusing array of often incompatible computer and communications equipment into a single integrated system. Another duty is explaining to managers how they

chief information officer

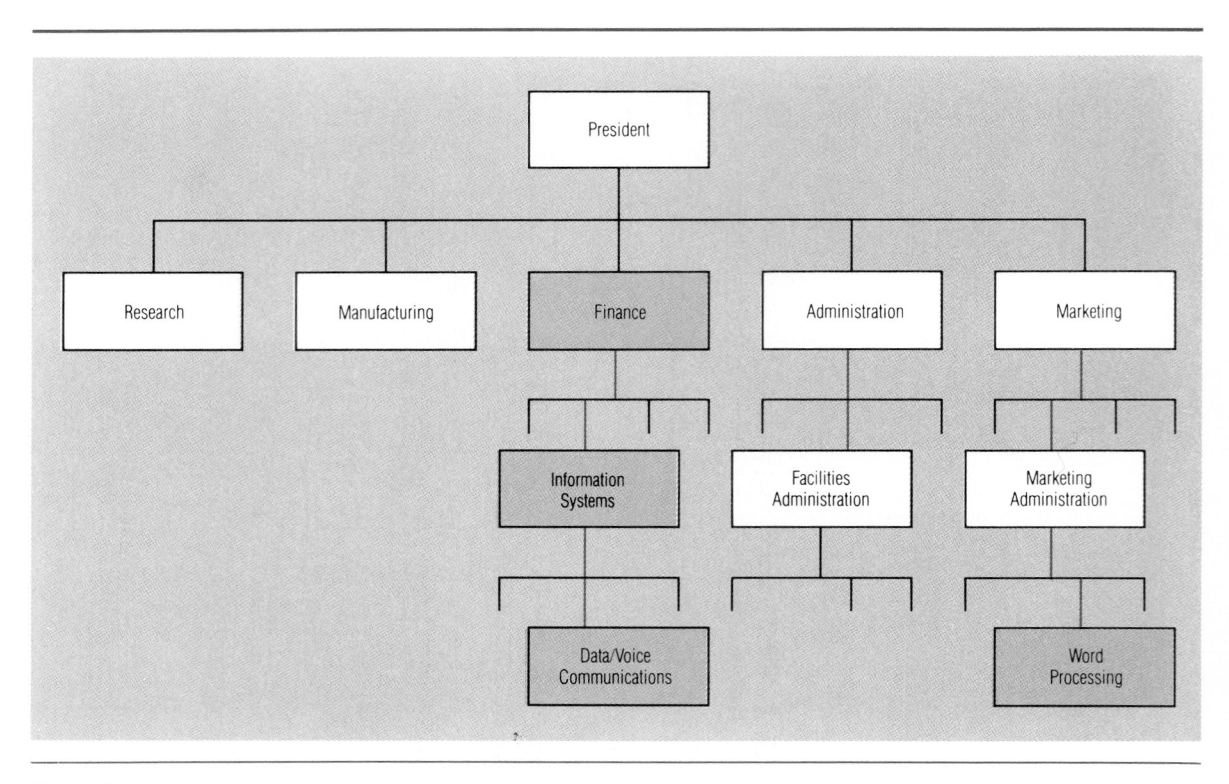

Figure 14–4
Organization after merger of data and voice communications.

can make the best use of technology while making the technical people understand what management wants to do.

Today, the IRM type of organization with a CIO exists in most companies in the information-intense industries of finance, banking, insurance, and transportation. Companies, such as Travelers Insurance, Bank of America, and United Airlines, have senior information officers with responsibilities like those just described.

Figure 14–7 shows an alternative organization that exists in a few companies where communications is of strategic importance to the company's success. Here, the telecommunications activity has been set up as a separate function, with the chief communications officer at the vice presidential level, whereas information systems reports at a lower level to finance or another function.

THE FUNCTIONS OF THE TELECOMMUNICATIONS DEPARTMENT

Regardless of where the telecommunications organization reports, it has a certain set of responsibilities and activities to perform. Parts of the telecommunications organization are analogous to the major functions

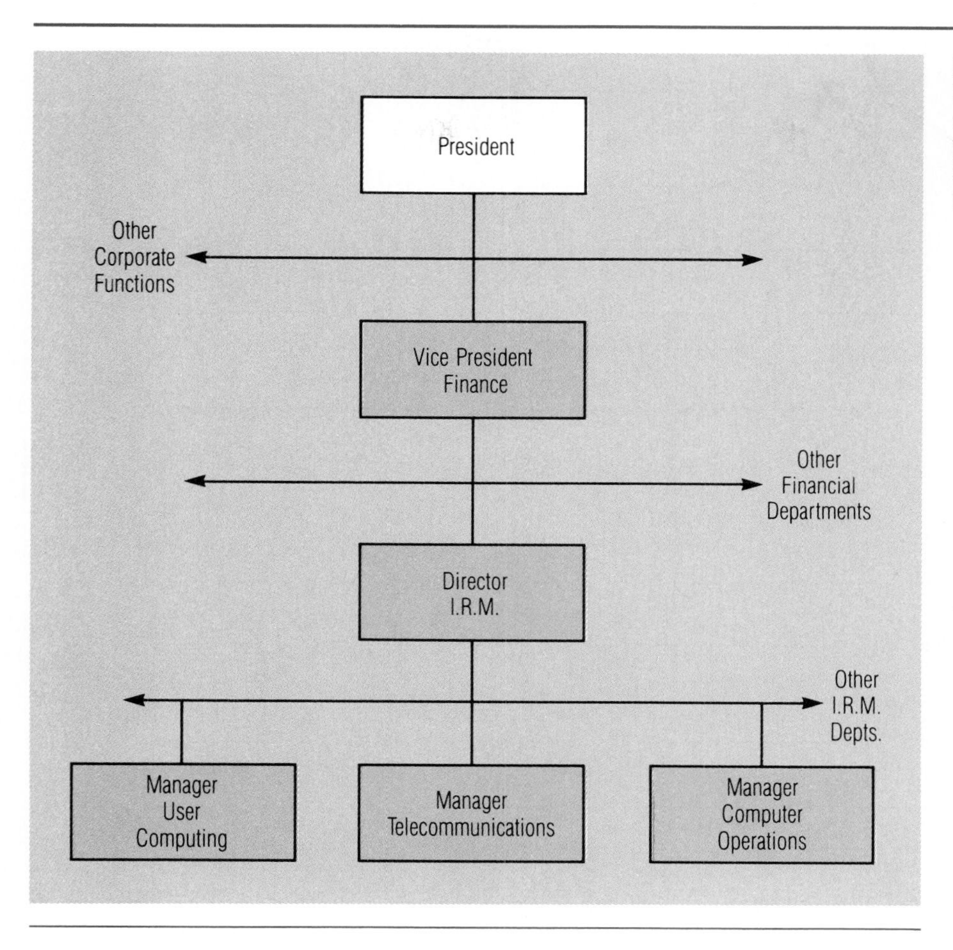

Figure 14–5
Information resources department with data processing, communications, and office automation responsibilities.

of the company, such as marketing, engineering, and production. The general principles of management that are taught in business management classes can and need to be applied to the management of the telecommunications department as well as to other departments in the company.

Following are descriptions of the basic business management functions that must be performed by the telecommunications department.

Design and Implementation of New Facilities and Services

All new communications facilities, including networks (both wide area and local area), building wiring, and equipment, and services must be designed and engineered to meet specifications. Facilities and services should be designed in response to a statement of requirements based on the user's needs. Installation activities range from simple repetitive installation of new telephones and computer terminals to infrequent but

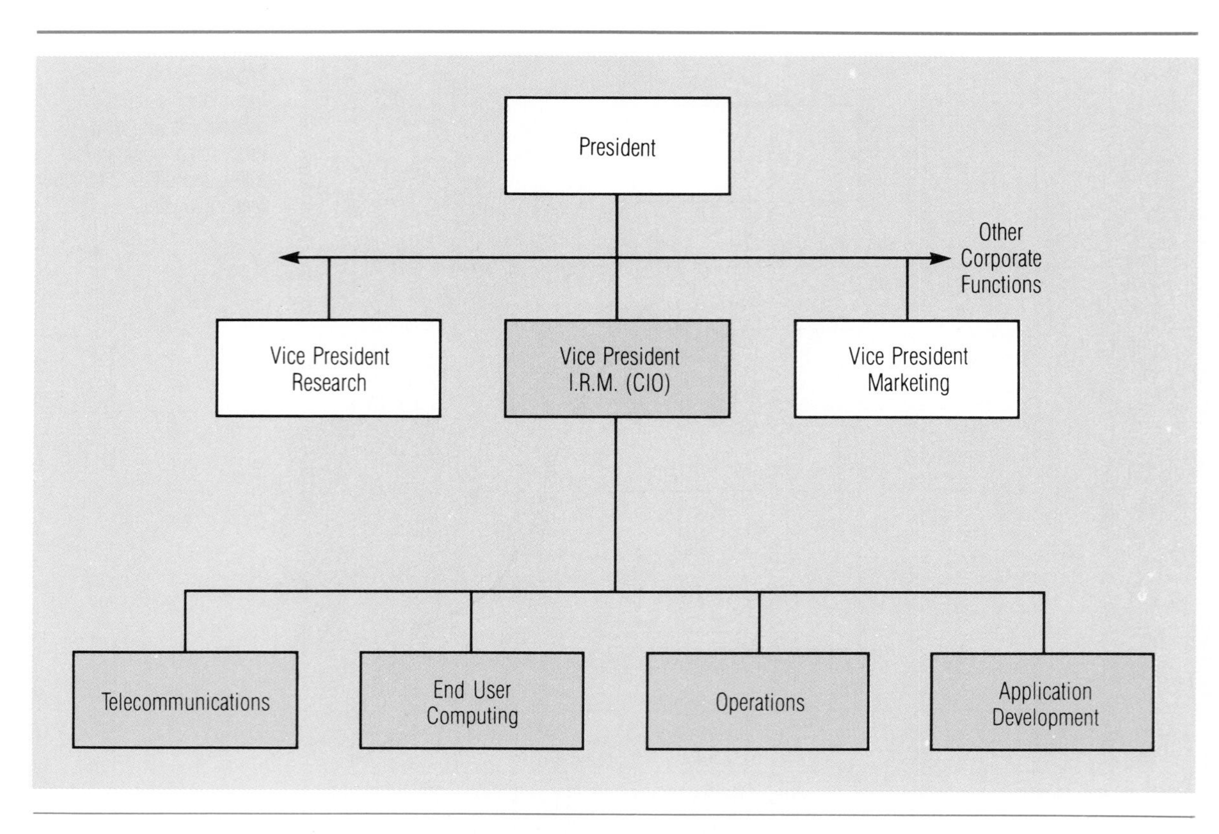

Figure 14–6
The IRM department as a separate function with the CIO at the vice presidential level.

complex tasks, such as wiring a new building or installing a new PBX, a front-end processor, or a new version of the software. Because most of the work in the design and engineering section is project oriented, project management skills and techniques are important. This activity is discussed in detail in Chapter 15.

Network Operations and Technical Support

The network operations group is responsible for monitoring the status of the communications network and services, performing first-level problem solving when trouble occurs and monitoring communications service levels. The central telephone operators may also be included in this group.

A help desk gives users a single point of contact for questions or problems relating to the use of telephone or data communications services. Another activity, telecommunications maintenance, is divided into two parts: routine maintenance, which includes activities such as moves and changes of telephones or terminals, and installations of new versions of software, and nonroutine maintenance, which is a response to trouble when something doesn't work. Problem determination, problem solving, and repair are all part of this activity.

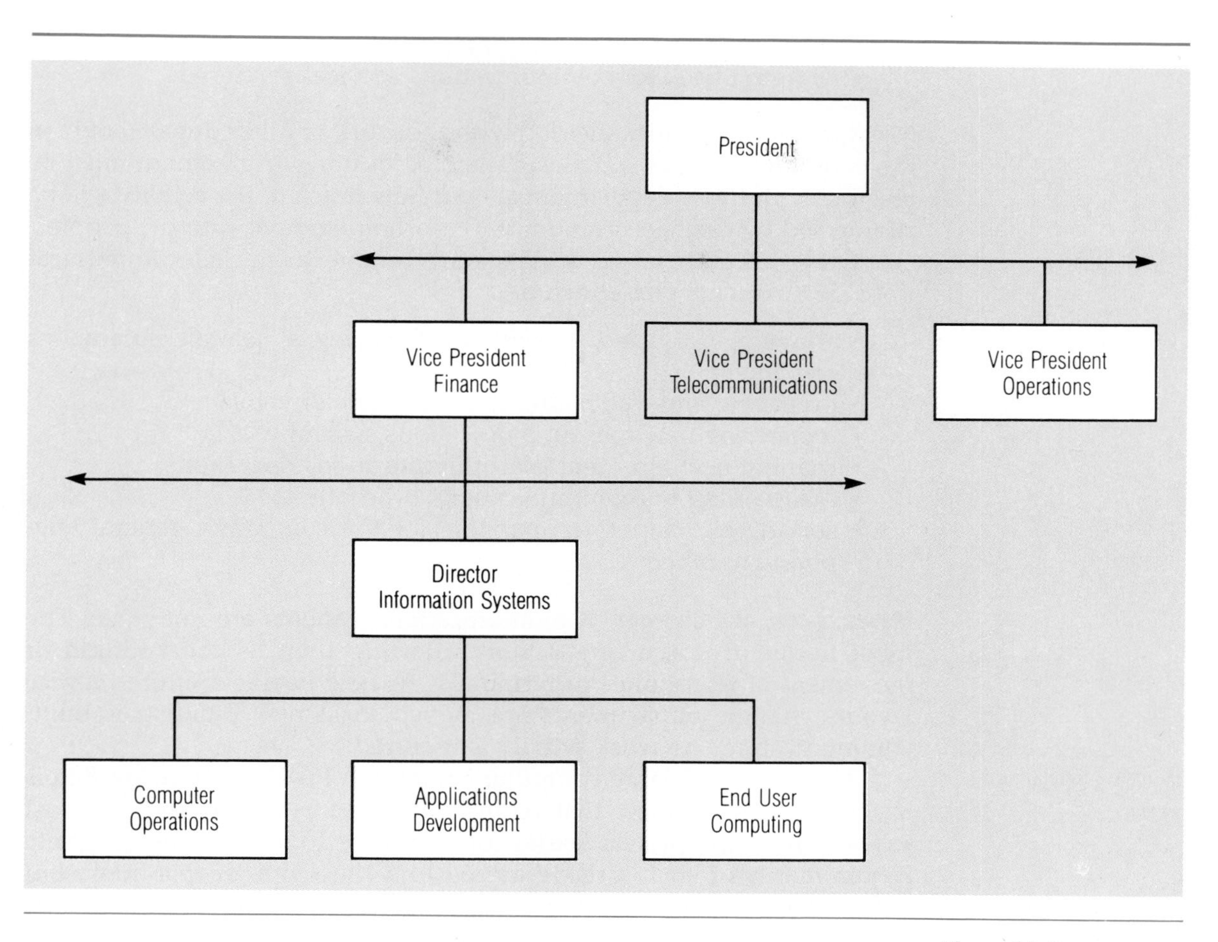

Figure 14–7
Alternative organization when telecommunications is of strategic importance to the company's success.

The technical support staff provides a high level of problem-solving capability for the most difficult problems. They also perform software maintenance and assist with other telecommunications activities.

Network operations and technical support are discussed in detail in Chapter 16.

Administrative Support

The administrative support group provides primarily clerical support to the telecommunications department. Familiarity with communications terminology and knowledge of at least the general concepts of the network and equipment are required, although detailed knowledge is not necessary.

The primary functions of the administrative support group are

- ordering/purchasing communications products and services;
- receiving and identifying products when they are delivered;
- inventorying all communications equipment;

- checking, approving, and paying communications bills;
- charging back for communications services rendered.

Many of these activities overlap with the work of other departments in the company, such as purchasing and accounting. Agreement must be reached with those departments about how much of the activity will be performed by the telecommunications administrative support people.

Other administrative activities are unique to the telecommunications department. These include

- arranging for adds, moves, and changes of telecommunications equipment;
- preparing and publishing the telephone directory;
- registering new telecommunications system users;
- training users in basic telecommunications operations;
- maintaining telecommunications procedures;
- serving as telephone operators for the company's general telephone number.

Moves, Adds, and Changes Communications systems are not static. The most frequently occurring activity affecting them is the addition or movement of telephones or terminals. As new people are hired, office layouts change, departments are moved, or a new building is built, communications network activity is required.

moving a telephone system

The move of a large computer center or a PBX telephone system is a very complex project that requires detailed planning and, in most cases, a full-time project leader for 6 months to a year. The complete requirements of such a move are beyond the scope of this book, but suffice it to say that numerous articles have been written on the subject, and consultants earn a living helping companies make a large move of this type.

Most moves, adds, and other changes, however, are relatively routine and can be coordinated by a person who is a good planner and communicator and who has some knowledge of the communications equipment. Despite the fact that moves can be handled routinely, they do require advanced notification so that the efforts of all of the parties involved in the move can be coordinated.

Preparing and Publishing the Telephone Directory Telephone directories in businesses range from simple typed lists to multipage books rivaling those of a small community. Although no general schedule can be defined that applies to all companies, most organizations find that they need to update their directories at least quarterly or semiannually.

In large companies, a mail number, a building number, or other location code usually is included in the directory. In organizations that use electronic mail, the mailbox name or user ID is also often included.

communications directory

The directory then becomes a *communications directory* rather than just a telephone book.

Registering Users Another administrative support function is the registration of new users to use the communications facilities. A part of the new employee orientation process may well be to assign the employee's telephone number and calling privileges. Registering people to use the data communications system and computers is another aspect of this responsibility. This often requires that the specific data transactions and files to which the employee will have access be identified and approved. Obviously there is a security implication to this process. Each company must develop procedures for authorizing an employee to access and use company data.

There may be a distinction between registering users to use the communications network and registering users to use a particular computer or computer application. It is most convenient if these two processes are combined into one, but for some organizations there is a clear distinction between them. In that case, the communications department registers users for the network, and the department that owns the computer uses a separate registration procedure to authorize access to it.

Training Users Sometimes the administrative support group has the responsibility for training people to use the communications system. Training can be simple, informal, and handled on a one-on-one basis, or it may be more highly structured and done in a classroom, depending on the size of the company and the complexity of the communications facilities. One good opportunity to do communications education is during new employee orientation. At that time, an introduction to the capabilities and features of the telephone system can be presented, and if appropriate, some basic terminal education can be given. With the proliferation of computer terminals and personal computers at all levels of our society, most people today have some familiarity with keyboards and VDTs, and therefore only a few comments about the unique characteristics of the company's terminals or computer systems may be required.

continuing education

Follow-up or refresher courses in communications are another aspect of the communications education process. Some companies find that having all employees attend a periodic class on telephone etiquette is desirable, and it also gives the communications staff an opportunity to discuss new capabilities or to remind people about other good communications practices, such as the use of WATS lines to help keep telephone costs down.

Maintaining Procedures The administrative support group may be responsible for writing and maintaining general procedures for the communications system operations. Procedures, such as how to logon, how to make a long distance telephone call, and how to report trouble, need to be provided to everyone who will use the communications system. It goes without saying that procedures of this type should be prepared on a word processor so that they can be easily maintained because, like the network itself, they will change with time and usually faster than expected.

Internal operating procedures for the various groups in the communications department are best written by those groups. For example, operational procedures for the help desk, such as the questions to ask users when they have trouble, how to record problems, and how to escalate problems, are best written and maintained by the help desk personnel and their supervisors. Procedures and project management techniques for implementing new communications circuits should be developed by network analysts.

Telephone Operators Telephone operators may or may not report to the communications department. Many times, they have other responsibilities, such as doing secretarial work or acting as a receptionist. When they perform the duties of a receptionist, they may report to an administrative or security function. In some companies, the telephone operators may be more customer service oriented, primarily handling calls from customers. If this is the case, they should report to the marketing or customer service department. Having the telephone operators reporting to the telecommunications department is not critical to its success.

In small companies, the telephone operator almost always does other work besides answering the telephone. In addition to the possible duties just listed, the operator may also send telex messages and keep the telephone list or book up to date. In fact, the job may be primarily secretarial, with only a secondary responsibility for general telephone answering.

In all cases, consideration must be given to backing up the telephone operator during breaks, illness, or vacations. Usually more than one person is trained to perform the functions of the job so that a pool of potential operators is available. As with any other personnel who are trained as backups, the backup telephone operators must be given a chance to use their training from time to time in order to refresh their memory and ensure that they are able to step in and perform the job when they are needed.

TELECOMMUNICATIONS MANAGEMENT RESPONSIBILITIES

Since telecommunications can be viewed as a business within a business, the basic responsibilities of general management can be applied to telecommunications management. These functions are

- staffing;
- organizing;
- planning;

- directing;
- controlling.

What is unique to telecommunications management is that only recently has the need to apply these management techniques to the telecommunications activity been recognized. Some very large organizations have done a good job of managing their telecommunications for years, but most companies have gotten along without paying much attention to the management skills and functions in the telecommunications organization. Technicians or clerical people with few management skills and little management training were promoted to supervisory positions and charged with the responsibility to run the department. Before deregulation and divestiture, this approach sometimes worked because many of the issues that had to be dealt with were technical. But the environment is more complicated now. Competition exists, and the telecommunications department has more responsibility for selecting among expensive alternatives, making trade-offs, and deciding to what extent the company will provide communications support itself as opposed to delegating the duties to suppliers or other outsiders. The department must have a leader with management and decision-making skills in order to select the best communications options for the company. We will look at the management activities in detail and see how they apply to the management of the telecommunications department.

Staffing

A management function that is particularly challenging for communications managers is staffing the department to effectively accomplish the work to be done. Staffing is especially challenging because for a number of years there has been a shortage of trained, experienced communications people. Only recently have colleges and universities introduced programs focused on educating people for careers in telecommunications. Historically, the three primary sources of communications people have been the telephone companies, the military, and internal people who learn telecommunications through on-the-job training.

sources of staff

Another complicating factor is the mix of skills required in the communications department. Building on the "company within a company" concept, it is apparent that planning, conceptualizing, operations, technical, and administrative skills are required by various members of the organization. When the department is small and has only a few people, it is difficult to obtain the right mix of skills. Larger departments have a better chance of obtaining people with the required skills or at least obtaining people with the aptitude to learn them. It may not be necessary for a telecommunications department to have all of the skills it needs. Outside consultants may be an effective supplement to the people within the telecommunications department, particularly when very specialized expertise is required.

qualities

The Telecommunications Manager The telecommunications manager must be a conceptualizer and visionary of how communications capabilities can be applied to solve business problems. He or she must be articulate and persuasive, able to understand company problems and opportunities, and able to envision how the application of communications technology might address these problems and opportunities. Ideally, the manager will be included in high-level planning discussions with senior management to keep informed about the factors that are of strategic importance to the company and to ensure that the mission of the telecommunications department is lined up in support of those strategies. The manager needs to have an ability to grasp technical subject matter, although he or she need not personally be a technician or former technician.

The telecommunications manager must be able to plan and implement projects at various scopes and levels. He or she must be a decision maker who can deal as effectively with today's operational problems as with a 3-year plan. A college degree in telecommunications, management, or engineering usually is required for such a position.

credentials

qualities

Designers and Implementers Telecommunications network designers and implementers must have good understanding of communications systems and product offerings. This knowledge usually is obtained through a combination of education and experience. Until recently there were no college courses one could take to learn how to do this work.

Good designers are creative and innovative and have strong analytical skills. Engineering training is very desirable. Since most design and implementation work is done in project form, a knowledge of project management techniques is required. Network designers must also have good verbal and written communication skills. They work with eventual users of the system, as well as with the vendors who provide the equipment or services. On large projects, it is feasible to take a strong project leader from another part of the company to manage the communications project if experienced people who have the specific communications systems knowledge are on the project team.

credentials

Generally, most large companies prefer that their network development people have a 4-year-college degree, although this is not mandatory. There is a feeling, however, that people with a 4-year degree are more flexible in their thinking and more likely to advance and move into other responsible positions within the company. Experience would, in general, support this notion.

Network Operations Staff Network operations is the manufacturing facility of the telecommunications department. Like many manufacturing people, network operations people often have more contact with the customers or users of the communications network than might first be expected. Therefore, they must be service oriented and strongly moti-

qualities

vated to keep the network operational and performing properly at all times. As in the previously described jobs, good verbal communications skills are required in order to deal effectively with the network users, many of whom experience problems with the network.

In addition to having good people skills, network operations people must thoroughly understand the network hardware and software that they are operating. There is a good deal of interchangeability between computer operations and network operations personnel. In most organizations, the ability to trade people back and forth is desirable, as it provides career path opportunities for people in both groups.

credentials

Normally, network operations people have at least a high school diploma. Some companies are requiring a 2-year or 4-year degree for these jobs.

qualities

Technical Support Staff Technical support personnel must understand the complexities of both software and hardware and the complex interactions between them. They are usually described as self-starting, analytical problem solvers.

credentials

A 4-year college degree normally is required, and it is almost always supplemented by advanced technical education conducted by the vendors of the products being used. The telecommunications technical support people and the data processing technical support people often have similar backgrounds and career paths. In many organizations, they are combined into a single technical support group.

qualities

Administrative Support Staff The administrative support required in the telecommunications department is primarily clerical in nature. Experience with accounting and purchasing activities is desirable. A professional accountant may be required to ensure that proper accounting procedures are followed. The administrative people need to be thorough and oriented toward handling a mass of detail.

credentials

Often, the administrative support people attend some specialized telecommunications courses to have a better feel for the products and services they are handling.

Staff Selection When considering telecommunications department staffing, several questions need to be asked:

- Should experienced communications people be hired from the outside, or should current employees be trained in telecommunications skills and technologies?
- What career paths are available for communications employees? Will they be able to transfer to other functions within the company?
- Should some, or all, of the telecommunications work be contracted to outside firms? For what expertise should consultants be considered?

Management must recognize that because the telecommunications industry is growing rapidly, the market for good, experienced communications people is tight and salaries are high. When trade-offs must be made, it is normally better to select management people based on their management skills and knowledge of the company. This may well bias the decision in favor of a person who is already an employee. For technical people, the trade-off should be made in favor of specific telecommunications knowledge, experience, and education. Knowledge of the company is less important, and therefore going outside to hire a skilled individual is feasible.

Use of Consultants Every department of every company occasionally needs advice or assistance in areas where it has little or no expertise. In telecommunications, this is particularly true because of the rapid technological changes. Telecommunications managers may find that they need help evaluating a new technology, such as fiberoptics, or the impact of a new regulation or tariff. Sometimes a vendor can help with this evaluation, but since the vendor has a product to sell, the information he or she provides may be biased. AT&T has good network evaluation and design tools that they make available to their customers, but obviously AT&T won't recommend a circuit provided by MCI or another carrier.

There are many telecommunications consultants in the field today. Some are independent, and some are associated with large consulting or public accounting firms. Before a consultant is engaged, it is important to determine the consultant's expertise in the particular field of interest. It is important to talk to other clients who have used the consultant to get their opinions about the consulting services provided. It is also important to find out exactly who will do the consulting work. The expertise and experience of the individual consultant is the primary determinant of the quality of the result and is ultimately more important than whether or not he or she happens to work for a "name" firm.

It is important to define the exact scope of the work to be performed in writing and to ask the consultant for a written proposal and cost estimate, as well as a list of all of the items or documents that will be delivered at the end of the consulting engagement. The proposal can be used as an indicator of how well the consultant understands the requirements of the job. Some consultants may suggest modifications to their proposal, either to expand the scope of the work to be done or to reduce the scope to hold costs down. In any case, it is important to reach an understanding of the scope before any work begins.

Organizing

Organizing is the grouping of people to accomplish the mission of the department. Obviously, telecommunications departments in large and in small companies are organized differently. In large companies, there are many specialists, whereas smaller organizations have fewer people, and they tend to be generalists. The structure of the department is also dependent on the sophistication and level of maturity of telecommunications in the company. As the company becomes more dependent on its telecommunications capability, the telecommunications department usually becomes larger and more structured in its operation. Different styles of organization are appropriate at different stages in the growth of telecommunications within the company.

One decision that management must make when organizing the telecommunications department is how much overlap there will be with other functions in the company. For example, the telecommunications department needs a purchasing procedure, and management must decide whether to set up its own purchasing group or use the company's general purchasing department. Another decision is to determine which telecommunications activities are going to be performed by employees and which will be contracted to outside services or performed by vendors. Usually, ongoing activities, such as communications operations, are performed by company employees, whereas occasional work, such as laying wires and cables, is contracted to outside specialists.

As in any department, there are many ways to organize the people to carry out their responsibilities. The grouping of activities discussed in this book is typical of what is occurring in industry today but not the only possibilities. Although each telecommunications department has certain basic activities to perform, there is always flexibility to tailor the organization to the talents of the individuals it comprises.

An organization chart of a typical telecommunications department is shown in Figure 14–8. This chart assumes that the voice- and data-related activities are handled by the same people.

Figure 14–8
Typical telecommunications department organization.

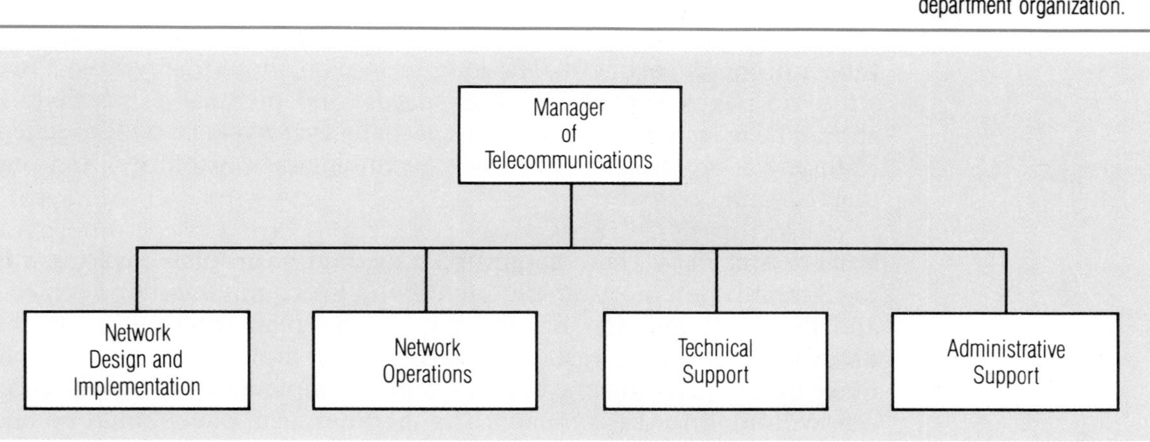

Planning

In any company, planning is required. The exact format and the frequency of the planning depends on the company and its culture. Some organizations have very formal, detailed, rigorously scheduled planning processes. Other organizations operate on a less formal basis, putting plans together only when required.

Before looking at specific types of plans, it is important to point out that the telecommunications plan cannot exist in a vacuum. It must relate to the strategic directions and plans of the company. In most cases, telecommunications projects are not undertaken for their own sake but are in support of some other activity within the company. For example, an expansion of the telephone system may be required to support a new telemarketing effort, or the data communications lines may need to be extended to support additional terminals for a new computer application. All telecommunications plans should relate to the needs of the company and should show the benefit that will accrue to the business. Management needs to understand that telecommunications capabilities can help break down constraints of geography and time on the product or the service that the company provides.

Long-Term Plan The long-term plan should look 3 to 5 years into the future and give a vision of what telecommunications facilities and services will be required by the company and implemented in that timeframe. Obviously, as a vision, it will never be 100 percent accurate, but it is important to put this type of plan together to give a sense of direction. The company may have a strong feeling, for example, that within 5 years it will begin operating in Europe. The 5-year plan may show the need to explore international communications options beginning in about 3 years in preparation for the anticipated expansion to Europe.

Often, the telecommunications manager puts this plan together. He or she has the most contact with senior management and others in the organization and may be more forward thinking and conceptual than others in the department. The long-term plan should be written, brief (10 to 15 pages at most), and as specific and pictorial as possible. It should be updated as appropriate (perhaps every year or two) to reflect changing company strategies, new technological capabilities, and regulatory implications.

Medium-Term Plan There should be a medium term plan covering a 6- to 18-month time horizon and identifying telecommunications projects and their priorities for that timeframe. This plan should identify the financial and people resources needed to complete the projects that make up the plan and the benefits that the implementation of the projects will bring to the company. The medium-term plan should be tied to the budgeting process to ensure that money will be available to implement the projects listed in the plan. It should show the starting and ending dates for each project but not the detailed tasks required for project implementation.

Like the long-term plan, the medium-term plan should be in writing. Each project listed in the plan should have information describing its scope, the reasons it is being done, its cost, benefits, starting and ending dates, and assigned people and equipment resources. One useful technique for medium-term planning is the *Gantt chart*. A Gantt chart, as shown in Figure 14–9, lists projects or activities on the left side and dates or other time periods across the top. A bar is drawn for each project showing when it starts and when it ends. The positions of the bars show how the projects or tasks overlap, and it is easy to see how they interrelate in time.

Gantt chart

Project Plan There is also a need for a specific task-oriented plan for each project showing the tasks, worker days, cost, and people resources required to complete it. This plan should set targeted completion dates and identify responsibilities for each task. A suggested format is shown in Figure 14–10. This task list may be supplemented by a detailed Gantt chart for the project. This detailed chart would list the project tasks in the left-hand column and would show how the tasks interrelate. The thought process that is required to put the task list and Gantt chart together is very worthwhile in itself because in the process of putting the chart together, problems, conflicts and previously unrecognized activities are uncovered. Detailed project plans should be updated as often as required. Depending on the size of the project, this may mean that monthly, weekly, or even daily updates are required. In any case, it is important to communicate changes in the plan to all of the individuals involved in its implementation, as well as to users and management.

Regular reviews with the affected users and key management personnel are desirable. Depending on the size of the project and the level of management, reviews may be held weekly or monthly. Like the discussions within the project team, these reviews help ensure that the project team is designing the telecommunications system correctly. The reviews also provide an opportunity to check progress and discuss any difficulties or unexpected circumstances that have arisen. The level of

Figure 14–9
A high-level Gantt chart of a medium-term telecommunications plan showing project start and end dates. A more detailed Gantt chart would show a finer breakdown of the tasks.

	1990 J	F	M	A	M	J	J	A	S	O	N	D	1991 J	F
N.Y. data circuit	1–15-	- - -	- - -	4–30										
Key system in Seattle					5–1 - -	- - 7–10								
Upgrade data to 9600 bps			3–1 -	- - -	- - -	- - -	- - -	- - -	9–30					
New terminals in plant					5–15-	- - -	- - -	- - -	- - -	- - -	11–30			
PBX in Atlanta								8–1 -	- - -	- - -	- - -	- - -	- - -	1–31

Install New York Multipoint Data Circuit

TASK	RESPONSIBILITY	TARGET DATE	COMMENT
Get exact addresses of the four locations	Jones	1–15	
Order the circuit	Jones	1–20	90-day lead time
Order the modems	Smith	1–20	
Contact each location about delivery date	Smith	3–1	Marketing has name of contact person
Modems delivered to each location	Vendor	4–1	
Circuit installation	Vendor	4–20	
Test circuit and correct problems	Jones/ Smith	4–25	With vendor help
Circuit operational	Jones	4–25	

Figure 14–10
A telecommunications project plan.

detail discussed in these reviews depends on the level of management that is participating. User line management usually is interested in hearing more of the details than does senior management, whose primary concerns are schedule and cost related.

Directing

Management's responsibility for directing is to ensure that the mission and plans of the telecommunications organization are executed on a timely and accurate basis. This is best achieved by first specifying the overall mission for the department and ensuring that it is understood by the members of the group. Furthermore, each subgroup within the department must know its responsibility for helping to achieve the overall objectives. Also, management must ensure that each individual is motivated and can see how his or her individual efforts help the department to meet its objectives and to be successful. To do this, specific objectives, goals, and standards of performance for each individual in the department are required. Objectives frequently are written annually, although there is significant benefit in reviewing and updating them more frequently. Other common techniques used in directing the department are requiring each staff member to write a monthly report and holding project review meetings regularly. The purpose of both of these techniques is to improve communications within the department, as well as to allow individuals to gain recognition for the results they are achieving.

Words that often are used to describe a successful telecommunications organization are *proactive, results oriented,* and *service oriented*. If these words are used by others in the company, it is a good sign that the department is being successful.

Controlling

Unfortunately, the word *controlling* often has a negative connotation, but it simply means ensuring that performance is taking place according to the plans. Control activities are very important to any management process. Even the best made plans are subject to change. Part of managing is ensuring that changes in plans are made consciously and with a full understanding of the implications. In the telecommunications department, two major controls are financial controls and the controls applied to ensure the quality of the service being rendered.

Financial Controls Financial controls take several forms. First, a financial plan or budget for telecommunications expenses is put together. Then, performance against the budget is measured. In some companies, telecommunications expenses are charged back to the users when services are rendered. If this is the case, chargeback rates must be established as a part of the budgeting process, based on the expenses that are expected to be incurred. By setting the chargeback rates higher than the expected expenses, the telecommunications department becomes a *profit center* generating more income than it spends. If the charge-out rates are lower than expenses or if no charge-outs are used, the telecommunications department is a *cost center.*

Having appropriate financial controls and overall financial management practices is important because the level of telecommunications expenditures continues to increase and become a more important part of a company's overall expenses. The Fortune 1000 companies are projected to spend about 3.8 percent of their operating budgets on communications in 1993, a 52 percent increase since 1983.

Expense Budgeting Expense budgets usually are put together for the fiscal year used by the company; this may or may not correspond to the calendar year. First, the expenses for all existing people and communications facilities must be identified. Then all expected changes, such as the addition of new people, circuits, or terminals, must be identified and their cost added to the budget. Typically, the budget is broken into expense categories, as shown in Figure 14–11 and described in the following list:

- Salaries—the total amount of money, including overtime, to be paid to employees in the department during the year;
- Employee benefits—the amount of money to be paid for insurance, retirement, dental, and other benefit programs. It is often calculated as a percentage of the salaries and wages amount;
- Rental/lease—the amount of money to be paid to vendors periodically for the use of equipment that has not been purchased;
- Maintenance—this amount of money is to be paid for the repair of equipment or for service contracts to cover the repair;

TELECOMMUNICATIONS BUDGET	JAN	FEB	MAR	APR	MAY	JUN	JUL	AUG	SEP	OCT	NOV	DEC	TOTAL
SALARIES	9600	9600	9600	9700	9700	9800	9800	9800	9800	9800	9900	9900	117000
BENEFITS	2688	2688	2688	2716	2716	2744	2744	2744	2744	2744	2772	2772	32760
RENTAL/LEASE	12150	12300	12300	12350	12500	12500	12700	12900	12900	12900	12900	13500	151900
MAINTENANCE	5250	5250	5400	5400	5400	5400	5700	5700	5700	5800	5800	5900	66700
DEPRECIATION	2200	2250	2300	2350	2400	2400	2600	2600	2650	2700	2700	2700	29850
SUPPLIES	150	150	150	150	150	150	150	150	150	150	150	150	1800
EDUCATION	400	400	400	400	400	400	400	400	400	400	400	400	4800
TRAVEL	500	500	900	600	600	400	700	200	1500	1000	400	0	7300
UTILITIES	350	350	350	350	350	350	425	425	425	425	425	425	4650
OVERHEAD (BURDEN)	700	700	700	700	700	700	700	700	700	700	700	700	8400
MONTHLY TOTAL	**33988**	**34188**	**34788**	**34716**	**34916**	**34844**	**35919**	**35619**	**36969**	**36619**	**36147**	**36447**	**425160**

Figure 14–11
The telecommunications department expense budget.

- Depreciation—money budgeted to cover the depreciation expense for equipment that has been purchased;
- Supplies—money in this category is for such items as ribbon and paper for printers, subscriptions to periodicals, general or special office supplies, and so forth;
- Education—money budgeted for vendor classes, general seminars, and other educational activities;
- Travel—money for the expenses of taking a trip, such as airline tickets, lodging, and meals;
- Utilities—money to pay the heating, air conditioning, and telephone bills;
- Building/corporate overhead (sometimes called "burden")—many companies assess a flat fee to each department for use of the office space and other corporate facilities. This is sometimes expressed as an overhead rate, which may be based on the number of square feet occupied, the number of people in the department, or the size of the rest of the budget.

The telecommunications manager may budget for literally hundreds of items. Sometimes it is desirable to break individual budget categories into finer detail. Figure 14–12 shows the items that make up the rent/lease item on a budget. Each of the items and its cost are listed. Items that will be added during the year show a zero cost until the month they are added. Totals are shown for each month and for each item.

A spreadsheet program operating on a microcomputer is an invaluable tool to use during the budgeting process. As additions or other changes are made to the budget, items can be added or numbers corrected quickly with the spreadsheet program recalculating the monthly and yearly totals instantly. Breaking the budget apart by months is desirable in order to easily account for items that are added or deleted in the middle of the year, and it also serves as a monthly milestone for checking actual expenses against the budget as the year progresses.

Capital Budgeting In some companies, it also is necessary to prepare a capital budget. The capital budget is a plan for money to be spent on items that are purchased and become assets of the firm. Whereas conservative accounting practice dictates that as many items as possible are charged to current expenses, the Internal Revenue Service guidelines state that certain items with a long life must be capitalized (become assets) and their expense spread over the asset's life. This expense is recovered in the form of depreciation. When purchased (as opposed to rented or leased), most telecommunications equipment must be capitalized and depreciated.

A capital budget is a list of the items to be purchased, their cost, expected useful life, and anticipated month of acquisition. The company may have special capital expenditure authorization procedures beyond the budgeting process so that specific approval for each major capital

RENT/LEASE BUDGET	JAN	FEB	MAR	APR	MAY	JUN	JUL	AUG	SEP	OCT	NOV	DEC	TOTAL
COMMUNIC. CONTROLLER	875	875	875	875	875	875	875	875	875	875	875	875	10500
CLUSTER CONTROLLER 1	295	295	295	295	295	295	295	295	295	295	295	295	3540
CLUSTER CONTROLLER 2	315	315	315	315	315	315	315	315	315	315	315	315	3780
CLUSTER CONTROLLER 3	0	0	0	280	280	280	280	280	280	280	280	280	2520
WEST CIRCUIT	1850	1850	1850	1850	1850	1850	1850	1850	1850	1850	1850	1850	22200
PLANT CIRCUIT A	470	470	470	470	470	470	470	470	470	470	470	470	5640
MODEM-WEST 1	165	165	165	165	165	165	165	165	165	165	165	165	1980
MODEM-WEST 2	165	165	165	165	165	165	165	165	165	165	165	165	1980
MODEM-PLANT A1	225	225	225	225	225	225	225	225	225	225	225	225	2700
MODEM-PLANT A2	225	225	225	225	225	225	225	225	225	225	225	225	2700
TERMINAL 1	47	47	47	47	47	47	47	47	47	47	47	47	564
TERMINAL 2	68	68	68	68	68	68	68	68	68	68	68	68	816
TERMINAL 3	110	110	110	110	110	110	110	110	110	110	110	110	1320
TERMINAL 4	0	0	68	68	68	68	68	68	68	68	68	68	680
TERMINAL 5	0	0	0	0	0	95	95	95	95	95	95	95	665
MONTHLY TOTAL	4810	4810	4878	5158	5158	5253	5253	5253	5253	5253	5253	5253	61585

Figure 14–12
Detailed budget for the Rent/Lease expenses category. Notice that some items are planned to be installed during the year; their budgeted amounts are 0 in the first few months.

purchase is obtained even though the budget had been previously agreed to.

verifying invoices

Cost Control During the month, invoices for installed equipment or new purchases come in from vendors. These invoices must be verified to ensure that they are correct before they are submitted to the accounting department for payment. As the bills are paid, they are charged against the telecommunications department budget.

checking expenditures

Periodically, it is necessary to check the total expenditures and compare them to the budget to ensure that costs are under control. Normally, reports are provided each month by the company's accounting department in a format similar to that in Figure 14–13. The left side of the report shows the actual expenditures for the current month, the budget for the month, and the variance from budget. The right side shows the same information on a year-to-date basis. With this type of report and the appropriate detailed information to back it up, the telecommunications manager can keep close track of the money his or her department is spending and, if necessary, take appropriate steps to keep the expenses in line with the budget.

Chargeback It is very common for some or all of the telecommunications expenses to be directly charged back to the users with some sort of a recharge system. It is desirable to have all of the invoices for telecommunications equipment and services pass through the communications department, where they can be checked and approved by knowledgeable people. However, there is also benefit in passing the costs on to the users of the communications system so that individuals realize that the service is not free. In most organizations, each department pays for its telephone equipment and long distance telephone charges through the internal chargeback system. It is also common for users of data terminals to purchase or otherwise pay for the terminal itself, and in many companies an internal bill for the use of the data network and other communications equipment, such as cluster controllers or front-end processors, also is generated. Many people believe that the only way to ensure that the costs of communications facilities are kept under control is to be sure that each user is paying his or her own way. If there is a perception that communications services are free, they are more likely to be abused.

Quality Control Telecommunications departments should have standards of performance for the services they provide. The standards should be agreed to with the users and should meet their requirements. Typical standards set a level of expectation regarding network availability, response time, ability to obtain long distance lines, and so forth. Responsibilities must be assigned for monitoring and reporting performance against the standards and for making changes in the standards to address users' changing requirements. Typically, network operations is responsible for this activity. This subject is covered in more detail in Chapter 16.

TELECOMMUNICATIONS EXPENSES
JULY 1990

CURRENT MONTH				YEAR TO DATE		
DETAIL ACTUAL EXPENSES	BUDGET	VARIANCE	TYPE OF EXPENSE	DETAIL ACTUAL EXPENSES	BUDGET	VARIANCE
9626.49	9800	174	SALARIES AND WAGES	66841.99	67400	558
2695.42	2744	49	BENEFITS	18715.76	18872	156
190.85	150	−41	SUPPLIES	1632.00	1050	−582
5219.38	5700	481	MAINTENANCE	36266.32	39900	3634
1119.21	300	−819	OUTSIDE SERVICES	'1119.21	2100	981
575.76	700	124	TRAVEL	3885.64	4900	1014
382.19	360	−22	TELEPHONE	2674.49	2520	−154
12519.38	12700	181	RENTAL/LEASE	82184.58	84300	2115
830.00	830	0	OVERHEAD	5810.00	5810	0
3019.00	2600	−419	DEPRECIATION	20874.00	19130	−1744
36177.68	35884	−292	TOTAL EXPENSES	240003.99	245982	5978
		−0.81%	VARIANCE AS A % OF BUDGET			2.43%

Figure 14–13
A monthly telecommunications department expense report.

The Telecommunications Audit Another type of control is the periodic audit of the telecommunications activity by internal or external auditors. The purpose of any audit is to review the activities of a department to ensure that weaknesses in procedures or controls do not exist. Audits normally are conducted by individuals not involved in telecommunications and need not be limited to financial matters. Audits also can be used very effectively in the operations area to review existing methods and procedures for possible improvement. Typically, internal auditors report to a high-level executive in the company in order to maintain their independence from any department. Therefore, their reports usually require a formal response by the management of the department that has been audited. A partial list of the areas that are examined in a telecommunications audit are shown in Figure 14–14.

Properly used, the telecommunications audit can be a very effective tool for checking the performance and upgrading the telecommunications activity. From senior management's standpoint, the audit provides an independent perspective of the operations of the department. From the telecommunications manager's perspective, the audit may provide valuable insights. If problems surface, it may give the manager some additional clout to get the problems corrected.

SECURITY

In addition to the traditional management responsibilities, telecommunications management has a very important security responsibility. Organizations have become dependent on information stored in computers and the data transmission facilities to access this information. No longer can they operate the business without having the information available. At the same time, terminals and data communications systems have made access to computers much easier. As information is increasingly viewed as a business asset, it must be treated like other assets and surrounded with proper controls and appropriate security to protect it. In order for a company to adequately manage information security either on the computer or on the network, it must have

- a network security strategy that clearly defines the reasons why security is important to the company;
- a security implementation plan that describes the steps to implement the strategy;
- clearly defined roles and responsibilities to ensure that all aspects of security are performed;
- an effective management review process to periodically ensure that the security policies and standards are adequate, effective, and being enforced.

Figure 14–14
Checklist for a telecommunications audit.

Administrative:
- ☐ Procedures in place and followed
- ☐ Separation of duties and responsibilities
- ☐ Comparison purchasing is done
- ☐ Purchases cannot be made without the approval of authorized persons

Personnel:
- ☐ Job descriptions exist and are up-to-date
- ☐ People are properly trained for their jobs
- ☐ Critical skills are backed up

Network Operations:
- ☐ Standards of performance exist and are up-to-date
- ☐ Service level agreements exist and are up-to-date
- ☐ Physical security is adequate
- ☐ A disaster backup plan exists and is tested periodically
- ☐ Software controls to prevent unauthorized changes or other tampering are in place

Each of these elements must be present. The strategy is management's statement of importance and commitment. The plan describes exactly what practices will be in effect. Usually, an information security officer is appointed and is responsible for carrying out the security plan. The security officer also investigates violations of the security policy and makes recommendations about additions to the security plan or changes to be made. The management review process is a periodic check that the security program is operating properly. The initial step in the review process may be a security audit performed by the company's inside or outside auditors. The audit report serves as the basis for the management review.

Communications security takes three primary forms:

- physical security of the telecommunications facilities, including the network control center, equipment rooms, and wiring closets;
- access control to prevent unauthorized use of the telecommunications terminal circuits, telephone systems, or computers to which they are attached;
- personnel security, such as security checks on prospective employees, training, and error prevention techniques.

Physical Security

The primary emphasis of physical security is to prevent unauthorized access to the communications room, network control center, or communications equipment that could result in malicious vandalism or more subtle tapping of communications circuits. It may be necessary or desirable to also inspect the facilities of the common carrier through which the communications circuits pass. Physical security can be thought of as the lock and key part of security. The equipment rooms that house

security measures

telecommunications equipment should be kept locked. Terminals should be equipped with locks that deactivate the screen, keyboard, or on/off switch. In lieu of actual locks and keys, magnetically encoded cards that resemble credit cards can be used to activate terminals or to unlock doors. These cards could also serve as a personnel identification badge.

One interesting facet of communications network physical security is the rising crime rate against ATM users. There have been an increasing number of robberies shortly after people have withdrawn money from an ATM. One significant legal question is how far banks must go to protect ATM users. Banks maintain that as long as they locate the machines in well-lighted areas or on major streets, they aren't liable when crimes occur. They also say that consumers should bear part of the risk when they use the machines after dark. Consumer advocates suggest that banks should have guards posted by the machines. This situation exemplifies the type of societal problems that will continue as communications systems become more widely used by the general public.

Network Access Control

Telecommunications access adds another dimension to the security concern. With telecommunications, data can be accessed from terminals outside the computer room and a new set of questions emerges:

- How do we know who is at the terminal?
- Once we know who is at the terminal, is the person authorized to access the computer?
- What operations is the terminal user authorized to perform?
- Is it possible that the telecommunications lines could be tapped?

Access control techniques include having a unique identification code for each user and a secret password to protect against three situations: unauthorized users, authorized people performing unauthorized activities, and authorized people using unauthorized terminals.

Users should be required to logon to the network each time they use a terminal by entering their unique identification codes and passwords. This information, along with the date, time, and terminal identification, should be automatically recorded by the software in a central file on a computer so that a complete record is kept of all users of the system. When someone tries to logon and enters an incorrect password a predetermined number of times, the user's ability to logon should be disabled, recorded in a security log, and a security officer notified. Reactivation of the user identification code should be done only when it has been determined that the user really is authorized. Usually, supervision is required to reauthorize the user, and a new password is issued. Similar control techniques should be put in place to limit the access to long distance voice facilities, such as WATS lines.

Dial-up data lines are especially vulnerable to unauthorized access. The most common security techniques used with dial-up data circuits are *call back* and *handshake*. With the call back technique, the user dials the computer and identifies himself or herself. The computer breaks the connection and then dials the user back at a predetermined number that it obtains from a table stored in memory. The disadvantage of this technique is that it does not work very well for salespeople or other people who are traveling and calling from various locations. The handshake technique requires a terminal with special hardware circuitry. The computer sends a special control sequence to the terminal, and the terminal identifies itself to the computer. This technique ensures that only authorized terminals access the computer—it does not regulate the users of those terminals.

In addition to these techniques, it is important that network operations management monitor the use of dial-up computer ports for suspicious or unusual activity. Usually, this is done by reading a printed log of all dial-in accesses or attempts. In particularly sensitive applications, such as the transmission of financial transactions or data about new product developments, encryption can be employed to scramble the data so that if unauthorized access to the computer or data is obtained, the data is still unreadable without further work to decrypt it.

Personnel Security

Personnel security involves using one or more of the following techniques:

- security checking or screening of prospective new employees before they are hired;
- the identification of employees and vendor personnel through badges or identification cards;
- having active security awareness programs that constantly remind employees about their security responsibilities and the company's concern for security;
- ensuring that employees are properly trained in their job responsibilities;
- having error prevention techniques in place to detect accidental mistakes, such as keying an erroneous amount.

All of these techniques are important because they provide an environment in which employees and others know that security is important to management.

OTHER MANAGEMENT ISSUES

Selling the Capabilities of the Telecommunications Department

The telecommunications manager needs to work constantly to keep communications-based opportunities and capabilities in front of senior company management and to look for opportunities for telecommunications to contribute to or improve the company's products, profits, or other capabilities. In some companies, telecommunications is still viewed as administrative overhead. The focus is on cost control, the emphasis being that less is better. Telecommunications department objectives are mainly related to saving money.

In many companies, the potential value of telecommunications is starting to be recognized. In these companies, management focuses first on the service required and then on the cost. Management in these companies is receptive to new ideas. Some ideas may, in fact, increase costs, but benefits can be achieved to offset the increased level of expenditure. The payoff of a new telecommunications project may occur over time, just as it does in other investments the company makes.

In a few companies, telecommunications is of such strategic importance that the companies could not survive without their communications facilities. In these companies, telecommunications is a money maker, not a cost center. To some extent, the more money spent on telecommunications, the more profit that is made by the firm. Financial institutions, transportation companies, and insurance companies are examples. In these companies, the telecommunications manager's job is one of prioritizing all of the possible projects that "could" be done rather than selling the idea that some telecommunications work "should" be done.

Project Justification Criteria

The criteria used for justifying telecommunications projects are closely related to management's perception of the strategic importance of telecommunications to the company's success. If telecommunications is viewed to be a major contributor to the company's success, it is a lot easier to get new projects approved. If the telecommunications project can be characterized as being of strategic importance to the company, such as helping to beat the competition or as an investment in the future, it will be easier to gain approval for its implementation.

In all cases, however, projects must go through some sort of justification process. This usually includes a look at the alternatives and a comparison of the anticipated costs of doing the project versus the benefits to be achieved. In most cases, the costs of a project are relatively easy to obtain and quantify. Benefits, on the other hand, often are less tangible and more difficult to quantify. There needs to be a continual effort made to attempt to quantify the benefits and put dollar values on them. On the other hand, the project sponsors should not be afraid to state what they believe the intangible or nonquantifiable benefits will be. Many times, the ultimate decision to do a project is based on a perception or belief that particular benefits will accrue to the company even though their value cannot be specified accurately.

Transnational Data Flow

Organizations involved in international operations need to be aware of the changing guidelines and laws affecting the transmission of data across international boundaries. As was discussed in Chapter 4, companies planning to transmit data across international boundaries need to understand and monitor the changing laws and guidelines that govern such transmissions. Depending on the degree of international communications, the telecommunications manager or one of his or her staff may have the responsibility to monitor these issues. Several consulting services and magazines are available to help a company keep up with the changes in this area.

Disaster Recovery

As the network becomes a vital part of a company's operation, plans must be made for recovery steps to be taken if a natural or other disaster destroys the network control center or part of the network. Organizations that have centralized computing capability must be concerned also about disaster planning for the computer center. The factors to be considered when planning for the recovery of a company's communications facilities are discussed in Chapter 2.

It has become evident in recent years that another type of disaster must be planned for—a disaster at a telephone company office that serves the company. In May 1988, a large fire struck a telephone company central office in Hinsdale, Illinois, leaving more than 35,000 customers, including many businesses, without telephone or data communications service for more than a week. Some companies that were dependent on telephone service actually went out of business. Others were very inconvenienced. In August 1989, workers of NYNEX were on strike for several weeks. Customers were not able to get new services installed, and service calls took far longer than normal. Clearly, these types of situations will occur again in the future. Companies must protect themselves by having adequate plans for emergency communications service if a disaster strikes their communications company. Management's responsibility is to ensure that the proper disaster recovery planning is done and that the telecommunications capabilities can be restored.

SUMMARY

Rapidly changing technology and regulatory environments combined with the growing use of telecommunications within business and government are forcing telecommunications managers to become more sophisticated in the way they run their organizations. In many ways, the telecommunications department can be viewed as a "business within a business," and traditional management techniques can be applied. The functions of the telecommunications department are analogous to the functions of the company itself. Telecommunications professionals must have skills in research development, product management, finance, marketing, administration, and performance measurement.

CASE STUDY

Dow Corning's Telecommunications Management

Telecommunications management at Dow Corning has been an evolutionary process. Data communications has always been managed in the Systems and Information Management (SIM) department. Until recently, data communications work was handled by the technical support or computer operations groups. Until 1982, voice communications were managed by the office services department, which also handled copying, printing, and other administrative functions. In 1982, the responsibility for planning the voice communications system was moved to SIM and merged with the data communications responsibility in the technical support group. The technical support manager had voice communications planning added to his list of responsibilities. In 1985, a new telecommunications department within SIM was formed. It was given the responsibility for planning, designing, and operating both voice and data communications networks.

High on management's list of concerns has been security. Internal users have had unique user identification codes and passwords for many years. In the last several years, there has been an increasing emphasis placed on making users aware of their security responsibilities. Most of the focus has been on ensuring that company data and computer applications are properly protected and on ensuring that users are properly authorized. The company maintains logs of all terminal and computer usage. Less emphasis has been placed on protecting the actual data transmissions themselves, but techniques for encrypting data transmissions are being studied.

A related concern and area of concentration has been disaster recovery planning. In 1986, Dow Corning signed a contract with Comdisco for disaster backup services. Testing of the disaster recovery plan is done regularly and focuses on ensuring that the company's computer and applications software can be restored to operation at the disaster backup site and that data communications can be established between the disaster site and all other Dow Corning locations. Testing of the backup capabilities is viewed as an ongoing process that occurs several times each year.

Telecommunications department planning is done at several levels. A 3- to 5-year vision is developed by the manager of telecommunications and his staff. This vision is updated approximately every year or when business conditions or technology change enough to warrant an update.

A 6- to 12-month project plan is put together late in the year to coincide with the company's budgeting process. Budgets are established for the fiscal year, which coincides with the calendar year. The list of planned telecommunications projects changes rapidly and must be updated about every 6 months. As specific projects are initiated, the project leader is responsible for developing a project plan that details the tasks to be performed and the timetable for completing them.

All vendor bills for telecommunications services are checked and approved by the telecommunications department. The costs for terminals, telephones, and long distance telephone charges are passed on directly to the users each month. SIM pays for the telecommunications network, including lines, modems, and cluster controllers. At the end of each year, all of SIM's costs are allocated to user organizations at a functional level. That is, a lump sum is charged to marketing, manufacturing, research, and so on. Ultimately, the user organizations pay the entire cost of the communications network.

Dow Corning has four people performing telecommunications administrative support activities. One person is responsible for working out the details of the data network moves, adds, and changes. He works closely with the users to ensure that when their terminal arrives, a port is available on the cluster controller where the terminal can be plugged in. He also coordinates with the Dow Corning people who arrange offices to ensure that proper planning is done for the related terminal moves.

A second person handles similar functions for telephones. She orders new telephone lines from the telephone company when required and coordinates moves and changes when offices are rearranged. As in most large companies, this coordination takes a significant amount of time, since many people change jobs and offices each year. Dow Corning has adopted a general policy of not changing telephone numbers when people change jobs. However, there are always exceptions and special circumstances that must be accommodated.

A third person handles the administration of the voice message system. She registers new users and trains them to use the voice messaging system. She also checks the system's statistics reports to ensure that all of the circuits are working properly and being used equally. When required, she sends voice messages to all users informing them about new features or upcoming service activities. In addition, she generates the monthly telephone bills that are sent to all Dow Corning telephone users by seeing that the programs are run and that the reports are accurate.

The fourth administrative support person performs the order processing and accounting functions for the entire SIM department, including the telecommunications department. She does the actual generation of purchase requisitions that are sent to Dow Corning's purchasing department. When bills are received, she checks them for accuracy and accumulates them before submitting them to the telecommunications manager for approval. After they have been approved, she sends them to the accounts payable department and then follows through to ensure that they are properly recorded in the Dow Corning accounting system.

Dow Corning's telephone operators are not a part of the telecommunications department. They report to the site security department and also serve as receptionists for the main entrance at the Dow Corning center. Since the Centrex telephone system has direct inward dialing (DID), most calls from the outside go directly to the person for whom they are intended. The only calls that the operators must handle are from people who call the general Dow Corning number because they don't know specifically who they want to talk to. Thus, the telephone operators are also information sources and must question callers to direct them to the proper person or department.

Review Questions

1. Discuss why it is important to manage the telecommunications activity within a company. Has the need increased or decreased since 1980? Why or why not?
2. What are the factors that favor having the telecommunications department report to the chief information officer? If a chief information officer does not exist, to whom should the telecommunications department report? Why?

3. What is meant by the statement in the text that political reasons have kept many companies from merging their voice and data telecommunications departments? Why is it important to overcome these stumbling blocks?
4. Describe some situations in which it would be appropriate for the telecommunications department to use consultants.
5. The Nova Company needs a good technical person to install and maintain a new data communications network. How would you recommend Nova go about locating such a person? What attributes should the company be looking for when the managers interview candidates for the job?
6. An up-and-coming systems analyst in the Photometrics Corp. was just promoted to be the telecommunications supervisor. She has excellent verbal communications skills and gets along well with people at all levels in the company. She has 3 years of experience in systems analysis and programming and has performed extremely well but has no specific telecommunications or supervisory experience. What difficulties might this person encounter in her new job, and what kinds of people should she try to hire to complement her strengths and weaknesses?
7. In a small telecommunications department consisting of one full-time and one part-time person, how do all of the different tasks, such as planning, designing network enhancements, solving day-to-day problems, and operating the network get handled?
8. What are the differences between a long-term, medium-term, and short-term telecommunications plan?
9. What are the consequences if a telecommunications department does not have a long-term, 3- to 5-year plan?
10. Why are financial management and controls important to the telecommunications department?
11. Explain the difference between expense budgeting and capital budgeting.
12. The text lists a number of types of telecommunications department expenses. Which of these do you think are most important to manage? Why? Which represent the largest dollar value for a typical telecommunications department?
13. The text gives several arguments in favor of charging out telecommunications costs directly to users. What might be reasons why some companies do not charge the costs back to the users?
14. What is the purpose of having a telecommunications audit?
15. What are some of the factors other than cost savings that can be used to help justify an expansion of the telecommunications network?
16. Why must the telecommunications manager be concerned about the physical security of the company's communications facilities?
17. Why is dial-in access an especially vulnerable point in a data communications network?

18. In most companies, several departments argue against having individual passwords for each user. The accounting department, for example, often wants all of the accountants to have the same user identification code and password because they all share the same data. As the security officer, what reasons would you give to defend your position that each employee should have unique identification codes and passwords?

19. In addition to planning for disasters at its own locations, why must a company plan for disasters at its communications companies' facilities?

Problems and Projects

1. Use a spreadsheet program to create a budget for a telecommunications department, using the expense categories listed in the text. Increase the personnel costs by 5 percent per year for 5 years and the rental expenses by 20 percent per year for the same time period. Observe what happens to the total budget for the department over the 5-year period. How would you present this kind of projection to senior management to gain support for a long-term telecommunications plan? What arguments would you use to support why the telecommunications costs are going up at a rate that greatly exceeds the general rate of inflation and the 7 percent growth rate for other administrative expenses in the company?

2. Call back and handshaking were two techniques described in the text for providing security against unauthorized access to a computer on dial-in lines. Invent another technique that would provide access control. Your technique could be implemented in hardware, software, or a combination of the two. Describe how it would work. See if your classmates can figure out a way to break your "security system."

3. Identify some of the different situations that would occur in a disaster caused by a tornado striking a computer and communications control center as compared to a fire striking the same location. How might these differences be reflected in the disaster recovery plan?

4. The Radix Supply company has a small but sophisticated online computer system providing most of the company's data processing capabilities. One of the people in the telecommunications group believes that the company telephone book should be put on the computer and made available online. In fact, she is advocating that Radix stop printing the telephone book every month and use only the online version. What are the factors the company must consider before the managers decide whether to eliminate the printed directory and have only an online version?

5. Visit a company with a telecommunications department. Find out where the department resides in the company's organization. To whom does the telecommunications manager report? How is the telecommunications department organized? How does the group recruit its people? Does the department use consultants and, if so, for what purposes? How often is an expense budget prepared? Are there regular reports of telecommunications expenses? What type of planning process does the department have? Are project leaders expected to prepare detailed project plans? How frequently is project progress reviewed? Who reviews projects?

Vocabulary

information resource management (IRM)
chief information officer (CIO)
communications directory
Gantt chart
profit center
cost center
call back
handshake

References

Blyth, John W., and Mary M. Blyth. *Telecommunications: Concepts, Development, and Management.* Indianapolis, IN: The Bobbs-Merrill Company, Inc., 1985.

Bock, Gordon. "Management's Newest Star," *Business Week,* October 13, 1986, p. 160.

Chester, Jeffrey A. "The Merging of Voice and Data," *Infosystems,* September 1985, p. 30.

"Executive Report on Telecommunications Management," *Computerworld,* March 17, 1986.

Fitzgerald, Jerry. *Business Data Communications: Basic Concepts, Security, and Design.* 2nd ed. New York: John Wiley & Sons, 1984, 1988.

Gottschalk, Karl D. "The System Usability Process for Network Management Products," *IBM Systems Journal,* Vol 25 (1), p. 83.

Harris, Catherine L. "Information Power," *Business Week,* October 14, 1985, p. 108.

Johnson, Bob. "The Almighty Telecom Budget Under Diebold Scrutiny," *Communications Week,* December 15, 1986, p. 18.

Kaufman, Bob. *Cost-Effective Telecommunications Management: Turning Telephone Costs into Profits.* Boston: CBI Publishing Company, Inc., 1983.

Moseley, Donald R. "Communications Security Network Management," *Telecommunications,* October 1985, p. 82.

Petersohn, Henry H. *Executive's Guide to Data Communications in the Corporate Environment.* Englewood Cliffs, NJ: Prentice-Hall, Inc., 1986.

Reynolds, George W. *Introduction to Business Telecommunications.* Columbus OH: Charles E. Merrill Publishing Company, 1984.

Schaevitz, Alan Y. "Managing the Future Network: Tactics, Technology, and Trauma," *Business Communications Review,* September 10, 1985, p. 20.

Sigler, Jerry L., and Gerard J. Cunningham. "Building a Financial Management Structure for Communications," *Business Communications Review,* August 1989, p. 18.

Swanson, Stevenson. "Phone Repair Woes Mount," *The Chicago Tribune,* May 18, 1988.

Xephon Consultancy Report: Network Management for IBM Users. Berkshire, England: Xephon Technology Transfer Ltd., 1985.

CHAPTER

Network Design and Implementation

OBJECTIVES

When you complete your study of this chapter, you should be able to

- describe the process of designing a telecommunications network;
- discuss each of the phases of the network design and implementation process in detail;
- explain the importance of understanding the requirements for a network before beginning its design;
- describe the differences between the design of a voice network and the design of a data network;
- describe several project management techniques that are relevant to the network design and implementation process.

INTRODUCTION

This chapter describes the process by which communications networks are designed and implemented. Proper design requires a detailed understanding of how the network will be used. Once the requirements are understood, the network designer can investigate alternative ways to design the network. The designer must understand different telecommunications technologies so that he or she can match them with the requirements to provide the optimal design at minimal cost. After the network is designed and the design is approved, the circuits and other equipment are ordered, and the network is implemented.

Network design and implementation are a structured process composed of several phases, each of which will be examined in detail in this chapter. In most phases, several techniques are used to accomplish the work. After reading this chapter, you will have a good understanding of the network design and implementation process, as well as the specific tasks required.

THE NETWORK DESIGN AND IMPLEMENTATION PROCESS

Communications network design and implementation exemplify the more general systems analysis and design process used to design and implement any computer-based system. In fact, computer systems analysis and design have many similarities to the architectural design and engineering processes followed when any new structure is built. The architectural and engineering processes for designing and building buildings has been in existence for many years. They are, therefore, widely understood and almost universally applied.

systems analysis and design

Computer systems analysis and design together form a much newer discipline. Although the process is becoming more scientific with each passing year, there are still many aspects that are somewhat "artistic" in nature. The methods and techniques are gradually becoming standardized as people in industry gain experience and understand the process and the work required. Communications network analysis and design have been performed within the telephone companies and other carriers since the invention of the telephone. These processes have been performed in other companies only since they began designing their own private data communications networks in the late 1950s.

Network analysis and design are a process for understanding the requirements for a communications network, investigating alternative ways for implementing the network, and selecting the most appropriate alternative to provide the required capability. Network implementation is the process of installing and implementing the network. The work is performed by a project team that has the responsibility to analyze the requirements, design the new network, and implement it.

Most aspects of network analysis and design are similar for voice and data networks; the basic process is the same. This chapter primarily describes data network design but includes a discussion of voice network design that examines the differences between voice and data network design.

In every network design, certain work must be performed, much of it in a particular sequence. That is why network design normally is broken into discrete phases that are completed in sequence. The phases are called by various names in different companies, and the dividing lines between phases vary a little from company to company. Overall, however, every network design requires the same general work. The goal of the network analysis and design process is to ensure that the network satisfies user requirements at an appropriate cost.

Project Phases

The phases of communications network analysis, design, and implementation are

1. the request;
2. preliminary investigation and feasibility study;
3. detailed understanding and definition of the requirements;
4. investigation of alternatives;
5. network design;

6. selection of vendors and equipment;
7. calculation of costs;
8. documentation of the network design and implementation plan;
9. management understanding of the design and approval of implementation;
10. equipment order;
11. preparation for network implementation;
12. installation of equipment;
13. training;
14. system testing;
15. cutover;
16. implementation cleanup and audit.

Although this may seem like a large number of steps, it bears repeating that on large and small projects alike, all of the steps are necessary. Obviously, some of the phases are quite simple and abbreviated for small systems. For example, in a small company, the manager who can approve the design and give the go-ahead to implement the project may be a part of the project team. Therefore, gaining approval to implement is a relatively minor phase. Nonetheless, at some point (whether implicitly or explicitly), management approval to implement is obtained.

Another example of a trivial activity may be the response when a business competitor comes out with a new, innovative communications-based service. In an effort to quickly provide a similar service, the preliminary investigation stages to respond to the competitor's edge may be very abbreviated. In addition, because the technical viability of the new system has already been demonstrated by the competitor, the feasibility study may be minimal.

Network Analysis and Design

This discussion explores the various phases of network analysis and design in detail.

The Request Phase For the network analysis and design process to begin, a person or department must ask for work to be done. The request phase is when the communications department is formally asked to perform some work. There are generally four sources of requests for projects. The most common is a request initiated by a user group or department that needs a communications service or facility to meet a business situation.

sources

A second type of request can come from senior managers who, through their industry contacts or other sources, get an idea for a new communications-based project. This type of request is relatively rare.

The third source of requests is outside organizations, such as customers, vendors, or the government. Vendors may request that all purchase orders be sent electronically. Customers may ask that shipping notices be sent via telecommunications. A change in the law may, in effect, be a government request for the transmission of certain types of data.

The fourth type of request comes from the communications department itself, which may initiate a project to improve or enhance the existing communications services. A typical request of this type is an increase in an existing network's capacity to handle increasing business transactions. A similar type of project might be aimed at improved response time or higher reliability. Requests generated by the communications department should be evaluated like any other project, and the benefits of the work compared to the cost.

form of request

Ideally, the request for work should be somewhat formal and in written form. Usually, the idea for a new project begins rather informally. Perhaps it comes from a conversation between two managers, a discussion over lunch, or an idea spawned by a comment made in a seminar. Often there are several meetings between the initiator of the idea and the communications department that are, in effect, preliminary feasibility discussions. The purpose of these meetings is to make an initial determination of whether the potential project has any merit. During these discussions, it may be determined that some formal work is required. If so, a formal request memo that documents the idea and the discussions should be written. The potential costs of the project and the potential benefits of the project to the company should also be outlined in the request memo.

project prioritization

In many companies, there are far more requests for work than can be satisfied by the staff of the communications department in a reasonable period of time. A prioritization process determines which projects the communications staff will work on. The most effective way to prioritize projects is to solicit the involvement of management from various departments, especially those who want to have work done. The management group should listen to a presentation that describes the potential projects and their potential costs and benefits in order to have information on which to base its decision. If properly conducted, this prioritization process should highlight projects that are potentially of the most benefit to the company. Some of the projects will drop from consideration, and the remaining projects will compete for the required dollar and staffing resources. If a prioritized list of projects cannot be established at the middle management level, the list of projects may have to be presented to senior management to obtain a decision.

outcome

Ultimately, the outcome of the prioritization process must be a list of projects for the communications department to work on. The communications department can estimate how long it will take to complete the prioritized list of projects with the existing staff. Frequently, the estimated time is quite long, and the departments that are sponsoring the projects are unhappy. The alternative may exist to add additional people or to allocate more money so that a greater number of projects can be worked on simultaneously and the elapsed time to complete projects may be shortened. The decision to apply these additional resources ultimately rests with senior management.

Preliminary Investigation and Feasibility Study The two primary objectives of the preliminary investigation and feasibility study phase are to gain an understanding of the current communications system and its problems and to determine whether some of the problems can be solved with a new or improved communications system. This phase usually is performed by a small number of people, perhaps an analyst or two from the communications department and one or two people from the user department. These people constitute the *project team*.

project team

The project team looks at how the existing communications system works and the problems that are prompting requests for changes or improvements. If, for example, the feasibility of a new telephone system were being examined, the team would look at the existing telephone system and its operation. The team members would talk to the people responsible for administering and operating the current systems, and they would meet with telephone users in various parts of the company to get their input about the existing telephone system. Similarly, if the feasibility of improving the response time on an interactive computer system were being examined, the team would talk to the users to determine what their perception of the response time was. They would also measure the actual response time, talk to the computer operations people about the response time problems, and try to determine whether the response time delays are really caused by the communications network or by other factors, such as slow processing in the computer.

study's objectives

The feasibility study attempts to determine whether implementing a solution to the identified problems makes sense from both a technical and a nontechnical standpoint. Clearly, if the cause of the response time delay in the timesharing example is slow computer processing, increasing the speed of the communications lines will have little or no effect on the response time. Similarly, the team investigating a new telephone system might find that the perceived problems with the existing system are not widespread. They may be primarily focused on one or two departments that have unique telephone requirements. In that case, the unique telephone requirements of those departments might be met with special equipment or circuits rather than by replacing the entire telephone system. Chances are that the customized solution for the two departments will be more effective and less expensive, too.

The project team might also determine that solving a particular communications problem is not feasible. There may be technical or cost reasons why a solution cannot be implemented. For example, providing dial-up 56 kbps data communications service to a disaster backup site may not be feasible if the local telephone companies do not have the service available. Although a solution using leased lines could be implemented, the costs are likely to be prohibitive. It is important to make an assessment of the feasibility of any potential communications project before recommendations are made.

creeping commitment of resources

A secondary objective in the preliminary feasibility stage is to use a minimum amount of people resources to study the situation. Properly performed, the phases of a telecommunications project require a *creeping commitment of resources* so that if, for one reason or another, the project is stopped, a minimum amount of personnel hours or dollars will have been spent. That is why the early phases of the project are performed by small numbers of people, and others are added to the team as the project progresses. This is also known as the *project life cycle*. Figure 15–1 illustrates this creeping commitment.

comparing costs and benefits

A preliminary estimate of the costs and benefits of completing the project should also be done in this phase. Since relatively little work has been done and the scope of the new or revised communications system has not been defined, the cost estimate may not be extremely accurate. It will be refined in subsequent phases, as a more thorough understanding of the project requirements is obtained. Identifying the benefits of the project work also is extremely important. Since communications systems frequently are installed for specific groups of users or departments in the company, their assistance in estimating the benefits should be obtained. Furthermore, benefit estimates made by the users are likely to be more credible with senior management than those made by the communications staff.

When the preliminary investigation and feasibility study have been completed, they should be documented with a written report that summarizes the findings. In many companies, the project team is also expected to make an oral presentation to one or more groups of management or users. Ultimately, the team needs to obtain a decision about whether the project will proceed to the next phase. The creeping commitment of resource concept suggests that it is far better to terminate infeasible projects that have marginal benefits early in the cycle before a lot of work has been done or money spent.

Detailed Understanding and Definition of the Requirements Assuming that the project team received a "go" decision at the close of the preliminary investigation and feasibility phase, it proceeds to the next phase. The team needs to gain a detailed understanding of the user's requirements. Often the project team is expanded in size at this point. As the phase name implies, the work is significantly more detailed and time-consuming than in previous phases. Additional people may be needed to help get the work done in a reasonable amount of time.

To help organize the requirements, it is common to break them into several categories. Organizationally, one member of the project team might be assigned to work on each category of requirements. Alter-

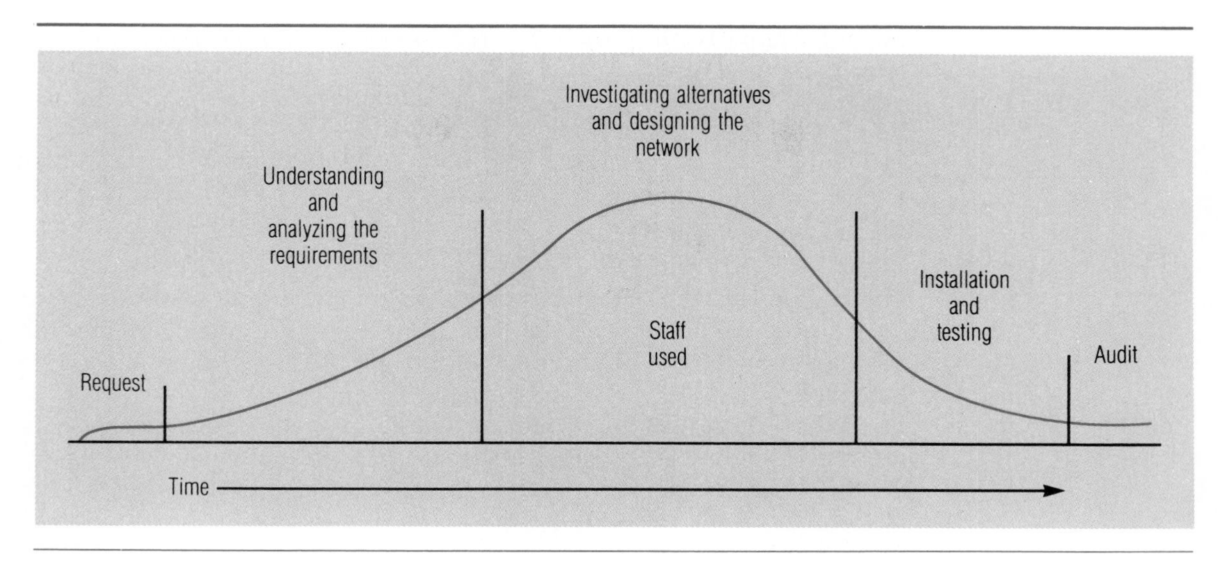

Figure 15–1
Creeping commitment of staffing as a project progresses. This concept is also known as the project life cycle.

nately, subgroups of the project team could be formed, each with responsibility for identifying the requirements in several of the categories.

Geographic Requirements The geographic scope of the network must be determined. It may be an international network serving company locations in several countries or a national network serving just one country. Regional networks serve a part of a country, a state, or perhaps a city. A local network serves a building or campus and is usually a candidate for private, company-owned circuits. Frequently, the network requirements don't fall cleanly into one category. An example of overlapping requirements would be a network connecting warehouses in many locations throughout the country to a central manufacturing plant. Within the plant there are many terminals also using the network. This system requires a combination of national and local networks.

It is quite probable that each geographic piece of the network has unique requirements. For example, there may be a requirement that terminals within the plant have a very fast response time. Some of the in-plant terminals may be production controllers that need consistently fast response from a computer in order to control machinery. These terminals might best be served by a local area network operating at a very high speed (such as 4 Mbps) and perhaps using the MAP protocol.

On the other hand, the terminals in the warehouses may be used by people to enter or access information about shipments. Standard leased telephone lines operating at 9,600 bps may be perfectly adequate to meet their requirements.

Traffic Loads One of the most important parts of this phase is to understand the amount of traffic that will flow on the network and develop definitive statements of requirements. A typical statement of requirements might be

> A circuit between Seattle and the computer center in St. Louis must be capable of carrying 15 million characters per day during normal business hours (8 A.M. to 5 P.M. Seattle time) with 3 million characters being transmitted during the peak hour of 10 A.M. to 11 A.M. Seattle time. The average response time for all transactions should be less than 5 seconds.

To develop this type of requirements statement, the network designer must know the anticipated number of messages and their lengths, the user's response time requirements, and the business hours at the two locations.

peak loads

When designers plan data networks, average message rates are most frequently used to calculate the traffic load. However, the designers must pay particular attention to those situations in which a high percentage of the message traffic occurs in a relatively short period of time. These peak periods can play havoc with the responsiveness of the network if they extend for very long. One of the most significant problems that can occur in communications systems is the inability to handle the transmission load during peak periods. The only way the problem can be avoided is by knowing what traffic the network must carry and designing a network with enough capacity to handle the peaks.

busy hour

It is especially important to determine how much traffic will be carried during the busiest hour of the day, called the *busy hour*. Busy hour traffic analysis is at the heart of both voice and data network design. It is not uncommon in business systems for busy hour peaks to occur

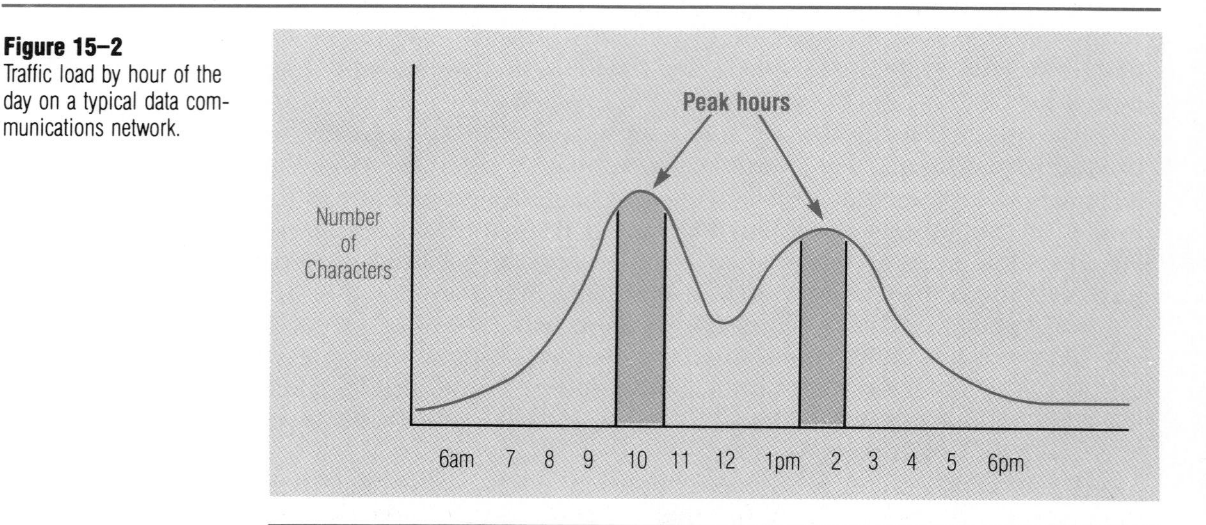

Figure 15–2
Traffic load by hour of the day on a typical data communications network.

Message Type	Average Length in Characters	Total Number of Messages per Day	Number of Messages in Peak Hour	Number of Characters in Peak Hour
Customer				
—Inquiry	30	170	45	1350
—Response	425	170	45	19125
Product				
—Inquiry	19	525	220	4180
—Response	1545	525	220	339900
Inventory Update	220	1750	175	38500
(list all of the transactions)				
Total		**3140**	**705**	**403055**

Figure 15–3
Analysis of expected traffic loads in a network.

around 10 A.M. and 2 P.M., as shown in Figure 15–2, although peaks may also occur at other time.

If, for example, a network carries 5,000 transactions a day and 80 percent of them occur between 3 P.M. and 5 P.M. each afternoon, the network must be designed to accommodate the volume during that peak period. The requirements are quite different if the same number of transactions is spread out over an 8-hour business day. An organization such as the New York Stock Exchange would have yet a different pattern.

Usually a table is prepared, like the one shown in Figure 15–3, to assist the analyst in gathering data about the expected traffic loads. If a totally new computer system is being designed, it may be difficult to get information about the number and length of transactions or messages. If exact information is not available, the analyst must make an estimate.

Traffic Flow Patterns Another aspect of requirements analysis is to determine message flow patterns. In a star network with a computer at the hub, it is obvious that all message traffic flows into and out of the central computer. In a mesh network, however, the traffic flows are not so simple. The network analyst must determine which message traffic goes to which location. If a map showing the locations on the proposed network has not been drawn, this is an appropriate time to do so. Depending on the geographic scope of the network, it may be necessary to have world, national, city, or building maps or a combination of all four. Typical maps of this type are shown in Figure 15–4. It is useful to connect the locations together with lines and identify the number of characters each must handle. At this point, the lines on the map do not

A.Map of U.S. network with hub in Denver

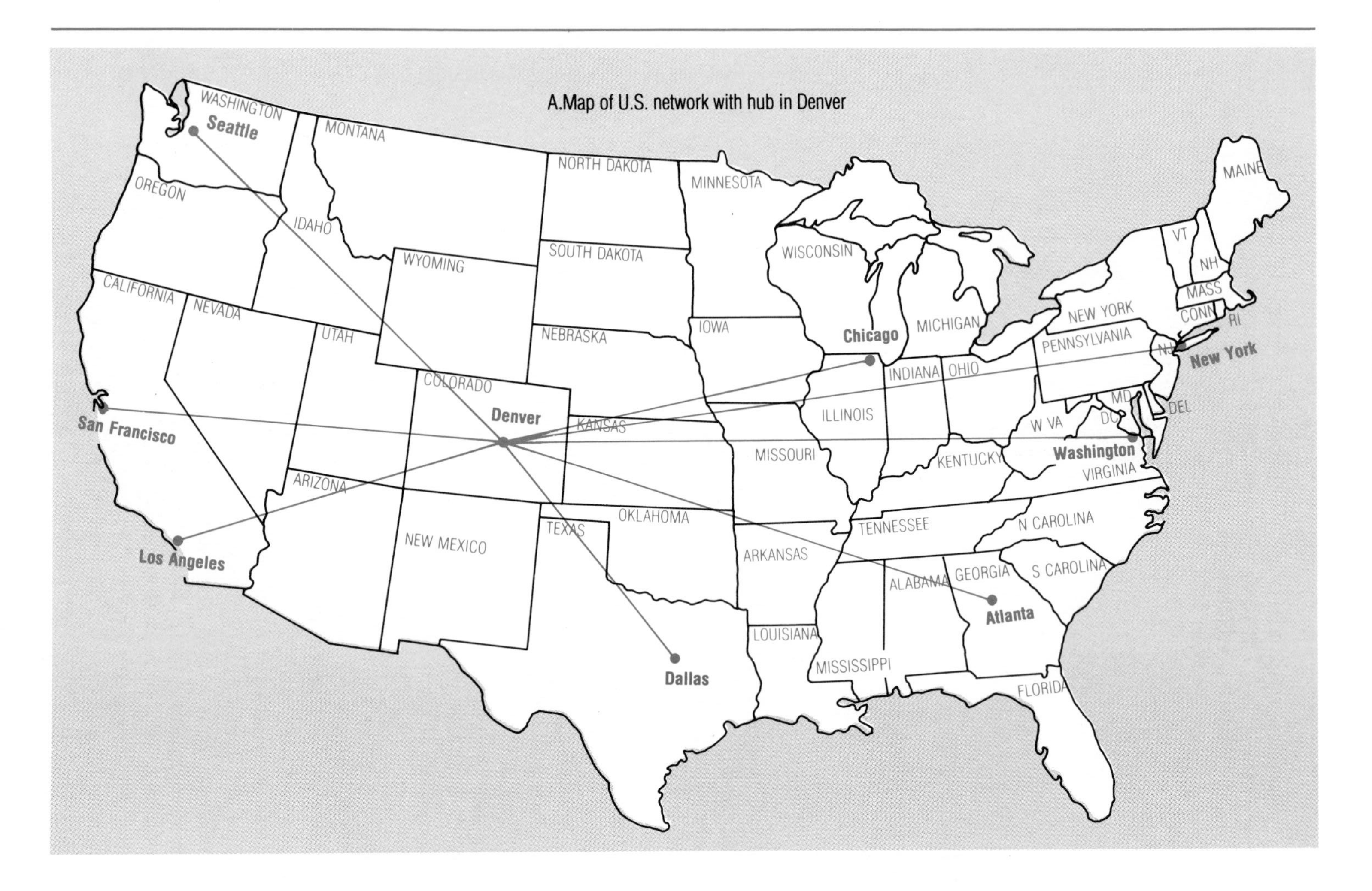

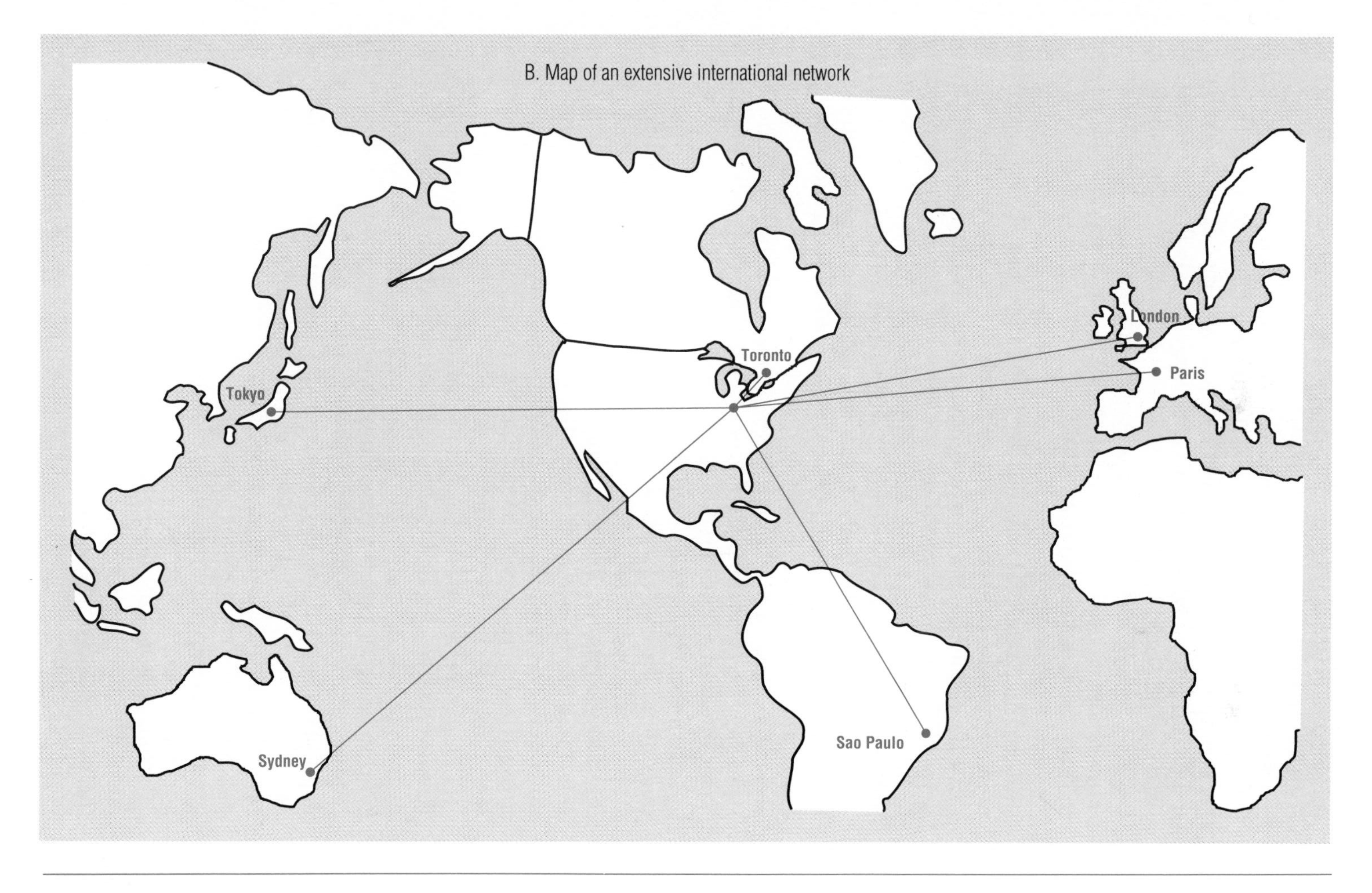

Figure 15–4
Typical network maps that are useful during system design and for network documentation.

correspond to circuits in the network but only serve to indicate the traffic patterns and volumes.

Depending on the geographic characteristics of the network, it may be necessary to make traffic estimates on a location-by-location basis. For example, a network with a hub in Kansas City might be connected to sales offices in El Paso, Phoenix, Los Angeles, San Francisco, and Salt Lake City. It would be important to know if the traffic volume from the Los Angeles office were significantly greater than the volume from the other offices. If so, it might be necessary to run a separate circuit just to carry the Los Angeles traffic load.

Availability Requirements Closely related to understanding the peak traffic period is knowing the range of hours during which the network will be operational and available to the users. The ability to carry 10 million characters in an 8-hour period requires quite a different network than one that must carry the same number of characters spread evenly over 24 hours. The problems created by having terminals in different time zones is even more severe in networks that handle traffic from Europe or the Far East. Japan's business day is 13 or 14 hours ahead of Eastern Standard Time in the United States, and Europe is 6 or 7 hours ahead, depending on the time of year. If a single network is serving locations in the United States, Europe, and the Far East, it must essentially be operational 24 hours a day, 6 days per week, in order to cover the normal business hours on the three continents.

Response Time and Reliability Requirements The project team must understand the true requirements for response time and reliability. Two studies, referenced in Chapter 3, have shown that for certain types of interactive computer-based systems, subsecond response time is desirable and leads to increased productivity by the users. In other applications, however, such as the delivery of electronic mail on a store-and-forward basis, it may be acceptable for the message to be delivered within minutes or even hours.

economics of reliability

Designing a network for high reliability can lead to some serious economic issues. It may be possible to achieve extremely high reliability by installing redundant hardware and circuits. It is important to assess from an objective point of view, however, whether the application really requires such high reliability. Although users like to think that their application is critical to ongoing company operations, it may turn out that the occasional loss of the network for several minutes or even several hours, though inconvenient and irritating to the users, is not critical to company operations. For most business applications, the greater than 99 percent reliability of standard leased circuits is perfectly adequate.

Type of Terminal Operators It is important for the network design team to understand the characteristics of operators who will be using the network's terminals. Full-time, highly trained operators who sit at terminals all day will put a more constant and overall higher demand on a network than will occasional users or the general public. The nature of the traffic load when the users are full-time operators is fairly easy to estimate.

Whereas the overall amount of traffic from consumer-operated terminals, such as ATMs, is lower, it tends to be "bursty" and to have high peaks at certain times of the day. A network that can handle the peak load from a collection of ATMs at lunch hour and right after work may be nearly idle the rest of the day.

Future Growth Projections—Capacity Planning Understanding the expected traffic growth over a 6-month to 5-year period is an extremely important part of network design. The single most common mistake made by network analysts and designers is underestimating the growth in traffic volume and failing to provide adequate extra capacity in the communications circuits and equipment. This causes great problems later when the network must be "unexpectedly" and rapidly expanded. One situation in which this occurs frequently is failing to adequately project the growth in the number of new employees who will require telephones when planning a telephone system. Inadequate cables may be run or a PBX that is too small might be purchased.

new employees

Data communications systems often get trapped by the *super highway effect* when a new terminal-based computer system provides an especially useful function. When super highways are built, it is often observed that traffic increases well above the level that existed on the old highway because travel on the new highway is so easy and convenient. More people travel to new destinations than ever before. As a result, a highway that was planned to last 15 or 20 years runs out of capacity in 5 years.

A similar phenomenon occurs in data communications systems when more users than were anticipated want to use the new system and they enter more transactions than were planned for. As a result, data circuits and other network hardware run out of capacity sooner than expected. Although extra circuits can always be ordered from the common carrier, all voice and data hardware, such as key systems, PBXs, and communications controllers, are designed to handle a certain maximum number of circuits, telephone calls per minute, or aggregate data rate. Once that capacity is exceeded, there is usually no alternative but to replace the hardware with a larger model. Although it may seem absurd, companies have spent thousands of dollars for telephone systems or communications controllers only to find that they are out of capacity within 6 months. Walking in to tell senior management that the company's "new" telephone system is now obsolete is not a pleasant experience!

planning for growth

With the explosion in the use of online terminals, many companies have found that their data communications traffic is growing at a rate of 40 percent or 50 percent per year, compounded! This means that the rate of traffic is doubling every 2 years. A communications controller that is only 50 percent used this year will be out of capacity next year.

Unless the network has been designed to handle such rapid growth and management has been primed to know what to expect, the company can be in for some nasty surprises. Of course, a 50 percent growth in the communications traffic usually has wider implications, such as a similar growth in computer use. If managers expect a computer to last 4 years, they will be shocked to find out they need to make major upgrades every year or two.

Capacity planning is not an exercise to be done only when a new communications system is designed. Traffic loads should be reviewed at least annually and more frequently in systems with a high growth rate. Many companies produce graphs of their communications traffic on a monthly basis. Firms that are highly dependent on communications may look at statistics hourly. Capacity projections should be made and presented to management regularly, even if network upgrades do not need to be made immediately. In some companies, capital expenditures must be planned ahead for more than 1 year. The capacity plan is an important input to the annual expense budgeting process. If major new facilities will be needed in the coming year, it is important to be sure that the financial plans (budgets) are made appropriately so that money is available when needed.

Date the New Service Must Be Available The network design team must understand when the new network or communications service must be installed and ready for use. In many situations, it takes at least 6 months from the time communications equipment is ordered until the service is operational. Yet equipment cannot be ordered until the network is designed and approved. Although there is a wide variability depending on exactly what work must be done, it is not unusual for communications projects to take a year or more from inception to implementation. If major new network capabilities are required in less time, it is sometimes possible to speed the ordering and installation process, but to try to do so is usually risky.

compatibility

Constraints The typical constraints implicitly placed on a network design are that an existing network must be expanded or that the new network must be compatible with the old one. Preexisting computer hardware or software presents a similar type of constraint. For example, if a Digital Equipment Corporation (DEC) computer is already in place and is going to process the transactions, the data network must be designed to interface to the DEC hardware and to work compatibly with it.

cost

Another type of constraint that may be placed on the network design is a cost constraint. Placing a direct limitation on the cost of the new network or an expansion occurs less frequently because the initial focus is usually on meeting new business requirements. Only after some design work has been done and the costs start to become evident are the costs usually given serious attention.

In general, the more detailed the information obtained in this phase, the better the network design will be. If the analyst is in doubt, it is better to err on the side of obtaining too much information rather than too little.

As the analyst gathers the data and analyzes the network requirements, inevitably he or she begins thinking about solutions to the communications problems. These "mental designs" are a natural part of the iterative analysis and design process, and these early, tentative thoughts or sketches may prove useful later. It is important, however, for the analyst to avoid the natural tendency to jump to conclusions and to design the network solution before thoroughly understanding the problem. Spending adequate time to ensure that problems and requirements are completely understood pays big dividends later in terms of a faster, more accurate network design and implementation.

Investigation of Alternatives Another facet of the work to be done before designing the network is to investigate alternatives for addressing the communications requirements. Everyone tends to do this work based on personal experience. Although experience is extremely useful, in a field that is changing as rapidly as telecommunications, it is important to continually assess options that may not have been available when the existing network was designed. This is particularly true since deregulation because the number of circuit and equipment offerings has multiplied rapidly.

Depending on the geographic locations to be served, certain vendors or communications service options may not be available. Obviously, if all of the locations are in major cities, more alternatives are likely to exist than if some of the facilities are in small towns or other less populated areas. Some companies like to have as few vendors as possible involved in their communications networks. An old rule of thumb says that the number of operational problems in a communications network increases as the square of the number of vendors. This is because at each point where two vendors' equipment interconnects there is a potential incompatibility. The fewer the number of vendors, the fewer the number of these interfaces.

Many vendors choose to market their products only in large metropolitan areas where there is a substantial potential customer base. They do not make their products available outside the areas where their marketing and support exist. The network designers must understand what products and services are available at the locations to be served by the network.

There are numerous other alternatives that must be investigated. Are public switched facilities available, or should private lines be used? Will the network be centralized in a star configuration or highly distributed? Is the use of multiplexers or concentrators appropriate? Do packet switching networks serve the network locations, and should they be used? Are satellite services appropriate?

When alternatives are being investigated, costs must be considered. Often there are trade-offs between costs and service. In some cases, an alternative that is technically viable may be eliminated because its cost is prohibitive. The basic cost information gathered while the analyst evaluates alternatives is useful for estimating the total cost of the network being planned.

It is helpful in this stage to eliminate some alternatives so that attention can be focused on the ones that are most relevant.

The objective of this phase of activity is to be sure that all relevant alternatives are considered and that inappropriate options are eliminated. After considering the alternatives, the design team must decide which alternatives to select. The products and services selected must meet the users' requirements.

Network Design The entire process of network design from information gathering through ordering the equipment is iterative, and the detailed design or layout of the network is, in itself, also an iterative process. Often new insights are gleaned or new information is obtained while analysts and designers go through the design process. When this occurs, it frequently is desirable to redo some of the work performed in previous phases to take the new insights or information into account. The result of the rework is usually a better network design.

If the locations the network must serve have not yet been laid out on a map, that should be done now. As the old saying goes, a picture is worth a thousand words. Many times the relative locations of the network nodes suggest certain network configuration possibilities when they are viewed on a map. Assuming for a moment that the network is not going to be a local area network with a ring or bus topology, the first cut at the circuit configuration should be made by connecting the nodes to the central computer (if one exists) in a star topology. If the network is configured around multiple computers instead of a single central computer, the initial configuration would normally have the computers connected together and the terminals connected to the nearest computer.

using a star network

From the traffic load figures developed in the data gathering and analysis phases, the traffic load should be applied to each route showing the number of characters to be transmitted in a given period of time—typically a business day. The number of characters during the peak hour should also be identified. A map of this type is shown in Figure 15–5.

When the number of characters to be transmitted and the available hours are both known, designers can calculate the speed of the line required to carry the traffic. This gives the capacity requirement but does not necessarily indicate the responsiveness. Although there are many factors that make up the overall response time of a communica-

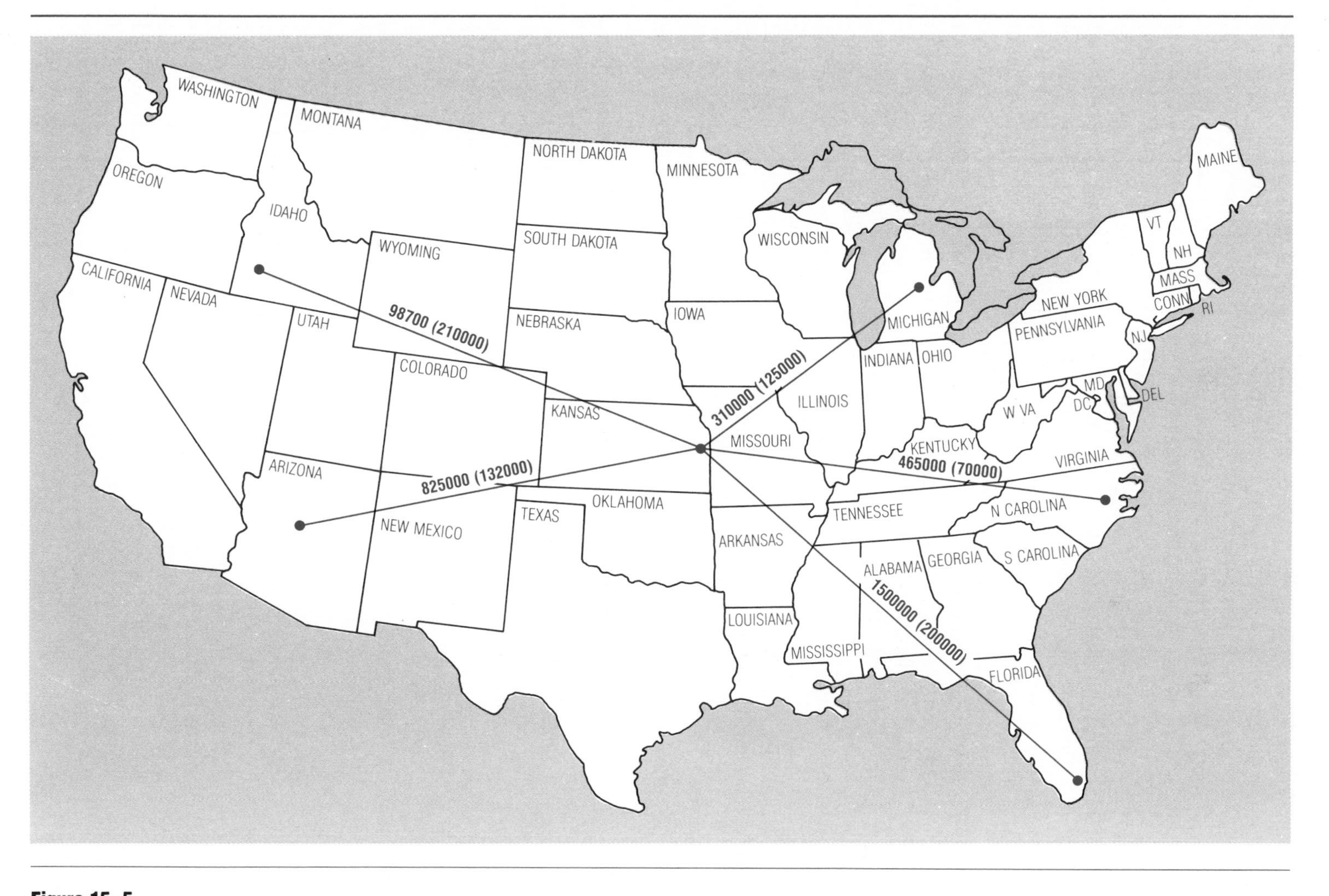

Figure 15–5
Typical star network map showing number of characters to be transmitted during the business day and during the peak hour (in parentheses).

tions computer system, the communications time to transmit the characters from the terminal to the computer and the response characters from the computer back to the terminal is, in most cases, a significant part of the overall response time. If the line is heavily loaded, the probability is high that queuing will occur somewhere in the network. Queuing occurs when the line is busy handling the data from one terminal and another terminal is waiting to send or receive data. The figures vary widely, but if the line use on a standard leased, multipoint, polled circuit is greater than 40 percent, it is very likely that some traffic is being delayed because of queuing. Queuing in itself is not bad, but it is important to understand that it introduces variability into the transmission time between a terminal and a computer. That is why most response time numbers are stated in probabilistic terms, such as "90 percent of the transactions will be processed in less than 5 seconds, 95 percent in less than 10 seconds, and 100 percent in less than a minute."

queuing

Detailed analysis of the effects of queuing in the overall performance of a network can be done with simulation tools. Several are commercially available. Creating a simulation model of a network, however, is a significant job, requiring a detailed understanding of both the network and the simulation tool. Once the simulation model is created, it must be validated for the set of assumptions used in its creation. Only then can it be used to project the performance of the actual network.

network simulation

A simulation model can give the answers to many "what-if" questions, such as "What if I change the line speed?" "What if I add more terminals to the line?" "What if I change the parameters of the network control program and increase the number of buffers?" Furthermore, after the network is operational, a simulation model can help predict when the network will reach capacity by increasing the simulated number of transactions until response time degrades to unacceptable levels. A simulation model not only helps to make the most of the available resources but also confirms that changes being planned for the network will have the desired effect.

If simulation is not used, it is still possible to estimate the responsiveness characteristics of the network. Response time to the user can be calculated by adding together the time for the transmission from the terminal to the computer, the processing time in the computer, and the transmission time of the response back to the terminal. This simple calculation assumes that no queuing exists and gives a "best possible" response time. If this simple model shows that the "best possible" response time is inadequate even though the circuit may have enough capacity to handle all the characters that will be transmitted in a day, two options exist. One is to speed the processing time on the computer, and the other is to speed the transmission time of the circuit. The option selected depends on many variables, but often the existing computer is one of the constraining variables put on the system design. That is, a computer of a given size is already installed and will be used. In that case, the only viable option is to increase the speed of the communications circuit to improve performance even though, for capacity purposes, the slower speed circuit is sufficient.

Let's look at an example. Suppose that a communications line must transmit 10 million characters in an 8-hour business day while providing terminals with no longer than a 1-second response time. Since this is a simple example, we will assume that there is only 1 terminal on a line and, therefore, that no queuing exists. As shown in Figure 15–6, a 9,600 bps circuit can transport 10 million characters in 2.3 hours, clearly meeting the capacity requirement. Suppose a typical transaction consists of 500 characters transmitted from the terminal to the computer, followed by .3 seconds of computer processing time and then 1,500 characters transmitted from the computer back to the terminal. The figure shows that the 9,600 bps circuit yields a response time of nearly 2 seconds and does not meet the response time requirement. A 56 kbps circuit cuts the response time to .58 seconds and clearly meets the response time requirement. At the same time, it greatly exceeds the capacity requirement of the circuit because it can carry all of the day's traffic in only .4 hours. The network designer faced with these circumstances would probably want to consult with the user to determine whether the additional cost of the 56 kbps circuit was warranted for this application.

The map of the network also needs to be viewed with an eye toward the possible use of concentrators or multiplexers. For example, as shown in Figure 15–7, if the computer doing processing for a network is located in Saint Louis, it may be desirable to concentrate multiple lines serving the west in Denver and then run a single high-speed line from Denver to Saint Louis. Similarly, circuits from the southeast might be concentrated in Atlanta and circuits from Texas in Dallas.

Line speed	9,600 bps	56,000 bps
Line capacity	= 1,200 characters per second	= 7,000 characters per second
	(at 8 bits per character)	
If 10 million characters per day are to be sent, the total transmission time would be	2.3 hours	.40 hours
500 characters in to computer	.42 sec	.07 sec
Computer processing time	.30 sec	.30 sec
1,500 characters out to terminal	1.25 sec	.21 sec
Total response time to user	**1.97 seconds**	**.58 seconds**

Figure 15–6
Response time at various line speeds.

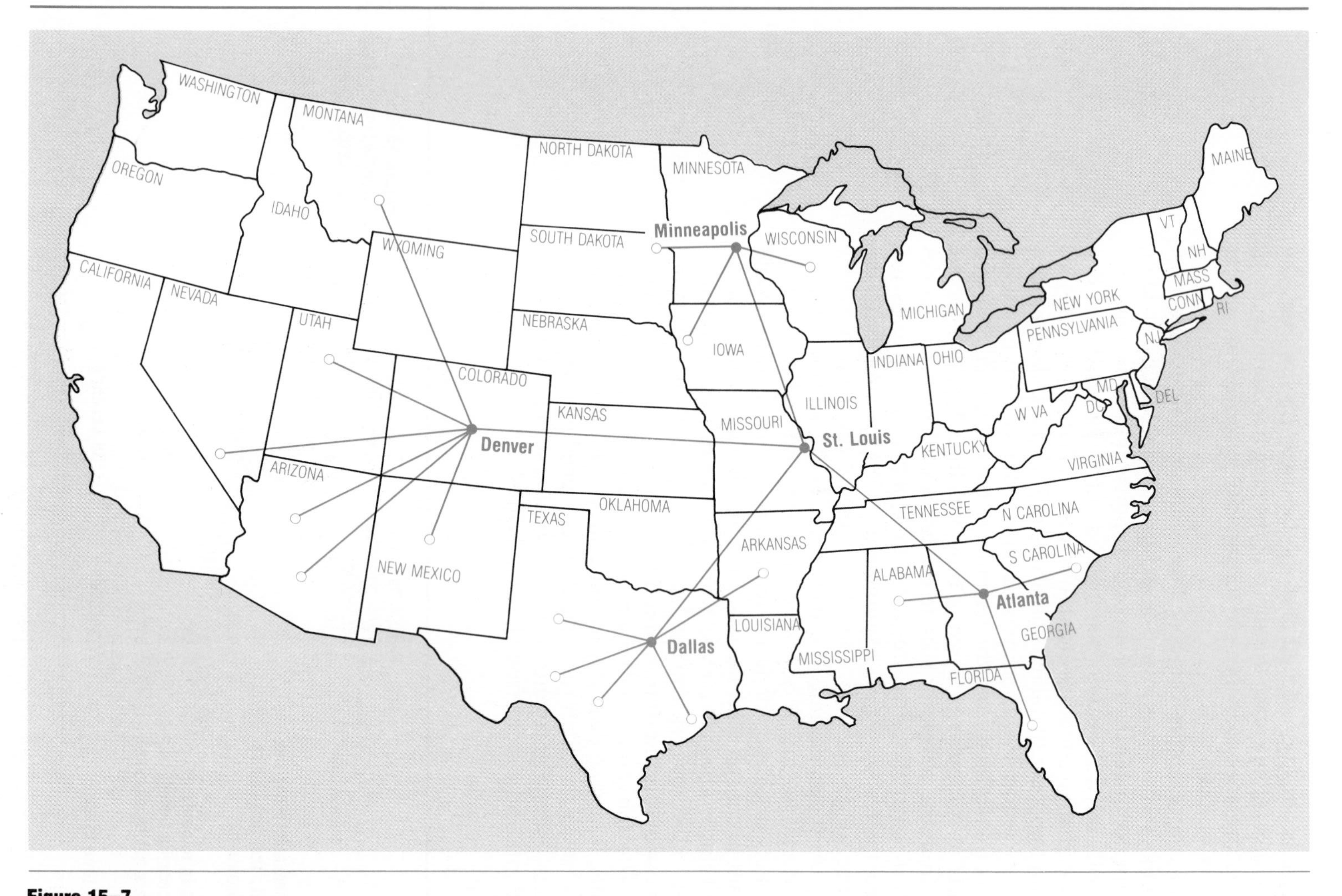

Figure 15–7
Network map showing the use of concentrators to consolidate traffic and reduce cost.

In the course of analyzing the data about network traffic and the configuration alternatives, it may be that more than one network configuration emerges as a possible candidate for the final design. In fact, it is a good idea to try to develop several alternative configurations. Each may have slightly different assumptions, use different equipment or vendors, and deliver different performance. In all likelihood there will also be cost variations between the alternative designs. Software is available to assist in evaluating alternative network designs. Communications carriers often make this software available to their customers free of charge. Similar software can be purchased from independent vendors.

Selection of Vendors and Equipment Once the network has been designed or when several viable alternative designs exist, the specific equipment and vendors must be selected. This may be done on a relatively informal basis in which known or preferred vendors are contacted and asked for price quotes and specifications for specific pieces of equipment. Alternately, a more formal approach can be used in which the company prepares a *request for proposal (RFP)*, also called a *request for quotation (RFQ)*, and sends it to several vendors.

RFP or RFQ

Before contacting vendors or sending out RFPs, it is best to get an overview of the types of equipment available on the market. There are many sources of information about data communications equipment. Advertisements in data communications magazines and to some extent data processing magazines are one possibility. Buyer's guides, such as those published by Auerbach Publishers, Inc., or Datapro Research, Inc., contain complete listings of data communications equipment by equipment type and vendor. The Auerbach and Datapro companies also publish equipment ratings based on information provided by users of the equipment. Salespeople normally are happy to provide information about the equipment they sell over the phone or by sending product literature. If at all possible, the salespeople usually like to make a direct sales call. Other sources of information about data communications equipment include data processing vendors, professional associates, and even friends. Most communications people are happy to share information and make recommendations about products they have used, especially if they are pleased with the results.

The RFP is a document that asks each vendor to prepare a price quotation for the configuration (or configurations) described in the RFP document. Some RFPs give the vendor considerable latitude in preparing responses or even ask for suggested solutions to specific design problems described in the RFP. Other RFPs are more structured and present a detailed outline for the vendors' response. In any case, certain basic information must be provided by the vendor. A list of this information is shown in Figure 15–8. Usually, the RFP is sent to several vendors so that a company can compare competitive responses.

Figure 15–8
Outline of typical information to be included by a vendor in a response to a request for proposal (RFP).

1. System design
2. System features
3. Growth capability
4. Installation and testing methods
5. Maintenance arrangements
6. System support
7. Installation schedule
8. Pricing and timing of payments
9. Warranty coverage
10. User training
11. Other recommendations

When the RFP responses are returned, the network design team must evaluate them against the requirements presented to the vendors. Silly as it may seem, it is necessary to check to be sure that each vendor is proposing the type of network requested. Some vendors don't read RFPs very carefully! Vendor responses can be compared on the basis of the proposed solution, the price, the sales support provided, technical support offered, and product maintenance and repair service. A vendor's financial viability should be considered, as there are always some vendors who are here today and gone tomorrow. Sometimes after analysis, pieces of one vendor's proposed solution are combined with parts of the proposal from another vendor to make a hybrid network that provides a unique combination of capabilities.

Calculation of Costs With one or more network configurations and actual cost information provided by the vendors in hand, the costs of the network designs or alternatives can be recalculated and compared. Just as there are technical trade-offs between different types of lines, modems, and other equipment, there is a cost/performance trade-off between vendor proposals and equipment that must be made. The objective of the cost analysis is to finalize the configuration of the network that will meet the capacity, performance, and other requirements at the lowest possible cost. The cost can then be compared to the benefits that the new communications system will yield, and a decision can be made whether to move ahead to the implementation phase.

Network costs can be grouped into the following categories:

- circuit costs;
- modem costs;
- other hardware costs;
- software costs;
- personnel costs.

Calculating the cost of leased circuits is a complicated process that usually is left up to the communications carrier. The details of the process are described in Chapter 9. Personal computer programs are available to assist with the calculation if a company wants to calculate the cost internally.

The costs of modems and other communications hardware and software can be obtained directly from the manufacturer or distributor of the equipment. Purchase prices usually are quoted by the vendor, and most companies buy (rather than lease) this type of equipment. Some vendors also provide rent or lease options for their equipment.

It is important to separate the costs of the personnel who develop the network from the costs of those who run it. The latter costs are ongoing operational costs, whereas the former are one-time development costs. The costs of the network analysts and designers are a part of the project costs but are not a part of the ongoing operational costs. Once the project is completed, the analysts and designers will work on other projects. Costs for the personnel to operate the network may be based on actual salaries and benefits or on a standard amount the company uses for all personnel cost estimation.

When all of the costs are calculated, they need to be assembled onto a worksheet, such as the one shown in Figure 15–9. It is helpful to build up a total cost for each circuit, including the cost of the circuit itself as well as all hardware associated with it. The cost for each of the circuits can be added together to give the total cost for all circuits and equipment. To this is added the cost of the network software and the operational personnel. The sum of all of these costs is a total that can be compared to the proposed benefits of the new network and analyzed according to the company's financial criteria.

worksheet

Documentation of the Network Design and Implementation Plan Once the network has been designed, it is important to document it in a form that can be understood by management for review and approval purposes.

Circuit Name	Circuit Cost ($ per mo.)	Modem Cost ($ per mo.)	Other Hardware Cost ($ per mo.)	Total Circuit Hardware Cost
Circuit 1	$ 565	$240	$310	$1115
Circuit 2	1884	465	920	3269
Circuit 3	140	120	–0–	260
Circuit 4	998	240	163	1401
	$3587/mo.	$1065/mo.	$1393/mo.	**$6045/mo.**
			Network software	**885/mo.**
			Operating personnel	**5100/mo.**
			Total Operating Cost	**$12030/mo.**

Figure 15–9
Network cost worksheet.

It is also necessary to document the design in technical terms for those who will implement it.

diagrams and maps

Network documentation takes two forms: words and pictures. The pictures may be maps that show the locations of nodes in the network, as previously discussed, or diagrams that show all of the nodes and the types of connections between them, although not in a geographical layout. The maps or pictures need to be accompanied by a narrative that describes the network. A circuit and equipment list, which shows a detailed breakdown of all of the network components, also should be included.

components list

specifications and model numbers

wiring diagrams

Detailed technical network documentation must show exact circuit specifications, including the types of conditioning and any special features. Hardware model numbers must be included, as well as any required accessories or cables. The equipment specifications must be detailed enough that they can be placed on a purchase order and understood by the vendor. In addition, wiring diagrams that illustrate equipment connections and where the wire and cable will run are included in the design package.

One other piece of required documentation is the implementation plan. This plan, which may be in the form of a Gantt chart, shows all of the activities required to implement the network once it is approved. Activities shown in the plan may include ordering equipment, writing software, training users, testing the network, and actual cutover. The plan should show the elapsed time required and the estimated completion date of each of the required activities.

Management Understanding of the Design and Approval of Implementation Once the network is designed and documented, the next step is to distribute the documentation to the users of the network, the network operations staff, and senior management. As with most complex designs, the documentation itself rarely provides enough information to give the reader a complete understanding. It is common to have a series of presentations and meetings in which the designers elaborate on the design and answer questions. The ultimate objective of this review and approval process is to gain the support of the users and operations staff and then the approval of management to implement the system.

review process

The users should understand the design well enough to be able to state that if the network is implemented as described in the documentation, it will provide the required communications capability and/or solve the communications problems they have been experiencing. The network operations staff need to understand the design of the proposed network well enough to ensure that they can operate it after it is implemented. Management must satisfy themselves that the implementation of the network is in the company's best interests and that the investment of staff and money will yield the claimed benefits.

Network Implementation

Once approval to implement the network has been obtained, the project moves into the implementation stage. Although the same project team that designed the network is likely to continue, the implementation of the network is a distinct set of activities in itself.

Equipment Order Ordering equipment is normally one of the first actions taken after implementation approval is obtained. Some equipment and services have lead times of up to 6 months, and therefore it is important to place specific orders promptly after the project is approved. The two most important provisions in the equipment agreement are the equipment specifications and the acceptance test. The specifications define the products and services being purchased, and the acceptance test provides an objective basis for determining whether the equipment installation, after cutover, meets the buyer's expectations. The order should specify the exact configuration of each piece of equipment, the date required, the location to which it is to be delivered, and any other terms and conditions the company's purchasing department requires. The more precise the specifications on the purchase order, the less likelihood of mistakes or misunderstandings when the equipment is built, shipped, or installed. Other important provisions of the purchase agreement should cover the vendor's warranty, maintenance arrangements, and the terms of the installation and cutover.

Preparation for Network Implementation There are many tasks that may occur after the equipment is ordered and before implementation. Depending on the nature of the work, there may be network operating policies and practices to establish. Standards may have to be written. Procedures for the network operations group may need to be developed. Purchased software may need to be modified to meet certain network requirements, or new software may have to be written.

Each of these activities should have been shown in the network implementation plan that was prepared earlier. Progress against the plan must be tracked to ensure that the target date for network cutover will be met.

tracking progress

Installation of Equipment Before equipment is delivered, the project team must prepare for the installation. Physical planning is the process for determining where the equipment is to be located, ensuring that adequate power and air conditioning are available, and seeing that large pieces of equipment can be conveniently transported from the shipping dock where they are delivered to where they will be installed.

Drawings should be made showing where all of the equipment will be placed, as well as how power and communications cables will be routed to it. Most vendors have specialists who can assist in the physical

layout drawings

planning, especially for large pieces of equipment, such as PBXs or front-end processors.

Once the equipment arrives, it must be installed. Depending on the nature of the equipment and the terms of the purchase agreement, it may be installed by the vendor or the company's own communications staff. Most frequently, the initial installation is performed by the vendor, who also handles any training that is required at the same time. Circuits are installed by the common carrier. When pieces of equipment from different vendors are connected together, it is important to ensure that the vendors work together and test the equipment while all are still on the premises. Then if there is a problem, they can work together to resolve it.

Training Both the users of the communications system and the operational and maintenance personnel must be trained in the characteristics of the new system. User training should be conducted by other users who have been trained by the network design team. Usually a combination of classroom instruction and hands-on experience is provided. The classroom instruction gives background and overview material, but the heart of the training is performed at the terminal, with the user actually entering transactions and receiving responses. This also gives the user a feel for how the network will perform, so that he or she gains a sense of its characteristics and responsiveness.

type of instruction

The operational and maintenance personnel will be trained by both the members of the project team and the vendors who install the equipment. Ideally, the operations staff will have been involved in the project since its inception and will already understand the design of the network. The final training, then, is a matter of getting hands-on experience with the commands and the various real-life situations.

Operations people can get good experience with the network while the users are using it in a training mode. If the users do a good job of simulating real operation, the operations staff will get a good feeling for how the network will operate and perform when it is finally cut into production. Of course, not every unusual situation that will occur during real operation will occur during the training period.

System Testing After all of the equipment and all of the circuits have been installed, the entire network must be tested. Ideally, each individual component of the network is tested under a variety of circumstances and conditions. However, with the size of some of the networks being installed today, this is virtually impossible. It is important that every terminal on the network exchange at least one message with another terminal or the host computer to verify that it is properly connected and working correctly. This type of test also exercises the modems, circuits, and multiplexers and gives some assurance that they are working properly.

Separately, the system software must be tested. Usually, this task is left to the application or system programmers. However, it is important to get users directly involved to be sure that the combination of network hardware and software is providing the required functions and capabilities. Sometimes, this testing can be accomplished during normal working hours. In other situations, it may be necessary to have people come in to do the testing in the evening or on weekends.

stress testing

Another type of testing required is *stress testing*. In stress testing, a heavy load is put on the system, usually by having a number of terminal operators simultaneously use the system. Stress testing indicates how the network will perform in real life and is also important as a final software test. Certain program bugs only show up when a high volume of transactions pass through the software in a short period of time.

One way to stress test a communications system is to use a tool called a *workload generator*, also called a *network simulator* or *driver*. The workload generator is a computer program that generates transactions. They are fed into the network to see how it behaves. To the network, the transactions appear to come from terminals. Workload generator programs often run on a computer that is attached to the front-end processor. They may also run in one or more personal computers attached at various points on the communications lines. Good workload generator programs, such as IBM's Teleprocessing Network Simulator (TPNS), allow the network designers to specify the type and length of transactions generated so that networks can simulate various transaction mixes. The advantage to the workload generator program is that if it is properly set up, it can simulate full network operation without requiring users to be involved. Furthermore, the transaction mix and rate at which transactions are introduced into the system can be controlled more precisely than they could be with human operators.

testing error handling capabilities

The most important oversight of most network testing plans is failure to test the error handling capabilities of the individual components and the system as a whole. The acid test of the error recovery capabilities built into each of the components is whether their error recovery procedures will work together. When a failed circuit is restored to service, do the modems automatically resume communications, or must they be restarted manually? What about the telecommunications monitor and other computer software? Do they recover from circuit or modem errors automatically?

Sometimes, the error recovery procedures of one device can create errors in other devices. It is important to identify these situations early so that proper procedures can be written for the network operations staff describing how to handle different types of errors. Some error recovery procedures will invariably be exercised during normal network testing. To thoroughly test the error recovery, however, it may also be necessary to introduce errors in the network by disconnecting circuits, powering off modems, or injecting noise on a circuit with an appropriate test instrument.

Cutover Cutting the new network into production use is a major milestone and an event that usually is very visible throughout the organization. If the planning and preparation have been done properly, all of the hardware and software components will have been installed and thoroughly tested. The users will have been trained and will be eager for the cutover to occur so that they can use the new network and accrue its benefits. Network cutovers frequently are scheduled to occur on a weekend. There may be a substantial amount of last-minute work to be done, and the extra time afforded by the weekend allows the work to be completed and a final round of testing to be conducted before all users come back to work on Monday morning. Sometimes, the cutover is scheduled for a Thursday night if it can be physically accomplished in a short period of time. This gives 1 day of live network operation before the weekend, and then the weekend is available for correcting problems or other unforeseen circumstances that became apparent on the first day of real operation.

vendor technicians

Despite the thoroughness of the testing, there may be some unanticipated startup problems. It is good practice to insist that technicians employed by the equipment vendors be on hand so that equipment problems can be resolved as quickly as possible. It is important that the help desk be in operation, and even overstaffed, in the first days of live operation so that incoming telephone calls can be handled promptly. The help desk should meticulously record all reported problems so that follow-up action and analysis can occur promptly.

changing displays

Implementation Cleanup and Audit When planning a network project, it is prudent to allow some time after the cutover for cleaning up problems that are identified but not quickly resolved. If, for example, it becomes apparent that despite careful analysis, a certain circuit is not performing up to expectations, a higher-speed circuit or an additional circuit will have to be installed. The network control operators may find that some minor changes to their status displays would help them identify problems before the user sees them. Making the required changes may take several days of analysis and programming, but many times, a few changes such as these can make the difference between a good network and a superior network.

revising documentation

Another activity that should occur after the implementation phase is updating the documentation of the network. In most cases, the network will have been implemented slightly differently from the way it was designed. These changes should be reflected in the final documentation so that it is up-to-date and provides an accurate description of the actual installation.

audit

Approximately 6 months after cutover, the new network should be audited by the company's internal auditors or a team of user and telecommunications management. The purposes of this audit are to determine whether the network is delivering the benefits that were promised, whether operational controls are in place and effective, and whether modifications need to be made to the network configuration, software, or application programs. The audit is conducted by interviewing network users and network operations personnel and by observing network parameters, such as response time at user terminals and the procedures in the network operations center. The audit team should write a formal report that describes its findings and makes recommendations for changes that need to be implemented. This report should be directed to senior management and appropriate telecommunications line management. Line management is responsible for writing a response to the audit report stating what actions will be taken based on the recommendations.

VOICE SYSTEM DESIGN

Most of the concepts and procedures that have been discussed thus far are applicable to both data and voice network design. After all, a circuit is a circuit, and both data and voice telecommunications needs are defined in terms of the amount of information transmitted. Voice conversations usually are quantified in terms of the minutes of conversation taken up by a call. Data transmissions are thought of as bit streams of a certain length that can be translated into transmission times.

When a designer creates voice systems, however, some unique techniques and terms commonly are used. Whereas a data traffic study measures the number of data messages from the terminal to the computer and back and the length of each message in characters, a voice traffic study measures the number of telephone calls and their average duration. The objectives of the studies are the same—to gather data about traffic volume and peak loads, so that enough circuits with sufficient capacity can be installed to meet peak demands. Many companies routinely gather voice traffic information and have graphs showing the call statistics extending back for several months or even years. As with the data traffic study, the purpose is to obtain data to use in predicting future requirements.

It is important to understand when the busy hours of each day occur, whether there are any days of the week that are busier than others, and whether there are any seasonal factors that cause unusual peaks during certain times of the year. There are some companies, for example, that do a heavy mail order business just before Christmas and employ extra telephone operators to handle the additional calls during the 2- or 3-month peak period.

holding time

The duration of a telephone call is called the *holding time*. Holding time is the time it takes the central office equipment to complete a telephone call plus the duration of the conversation. The number of telephone calls multiplied by the holding time measured in seconds is the number of seconds that the equipment is in use. For example, if there were 200 calls per hour each lasting 4 minutes, or 240 seconds, the equipment would be in use 200 × 240 or 48,000 seconds. Dividing this number by 100 gives the number of hundreds of call seconds, which is abbreviated *CCS*. CCS stands for *centa call seconds*, where *centa* is the common designation for hundreds. Another measure of equipment or circuit usage is the *Erlang*. One Erlang is 36 CCS, and because 36 CCS equal 3,600 seconds, both 1 Erlang and 36 CCS equal 1 hour of equipment usage. Both CCS and Erlangs are used frequently by telephone engineers when discussing telephone traffic.

Telephone systems are designed to provide sufficient equipment to handle traffic during the busiest hours with a certain grade of service, as was discussed in Chapter 5. The grade of service is the probability that a call cannot be completed because all of the equipment or circuits are busy. This probability is expressed as a percentage, such as P.02, which means that 2 percent of the calls during the busy hour are likely to be blocked. Telephone systems can be designed to provide any grade of service.

Determining the number of circuits required to carry the traffic in a voice communications network and provide a given grade of service is done using capacity tables. The *Poisson capacity table* is based on the assumption that there is an infinite source of telephone traffic and that all blocked calls will be retried within a short period of time. A sample Poisson capacity table is shown in Figure 15–10. The table is used by

Figure 15–10
Sample Poisson capacity table.

	Grade of Service			
Trunks Required	**P.01**	**P.02**	**P.05**	**P.10**
2	5.4	7.9	12.9	19.1
4	29.6	36.7	49.1	63.0
6	64.4	76.0	94.1	113.0
8	105.0	119.0	143.0	168.0
10	148.0	166.0	195.0	224.0
15	269.0	293.0	333.0	370.0
20	399.0	429.0	477.0	523.0
25	535.0	571.0	626.0	670.0
30	675.0	715.0	773.0	636.0
40	964.0	1012.0	1038.0	1157.0
50	1261.0	1317.0	1403.0	1482.0

Usage in CCS

Trunks Required	Grade of Service P.01	P.02	P.05	P.10
2	.153	.224	.382	.6
4	.870	1.093	1.525	2.0
6	1.909	2.276	2.961	3.8
8	3.128	3.627	4.543	5.6
10	4.462	5.084	6.216	7.5
15	8.108	9.010	10.63	12.5
20	12.03	13.18	15.25	17.60
25	16.13	17.51	19.99	22.80
30	20.34	21.93	24.80	28.10
40	29.01	31.00	34.60	38.80
50	37.90	40.25	44.53	49.60

Usage in Erlangs

Figure 15–11
Sample Erlang B capacity table

finding the CCS usage in the appropriate grade of service column and reading the number that appears in the trunks column.

Another commonly used capacity table is the *Erlang B capacity table.* The Erlang B capacity tables are based on the assumption that the sources of telephone traffic are infinite, but that all unsuccessful call attempts are abandoned and not retried. A sample Erlang B capacity table is shown in Figure 15–11. This table is read by finding the traffic load measured in Erlangs in the appropriate grade of service column and reading the number in the trunks column. If the usage figures are converted to common units (either CCS or Erlangs) and the tables are compared, it will be seen that the Poisson table is slightly more conservative than the Erlang B table. That is, for a given traffic load and grade of service, the Poisson table will suggest that a slightly higher number of circuits is required to carry the load. The reason for the difference is the slightly different assumption the two tables make about blocked calls.

Considering the total costs of the people and machines that use a communications network, compared to the costs of the lines themselves, there is some argument for using the most conservative capacity tables (Poisson) and slightly *overdesigning* the network. Overdesigning means that additional lines or trunks are added so that users will almost never experience a delay or a network-busy condition. The Bell System has overtrunked the public switched telephone network for years, and that is at least part of the reason why the network is so reliable and always available.

In some cases, telephone systems are purposely designed so that some calls will experience a delay. The most common examples are reservation systems and the telephone systems of organizations that dispense information. For example, when people call the Internal Revenue

Service to check on their tax returns, they frequently receive a recorded message saying that all tax agents are busy and that the call will be handled as soon as possible. Since the IRS is the only organization that has the information the caller wants, he or she will wait, even though the delay may be annoying. In a system such as this, the cost of providing a better grade of service is traded off against the probability that the call will be lost. In the case of the IRS, the probability that the caller will hang up and never call again is low. For an airline reservation system, however, the trade-off is not so clear because the caller may call another airline to make the reservation. Therefore, the balance between the cost of the network and the grade of service may be viewed from a competitive perspective.

Calls waiting for service are queued. Most telephone queuing systems queue the calls in the order that they arrive, and the customer gets service when his or her call reaches the head of the waiting line. This is known as *first-in-first-out (FIFO) queuing*. Some PBXs have the ability to queue calls, but in most cases, a separate piece of hardware—the automatic call distribution (ACD) unit—does the queuing and passes the call at the head of the queue to the next available agent.

Telephone systems with the queuing capability are designed using a different traffic capacity table, the Erlang C table. This table takes into account the number of calls to be held in the queue as well as the grade of service and predicted traffic load.

Other traffic capacity tables have been developed by communications engineers and consulting organizations. They are all based on the same concepts and probabilities, and they vary only in the assumptions they make and in their degree of conservatism in suggesting the number of lines to be used to carry a given traffic load at a given grade of service.

PROJECT MANAGEMENT

Implementing a new communications system is a project made up of many phases or steps. As we have seen, the steps are interrelated and are sometimes repeated as more information is gathered. It is extremely important that the project team use project management tools that are appropriate to the size of the work to be done.

A detailed implementation plan is mandatory for any size project. Gantt charts, which were discussed in Chapter 14, are appropriate for all but the largest projects. Large projects may require the use of a computerized *program evaluation review technique (PERT) chart*. The PERT chart shows the sequence in which tasks must be executed and the interrelationships between tasks. Software is available for personal computers and mainframes to assist in constructing PERT charts and keeping them current.

PERT charts

Regardless of which project management tools are used, it is important that an initial plan for the project be prepared, key milestones of project activity be identified, and project progress be reviewed on a regular basis. Frequently, the project team will have regular review meetings, typically weekly, in which team members review their activities for the past week, problems that they have encountered, and plans for the coming week. The frequency of these meetings depends on the size of the project, the number of people involved, and the technical complexity of the network that is being implemented.

Most project teams find it appropriate to review their progress with management at least monthly. The purpose of such a review is to keep management apprised of the progress the team is making as well as any difficulties encountered. Properly done, the management review meetings ensure that no one is surprised by changes to the schedule, scope, or resource requirements.

SUMMARY

Planning a communications network is a multistep, iterative process. After a formal request for work, the team studies the feasibility of a new communications system and then gathers more detail so that members grasp the problems with the existing system and the requirements for the new system. Then they examine the various alternatives for the new network and prepare one or more network designs. Each alternative is subjected to a cost analysis, and one or more of the alternatives is presented to management for review and approval. After approval is given, the equipment is ordered and installed. After the network is tested, the users are trained and the system is cut over into production use. Time should be allotted for implementation cleanup activities, and an audit of the system should be conducted approximately 6 months after the network is installed.

Designing voice systems is a similar process, but special terms and traffic tables are used. Based on the traffic load and the grade of service to be provided, these tables help determine the number of circuits required.

In any network design and implementation activity, project management tools need to be applied. The number and complexity of the tools depends on the size of the project.

CASE STUDY

Dow Corning's Communications Network Design

Dow Corning's communications network design has been done much less formally than the method described in the text. In the past, data network design was done by a data communications specialist based on the requirements for new terminals and the desire to upgrade line speed or to take advantage of changing technologies. In those days, the network was small enough that a single specialist could keep track of the various parts of the network and how they were performing. He knew when it was time to look at a particular piece of the network and evaluate options for upgrading it. Because of his expertise and knowledge of product offerings, he was able to do the design "in his head," gain management approval, order the required equipment, and install the changes relatively rapidly. Data network design was handled more as a project than an ongoing process.

As Dow Corning has grown and the networks have become more complicated, it has become necessary to handle data network design in a more sophisticated way. LANs and manufacturing networks have placed new requirements on and added complexity to the task of planning for network growth and expansion. Also, the telecommunications department is larger than it was a few years ago, and several people are involved in most of the projects. Therefore, there is a greater need for formal planning and good communications among all members of the telecommunications department and users than there was in the early 1980s.

Regular meetings are held to discuss all projects that are underway and to review their progress. Target dates are set, and responsibilities are assigned to the members of the project team. Project leaders on the larger projects prepare activity lists and Gantt charts so that progress can be tracked and measured. The network operations supervisor attends these meetings so that he will be aware of the changes planned for the network before they occur. He is able to ensure that the help desk operators are not surprised by changes and that new procedures are written when necessary.

Voice network design has been largely handled by Michigan Bell Telephone and AT&T. These vendors receive input from Dow Corning about actual or anticipated changes in facilities or personnel. In addition, the telephone companies perform periodic studies of actual traffic patterns and recommend changes to Dow Corning's telecommunications management.

The telecommunications staff is aware that these simple project management techniques will have to be improved as further growth occurs. The staff is trying to maintain the proper balance between too little project management and too much.

Review Questions

1. Why is a phased approach to network analysis design used? What does the text mean when it says that network analysis and design are an iterative process?

2. Why is it necessary to have a formal request document for network analysis and design work to be done?

3. Compare and contrast the jobs of a computer systems analyst and a network analyst.

4. Discuss the criteria that might be used for prioritizing the list of network projects that could be undertaken, assuming that there are not enough people to work on them all.

5. Explain the concept of the creeping commitment of resources to a project. Why is it important?

6. What information can network simulation give the network designer?

7. Why is it so important to understand the requirements for a communications network before a designer begins the design work?

8. Why is it important to understand the geographic requirements for a network?

9. What are the components of the total cost of a network?

10. Why is important to have users involved in the network design activity?

11. Why might communications software have to be written?

12. Explain the concepts of *peak load* and *busy hour.*

13. Why is it important to investigate alternatives for a network design?

14. Why is it important to test error handling and error recovery procedures?

15. What is stress testing? Why is it important?

16. What is the purpose of the implementation cleanup phase?

17. What are some examples of telephone systems that are purposely designed so that some users experience a delay in getting their call answered?

18. What is a CCS? What is an Erlang?

19. Explain how telephone grade of service is calculated.

20. What is telephone call holding time?

Problems and Projects

1. A point-to-point circuit from San Diego to Miami is being designed. The projected traffic on the circuit is four million characters during the business day, with 15 percent of the traffic occurring during the busy hour. If less than a 2-second response time is required, what speed circuit would you recommend be installed? Why?

2. Determining the proper amount of documentation for a network is not an easy task. Describe several different levels of detail at which the documentation could be created. What are the advantages and disadvantages of the most detailed level? The least detailed level?

3. Management understands the economic implications of a new or expanded network better than the technical aspects. How would you quantify the benefits of a new network that would provide electronic mail capability to your company's 35 locations throughout the United States? Assume that the network would deliver any piece of electronic mail within 30 minutes after it is sent. What might be the benefits of tying your customers to the network and letting them use it to communicate more effectively?
4. The holding time for a voice call is typically longer than the holding time for the transmission of a data transaction and its response. Why?
5. Telephone companies design the public telephone equipment and networks to provide a P.01 grade of service, whereas business telephone systems often are designed to provide service in the P.01 to P.05 range. Is the design level for the business system better or worse than the public system? If better, how can private business afford the better grade of service? If worse, how can business afford to have inferior telephone service?
6. The impact of a 40 percent or 50 percent compound growth rate is often not realized by business people. Assume a network is handling 2,000 messages per day this year but that the traffic will grow by 50 percent for each of the next 5 years. How many messages will the network need to carry 5 years from now?
7. As a further exercise in compound growth, take the starting salary you expect to earn when you graduate from college and increase it 5 percent for every year you expect to work to see what your salary would be when you retire. Try it at 8 percent. (Hint: It's an easy program to write if you have access to a computer.)

Vocabulary

project team
creeping commitment of resources
project life cycle
busy hour
super highway effect
request for proposal (RFP)
request for quotation (RFQ)
stress testing
workload generator
network simulator
driver
holding time
centa call seconds (CCS)
Erlang
Poisson capacity table
Erlang B capacity table
first-in-first-out (FIFO) queuing
program evaluation review technique (PERT) chart

References

Beshai, Maged, and Ray Vilis. "Meeting the Challenge of Network Configuration," *Telesis,* 1987 (2), p. 57.

Blythe, W. John, and Mary M. Blythe. *Telecommunications: Concepts, Development, and Management.* Indianapolis, IN: The Bobbs-Merrill Company, Inc., 1985.

Doll, Dixon R. *Data Communications: Facilities, Networks, and Systems Design.* New York: John Wiley & Sons, 1978.

Effron, Joel. *Data Communications Techniques and Technologies.* Belmont, CA: Wadsworth, Inc., 1984.

Fitzgerald, Jerry. *Business Data Communications: Basic Concepts, Security, and Design,* 2nd ed. New York: John Wiley & Sons, 1984, 1988.

Gurrie, Michael L., and Patrick J. O'Conner. *Voice/Data Telecommunications Systems.* Englewood Cliffs, NJ: Prentice-Hall, Inc., 1986.

Jarrett, C. Douglas. "A Practical Perspective on PBX Contracts," *Data Communications,* November 1986, p. 171.

Jewett, J., J. Shrago, and B. Yomtov. *Designing Optimal Voice Networks for Businesses, Government, and Telephone Companies.* Chicago: Telephony Publishing Corp., 1980.

Kaufman, Bob. *Cost-Effective Telecommunications Management: Turning Telephone Costs into Profits.* Boston: CBI Publishing Company, Inc., 1983.

Leeson, Marjorie. *Systems Analysis and Design.* Chicago: Science Research Associates, Inc., 1985.

Martin, James. *Systems Analysis for Data Transmission.* Englewood Cliffs, NJ: Prentice-Hall, Inc., 1972.

Salinger, Anthony W., and Leigh Gerstenmaier. "How to Pin Down User Requirements," *Data Communications,* August 1985, p. 155.

Schaevitz, Alan Y. "Network Design Issues for the 1990s," *Business Communications Review,* November–December 1988, p. 49.

Stamper, David A. *Business Data Communications,* 2nd ed. Menlo Park, CA: The Benjamin/Cummings Publishing Company, Inc., 1986, 1989.

Thompson, H. Paul. "Increasing Data Communications Control Through System Simulation," *Information Systems Management,* Vol. 4 (1), Winter 1987, p. 29.

Xephon Consultancy Report: Network Management for IBM Users. Berkshire, England: Xephon Technology Transfer Ltd., 1985.

CHAPTER

NETWORK OPERATIONS AND TECHNICAL SUPPORT

OBJECTIVES

When you complete your study of this chapter, you should be able to

- explain why it is necessary to manage the network;
- describe the functions of the network operations group;
- describe how network problems are diagnosed and repaired;
- explain why problems are escalated;
- explain the three levels of problem resolution;
- describe how network performance is monitored and measured;
- explain how the network configuration is documented and controlled;
- discuss the importance of having change management procedures;
- describe the functions of the communications technical support group.

INTRODUCTION

This chapter deals with the tasks required to operate and manage a communications network. The operation of a communications network is a complex, interrelated set of activities, some oriented toward normal or routine operations, and others that are performed when problems occur. The various facets of network operations must work together cohesively to provide the user with consistent, reliable service.

The communications technical support group is often a part of network operations. This group has the responsibility of supporting the communications software and solving the most complex network problems. Technical support specialists must understand the interrelationships between the network hardware and software.

DEFINITION OF NETWORK OPERATIONS

Network operations, sometimes called *network management,* is the set of activities required to keep the communications network operating. The proper view of the scope of network operations is the same view that the user has of the network. It is an all-encompassing view in which all of the elements that are required to deliver communications or computing service to the user are equally important. If any of the elements are not functioning properly, from the user's perspective, "the network is down." If the central computer or one of the application programs running in it is not operating, the user cannot receive the service he or she expects. The network operations group should be concerned about the problem and involved in getting that service restored. From the user's perspective, anytime he or she wants to use a terminal on the network, it should be available.

Communications users have become accustomed to this level of service because of the extremely good reliability of the telephone system. For most of us, waiting for a dial tone when we lift the handset rarely occurs. We expect that when we want to make a telephone call, the network will be ready to serve our needs. Unfortunately, most business data communications networks do not operate at that same high level of reliability. In many companies, several network failures per day or erratic response time is considered normal or "good enough."

This implies that all business communications networks should operate at the same level of availability and reliability as the public telephone network. From a practical standpoint, the service requirements of the business must be defined and the network designed to meet those requirements. If 24-hour-per-day availability of the network is not required, it is a waste of money to design the network for 24-hour operation. If an outage or two each week is not critical (albeit inconvenient for the users), perhaps it is not worth spending money to provide more reliable service. The key concept is that each company must define the service level it expects from its communications network.

Network operations people have the challenging responsibility of ensuring that the defined requirements for availability and reliability are met. In most companies the service requirements also include requirements about consistently good response time and fast problem resolution. The network operations group can only meet these service objectives if the scope of its responsibility is defined broadly.

In some companies, however, the network operations group is viewed more narrowly as only having the responsibility for lines and modems. By limiting the scope, major pieces of the network may be left with no management. The proper scope of network operations responsibilities includes:

- user workstations;
- cluster controllers;
- modems;
- communications lines;
- line concentrators;
- multiplexers;
- front-end processors;
- communications software.

In addition, the network operations personnel need to be familiar with the computers and applications software that are attached to and use the network. Admittedly, there is some potential overlap with the functions of the computer operations staff, and in each company, the lines of responsibility must be drawn. However, the network operations people are in the best position to view the entire process of delivering information from the computer to the user and are, therefore, in the best position to communicate with the user and to see that problems get resolved.

The same network operations staff should also be responsible for the proper operation of the voice network and the services it provides. Although historically the voice and data operations in a company have been handled by two different groups, there is no organizational or technical reason why they shouldn't be brought together. There may be strong political reasons for not combining the two groups, but it is in the best interest of the company to overcome these objections and move toward a single network management organization.

WHY IT IS IMPORTANT TO MANAGE THE NETWORK

One might ask why it is worthwhile to spend any time or effort managing the communications network. There are three primary reasons:

- the network is a corporate asset;
- the network is a corporate resource;
- the network is growing rapidly.

The communications network, as well as computers and applications, are assets of the business that in most cases are becoming increasingly vital to the company's operation. Perhaps this can be most easily understood by thinking about the importance of the telephone to businesses and what would happen if it were not available. Similarly, companies with computers and terminals are finding that they are increasingly dependent on the reliable operation of their data communications system in order to do business. This dependence is similar to the company's need for good people, sufficient financial resources, suppliers, and customers. Each of these assets must be managed, and the communications network is no exception. It must be managed, too.

Viewing the communications network as an asset implies that its creative use can give the company advantages in its business operations. The communications facilities might be employed to make it easier for customers to contact the company and get product information. They might be used to give production people instantaneous access to material requirements or to give service people access to a database containing solutions to customer problems. Communications facilities also may be used to improve the productivity or effectiveness of employees by making it easier for them to exchange information, schedule meetings, or gain approvals. Imagination is the major factor limiting the creative use of communications facilities.

network growth versus budgets

Computer and communications networks in many companies are growing at a rate of 15 percent to 50 percent per year, compounded. The average communications budget for companies in the United States is growing 10 percent to 20 percent each year. The breadth of service being provided and the amount of money being spent on communications makes network management virtually mandatory. Without management, there would be chaos. In particular, network operations management must ensure that the service provided to existing users of the network does not degrade as new users are added. Furthermore, they must ensure that the change (growth) is implemented in a controlled manner.

Even in small companies, communications networks should be managed. Although the small company cannot afford to have very many people working in the network operation area, at least one person must be given the responsibility to ensure that the company's communications facilities operate properly. Many of the tasks described in the rest of this chapter are handled very informally in a small company, but the basic network operations and problem-handling techniques are as relevant to the small company as they are to the large organization.

THE FUNCTIONS OF NETWORK MANAGEMENT

The business objective of network management is to satisfy users' service expectations by providing reliable service, consistently good response time, and fast problem resolution. Network management is made up of the following six activities:

- network operations;
- problem management;
- performance measurement and tuning;
- configuration control;
- change management;
- management reporting.

Each of these activities interrelates with the others. In large companies, each of the activities might be handled by a separate department, whereas in small organizations, one person might perform all of the functions. We will study each activity separately as though it were done by a separate department, and we'll examine where the interrelationships occur.

Network Operations Group

The network operations group is the heart of network management. It is most easily visualized as a group of people who reside in a place called the *network control center.* The network operations group has many similarities to a traditional computer operations group. In some companies, the two groups are combined or at least report to the same supervisor. The network operations group is responsible for the management of the physical network resources, activation of components, such as lines or controllers, rerouting of traffic when circuits fail, and execution of normal and problem-related procedures.

starting and stopping the network

If the network does not operate 24 hours per day, 7 days per week, there is a daily routine of starting and stopping the network. To start the network, communications software is loaded in all of the computers and communications controllers that are software-driven. Then commands must be issued to instruct the software to activate each circuit. A command such as Start Line might, for example, instruct the software to begin polling the terminals on a line. In some cases, individual terminals may need to be activated with a Start Terminal command. This might occur when terminals are in different time zones, and the terminals in each time zone are activated just before the users come to work.

At the end of the day, there is a similar process for shutting down or deactivating the network. Individual terminals may be stopped as people leave work, circuits may be individually deactivated, and finally, after the entire network is shut down, the communications software may be removed from the computers, freeing the memory for use by other programs. In most communications systems, the individual commands required to start up and shut down the network are aggregated into a macrocommand that might be called Startnet or Stopnet. When these macrocommands are entered, all of the individual line and terminal commands are executed automatically. In some systems, even the issuance of the Startnet and Stopnet commands can occur automatically when triggered by a clock at a certain time each day.

monitoring the network

When the network is operating, the network operations staff is responsible for monitoring its behavior. Because communications circuits and terminals exist in relatively uncontrolled environments (as compared to the controlled environment of a computer room), unusual conditions and problems are a certainty. Good telecommunications software provides regular status information for the network operations group. It identifies which components of the network are operating normally and which are having problems. Often the status is in the form of a visual display on a VDT and is updated every few seconds, giving a current, realtime picture of the network's operation. Well-designed network monitoring hardware and software should quickly alert the network operations group when problems occur so that appropriate actions can be taken promptly. Problem identification and resolution are discussed extensively in the next section.

collecting statistics

Another responsibility of the network operations group is to collect statistics about the network's performance. The statistics-gathering process

The help desk serves as a single point of contact for users with network-related problems. Help desk operators have access to a variety of network status displays and other tools, which allow them to monitor the network's performance and assist in problem diagnosis.

should be a routine part of the operation, perhaps aided by software or hardware monitors that automate the process. When performance or other problems occur, additional information should be gathered. Although the first responsibility is to get the problem solved and the network operation back to normal, the secondary responsibility is to gather other data that can be analyzed later to prevent the problem from recurring. In networks where the performance data is gathered automatically by the hardware and software, the network operations group must ensure that the data-gathering mechanisms are working properly.

The network operations staff normally has other responsibilities, such as problem management, configuration control, or change management, each of which will now be discussed in detail.

Problem Management

Problem management is the process of expeditiously handling a problem from its initial recognition to its satisfactory resolution. One of the important subgroups of network operations is the *help desk*. The help desk is the single point of contact with users when problems occur. Ideally, a single telephone number is established, and users are instructed to call the help desk whenever they have problems with any of their communications equipment.

The first responsibility of the help desk personnel is to log each problem reported. A sample log sheet is shown in Figure 16–1. The information may be kept manually on a log sheet or entered into an automated system using a terminal or personal computer. The types of information that are recorded include the date and time, the name of the user reporting the problem, the type of terminal being used and its

HELP DESK PROBLEM LOG						
Date	Time	Name of Caller	Phone No.	Terminal Type	Symptoms	Resolution

Figure 16–1
Help desk log of each reported problem.

identification, and the symptoms of the problem being reported. The advantage of an automated problem logging system is that the information recorded in it is available to those who have a need for it. Another advantage is the system's ability to easily sort and report the problems in various sequences for later analysis.

In many cases, the help desk operator will be able to offer immediate assistance while the user is on the telephone. If, for example, a communications circuit with 10 terminals on it has failed, it is likely that the help desk will receive several calls from users at terminals on the circuit. After the first call, the help desk operator will be familiar with the problem and will be able to assure subsequent callers that action is being taken. Other types of problems, such as certain types of terminal errors, tend to be repetitive in nature. The help desk operator, on hearing the symptom, may be able to tell the user a particular sequence of key strokes that will clear the problem or correct the error. Even in such simple cases, the problem should be logged. Later analysis may show that many users are having the same difficulty and that additional training or documentation is required.

Help desk operators should be provided with a script or at least a standard list of questions to ask all callers. Some of these questions are designed to ensure that the proper data about the problem is gathered, such as "What is your name and telephone number?" or "What type of terminal are you using?" Other questions are diagnostic in nature, such as, "Is the green light on your terminal lit?" or "Have you checked to be sure your terminal is plugged in?" or "Are any other people sitting near you having a similar problem?" Some companies have developed flowcharts or decision trees of questions that assist the help desk operator in diagnosing even relatively complex problems. The help desk person asks the user a series of questions, and the answer to one question narrows the range of possible causes and determines which question will be asked next. Diagnostic tools such as these are a simple form of artificial intelligence, and they themselves are subject to automation.

trouble ticket

In large organizations, a *trouble ticket* is opened for each problem reported. The trouble ticket is a form that is filled out by the person taking the call and then passed to other people or groups who work on the problem and resolve it. The information on a trouble ticket is similar to the information recorded on the help desk log. When a problem is resolved, the trouble ticket is completed with the time, date, and action taken to correct the problem. The person who corrects the problem is usually the one who closes the trouble ticket and completes the form or log.

An automated system that records the information about the problem and produces trouble tickets can simplify this entire process because the information about the problem can be accessed by people at the help desk, by network technicians, and even by users. Each person can be given the authority to view or update certain parts of the record about the problem, and the system can be programmed to notify the help desk and the user when the problem is resolved.

Problem Resolution Levels One approach to problem tracking and resolution uses the notion of *levels of support*. The exact definition of the levels varies from organization to organization, but the idea is that the person who initially takes the telephone call about the problem has enough knowledge to be able to quickly resolve a high percentage—say 80 percent to 85 percent—of all problems reported. Problems that cannot be resolved within a few minutes at this level, called level 1, are passed to level 2, which is made up of technicians who have a higher degree of skill, experience, or training. Level 2 handles all of the problems remaining from level 1 and those where a vendor must be involved. It is expected that level 2 would solve 10 percent to 15 percent of the total problems that are reported.

All problems not solved at level 2 are passed to level 3, which is made up of communications technical support specialists or vendor specialists. Level 3 usually receives only about 5 percent of the problems—those that are extremely complex, difficult to identify, or difficult to solve. Problems that reach this level may require software modifications, hardware engineering changes, or other corrective actions that take a long time to put in place.

The advantage of the leveling approach to problem solving is that the problem is solved by the person with the lowest skill level who is able to do so, and highly skilled technical people are reserved to work on the most difficult problems.

Escalation Procedures Procedures need to be in place to escalate the status of a problem if it has not been resolved within a predetermined period of time. Certain problems are more critical than others, and these are the ones that need more formal and rapid escalation. A single inoperative terminal might be escalated if it has not been repaired within 24 hours, whereas a front-end processor outage that causes the entire network to be down may be escalated immediately.

Problem escalation takes two forms. One type of escalation is to bring additional technical resources to bear in order to help solve the problem. The other type of escalation is to make users and management aware of the actions being taken to resolve the problem. Often these escalations proceed at different rates. Figure 16–2 shows a sample of a generic, escalation procedure. It is generic because it does not distinguish between types of problems that might be escalated at different rates. In many companies, problems are first ranked according to their severity. Severe problems are escalated very quickly, whereas problems with limited impact or that only affect one or a small group of users are escalated more slowly. The example in Figure 16–2 shows the differences between technical and management/user escalation.

Technical:

1. Level 1 (Help Desk) works on problem for a maximum of 15 minutes. If problem is not resolved, pass it to level 2.
2. Level 2 (technician) works on problem for up to 1 hour. If the problem is not resolved, notify the help desk supervisor and continue working on the problem.
3. If the problem is not resolved in 4 hours, get the appropriate level 3 (network specialist) involved. Level 2 retains "ownership" of the problem. It is level 2's responsibility to monitor the progress and to keep the user and the help desk supervisor informed about the status of the problem every 2 hours after level 3 gets involved.

Managerial:

1. The help desk supervisor is notified by level 2 if problem has not been resolved in 1 hour.
2. If the problem is not resolved in 2 hours, the help desk supervisor notifies the supervisor of network operations.
3. If the problem is not resolved in 4 hours, the supervisor of network operations notifies the manager of telecommunications. The manager of telecommunications calls the manager in the user department to discuss the situation and decide on any extraordinary action to be taken.
4. If the problem is not resolved in 8 hours, the manager of telecommunications notifies the chief information officer (or equivalent) and discusses the actions taken to date, and the future plans to get the problem resolved. The discussion should also include the possibility of contacting vendor management if appropriate.

NOTE: These sample procedures do not account for actions to be taken if the problem continues after normal working hours. The actions to be taken depend on the nature of the problem and the criticality of telecommunications to the company.

Figure 16–2
Problem escalation procedure.

From time to time, all open problems or trouble tickets need to be reviewed by the network operations supervision. Typically, this is done each day with an eye toward spotting unusual problems or those where the normal problem escalation procedures may not be applicable. On an exception basis, supervision can make decisions about the relative priorities of outstanding problems and whether to take extraordinary steps, such as accelerating the normal escalation procedures, in order to resolve the problem.

Another technique used by many companies is a periodic *problem tracking meeting*. In this meeting, people from network operations, software support, vendor organizations, and perhaps computer operations and user groups review all outstanding problems. This type of meeting is an excellent communications vehicle, and if the computer operations people are involved, it gives an opportunity to prioritize all of the unresolved problems, whether they are computer- or network-related.

Bypassing the Problem If a problem is caused by a piece of equipment or circuit failure, the ideal solution from the user's point of view is to bypass the problem, enable the user to keep working, and then diagnose and fix the problem later. A failing terminal or modem might be "fixed," from the user's perspective, by replacing it with a spare that is available for just such a purpose. The failed equipment can then be repaired or returned to the vendor for service. When it is repaired, it can either become a spare itself or be returned to the original user.

substituting equipment

Reconfiguration In the network control center where all of the circuits come together, spare equipment can be substituted by the use of switches and *patch panels*. A patch panel is a piece of equipment on which each circuit has one or more jacks. Using cords with plugs, the network technician can connect spare equipment into the circuits temporarily by inserting a plug into the appropriate jack at the patch panel.

Switches also can be used. Depending on what equipment needs to be substituted, analog or electronic switches can be used that switch a circuit from a failing modem to a good one or switch a modem from one port on the front-end processor to an alternate port. Usually, software switching must also occur to allow terminals to be addressed on the alternate ports, modems, or circuits.

dial backup

Another technique for bypassing a failure is the use of a dial-up line as the backup for leased circuits. This technique is called *dial backup*. Certain types of modems can handle both leased and switched connections and are designed to automatically make a dial-up connection if the leased circuit fails. Often the dial-up connection operates at a slower speed, but for most applications, slower speed operation is preferable to being totally out of service.

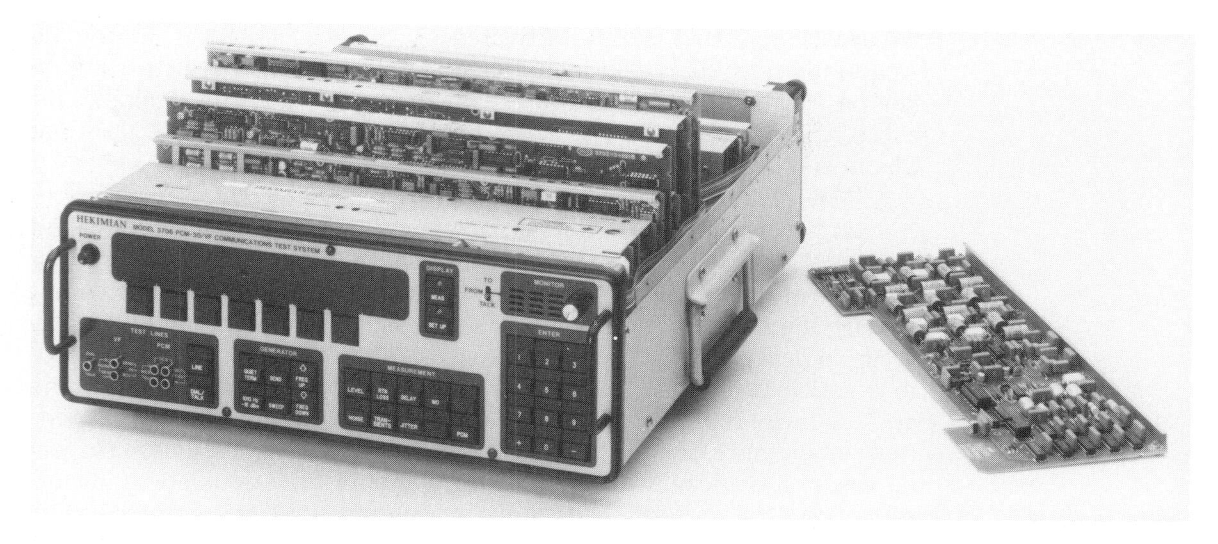

The Hekimian Laboratories 3705 test system can perform tests on analog or digital circuits. It is connected to the circuit using patch cords that plug into the jacks on the front panel and similar jacks on a patch panel. (Courtesy of Hekimian Laboratories, Inc.)

Problem Diagnosis and Repair Because it is not so easy and usually not economically attractive to have spare circuits, front-end processors, or other expensive equipment available, diagnosis and repair of many types of equipment problems are essential. Front-end processors do not normally fail often, and because the diagnosis and repair requires specialized equipment and training, it is almost always left to the vendor's maintenance people. On the other hand, since communications circuits are more prone to failure, many companies find it desirable to have some testing equipment available to assist in diagnosing circuit problems.

Depending on whether an analog or digital circuit is to be tested, different types of equipment are required. Analog signals are analyzed with simple speakers, tone generators, butt sets, and conductive probes. The techniques and equipment are similar for all analog lines, whether they are used primarily for voice or for data. Digital circuits or the digital sides of analog circuits require the use of breakout boxes and oscilloscopes. Some modems are capable of running tests on the communications circuits to which they are attached. The types of problems that this equipment can identify include frequency response, dB loss, and various types of distortion (problems described in Chapter 9).

The carrier that provides the circuit obviously has extensive testing equipment as well and can be relied on to diagnose and resolve circuit problems. There are many cases, however, in which additional information provided from tests at the user's end of the circuit greatly assists the vendor in problem resolution. Test data can pinpoint the exact location of the problem and help resolve it.

Most modern communications equipment has diagnostic capabilities built into its basic design. Modems usually have self-test routines that can be activated manually through the front panel or, under certain circumstances, that will be activated by the modem itself. Terminals frequently have diagnostic routines that the user can activate through a key on the keyboard. One of the first questions the help desk should ask when a problem is reported is whether the appropriate terminal diagnostics have been run.

When all hardware appears to be working correctly but communications is still impossible, it may be necessary to employ a device called a *protocol analyzer.* A protocol analyzer is an electronic instrument that can look at the actual bits or characters being transmitted on the circuit to determine whether the rules of the protocol are being followed correctly. Some protocol analyzers are designed to handle a specific protocol, such as BISYNC or SDLC, whereas other equipment can examine and diagnose multiple protocols. When protocol analysis is performed, it is usually necessary to have software specialists involved because few hardware technicians are trained to understand the detailed characteristics and sequences of the protocols and data being transmitted.

A basic decision that companies must make is how much problem diagnosis and testing they will perform themselves and how much they will leave to vendors or other maintenance organizations. Often the answer to the question is to use a combination of people, some from inside the company and others from the vendor. In that case coordination of the work is important.

One reason why it is important to have trained technicians and appropriate diagnostic equipment is that in today's multivendor networks, it is not always clear which vendor should be called when a problem occurs. For example, a line from Colorado to California has two Regional Bell operating companies, a long distance carrier, modem manufacturers, and terminal manufacturers involved. Sometimes, technicians must perform tests on the network to determine exactly where the problem lies. Customers may be paying for service of a certain quality. They may want to test the circuit periodically in order to ensure that they are receiving the quality of service they are paying for. Even if a vendor's equipment is clearly at fault, it may be possible to perform tests that further pinpoint the problem while the vendor's technician is on the way. This may greatly speed the process of repairing the failing component and keep the cost of the technician's time to a minimum.

Performance Management

Performance management is the set of activities that measure the network performance and adjust it as necessary to meet users' requirements. The people responsible for network performance management use the statistical data gathered by the network operations staff. The data is analyzed and summarized, and the network performance is reported to management. This work also triggers performance adjustment activities or performance problem resolution, which is normally done by the communications technical support group. It also provides utilization

The Hewlett-Packard HP 4954I wide area network protocol analyzer allows the network specialist to examine characters transmitted on a circuit to determine whether the rules of a protocol are being followed. This analysis is useful to isolate and identify installation and maintenance problems. (Courtesy of Hewlett-Packard Company)

data to network capacity planners. The location of the performance management responsibility depends on the size of the telecommunications department, the skill levels of the staff, and the preference of the telecommunications manager. Performance tuning will be discussed in more detail later in the chapter because it is most frequently performed by the technical support people.

In order to determine whether the performance of a network is adequate, as measured by its reliability and response time, preestablished performance objectives are required. Performance objectives may be established for the entire network or for a specified set of equipment or users. When the objectives are established for a group of users, such as a department, the objectives are commonly called a *service level agreement*. Service level agreements often involve stating that of the X hours of the day or week that a user's terminal and applications will be available, the reliability of those applications will be measured as a certain percentage of X. It also contains a statement about the response time the user can expect to receive at the terminal. A typical service level agreement might include the following:

service level agreement

- the order entry application and customer service terminals will be available from 7 A.M. to 8 P.M. EST each business day;
- the reliability will be 98 percent as measured at the user terminal. Downtime will not exceed 2 percent;

- response time, measured from the time the user presses the ENTER key until the response is received at the terminal, will be less than 1.5 seconds 90 percent of the time and less than 10 seconds 100 percent of the time.

This type of service level statement must be negotiated between the user and the network and computer operations people. The starting point should be to identify the user's requirements. The network and computer people may have an option to provide a better grade of service at an increased cost if the user is willing to pay for it.

The other part of performance measurement is the actual tracking of the network's performance. Performance data about a data communications network can be gathered by hardware, software, or a combination of the two. Recording the time that the network actually started in the morning and closed at night is normally a software function. Measuring the availability and response time from the user's point of view must be done at or close to the terminal. A common way to gather the data is with hardware in the cluster control unit. This hardware can capture the response time and all outages except the actual failure of the user's terminal. Some types of terminal failures can be detected at the host. For example, the failure of a terminal to respond to a poll may mean that it has failed, or it may mean that the user simply has turned the terminal off and gone home. Whether that type of "failure" is counted in the network statistics needs to be determined by management.

For voice networks, statistics are gathered by the carrier that provides the circuits, as well as by PBXs and many key systems. The carrier is usually willing to provide the statistics collected to the customer. Working together, the carrier and the customer can ensure that the network is properly configured for the actual traffic loads. In some PBX systems, the statistics collection can be tailored to meet the particular requirements of the company. Key systems usually have less flexibility and provide a predefined set of standard statistics about a call: its duration, the extension number of the caller, and perhaps the outgoing circuit or trunk identification if one was used.

The most useful type of performance measurement hardware or software on either voice or data networks provides a means for forwarding the information to a central database that is accessible to the network operations staff. One such system for a data network collects data at the cluster controller and automatically forwards it to the host computer when the communications line is otherwise idle. The data in the central database can be examined on a realtime basis or processed periodically by programs that produce summary or statistical reports showing the entire network's performance. Once the performance data is available, it can be compared to the service level agreement to see if the objectives were met.

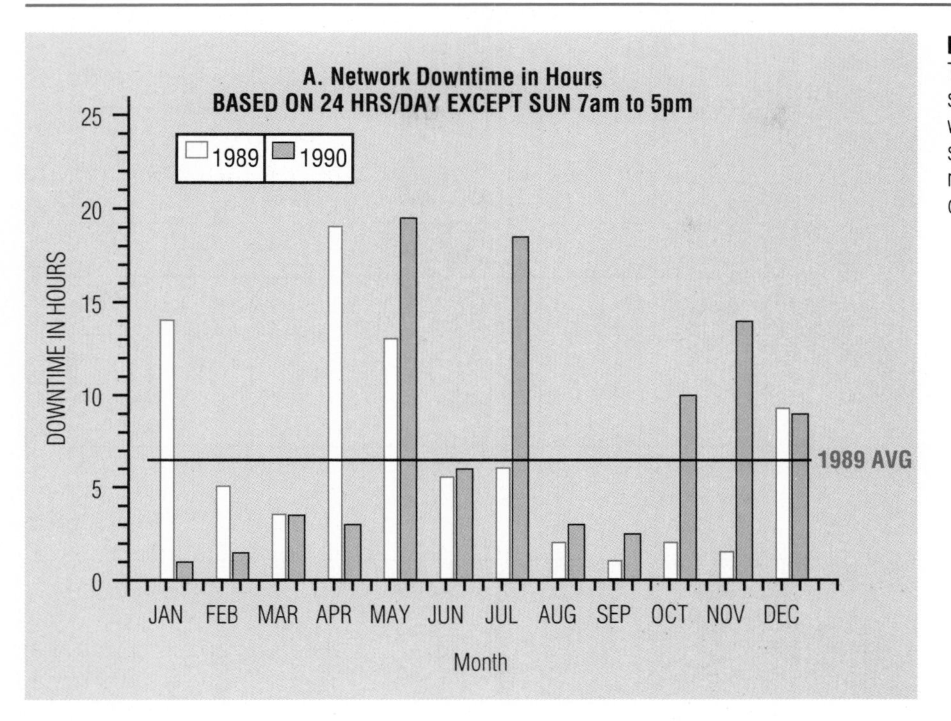

Figure 16–3
These trend charts show several parameters of a network's operations. The standard level of performance is shown on many of the charts.

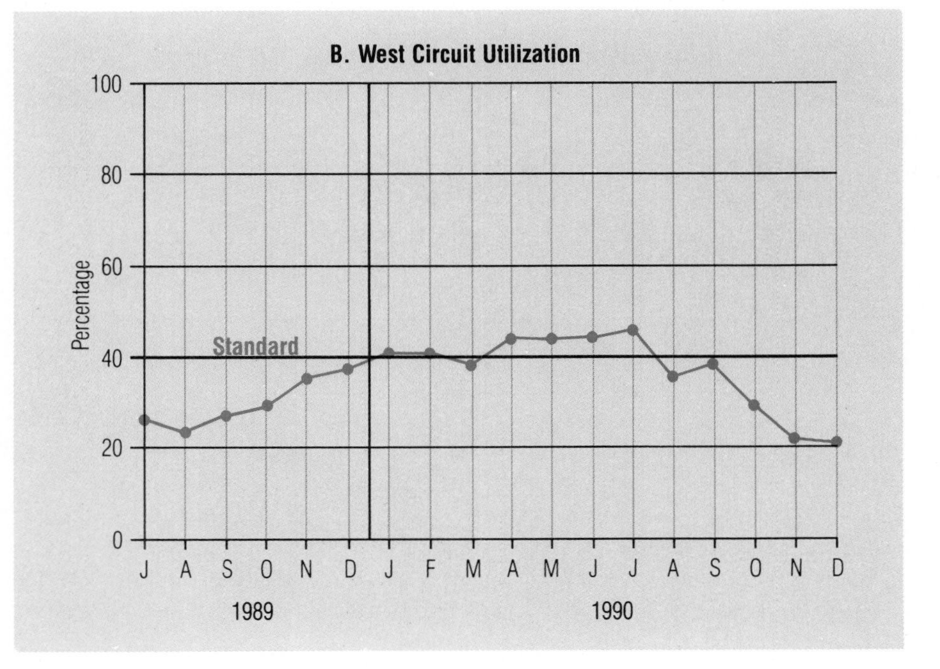

Figure 16–3
(Continued)

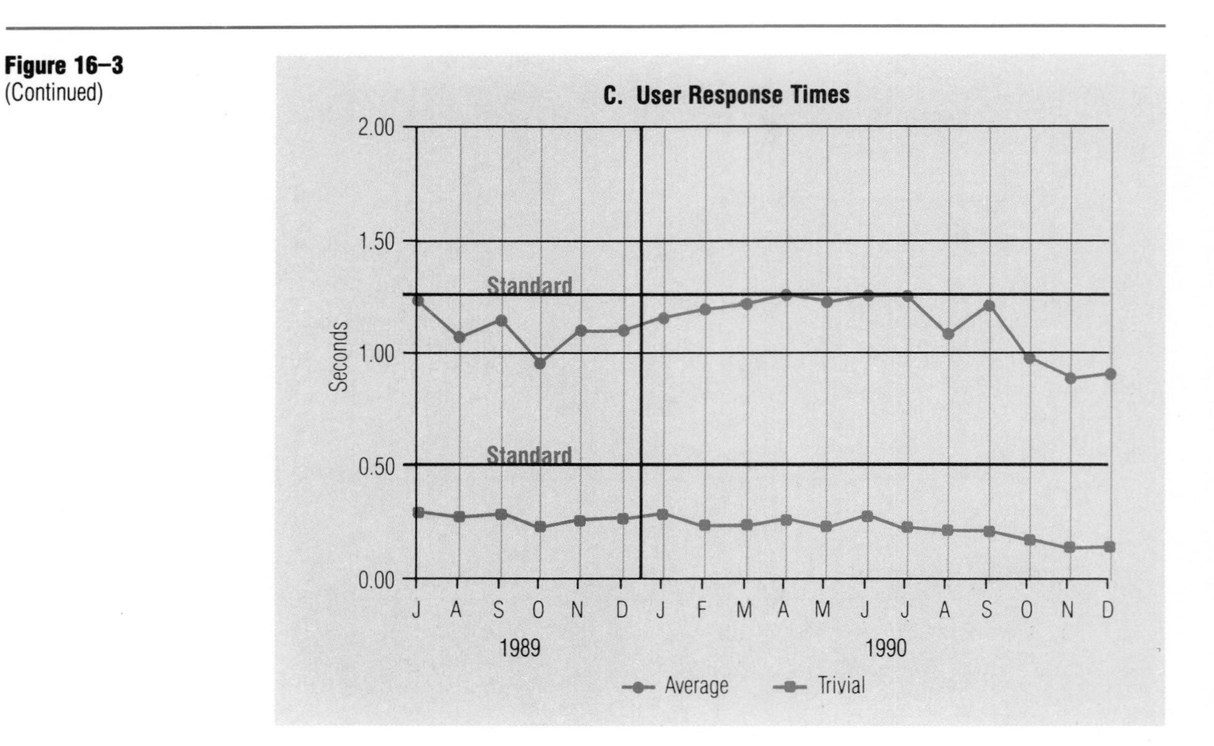

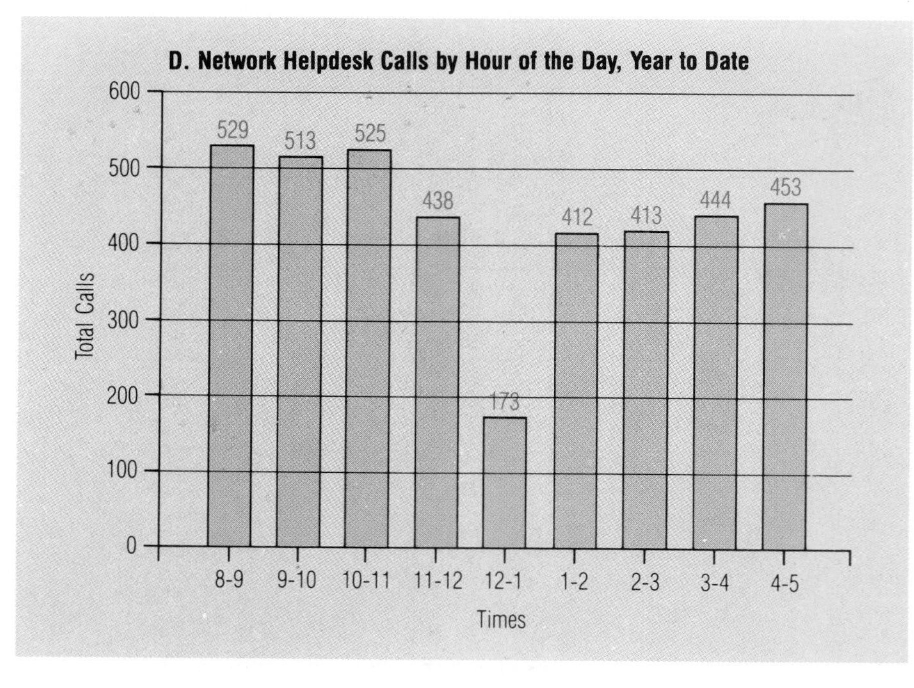

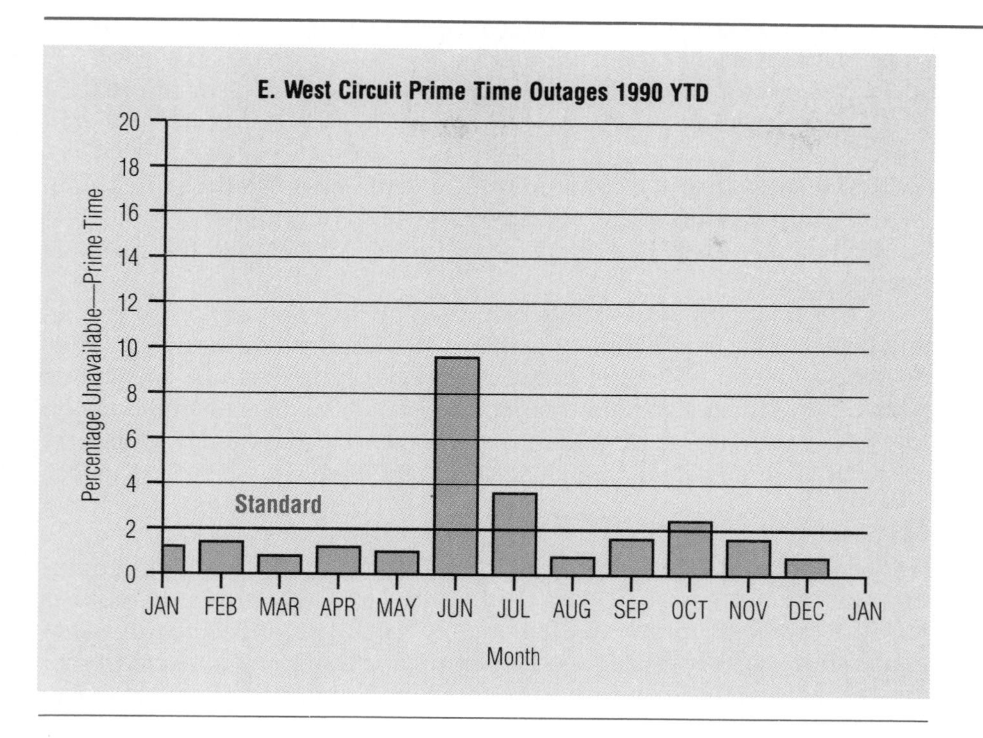

Figure 16–3
(Continued)

In any type of performance measurement system, it is important to keep a history of performance statistics so that trends can be examined. One useful technique is to plot trend charts that show how certain network parameters, such as availability, response time, or circuit utilization, are changing over time. Some examples are shown in Figure 16–3. The trend charts should also show the standard or desired level of performance. Trend charts are particularly useful for watching parameters that change gradually over time. A gradual increase in response time might not be noticed for a long time if the daily performance reports were the only information available. On the other hand, a trend chart would show the increase in response time more quickly.

Some of the parameters of the network that need to be tracked are

- response time—by hour of the day, by application, by circuit, all measured from the user's point of view;
- circuit utilization—by circuit and by time of day;
- circuit errors—by circuit;
- transaction mix—by time of day, over time;
- routing—utilization of each circuit when multiple routes to the same destination are possible;
- buffer utilization—in the FEP, in cluster controllers, and in the computer;

- queue lengths—in the cluster controller, in the multiplexers, in the FEP, and in the computer;
- processing time—the time to process a transaction or message in the computer.

Each of these statistics gives a unique view of network operation. Often two or more will correlate, as, for example, when response time gets longer because there is a shortage of buffers or because the message queues are long.

Over time and with experience, the network operations personnel will begin to know what the normal values are for their network. They develop a feel for when the network is performing normally and when it isn't. More formal statistical techniques can establish control limits on the values so that it is easy to determine whether day-to-day variations are normal or indicate that a statistically significant change has occurred.

Configuration Control

Configuration control is the maintenance of records that track all the equipment in the network and how it is connected. One required record is an up-to-date inventory of all network hardware and software. Diagrams showing how the pieces of equipment and circuits connect together are another useful form of documentation. Configuration control may be performed by the network operations group or the administrative support group. As with all of the organizational alternatives, the definition and assignment of responsibilities are a management function.

Equipment Inventory The network equipment inventory is like any other inventory system in that it is updated to keep track of the installed equipment and circuits as well as all additions and deletions. If the inventory is kept in a computerized system, the data can easily be manipulated, sorted, or reported in different ways. Network operations personnel might want to see a report of all of the terminals of a certain model or to find the location of the cluster control unit with a certain serial number. Management might be interested in data about the total value of all of the terminals or the total costs of all circuits. Such information is obtained easily from a properly designed, computerized inventory of network equipment. An example of a report from a network equipment inventory system is shown in Figure 16–4.

inventory software

Software is available for personal computers to keep these types of inventory records, or a simple system may be developed in-house. As an alternative, the company may decide that its network equipment inventory will be kept in the company's property ledger, the accounting system that tracks all of the company's physical assets.

maps

Network Diagrams One of the most useful forms of displaying the topology of the network is a map that shows all of the network locations,

Figure 16–4
Inventory list of telecommunications equipment.

ACCOUNT 7045-336 DETAIL BY DEVICE CODE
RUN DATE: 90/07/02

DEVCD	TYPE	MODEL	SERIAL	SITE CODE	ACCOUNT BILLED	VENDOR CHARGE
COMMCTL	3745	CONTROL	00951A	CMD	7045-336-4X3	5228
	3720	CONTROL	03498	MID	7045-336-4X3	1424
CONTROL	3274	31A	12506	CMD	7045-336-4X3	1182
	3274	41C	D0055	CMD	7045-336-4X3	338
	3274	41C	D0056	MID	7045-336-4X3	338
	3274	41C	D4528	MID	7045-336-4X3	331
	3274	41C	D5477	CMD	7045-336-4X3	338
CRT	3101		G1843	CMD	7045-336-4X3	–
	3178	C10	W3090	CMD	7045-336-4X3	36
	3178	C20	BH624	MID	7045-336-4X3	38
	3178	C20	BH629	MID	7045-336-4X3	38
	3178	C20	BR715	CMD	7045-336-403	38
	3178	C20	DL634	CMD	7045-336-4X3	35
	3178	C20	EU334	CMD	7045-336-4X3	36
	3178	C20	P0319	MID	7045-336-4X3	38
	3178	C20	P7802	CMD	7045-336-403	38
	3178	C20	Q1611	CMD	7045-336-403	38
	3178	C20	T0173	CMD	7045-336-4X3	36
	3178	C20	UN757	MMS	7045-336-4X3	36
	3180	110	AE958	CMD	7045-336-403	53
	3180	110	V1904	MID	7045-336-4X3	48
	3180	110 DE	T4790	MID	7045-336-4X3	47
	3277		B1140	MMS	7045-336-4X3	–
PRINTER	2700		32255	CMD	7045-336-4X3	822
	2700		32305	CMD	7045-336-4X3	822
	2700		32431	CMD	7045-336-4X3	822
	2700		40435	CMD	7045-336-4X3	723
	2700		40560	CMD	7045-336-4X3	723
	2700		40938	CMD	7045-336-4X3	723
	2700		41039	CMD	7045-336-4X3	723
	2700		41416	CMD	7045-336-4X3	723
	2700		41434	CMD	7045-336-4X3	723
	2700		41676	CMD	7045-336-4X3	723
	3262		46878	MID	7045-336-4X3	499
	3262		47021	CAR	7045-336-4X3	417
	3262		57195	HLK	7045-336-4X3	583
	3262	2	50563	GRN	7045-336-4X3	404
	3262	2	64770	ELI	7045-336-4X3	583
	3287	1	N0587	DAL	7045-336-4X3	0
	3287	1	P9777	CMD	7045-336-4X3	160
	3287	1	Q1753	CMD	7045-336-4X3	160
	3287	1	Q1986	SFD	7045-336-4X3	160

controllers, and computers, and the circuits connecting them. Examples of this type of map are shown in Figure 16–5. Such maps can be produced in many sizes. Wall-sized versions can be hung in the network control center. Versions of an 8½ by 11 inch size can be put in notebooks or made into transparencies for use in presentations or at meetings.

circuit charts

Another level of detail, as shown in Figure 16–6, is a listing of each circuit and the devices attached to it. Charts of this type usually indicate the circuit number assigned by the carrier as well as the names or network addresses of the controllers and terminals. Equipment model numbers and serial numbers are also generally included.

wiring diagram

A third level of detail is the wiring diagram. The wiring diagram is drawn for a single building or floor of a building. One version is a type of map that shows the actual locations of the devices on the floor or in the building as well as the connections between them. Another type of wiring diagram has no bearing on the actual location of the equipment but shows in detail the exact cable runs and types of wire, twisted pair numbers, and pin numbers of each component in the network.

Drawings and diagrams of the network are most easily developed and maintained on a computer-assisted drafting (CAD) system that captures the drawing in a computer database and simplifies the maintenance. CAD software with enough capacity to handle most networks is now available for personal computers, so the tool is easily affordable by most telecommunications departments.

Other Documentation Other network documentation includes

- software listings for all telecommunications software;
- the names and telephone numbers of a contact person at each location on the network;
- vendor manuals for all hardware and software;
- disaster recovery plans;
- vendor contact names and telephone numbers for all equipment and software;
- problem escalation procedures;
- routine operating procedures.

The network documentation should be backed up. Whether it is kept online in electronic form or printed, duplicate copies of all lists and procedures should be kept, preferably off-site. Furthermore, a procedure to keep the backup copies up-to-date must be in place.

Change Management

Change management is an activity for monitoring all changes to the network and coordinating the activities of various groups in such a way that there is minimal impact on network operations and user service when changes are made. The change management activity is often called *change control*. However, the word *control* tends to imply that many net-

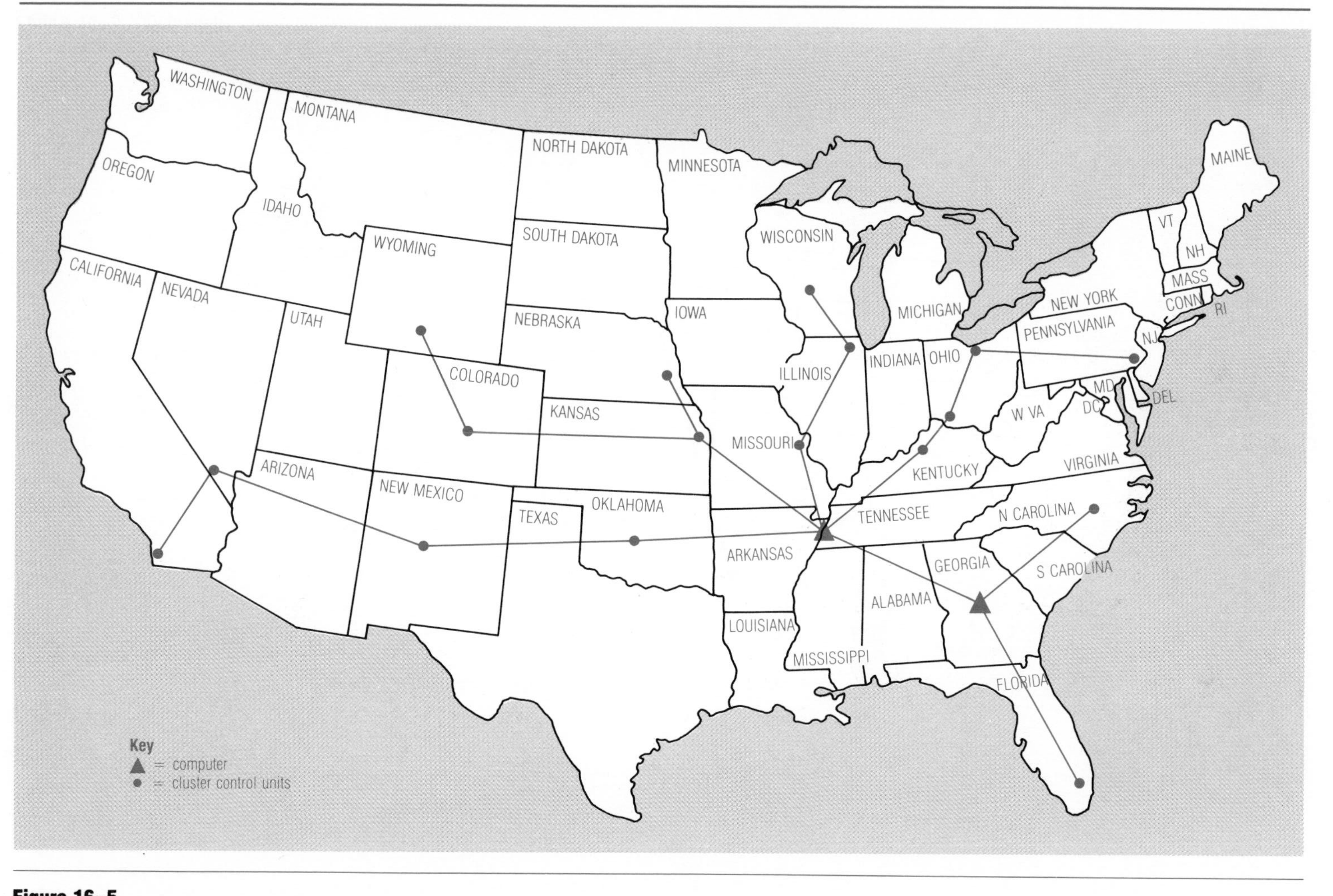

Figure 16–5
Maps of typical communications networks.

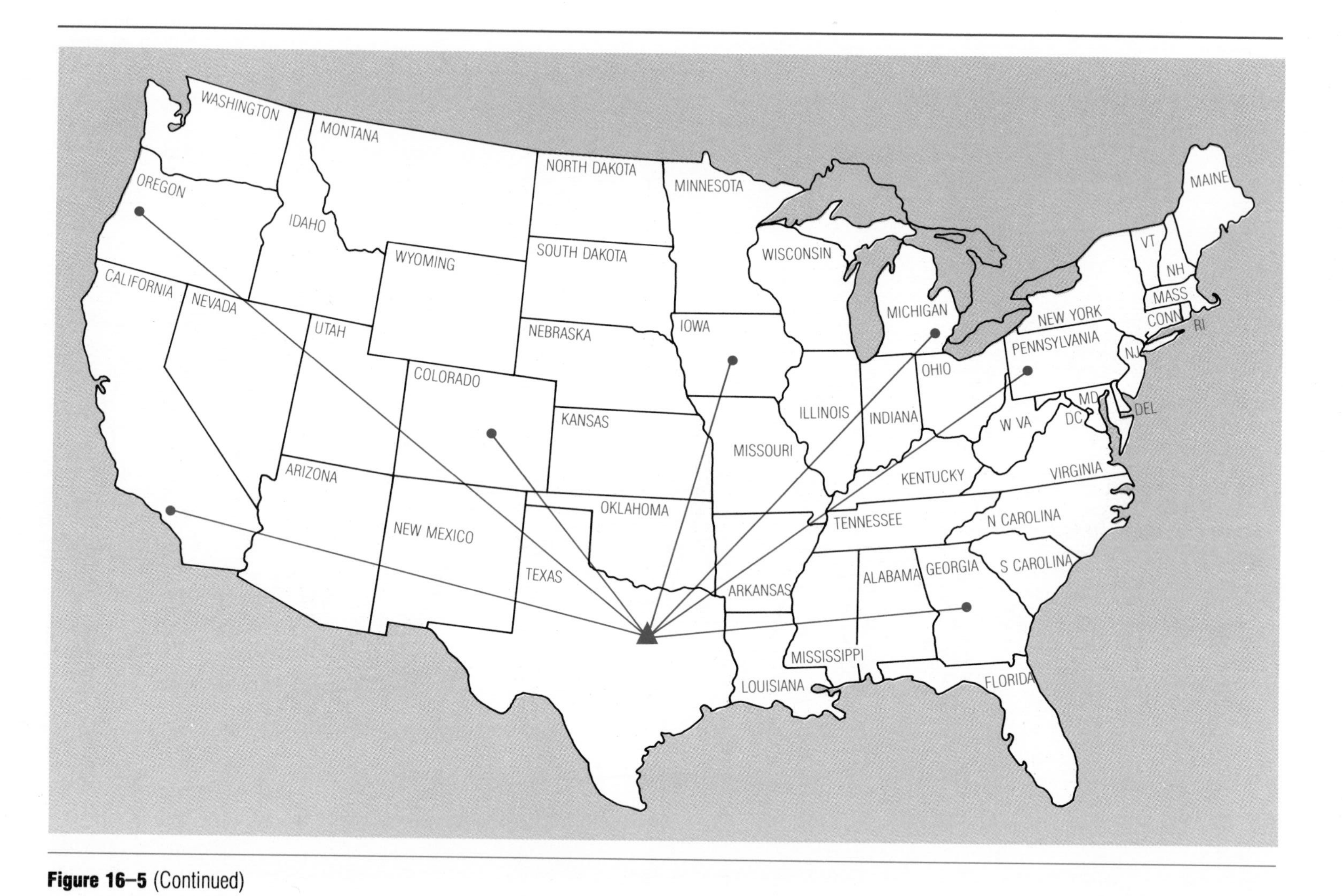

Figure 16–5 (Continued)

Circuit Name	Circuit Number	Controller Type	Model No.	Serial No.	Network Address	Location
West	FDA3856B	3725	N.A.	135269	3E1C	Midland
		3174	3	426578	3E9D	Chicago
		3174	2	890124	3E1A	Denver
		3274	41C	182930	3E4F	San Fran
South	FDA3472	3725	N.A.	135269	3E1C	Midland
		8130	30	42457	1A46	Toledo
		3174	2	675645	1A3D	Memphis
		3274	61C	276578	1A88	Mobile
		3174	2	764190	1A6A	Miami

Figure 16–6
Listing of circuits and attached controllers.

work changes are unnecessary and could be eliminated. Although many operations people would like to eliminate change to the network to preserve stability, it is hardly realistic to do so in a telecommunications environment. However, change must be implemented in a planned, coordinated way.

Most change management systems require that a "request for change" form, such as the one shown in Figure 16–7, be filled out by the person who is sponsoring the change. A change may be the installation of a new terminal, the reconfiguration of a circuit to support a new location, or the installation of an updated version of the telecommunications software. In most companies, all of the change request forms go to a single person, the change coordinator, who screens the requests to find any obvious conflicts between changes or other inconsistencies. The change coordinator can also keep a master list of all planned changes, such as the one shown in Figure 16–8.

change coordination meeting

From time to time, the change coordinator convenes a *change coordination meeting* that is attended by representatives from network operations, software support, computer operations, vendors, users, and others who may be interested in or affected by the proposed changes. At the coordination meeting, the changes are reviewed and discussed. The plan, which shows when each change is to occur, the nature of the change, and who is responsible, is updated. Assuming no major obstacles are discovered, the plan is approved and the changes are implemented according to the plan.

In many organizations, the change coordination meeting is held weekly. Many network changes are implemented on the weekend to avoid disrupting the network during normal business hours. Software changes in particular often require deactivating the entire network so that the new software can be installed and tested, and usually the only time this can be done is at night or on the weekend.

When major changes to the network are anticipated, such as a complete reconfiguration of the circuits or the provision of a totally new

Figure 16–7
A change request form that would be completed by the person requesting the change.

Change Request Form

Request
Date: ____________ Request No. ____________

Implementation
Date: ____________ Requestor: ____________

Type of change: Addition __ Modification __ Replacement __
Removal __ Correction __

Description of the change: ____________

Assigned to: ____________

Approvals:

Supvr. Network Operations: ____________

Supvr. Technical Support: ____________

Implementation Data:

Date Change Made: ____________

Comments: ____________

service, the network operations people should participate with the network analysts and designers in the planning and design process. By participating in the development project, the operations people can stay in touch with coming events that will affect the work they do or the procedures that are in place.

documentation updates

It is extremely important to keep the network documentation up-to-date as changes are made. Of course, if the documentation is kept online in a word processing system, the maintenance process is simplified. The proper way to keep network documentation up-to-date is to make it a part of the change management process so that it is done routinely at the time that the changes are implemented. Although this statement may seem obvious, it is surprising how many organizations ignore the documentation in the rush to implement changes.

Management Reporting

Management reporting of network operations activities and statistics is discussed here as a separate topic, although a piece of it is likely to be done by each group within the network operations department. Several different types of reporting need to be done for different levels of management. Generally, the higher the level of management, the more the data should be summarized and the more the reports should focus on

Change Coordination Plan
As of 10–25

Change No.	Description	Planned Date	Responsibility	Date Completed
1047	Add Dallas to network	10/27	Vandewoe	
1048	Maint. to NCP software	10/29	Morris	
1049	Add 5 VDTs in marketing	11/15	Vandewoe	
1050	Upgrade west circuit to 9,600 bps	11/28	Aymer	
1051	Add cluster controller in research department	12/10	Aymer	
1052	Maint. to access method software	12/12	Morris	

Figure 16–8
A change coordination plan for a small network. Large networks would have a much more extensive list of changes.

the exceptions rather than routine detail. User management is primarily interested in data about the network performance, particularly when it is compared to agreed upon service levels. Network operations management is equally concerned about the performance compared to the service levels because their objectives are based on providing the agreed to service. They are also interested in basic operational statistics, such as the number of problems that were logged, the length of time it took to resolve problems, and the overall use of the network facilities. Graphs normally are the most useful way to convey this type of information.

Different types of information are reported on different time schedules. In most networks other than the smallest, a daily status review meeting is held with the computer operations people to discuss the results of the previous day and any outstanding problems. Plans for the coming day also may be discussed. This type of meeting is held most frequently early in the morning and is kept quite brief—normally 15 minutes or less. Some companies connect remote locations into these meetings via audio teleconferencing so that status information can be distributed widely and quickly.

Network performance and utilization statistics are usually calculated and distributed monthly. A useful variation of this approach, however, is to calculate performance numbers daily. If service levels are not being achieved, it is considered a problem and handled through the problem tracking system. Since open problems are discussed at the daily status meeting, the performance problems get immediate attention.

Statistics in a voice network are similar to those in a data network if a PBX is installed. PBXs can generate statistics and measurements, such as trunk utilization and average call length. Much of the focus of telephone statistics is on long distance calling, which is frequently the most expensive part of the voice network. The number and duration of long distance calls, evidence of call abuse, utilization of FX lines or tie lines, and call queuing times, where appropriate, are all relevant to the efficient and effective operation of a voice communications network.

NETWORK MANAGEMENT SOFTWARE

Software packages that are designed to help the telecommunications staff manage a data network are becoming available. These packages provide an array of status displays, commands, databases for storing data, error analysis programs, performance measurement, capacity management, and reporting tools. Two of the best known are Netview from IBM and Netmaster from Systems Center, Inc., both programmed to support IBM SNA networks. DEC has a set of products called Enterprise Network Management Architecture (EMA) for its networks. AT&T has a network management architecture called Unified Network Management Architecture (UNMA). UNMA is a combination of concepts, hardware, and software with which AT&T hopes to build network management capability for data and voice networks that include products from many vendors.

There are also many software packages on the market to assist the telecommunications staff with such activities as keeping equipment inventory, drawing network diagrams and maps, keeping track of changes, and charging costs to users. Software is available that runs on either mainframes or personal computers, depending on the customer's requirement. Some vendors have integrated but modular packages, so that the customer can buy only the pieces that are needed. Most organizations should be able to find commercially available network management software that meets their needs.

THE PHYSICAL FACILITY

network control center

The network technicians are normally housed in the network control center (NCC). The network control center contains the consoles or terminals used to operate and monitor the status of the network and other telecommunications equipment such as modems, multiplexers, and front-end processors. Because it contains this equipment, much of which has the same environmental requirements as mainframe computers, the network control center is often located in or adjacent to the company's computer room. Depending on the needs of the particular equipment being used, special power and air conditioning may be required.

flooring

A raised floor, at least 12 inches high, is extremely desirable because of the large number of cables and other wiring that connects all of the telecommunications equipment together. If the wiring cannot be run under the floor, it can be run overhead above the ceiling, but it is less accessible there. If that option is not available, special cable trays or racks can be obtained. Without proper planning, it is very easy to end up with a "rat's nest" of wires and cables that are difficult to trace and troubleshoot.

Another similarity between the network control center and the computer room is that smoke and fire detection and extinguishing equipment must be installed. Usually, smoke, water, and heat detectors are mounted under the raised floor and on the ceiling of the room. They are designed so that if any two detect smoke or heat, the fire extinguishers are set off. The fire extinguishers can use water, carbon dioxide, or a gas called Halon as an extinguishing agent. Water sprinklers are the least expensive, but a discharge of water is almost certain to damage some of the electronic equipment. Carbon dioxide is effective, but messy. Halon is effective against all types of fires except smoldering paper and does not damage electronic equipment. In the concentration needed to extinguish fires, Halon can be breathed by humans. Its biggest disadvantage is that it is very expensive.

smoke and heat detectors

A key attribute of the network control center is that it must be flexible so that it can be expanded and rearranged easily. Given the dynamic, growing nature of most telecommunications networks, the configuration of equipment in the network control center seldom stays the same for long. The layout that seems ideal today will seem hardly workable in six months. Network operations people should be involved in the design of the network control center because they will have the best understanding of the requirements and the implications of seemingly small changes.

configuration

Most communications equipment is designed to be mounted in standard 19-inch wide racks and cabinets. Some small equipment, such as modems, if not designed to be rack mounted, can sit on shelves in the cabinets. Many communications equipment cabinets have glass doors in the front so that the operators can see the lights and meters on the equipment. There are also doors in the back for access to the wiring that interconnects the various pieces of equipment. There is some debate about whether individual pieces of equipment should be labeled, particularly if they might be swapped for a spare piece of equipment when a problem occurs. If labels have to be changed, it delays the problem resolution. An alternative approach is to label the shelves on which the equipment sits or into which the rack-mounted equipment slides. Many cabinets have a space for such labels on the front of each shelf.

labels

Cables should be labeled at both ends, and the cable identification should be typed or written twice on the label so that it can be read without twisting the cable to an unnatural position. This approach also ensures that the label is readable if one of the identification codes becomes smeared or otherwise unreadable.

Other equipment that should be located in the network control center are the PBX, patch panels and switching equipment to bypass failing components of the network, and the network test equipment.

These modem cabinets contain many modems. Lights on each modem display its status. The cabinet also has a door in the back for easy access to the power cables and communications circuits.

The help desk does not need to have access to all of the equipment. The help desk can be located in a traditional office environment outside of the network control center in less expensive floor space. Help desk operators need telephones, perhaps with headsets, access to one or more VDTs for monitoring network status and problem conditions, and access to the tool (perhaps the same VDT or another one) on which all incoming problems are recorded. If trouble tickets are filled out, the help desk operators must have a way of transmitting or transporting them to the technicians who will work on the problems.

Depending on the hours of operation, the network control center and help desk may need to be staffed for a long shift, say from 7 A.M. to 8 P.M., or for multiple shifts. If a long shift is required, the problem may be solved by having some people start early, for example, at 7 A.M., and work until 4 P.M., and others start later, say at 11 A.M. and work until 8 P.M. If second or third shift staffing is required, it may be possible to share people with the computer operations group. This can be done if the network control center, help desk, and the computer room are located in close proximity to one another.

STAFFING THE NETWORK OPERATIONS GROUP

The staff of the network operations group is made up of people with a variety of skills. There are two philosophies of staffing the help desk. One suggests that it be staffed by the most senior, most experienced people available, such as experienced technical support people. These people can resolve virtually all of the reported problems on the spot. Proponents of the other staffing philosophy believe that the help desk can be staffed by relatively new, inexperienced people who, with a minimum of training, can solve a high percentage of the problems and, by asking a set of predetermined questions, can properly route the other problems to the right person. The advocates of this approach say that recent high school graduates with good verbal communications and telephone skills can be trained to perform well on the help desk in a few weeks. Proponents suggest that this is a good entry-level job into the telecommunications or data processing organization.

Routine network operations work, such as starting and stopping the network, circuits, or terminals, also can be handled by help desk personnel in all but the largest networks. Since they are handling trouble calls, the help desk operators are usually the most closely in touch with the state of the network at any time.

Hardware technicians, the second level in the problem-solving hierarchy, are people with electrical and electronics training and aptitude. Often they are recruited from the local telephone company or the military, but they might also come from a local electronics school. They typically do not have a 4-year degree. The hardware technicians can also be used to install equipment, such as terminals or personal computers, where some initial setup as well as connection and checkout are required.

Network operations supervisory personnel should have a production orientation. They should be focused on meeting the numerical service objectives and constantly on the lookout for techniques to improve the efficiency and effectiveness of the network operations group. In addition, they should have good verbal communications skills because they may have to deal with users under stressful situations. Ideally, they should have education and experience in data processing and data communications, as well as some general supervisory experience. The network operations supervisor and computer room supervisor are in many ways similar jobs, and each may provide a backup for the other.

It is important that network operations personnel at all levels spend 5 to 15 percent of their time each year in formal or on-the-job training. Because of the rapid changes in the telecommunications field and the changes in most companies communications network, this type of educational investment is necessary to keep the staff's knowledge and skill level current.

NETWORK MANAGEMENT SERVICE COMPANIES

A relatively new service becoming available throughout the country is the management of a company's network by an outside service organization. For a fee, a service company will contract to operate the network, around the clock if necessary, providing all liaison with vendors and service organizations. Some businesses would probably not feel comfortable turning over the operation of their network to a service company, but for companies that have difficulty hiring skilled technicians or who simply don't want to manage their own network, help is now available.

The State of Wisconsin has a contract whereby AT&T designed, built, and now manages the state's communications network. Patterned after the concept AT&T uses to manage its nationwide long distance network, the Wisconsin Network Management Center (NMC) is the nerve center of the network that ties together more than 1,800 state, county, and municipal government locations and serves 50,000 employees. The NMC is staffed around the clock by AT&T personnel and is actually located in an AT&T building near the state capitol in Madison.

COMMUNICATIONS TECHNICAL SUPPORT

Communications technical support specialists manage the communications software and handle other nonroutine situations or problems that arise during the design or operations of a communications network. To be most effective, they require a broad understanding of all facets of the communications system, both hardware and software.

In some companies, the communications technical support group is a separate department reporting to the manager of telecommunications, as was illustrated in Figure 14–8. In other companies, communications technical support is a part of the network operations group. The actual organization depends on the size of the company, the skills of the particular individuals, and the preference of the telecommunications manager. Either organization can work effectively.

THE FUNCTIONS OF COMMUNICATIONS TECHNICAL SUPPORT

The major functions of communications technical support are

- supporting communications software;
- providing third-level problem solving;
- assisting in network analysis and design;
- network performance analysis and tuning;
- hardware evaluation;
- programming;
- consulting and general problem solving.

Each of these activities is discussed in the following sections.

Support Communications Software

The communications technical support group is responsible for installing and maintaining the communications software that runs the network. This includes the network control program that operates in the front-end processor and the communications access method and communications monitor software that reside in the mainframe computer. Communications software is generally delivered by the vendors in various levels or releases. Each succeeding release contains additional capability and may correct software problems (program bugs) of prior releases. The technical support staff and its management must decide whether to install each new release of the software as it becomes available. This decision largely depends on whether the new release cures problems the network has experienced and whether the new capability is required.

Testing communications software on a personal computer is relatively simple, but testing the network control program, which runs a large multiline network with several thousand terminals, is extremely complex and difficult. Because it is usually impossible to test all of the terminals and combinations of communications possibilities, the communications technical support specialist's experience is critical in determining which parts of the software have changed and should be tested.

Another aspect of software maintenance is to reconfigure the software as necessary to support new circuits or terminals as they are installed. In some systems, this process must be done each time a circuit or terminal changes. In other systems, the software itself senses new equipment and adapts to the changes.

A third aspect of software support is diagnosing problems that occur with the software, obtaining fixes for the problems from the vendor, and applying and testing those fixes to solve the problems. Communications software problems are frequently very difficult to diagnose because it is impossible to specify or repeat the exact conditions under which a problem occurred.

Vendors have different approaches to software maintenance. Some vendors take the approach of sending out "fixes" for all known software problems and asking the users to apply them as *preventive software maintenance.* Other vendors adopt the strategy "if it's not broken, don't fix it"—fixes are only applied for the specific bugs that occur at each site where the software is installed.

Provide Third-Level Problem Solving

Usually, the help desk and network technicians solve approximately 95 percent of all network problems. Some problems are so unusual or difficult that they require a higher degree of skill to resolve. The communications technical support group serves as the third level for problem resolution. In this role, the group stands behind the help desk and network technicians ready to jump in and work on the most difficult problems.

Whereas the network technicians are very familiar with communications hardware, they are not usually trained in communications software. The communications technical support specialists' knowledge of both hardware and software becomes very important because in many of the most difficult problems, it is hard to tell whether hardware or software is at fault. The technical support person may also call on the vendor's technical specialists to work on the problem because in many cases, a problem experienced at one company has been seen and solved elsewhere.

Assist in Network Analysis and Design

Communications technical support specialists often are called upon to assist when network designs are being reviewed and updated. Depending on the size of the communications department, the technical support specialist may even perform the network analysis and design. In any case, the detailed knowledge of hardware and software capabilities make the technical support specialist a valuable member of the network design project team.

Network Performance Analysis and Tuning

The regular gathering of performance data can be performed by the network operations staff according to procedures established by the technical support specialist, but the analysis and interpretation of the performance data require considerable skill, judgment, and experience. Usually, the technical support specialists are charged with the responsibility of making sure that the network is delivering the performance for which it was designed and for spotting trends, such as increases in message or transaction volume or error rates, that could lead to capacity problems and performance degradation. The technical support specialist sometimes can make adjustments in the software, such as varying the number of buffers in the network control program or changing the priority of certain transactions in order to improve performance. In other cases, he or she may recommend that existing circuits be reconfigured or additional circuits be added.

tuning

The process of *tuning* a network consists of monitoring its performance, making adjustments to improve the performance, and then monitoring the results of those adjustments. This is illustrated in Figure 16–9. The results of the initial adjustments may suggest that further adjustments are necessary or may show that performance has returned to the desired operating range. Network performance monitoring and tuning should be an ongoing process that is a routine part of the technical support job.

Hardware Evaluation

Another function of the communications technical support group is testing and checkout of new equipment being considered for installation

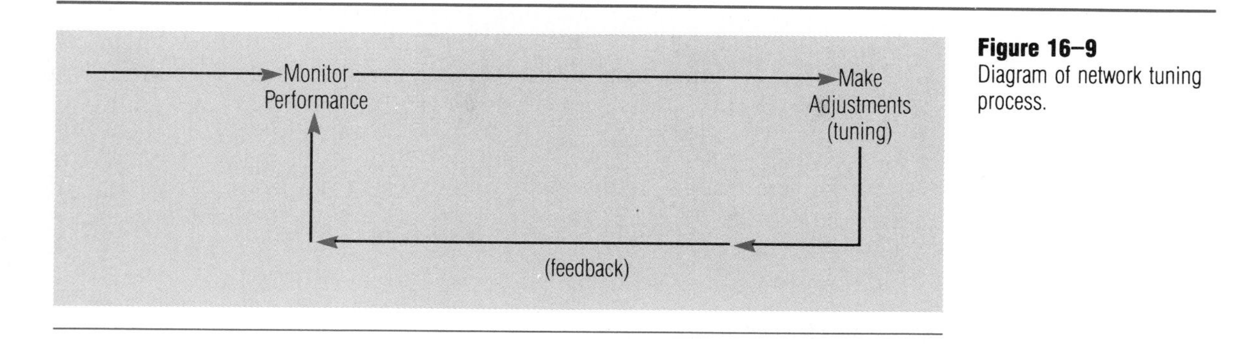

Figure 16–9
Diagram of network tuning process.

on the communications network. Most manufacturers claim that their equipment conforms to certain specifications or standards, such as RS-232 or SNA. However, there is enough variation in the way the standards are implemented that testing in the actual environment where the equipment is to be used is desirable. Experienced telecommunications people insist that the vendor provide a piece of equipment for testing purposes. The job of connecting the equipment to the network and testing it falls to the technical support specialist.

When the specialist tests new equipment, it is important to test the equipment's error handling capability as well as its ability to operate under normal conditions. The device may work fine when no errors occur, but it is important to know how it will perform in all situations.

Programming

In some companies, communications technical support specialists are communications programmers who design and write communications software. This type of programming is complicated and difficult, and most organizations believe it is best to avoid writing such programs if possible. Occasionally, however, a unique capability or function not provided by any vendor's software is needed, and the only way to get it is to do the programming in-house. Software of this type needs to be especially well documented because very few people in the organization will understand how it works well enough to be able to maintain or debug it.

Consulting and General Problem Solving

Good communications technical support specialists get called on to help solve a surprising variety of communications and data processing problems in an organization. This is because they are highly skilled, analytical, and by nature good problem solvers. They may find that a high percentage of their time is spent on ad hoc or spur-of-the-moment activities that are outside of their control. When managers plan the specialists' time, it is important to recognize that some must be left unscheduled and available for such activities.

STAFFING THE COMMUNICATIONS TECHNICAL SUPPORT GROUP

Communications technical support specialists have many of the same characteristics, and much of the same training and experience, as the system programmers in the data processing organization. In fact, in many companies there is a single technical group for both communications and data processing.

Communications technical support people should be

- intelligent;
- analytical;
- technical;
- curious;
- self-motivating;
- independent.

A study, recently performed by IBM, has shown that communications technical support specialists who work for IBM's largest customers have the following education and experience:

- 100 percent have a high school education;
- 52 percent hold a bachelor's degree or higher;
- 84 percent have formal software training;
- an average of 8.1 years experience in data processing or communications and 3.1 years experience on their current job;
- 50 percent are familiar with four or five IBM software products.

Good technical specialists are excellent problem solvers who often prefer to work independently. Their independence and self-motivation serve them well when they are called on to solve difficult problems. They have a "can do" attitude and enjoy the challenge and self-satisfaction that comes from diagnosing and solving a problem when others have failed.

SUMMARY

This chapter studied the network operations and communications technical support organizations that must be in place to properly operate the communications network of a company. We have looked at why it is important to manage the network, and we have seen the six major activities that the network operations function comprises. The chapter discussed the different skills required of the network operations group and described the physical facility, called the network control center, which is the focal point for network operations and trouble shooting.

Technical support people install and maintain the communications software that drives the network. The communications technical support group provides detailed technical assistance and consultation, heavily focused on software but with a large amount of hardware expertise required as well. Because of the heavy education and experience requirements, good technical support people usually are in great demand.

CASE STUDY

Dow Corning's Network Operations and Technical Support

Dow Corning has had a network control center and help desk since 1968. The original help desk evolved from the message center in the days when the company had a torn tape message switching system. When the message switching was taken over by the computer in 1968, the message center became the company's first help desk, and the staff also monitored the operation of the communications circuits.

The current help desk is located in the same area as the mainframe computer operators. The network is operational from 3 P.M. on Sunday afternoon to serve Sydney, Australia, until 6 A.M. the following Sunday morning, when the computer is taken down for hardware and software maintenance. The help desk is currently staffed only from 8 A.M. to 5 P.M., Monday through Friday. At other times, the help desk activities are performed by the computer operators. At some time in the future, it may be necessary to expand the help desk's hours of operation. This situation is continually monitored by telecommunications management.

Incoming calls to the help desk are taken by an operator, or if all operators are busy, a recorded "all lines are busy" message is played. It reminds the users to have their terminal name available to give to the operator when he or she answers the call. The operator logs the call, the problem symptoms, and the user's name and telephone number. Approximately 90 percent of the calls are relatively routine and can be handled by the person taking the call within a minute or so. Callers who have more difficult problems are passed to a network technician or to the help desk supervisor for handling. In the most difficult cases, communications technical support or vendor personnel are contacted and asked to help resolve the problem.

Dow Corning has been very successful using high school and college students to staff the help desk. These students typically work 4 hours per day and are trained to handle calls just like the full-time staff. Currently, there are two students employed. One works in the morning and one in the afternoon.

Dow Corning has a problem tracking and change control system in place for both the network and the Midland computer center. A meeting is held each Tuesday morning at 8:30 A.M. to discuss all outstanding problems and upcoming changes. This meeting is intended to ensure that difficult problems are getting the proper attention and that all changes to be made to the computer or network are communicated to network and computer personnel and vendors.

Central communications equipment at Dow Corning is primarily located in the computer centers. A conscious effort has been made in the past few years to get the communications equipment relocated into a few secure areas rather than being scattered throughout the headquarters building. Moving communications equipment takes a great deal of planning to minimize the disruption to users and unscheduled outages. In plants and other outlying locations, communications equipment normally is kept in an equipment room.

Dow Corning's communications technical support group consists of two people. One of these people has over 20 years of telecommunications and computer technical support experience. The other person has been working in telecommunications technical support for nearly 5 years and has a strong technical background that he is now applying to the telecommunications environment. These two people handle the communications software maintenance, get involved in difficult communications problems, and act as consultants on communications matters. One of them specializes in the IBM SNA software and changes to the network of IBM terminals. The other person specializes in all of the non-SNA communications hardware and software. There is also a technician, a former telephone company employee, who handles the physical connection of equipment and who has become a specialist in the T-1 and fiberoptics multiplexing equipment. For the most part, voice technical support is handled by Michigan Bell, AT&T, and Siemens-Rolm technical personnel.

Since most of the network changes have to be made during the short time when the network is down on Sunday, one of the technical people normally takes part of Friday off and works part of the day Sunday. The computer technical support people work a similar schedule, so having a plan for the Sunday activities is mandatory to ensure that the staff can get their work done and that they don't interfere with one another. The technical support people are also on call when severe network problems develop at night or on weekends. Dow Corning has found that having talented people who are flexible about their schedule is a necessity for this type of work.

Review Questions

1. What is the function of the network help desk?
2. Distinguish between the help desk and the network control center.
3. In a star network, why should the network operations group be concerned about the reliability of the computer processing?
4. What are some of the startup and shutdown activities the network operations group performs if the network is started and stopped each day?
5. What is the purpose of having a daily status meeting?
6. Is it possible to design a network that is totally redundant and, therefore, does not need to be managed? Why?
7. What are some ways that problems in a network can be bypassed?
8. Who should attend a problem tracking meeting?
9. Describe the process of problem escalation.
10. Explain the level 1, level 2, and level 3 method for problem resolution.
11. What is a service level agreement?
12. What are some of the statistics about network operations that the network operations supervisor would be interested in seeing each day?
13. Some companies institute a moratorium on all network and computer changes at the end of the year. Why might such a moratorium be desirable?

14. What information should be kept in an inventory file of telecommunications equipment?

15. List several techniques to ensure that when telecommunications change is introduced in a company, it is implemented in a controlled, well-managed way.

16. What is the purpose of telecommunications management software packages?

17. Frankly, preparing and updating network documentation is tedious. No one likes to do it, and few companies do it well. If you were the manager of telecommunications, what incentives can you think of that would motivate your people to keep the documentation up-to-date? What guidelines would you develop to ensure that the documentation is kept up-to-date?

18. Prepare a list of management reports that would be useful for the help desk supervisor, network operations supervisor, and telecommunications manager to keep them informed about network operations. Describe what information would be included in each report and how frequently it would be created. Consider also the possibility of providing some of the information via online inquiries.

19. What are the factors to consider when a manager decides whether or not to apply maintenance to the network software?

20. What are the advantages and disadvantages of using an outside service organization to manage a company's telecommunications network?

21. How can the communications technical support people assist with network analysis and design?

22. Discuss the factors to be considered when deciding whether to include the communications technical support group in the network operations group or create a separate group in the telecommunications department.

23. Joe Brown, a senior network technician, has expressed a strong interest in getting into the communications technical support group. He has 9 years of communications experience with the telephone company and 4 years with your organization. He graduated from high school and completed 2 years of college, where he studied data processing and programming. He has been a top-rated performer and has shown a great deal of initiative in solving hardware problems. What attributes and skills would you look for in assessing whether Joe has the capability to work successfully with the other technical support specialists and deciding whether to give him an opportunity in technical support?

Problems and Projects

1. Describe some advantages of having all telecommunications personnel spend some time working in the network operations group.

2. Write an appropriate service level agreement for

   a. an airline reservation system;

b. an ATM application;

c. an online inventory system in a chemical company.

3. To some extent, making decisions about future network growth and configuration by looking at past network statistics is like steering a boat by looking at the wake. Discuss how the statistics about past performance can be properly used and misused for forward planning.

4. Draw a floor plan of a network control center containing a central console with three VDTs, four modem cabinets, a front-end processor, two cabinets of test equipment, and any other equipment that would be appropriate to have in the room.

5. When a three-level approach (as described in the text) is used for problem resolution, it is possible that none of the people that have worked on a problem feel totally responsible for it. Level 1 passes the problem off to level 2, and level 2 may pass it to level 3. As the manager of network operations, what procedures would you institute to ensure that the responsibility for a problem is always well defined and that the help desk always knows its status in case it gets questions from the users?

6. What types of information would a good communications line trace program capture?

7. Most vendors keep a database of known problems with the hardware and software it sells. Do you think a vendor would ever let its customers have direct access to such a database? What would be the advantages and disadvantages of giving customers such access?

8. The text stressed the importance of doing ongoing network performance monitoring and tuning. Why do you think many companies don't do this important activity regularly?

9. Good technical support specialists find that communications programming is challenging work. What are the trade-offs for a company in letting them develop communications software? As the manager of technical support, how would you convince the specialists that writing communications software is generally not in the best interests of the company?

Vocabulary

problem management
help desk
trouble ticket
levels of support
problem escalation
problem tracking meeting
patch panels
dial backup
protocol analyzer
performance management
service level agreements
configuration control
change management
change coordination meeting
network control center (NCC)
preventive software maintenance
tuning

References

Benkoil, Dorian. "Testing Equipment—Testing Testing," *Teleconnect*, September 1986, p. 36.

Chorafas, Dimitris N. *The Handbook of Data Communication and Computer Networks.* Princeton, NJ: Petrocelli Books, Inc., 1985

Effron, Joel. *Data Communications Techniques and Technologies.* Belmont, CA: Wadsworth, Inc., 1984.

Fitzgerald, Jerry. *Business Data Communications: Basic Concepts, Security, and Design,* 2nd ed. New York: John Wiley & Sons, 1984, 1988.

Kasperek, Gage. "Network Control Center Structure," *Network World,* August 25, 1986, p. 13.

King, Julia. "Does Anyone Have a Line on Network Management Systems?" *Information Week,* November 3, 1986, p. 22.

Kreager, Paul. *Practical Aspects of Data Communications.* New York: McGraw-Hill, Inc., 1983.

Leonhart, James. "Wisconsin is Wired," *Communications News,* August 1989, p. 42.

Mulqueen, John T. "Running Other People's Networks: $800 Million for Off-Loading Headaches?" *Data Communications,* March 1989, p. 77.

Petersohn, Henry H. *Executive's Guide to Data Communications in the Corporate Environment.* Englewood Cliffs, NJ: Prentice-Hall, Inc., 1986.

Scheavitz, Alan Y. "Managing the Future Network: Tactics, Technology, and Trauma," *Business Communications Review,* September–October 1985, p. 20.

Stamper, David A. *Business Data Communications,* 2nd ed. Menlo Park, CA: The Benjamin/Cummings Publishing Company, 1986, 1989.

Thoenen, David. "An Executive Overview of Information Network Management," *IBM Manual,* GG22-9106-00, November 1986.

———"Managing Connectivity: The IBM Network Management Architecture," *IBM Manual,* GG22-9121-00, September 1986.

Xephon Consultancy Report: Network Management for IBM Users. Berkshire, England: Xephone Technology Transfer Ltd., June 1985.

CHAPTER

17

Future Directions in Telecommunications

OBJECTIVES

After studying the material in this chapter, you should be able to

- describe many of the trends that are shaping the future capabilities of telecommunications systems;
- describe applications of telecommunications that are "just arriving" or are on the horizon for the near future;
- understand why these new telecommunications capabilities are likely to have a profound impact on our society.

INTRODUCTION

One of the aspects of telecommunications that makes the field so exciting is the constant and rapid technological change, a characteristic of today that is predicted to continue for the foreseeable future. As new technology becomes available, it brings with it the capability of applying communications to situations where it was previously not feasible, justifiable, or even possible in a way that meets user requirements. The constant technological change is one reason why some people have chosen to work in the field for many years.

Often, as new technology becomes available, the potential uses of it are not immediately evident. People must become familiar with the technology and figure out how it might be applied to solve a problem or to do something that has not been done before. Even when a use of technology seems evident, skeptics are likely to pooh-pooh the idea. When the telephone was first invented, one futurist in England declared there would never be a market for the telephone in England because there was a perfectly adequate supply of messenger boys!

In this chapter, we will look at some of the discoveries, events, and ideas that are giving us insight into the uses of future communications capabilities. Some of the ideas will be great successes, whereas others will fail. Furthermore, for every idea discussed in this chapter, there are probably at least 10 that are not covered, and more will emerge in the coming years.

TRENDS IN ELECTRONICS

The capabilities of electronic circuit chips have been increasing exponentially and are continuing to do so. Whereas a few years ago, memory chips were capable of storing 4 k, 8 k, or 16 k bits of information, today's chips store 256 k, 1,000 k, and 4,000 k. The microprocessor chips that drive computers of all sizes allow the computing power of a room-size mainframe computer of 10 years ago to reside in today's desktop personal computer. Microprocessor chips are becoming ubiquitous. It has been predicted that by 1995, the average home will have approximately 25 microprocessors (about the number of fractional horsepower motors found in the home today). These chips will—and many already do—reside in common items found around the home: thermostats, appliances, entertainment systems, telephones, automobiles, and the like.

The price-performance of memory chips, microprocessors, and other related electronics has been coming down at a rate of 20 percent per year, compounded. This trend has existed for the past 15 to 20 years and is expected to continue at least until the turn of the century. Of course, this does not mean that every single chip decreases in price by 20 percent each year or that the decreases are uniform, but the trend is clear. As shown in Figure 17–1, at 20 percent per year, the price of a particular chip reduces to half its original cost in just over 4 years. If such chips represent a substantial part of the cost of a product sold in a highly competitive market, say a personal computer, the price of the end product is likely to drop at a similar rate.

In communications equipment, microprocessors and memory chips find many uses: adding intelligence to telephones and modems and allowing more memory buffering capability in PBXs and front-end processors are two examples. One type of microprocessor, called the *digital signal processor (DSP),* is having an especially significant impact. Digital signal processing is the technique of using microprocessors to analyze, enhance, or otherwise manipulate sounds, images, and other real world signals. A DSP chip processes a digitized signal through a series of mathematical algorithms over and over again to manipulate the signal. Before DSP chips became available, analog techniques that processed

digital signal processor

Figure 17–1
Declining cost of circuit chips, assuming 20 percent per year cost improvement.

Year	$100 chip	$20 chip	$5 chip
1	$100.00	$20.00	$5.00
2	80.00	16.00	4.00
3	64.00	12.80	3.20
4	51.00	10.24	2.56
5	41.00	8.19	2.05

signals in their original wavelike form were used, but analog signal processing is much slower and less accurate than the digital method.

Applications of DSP that are of particular interest in communications are *voice compression* and *video signal compression*. The objective of compression is to reduce the number of bits required to carry a digitized signal while still maintaining its unique characteristics and quality. In voice signal compression, the objective is to strip out redundant bits of information, yet maintain tone and inflection so that the listener can recognize the speaker. Twenty to one compression of a voice signal is possible while still meeting the recognition objective.

Of even more interest is video signal compression. A standard broadcast-quality, color television signal requires 92 million bits per second when transmitted digitally. With sophisticated digital signaling processing techniques, it is possible to get good quality (although not broadcast quality) color television signals compressed into 1 million bits per second. The compressed signal can be carried on a standard T-1 circuit leased from a common carrier. It yields a color television picture that is adequate for most business teleconferencing needs.

speech recognition

Another application for digital signal processing techniques is speech recognition. The idea of talking into a microphone attached to a device, such as a personal computer, and having the speech converted to text and a typewritten page emerge from the printer has long been the researcher's dream. Until recently, the speeds of the DSP microprocessors were simply not adequate to keep up with the rapid flow of normal human speech. Speech recognition systems have been available for several years that perform adequately if the speaker talks slowly, enunciates clearly, and limits his or her vocabulary to a predetermined set of words. Such devices have found application in factory applications where simple instructions, such as "stop," "left," or "forward," are given to a machine. Ford Motor Company has installed such a system for its automobile paint inspectors. They speak into a headset and identify flaws in paint jobs of new cars as they roll down the assembly line. The inspectors use simple terms such as "right front fender, sag" or "right quarter roof, dirt." The information is fed directly into a computer where it can be easily manipulated to stop aberrations in the painting process.

Recognizing a few words, spoken slowly, is much simpler than recognizing continuous speech, however. In addition to the speed difference, other complications in continuous speech recognition include recognizing the difference between homonyms, such as *to*, *too*, and *two*, and recognizing the same word spoken with different regional accents, such as *drawer* versus *drawh*. The latter problem is solved in today's systems by "training" the speech recognition processor to understand a certain voice. The prospective user of the system begins by speaking a few predetermined words into the microphone. The speech recognition equipment analyzes the wave forms of the speaker's voice and compares them to standard wave forms for those words stored in its memory. It then generates conversion rules that allow a successful comparison between the wave form of the speaker's voice and the standard wave form for all words in its vocabulary.

Speech recognition systems exist today that can recognize 5,000 words spoken clearly and at a moderate rate. Because there are so many potential applications for speech recognition systems, research is being conducted by many companies, and progress is rapid.

TRENDS IN COMMUNICATIONS CIRCUITS

The single most universal trend with communications circuits is the rapid advance toward end-to-end digital capability. The advantages of digital transmission are clearly recognized, and most of the common carriers are moving as rapidly as possible to install the electronics that can handle digital signals in their central offices. The carriers are also converting microwave radios from analog to digital transmission, and installing fiberoptics circuits to provide the high bandwidth that digital transmission requires.

As shown in Figure 17–2, in the United States today, 80 percent of the common carriers' local loops are capable of handling digital signals. Thirteen percent of the local loops connect to digital switches in the telephone company's central offices. Eighty percent of the links connecting the local telephone companies with the long distance carriers are using digital transmission. Eighty percent of the long distance switching centers have digital equipment installed. The long haul links connecting long distance switching centers for the most part use analog microwave technology. Although these circuits are being converted rapidly, only 50 percent of them were using digital technology by 1989.

Because of the inherent delay associated with satellite transmission, since the early 1980s there has been a movement away from using satellite circuits for voice transmission. Satellite circuits are perfectly adequate for batch data transmission, television broadcasting, or the distribution of programs for entertainment. They are less satisfactory when voice or interactive data transmission with rapid turnaround is required.

Two major trends in satellite technology will have an impact on the way satellite transmission is used in the future: higher bandwidth and more powerful satellite transmitters. The higher bandwidth is needed to help keep the cost of satellite transmission competitive with fiberoptics. The more bandwidth that can be placed on a single satellite, the fewer satellites that are needed and the lower the cost. Higher-powered transmitters in the satellite allow the receiving dishes on the ground to be smaller, but the trade-off is that larger solar panels and batteries are required on the satellite. It had been predicted that by 1990, it would be possible to receive satellite television signals using a small dish antenna

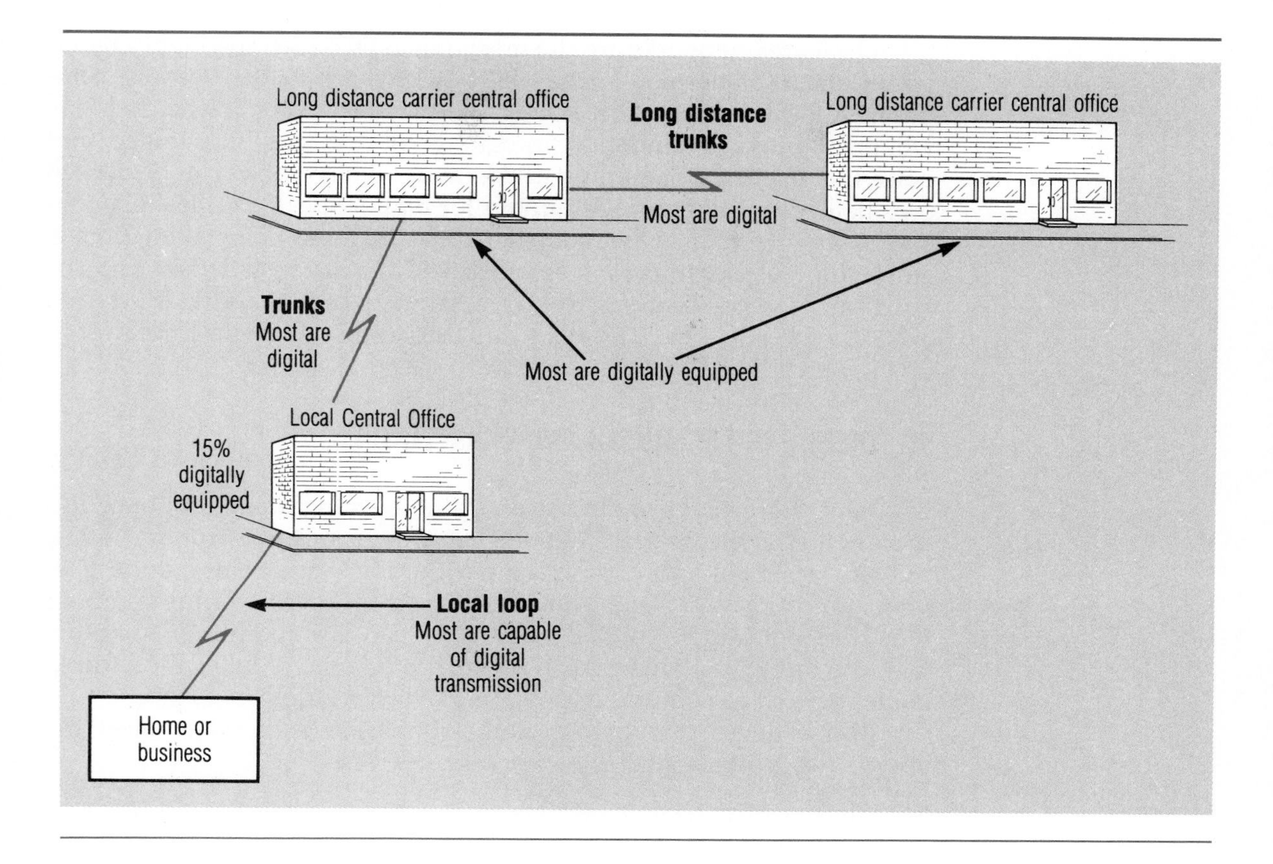

Figure 17–2
Amount of U.S. telephone system that is digitally equipped.

that sat on top of the television set inside the house. Although this was technologically possible, other forces prevailed and kept the capability from becoming generally available. Cable television operators obviously were opposed to the idea, and the companies that distribute entertainment programming via satellite were concerned as well.

Optical fiber technology is making the biggest impact on communications circuits today. Common carriers, as well as many private companies, are installing fiber cables at an unprecedented rate. There is expected to be a glut of fiber capacity in the early to mid-1990s, at least in the major metropolitan areas in the United States. Although 45 million bits (45 Mbps) per second is the most common rate for digital transmission on a fiberoptic circuit today, transmission rates of 135 million bits (135 Mbps) and higher are coming into use. Rates of billions of bits per second have been tested under laboratory conditions. Although production systems lag behind laboratory experiments by several years, actual progress has exceeded even the most optimistic forecasts.

Another factor that affects the economic viability of fiberoptic transmission is the distance between repeaters. Today, using single mode fiber, distances of 100 to 120 kilometers between repeaters are common.

Local area networks are the mechanism by which the higher bandwidth and transmission speed will be delivered to the desktop computer. A local area network carrying data at a speed of at least 16 Mbps will be required to handle the integration of text, graphics, image, and voice that users are beginning to demand. Sixteen Mbps can be carried on shielded twisted pair wiring, but for higher speeds, optical fibers or some other media may be required. Today, the cost of installing fiber is still not low enough to consider such widespread installation, and the hardware and applications to take advantage of its bandwidth do not exist. By the mid-1990s, however, this situation will begin to change.

TRENDS IN TERMINALS AND WORKSTATIONS

For years, there has been discussion about the integrated, desktop workstation, a device that combines the capabilities of the telephone, VDT, printer, facsimile, and copy machine. Products that combine the telephone and the VDT have been announced by several vendors. So far they have not caught on, and sales have been disappointing. Nonetheless, the integrated workstation—which has some or all of these capabilities as well as computing power—is a safe bet for the future.

IBM believes that special workstations are required to meet the unique needs of six groups of workers:

- business/professional workers;
- engineers/scientists;
- clerical workers;
- production workers;
- data processing professionals;
- secretaries.

IBM believes that the requirements of these job types are different enough that they each require unique workstations, and IBM continues to develop products aimed at those job families. For example, a terminal used on a production line needs to be extremely easy to use, perhaps voice-driven, but may not need heavy computational power. On the other hand, a terminal used by the engineer/scientist needs strong computational and graphics capability.

For high-quality graphics display, much higher resolution than is commonly available today is required. Chapter 6 discussed the fact that VDT resolution is measured by the number of picture elements, commonly called *pixels* or *pels*, per square inch. To obtain "photograph quality" displays on a VDT, a display density of 400 × 400 pels per square inch is desirable. Figure 17–3 compares the number of bytes of data that are required for an 8½ by 11 inch page of typewritten and graphics data. The typical typewritten page contains between 2,000 and 5,000 charac-

Figure 17–3
Comparison of data that needs to be transmitted to display an 8½ × 11 inch page of typewritten data versus a graph of the same size.

Typewritten Data

2,000 to 5,000 characters per page
at 8 bits per character
= 16,000 to 40,000 bits
= 2,000 to 5,000 bytes per page

Graphics Data

$8\frac{1}{2}$ by 11 = 93.5 square in.
At 400 × 400 pels per sq. inch
= 160,000 pels per sq. inch
160,000 pels × 93.5 sq. in.
= 14,960,000 pels per page

Black and White:
= 1 bit per pel
= 14,960,000 bits per page
= 1,870,000 bytes per page

Color: (256 possibilities)
= 8 bits per pel
= 119,680,000 bits per page
= 14,960,000 bytes per page

With 20 to 1 Compression
Black and White = 93,500 bytes
Color = 748,000 bytes

ters, which, at 8 bits per character, requires 2,000 to 5,000 bytes to represent. That same page with graphics data at 400 × 400 pels per square inch contains 14,960,000 pels. If the graphics are to be displayed in black and white, each pel can be represented by 1 bit, and the page requires 1,870,000 bytes. On the other hand, if a color graph is to be displayed and we assume that 256 colors are required, 8 bits per pel are required to code the data. The page of colored graphics requires 14,960,000 bytes to represent the data.

Fortunately, with digital signal processing, 20 to 1 data compression is possible. The 1,870,000 bytes for the black and white graph can be compressed to 93,500 bytes for transmission, and the 14,960,000 bytes for the color graph can be reduced to 748,000 bytes. Even with this reduction, there is still a lot of data to be transmitted for one graph. In order to deliver reasonable response time, transmission speeds that only a local area network can deliver are required. This is shown in Figure 17–4, which shows how long it would take to transmit the typewritten page, the black and white graph, and the color graph at various transmission speeds.

There is some indication that the power of desktop workstations will change the basic characteristics of the way data processing is performed in business organizations. With adequate computing power in the desktop workstation, most workers may find that a large portion of their data processing is done at their terminal. The corporate mainframe would become a back-end processor that would provide extra computing capacity when heavy duty processing was required. The mainframe

Figure 17–4
Time required to transmit a page of data at various transmission speeds.

Transmission Speed	Number of bytes Transmitted per second	Time to Transmit (seconds): Average Typewritten Page (4,000 bytes)	Compressed Black & White Graph (93,500 bytes)	Compressed Color Graph (748,000 bytes)
1,200 bps	150	26.7 secs.	623.3 secs.	4,986.6
2,400 bps	300	13.3	311.7	2,493.3
4,800 bps	600	6.7	155.8	1,246.7
9,600 bps	1,200	3.3	77.9	623.3
56,000 bps	7,000	.6	13.4	106.9
1 Mbps	125,000	.032	.75	6.0
4 Mbps	500,000	.008	.05	1.5
16 Mbps	2,000,000	.002	.05	.4

would also store vast amounts of data, and the desktop computer user would pull subsets of this data to the workstation for local work. Then, at night or during other nonbusiness hours, the corporate mainframe would connect to each workstation and pull copies of all of the data files that have been changed back to the mainframe, providing an automatic backup of the data. It should be obvious that the use of desktop workstations in this way will require high-capacity data transmission facilities to transmit data files back and forth rapidly.

TRENDS IN STANDARDIZATION

In any new industry, standardization comes slowly. Initially, it is not clear which facets of the industry or its products should be standardized. Furthermore, each company producing products believes that it has better ideas than its competition. Since standards often represent compromises and, therefore, not necessarily the best technical solutions, companies must be convinced that they can still maintain their unique product identity and advantages even while meeting the standards.

In the communications industry, standards have been stronger and better developed in the telephone segment than in the data segment. This is because the telephone segment of the industry is older, and the need for any-to-any communications between telephones has been obvious from the start. Data communications standards are now evolving rapidly because users want the capability to connect any terminal to any computer and to have more flexibility in the connection than they have enjoyed in the past. The CCITT, ANSI, and other groups are working hard to develop new standards, but the process is slow. Standards for the bottom three layers of the ISO-OSI model for data communications were approved in 1978 but did not start to be widely used until the mid-1980s. The international ISDN standard for digital networks, which has been under discussion since the early 1980s, will in all likelihood not receive final approval until 1992. Despite the slow starts, the OSI model is gaining broader support each year, and ISDN capabilities are now being implemented in countries throughout the world.

Because of the slow process and the fact that standards represent a technical compromise, nonstandard and proprietary products and solutions to communications problems will continue to exist. Sometimes, these products provide a unique ability for a company to solve a business communications problem. The fact that the product falls outside of any standards may be irrelevant. Sometimes, products evolve in areas where no standards work is under way. For example, each vendor of voice messaging systems has its own proprietary method of coding the voice and storing it in the system's disk. VMX voice mail products cannot exchange voice messages with Rolm Phonemail systems. In the future, the need to standardize the systems so that they can exchange voice messages may be identified.

A company that can solve an existing business communications problem can be highly successful, even if its product is nonstandard. By installing the proprietary solution, a user of the product is able to solve a problem months or years earlier than if it waited for the standardization process to occur and standardized products to become available. Both the vendor and user face the risk of potentially having to convert to standardized products in the future. This risk must be weighed against the potential benefit that the product/solution can bring in the meantime.

TRENDS IN REGULATION

In the United States, there is a general trend toward continuing deregulation. Competition has opened the communications industry and allowed many new companies to market products. Many people believe that in the early 1990s there will be a number of failures and mergers of communications companies because of the continuing high cost of staying competitive and keeping up with the rapidly changing technology. It would not be surprising by the mid-1990s to see only five or six companies dominating the industry in a classic oligopolistic fashion.

Outside of the United States, there is somewhat of a dichotomy between those countries that are maintaining tightly regulated, monopolistic control over their communications industry and those countries that are taking some steps toward deregulating it. The United Kingdom and Japan are two countries that have made significant moves toward deregulating their industries. It is thought that they will continue to follow the United States' lead, perhaps lagging in the deregulatory process by 5 years or so. Other countries, such as Germany and Brazil, are keeping tight control of communications by centralizing it in a government agency. They are not expected to change this posture in the foreseeable future.

TRENDS IN THE APPLICATION OF COMMUNICATIONS TECHNOLOGY

The deregulated, competitive environment, coupled with the rapid technological advances, act as an enabler for new communications applications. The application of communications technology is the most exciting aspect of communications. This is where the benefits to business and individuals can be seen. This section discusses some of the things that can be done through the application of new communications technology and developments. Some of the applications are being performed today on a limited basis, whereas others will be coming available in the near future.

Telecommuting

Telecommuting is the use of an alternate work location for employees who are normally based in the office. Telecommuters typically work at home and use the telephone, modem, and a personal computer connected via telephone lines to the mainframe computer at the office. Telecommuting works best for people who are used to having a certain amount of freedom and who work well alone. By working at home, they can have an unlimited amount of quiet time with no interruptions (assuming that the worker doesn't have three small children or a spouse who gives music lessons!). On the other hand, communications with associates may suffer, particularly the informal communications that are a vital part of any business environment. Too much telecommuting can be professionally risky if an employee gets out of the mainstream. Usually, telecommuters schedule regular times to go into the office for face-to-face contact with their peers and superiors and for meetings.

Telecommuting can also work well for part-time employees who only want to work a few hours each day. They can work out a schedule with their employer so that they work at home a certain number of hours per week and go into the office one or two half-days to keep in touch with their peers and attend meetings. Usually, the employer covers the cost of the terminal and an additional telephone line in the employee's home.

It should be emphasized that telecommuting is not a future application of communications technology but is in widespread use today. Many employees of the city of Fort Collins, Colorado, regularly work at home using a terminal. A lawyer who works for a Chicago law firm telecommutes from Colorado. He consults with his partners by telephone and sends reports to the office using his personal computer and a modem. A speechwriter works in his home and sends speeches to his clients electronically. Many more examples abound, and as our society gets used to the concept of having workers at home, telecommuting will become more widespread.

Electronic Document Interchange (EDI)

Electronic document interchange (EDI) is the use of communications techniques to transmit documents electronically. The term is used most frequently in a business context today, but there is no reason why EDI techniques cannot be used by consumers as well. EDI relies primarily on the development of standard formats for various business documents, such as invoices, purchase orders, and acknowledgments. Furthermore, there must be an agreement between businesses (or other parties) to use these formats and transmit the data electronically using a standard communications technique. The benefits of EDI are the more rapid transmission of the documentation supporting the business activities and transactions and the elimination of paper. These can lead to other benefits, such as reduced inventory levels or faster collection of money from creditors.

All of the technology required for EDI is in place today. The banking and grocery industries have been leaders in developing the EDI capabilities, and the techniques are now spreading to other industries and across industry boundaries. The chemical industry is very active in developing standards for the electronic exchange of business documents. General Motors has been very active in promoting the use of the EDI with its suppliers and has made it a requirement for the suppliers of certain commodities to use EDI to do business with General Motors.

Public Voice Messaging

Voice messaging is another application area where all of the technology is available. To date, voice messaging has been implemented primarily by companies on a private basis for the use of their employees. A few voice messaging service bureaus have begun operation to provide voice messaging capability to subscribers in several companies. There is no technological reason why a telephone company could not offer its customers a voice messaging capability as an adjunct to standard telephone service. This is exactly the type of capability that is now permitted in the deregulated communications environment. In the mid-1980s, voice messaging computers were somewhat limited in capacity and could not handle more than 1,000 to 2,000 subscribers. However, the capacity barrier has been overcome, and it is expected that public voice messaging will become a reality in the near future.

Telediagnosis

Telediagnosis uses microprocessor chips and standard communications facilities to allow a device, such as an appliance or a computer, to automatically place a telephone call to a diagnostics center. The IBM 3090 computer, for example, continually monitors its own operation. If it detects a problem, the computer places a telephone call to a diagnostics center and transmits relevant data about the problem it is experiencing. At the diagnostics center, engineers can examine the data, request other information, and essentially operate the IBM 3090 in diagnostic mode. Assuming that the problem is not disabling the computer, normal data processing can continue while the diagnostic work is under way. In a high percentage of cases, the remote diagnostic center can determine which part of the computer is failing and can dispatch a local technician with the correct replacement part.

This concept could just as well be implemented in home appliances, such as televisions, washing machines, or furnaces. Similarly, an automobile might be driven to the local dealer's garage, where it could be connected to a diagnostic computer that communicates with a central diagnostic computer at the automobile manufacturer's headquarters. The technology exists; economics will drive the general availability of such services.

Medical Telediagnosis

A variation of telediagnosis will be used in the medical profession. A computer could give a doctor recommendations about how to assess a particular patient complaint. Using audio and video signals and diagnostic medical equipment with communications capability—such as a communicating X-ray machine—a doctor at one location could converse with a patient in another location, conduct tests, examine the results, diagnose the condition, and prescribe a treatment. The telecommunications connection would allow patients in remote geographic locations to consult with specialists in medical centers and provide doctors the capability to consult with one another and share information about particularly unique or difficult cases.

Automobile Navigation by Satellite

In this application, a car equipped with a small television-like device is capable of displaying a map of the local area. A transmitter in the car would send a signal to a satellite, which would calculate the location of the automobile and send a signal back to the car. A microprocessor in the car would translate the signal from the satellite and display the location of the automobile on the map. Ocean-going ships have been using a similar technique with the Inmarsat satellites for years, but until now the equipment has been expensive and satellite capacity far too limited to make the technology available to consumers. With the increasing bandwidth of satellites and the decreasing cost of the electronics required in automobiles, the application is coming closer to being affordable in the consumer market.

Electronic Journalism

Electronic journalism involves communications in several ways. Examples best illustrate the point:

- For a number of years, *The Wall Street Journal* and *USA Today* have been composed in a central location and transmitted via satellite circuits to several printing companies located around the world. The papers are printed and distributed on a regional basis, thereby giving simultaneous availability nationwide.

- Recently, the *Phoenix Gazette* began allowing letters to the editor to be transmitted to the paper from personal computers using a modem and telephone line.
- Soon, photographers will use electronic cameras that send images over telephone lines directly to computers at the publisher's office.
- When President Lincoln was shot, the news reached Europe via ship in about 2 weeks. When President Kennedy was shot, viewers in Europe watched the funeral live via satellite transmission. When President Reagan was shot, a reporter for a London newspaper, who was located in his office three blocks from the scene of the shooting, first heard about it via a telephone call from his editor in London, who had watched the shooting occur on live television!

Communications and microprocessors will combine TV and newspaper journalism in the future. It won't be long before we will have the option to have our newspaper(s) delivered electronically to the disk of our computer. The computer will scan the newspapers looking for topics of interest to us and will check the advertisements looking for items that we have told it we want to buy. Then we can select which, if any, parts of the paper we want to print. Similarly, our television will scan all programs on all channels and record those that it knows are of interest or importance to us. All of this information will be available to us to scan at our leisure, saving what we want to keep and deleting the rest.

A precursor of this capability is here today. Companies, such as CompuServe, contain regularly updated news stories, stock prices, and other articles of interest. These stories and articles are categorized so that subscribers can readily find information of interest and read it on their terminal.

Optical Fiber to the Home

Telephone companies are working hard to find economical ways to replace the wiring that runs to virtually every home and business in America with optical fiber. It will take years for this transition to take place because of the hundreds of millions of dollars that are invested in the exisiting cabling. However, on a gradual basis, beginning in new housing developments, optical fiber will be installed for communications instead of copper wire.

High bandwidth optical fiber will allow innovations in home or business communications. Most current examples of the potential relate to broadcasting, but they include the possibility for advertisers to direct different messages to viewers watching the same program, based on a viewer profile. Families with children might see a different advertisement than a retired couple. Viewers might be able to select their own schedule of television programs, watching the programs they want to see at convenient times—without videotaping.

In Heathrow, Florida, Southern Bell and Northern Telecommunications have been working on a project to explore the technical and economic problems of wiring a community with optical fiber. The system the companies are installing will initially carry 54 television channels, regular telephone service, and ISDN to homes in the community. The first phase of the system was put into operation in mid-1988, and services are continuing to expand. The telephone companies say they have identified a number of challenges that must be met before such a system could be widely deployed, but there is no indication that the challenges cannot be overcome.

Smart Homes

A *smart home* is a house that has microprocessors to help control home functions, such as heating, lights, and security alarms. When coupled with telecommunications, the computers can be instructed from a remote location to perform certain functions. For example, one could call home just before leaving the office and tell the computer to turn the heat up or the oven on. A smart home might also contain smart appliances that have the built-in diagnostic capability discussed earlier in the chapter. Some prototypes of these homes are scattered around the country. Xanadu House in Kissimmee, Florida, and Ahwatukee House in Arizona have been designed to be energy efficient and computer controlled.

common wiring system

Another aspect of the smart home of the future is that it may have a common wiring system. Every piece of wiring will be connected to the same control center. This means that every plug in the house will be interchangeable. You will be able to plug in a hair dryer, a stereo speaker, a telephone, or a security system to any outlet, and each of those appliances or devices will send a coded signal to the control center to tell it what kind of electrical power or other signal it needs. The blender will send a signal requesting electrical power, and the audio speaker will send a signal requesting music from the entertainment center. A child who sticks a finger into an empty socket will not be shocked because fingers don't carry the coded signal asking for electricity. As houses with this type of wiring system become available, manufacturers will begin to offer smart appliances with the coded signals built in. In the meantime, there will be adapter plugs available so that the appliances that people own today can be used with the new wiring system.

With a common wiring system, it will be possible to control and monitor any appliance or any function from anyplace in the house or, via telecommunications, from anywhere else. Furthermore, since the appliances will all be interconnected through the central control system, they will be able to talk to each other. For example, when the clothes washer finishes a load, it will be able to flash a message on the television screen you are watching to tell you that it's time to put the clothes in the dryer. The alarm system will be able to call you by telephone and tell you if someone is in the house when you are away. A fire detector, in addition to calling an emergency number, could turn on televisions and radios and have them speak a fire warning message. These capabilities and others are expected to be available in houses in the 1990s.

SOCIOLOGICAL IMPACT

With all of the trends pointing toward more information being available, faster, and at any location, thanks to the rapidly advancing microcomputer and communications technologies, an interesting question arises. One must ask whether individuals will be able to absorb and take advantage of all of the data and information that are becoming available. Studies have shown that our society is more stressful than in years past because our senses receive so many stimuli every minute of every day. Are our human data acquisition and processing skills up to the challenge of dealing with the rapidly growing output of an ever increasing number of microprocessors?

Clearly, one approach that will help us deal with all of the information available to us is the use of the computer itself. Using artificial intelligence techniques, the computer will digest, cull, and summarize data and present it to us in manageable forms and quantities. With proper instruction, the computer should be able to provide us with information that is meaningful and of interest while serving as a buffer to protect us from the millions of other information stimuli that will come our way.

It must be recognized that the trends described in this chapter are likely to bring about some new problems that will have to be solved. The invasion of privacy and the ability for someone to more easily monitor our daily activities are two simple examples that will have to be dealt with. It is a safe bet, however, that human intelligence and people's natural adaptability will ensure that ways will be found to deal with the problems. Furthermore, it is virtually certain that new communications technologies and capabilities will be put to use in ways we have never imagined.

SUMMARY

This chapter explored many emerging communications and computing trends and directions. The new capabilities that technology affords will shape our lives and lifestyles. As new technologies become available, there is always a transition period when the old and the new technologies are mixed. Organizations and individuals have large investments in installed equipment and capability, and they are only able to afford and take advantage of new technology at a certain rate. Therefore, compatibility of new with old is important so that we can take advantage of new developments while protecting the investment we have made in older equipment. We want to be able to use a new compact disc player but are not willing to throw out our record collection. Similarly, we want to add optical disks to our computers without making obsolete our magnetic disk. It is extremely important that vendors consider these compatibility issues when designing new products. Assuming that new technology can be implemented in an evolutionary fashion, the rapid growth of the computer and communications industries should continue indefinitely.

CASE STUDY

The Dow Corning Communications Network Vision

Dow Corning's vision of the future for telecommunications in the company has several facets. There is a strong belief that the terminals used by its employees will become more intelligent and that they will be used for an increasing variety of activities. Management also believes that the terminals and personal computers will not replace the mainframe computers. Rather, both will continue to grow at rapid rates. In order to take advantage of the increasing capability of the equipment and software, employees must undergo continuous education and renewal.

To support the growth in terminal sophistication, higher transmission speeds to the desktops will be required, especially as it becomes possible to integrate voice and image transmission with data. LAN technology, especially fiberoptics and the new FDDI standard, holds the promise of being able to deliver the needed speed, and the company is actively studying the technology and cost/benefit trade-offs.

Telecommunications management recognizes that network and computer systems are going to have to be of higher reliability. They must approach the reliability of the public telephone system and essentially be operational all the time. In order to achieve this reliability, the network management procedures and practices must be improved so that the help desk knows of problems before the users do. More redundancy must be added to the existing network to make it more fail-safe so that there is no single point of failure.

Finally, Dow Corning believes that the network can be used to help it communicate better with other companies with which it deals, most notably its customers and suppliers. Dow Corning believes that it can improve its customer service and add value to its products by providing new types of information and new ways of doing business.

The implications of all of these changes for the telecommunications department are significant because the number of locations served and the capacity of the network are likely to far surpass today's capabilities. The department's people will need to acquire new technical and management skills in order to be ready to implement and operate new telecommunications systems. Having innovative and creative people will continue to be a critical success factor for the telecommunications department.

Problems and Projects

1. Count the number of fractional horsepower motors you have in your home. Does the number surprise you? Can you imagine that number of microprocessors being in your home in the near future? Identify some novel places where those microprocessors will be used.

2. If sophisticated speech recognition equipment were available, it would theoretically be possible for you to dictate your college term papers into your personnel computer and have the result printed immediately. Is your thought process organized well enough so

that this would really work for you? What other obstacles will you have to overcome so that you can take advantage of the speech recognition capability when it is available to you?

3. Based on your knowledge of fiberoptic technology, what are some of the practical problems of running a fiber to every desk?

4. Thinking about the differences in their jobs, identify some of the differing characteristics that would be required in a terminal designed for clerical workers and one designed for managers.

5. Engineers and other technical people tend to push for standardization in all parts of our society. What are some of the advantages of not having all communications standardized?

6. The banking industry has been using EDI for years to transmit money between financial institutions. If you had EDI capability available to you personally, can you think of any uses you could make of it?

7. If automobile navigation by satellite were available, as described in the text, how else could the communications capability be used by the driver or other occupants of the car?

8. The text mentioned that telecommuters must go into work periodically for meetings. What would be required to allow them to participate in the meetings from their homes?

9. What are some sociological problems that may occur because of ubiquitous, high-speed data communications being readily available?

Vocabulary

digital signal processor (DSP)
voice compression
video signal compression
telecommuting
electronic document interchange (EDI)
telediagnosis

References

Atchison, Sandra. "These Top Executives Work Where They Play," *Business Week,* October 22, 1986, p. 132.

Balmes, Mark, John Bourne, and Jung Mar. "The Technology Behind Heathrow," *Telesis,* Volume 2, 1989, p. 31.

Becker, Hal B. "Can Users Really Absorb Data at Today's Rates? Tomorrow's?" *Data Communications,* July 1986, p. 177.

Blyth, W. John, and Mary M. Blyth. *Telecommunications: Concepts, Development, and Management.* Indianapolis, IN: The Bobbs-Merrill Company, Inc., 1985.

Gordon, Gil E. "Telecommuting: Planning for a New Work Environment," *Journal of Information Systems Management,* Summer 1986, p. 37.

Harb, Joseph A. "No (Beep) Place Like Home (ZZZT)," *Nation's Business,* February 1986, p. 34.

Martin, James. *Future Developments in Telecommunications,* 2nd ed. Englewood Cliffs, NJ: Prentice-Hall, Inc., 1977.

Mazur, Thais, and Ives Brant. "The PC Hear and Now," *PC World,* November 1986, p. 226.

———. "Telecommunting Benefits Business with DP's Help," *Computerworld,* February 17, 1986.

Miller, Thomas. "High Performance Telework Helps Companies Compete," *Computer World,* February 24, 1986.

Payne, Seth. "It's 10:30, Do You Know Where Your Truckers Are?" *Business Week,* May 23, 1988, pp. 116H, 116L.

Rohan, Barry. "Computers That Listen Join the Work Force," *Detroit Free Press,* November 10, 1986.

Stamper, David A. *Business Data Communications,* 2nd ed. Menlo Park, CA: The Benjamin/Cummings Publishing Company, Inc., 1986, 1989.

"Visions of The Future," *Communications News,* 25th Anniversary Edition, 1989, p. 16.

Wilson, John E. "A Chip That Lets Computers Tune in on the Outside World," *Business Week,* July 21, 1986, p. 118.

APPENDIXES

Appendix A

Readings in International Telecommunications

This series of articles provides more information about the status of telecommunications in countries other than the United States. The international telecommunications situation is changing rapidly, as these articles show, but the overall, worldwide trend is toward deregulation. These articles were originally published in the magazine, *Communique,* a publication of the International Communications Association of Dallas, Texas. Articles in this appendix describe the telecommunications situation in

Australia
Austria
Belgium
Canada
The Caribbean
Finland
France
Hong Kong and the Far East
The Netherlands
Norway
Spain
Sweden
Switzerland
United Kingdom

Australian Telecommunications at the Crossroads

by Walter E. Rothwell
Executive Director
Australian Telecommunications Users Group

Following a number of significant events over the last eighteen months, the telecommunications industry in Australia appears set for major change.

Presently, regulation of telecommunications is the province of Telecom Australia, a monopoly government business enterprise, much in the mold of British Telecom preprivatization. Telecom is the sole permitted carrier of domestic public switched voice and data traffic, no other organization being allowed to carry third-party traffic or to sell spare capacity on private networks. The only exceptions are the Overseas Telecommunications Commission (OTC), which has a virtual monopoly on overseas traffic, and AUSSAT, which may provide interconnected private satellite networks but may not compete for public switched traffic.

Telecom's monopoly extends to maintenance of PBXs and interconnected private networks, the lease and installation of the first telephone in all households, and the sale of small business or key-phone systems. It is also the setter of specifications and standards and is conformance testing authority while also being the virtual monopsonist purchaser of telecommunications equipment in Australia.

Telecom is an organization of 91,000 employees, and revenue last year was $6.04 billion. Telephone service is now available to 92% of Australian homes.

However, in spite of a well-evident social conscience, there is a wide feeling that the benevolent dictator that is Telecom Australia is often seen as a lumbering giant, providing inconsistent and frequently slow service, with policy being decided more with an eye to protecting its revenue base and marketing position than encouraging more liberal use of the telecommunications medium.

Telecom is probably as advanced technologically as any of the world's PTTs—ISDN trials are underway, fiberoptics are in fairly wide use, it is a world leader in EFT/POS, and QPSX has just been accepted as a world standard. But it is simultaneously involved in policy making, regulation, competition, and arbitration. It is felt strongly that Telecom should not be rule maker, player, referee and "Judge Greene" all at once and that these areas need to be separated out. There is even evidence that Telecom itself would like to be relieved of some of these responsibilities.

In the last year, Telecom has come under fire for alleged predatory pricing in the competitive telephone market and for some of its other trading practices. Its activities in the marketplace and some rather arbitrary tariff setting, not to mention the commercial opportunities seen by business, especially in the Value Added Service area, have heightened the calls for change.

As a consequence and to the loud acclaim of users, Transport and Communications Minister, Senator Gareth Evans, has set up a task force to inquire into the need for change.

After wide consultation, that task force will be making its final recommendations in February.

Hopes and expectations are high that the result will see considerable liberalization. Two things seem to be almost certain: first, policy making will largely return to government, where it belongs, and second, Telecom will be relieved of most of its regulation and arbitration roles by the appointment of an independent regulatory body, possibly similar to Britain's OFTEL. But users are seeking much wider reforms than these two, both of which seem to be self-evident necessities for a healthy life.

The task force has been asked to look at other proposals, such as

- restricting Telecom's monopoly, for the time being, to the basic telephone network;
- allowing open and equitable competition in all other areas, particularly in the supply, installation, and maintenance of Customer Premises Equipment;
- providing free access to public networks by allowing full interconnection of private networks and allowing carriage and third-party traffic and resale of spare capacity on those networks;
- freeing all three carriers from some government and public service constraints so that they can compete equally;
- allowing AUSSAT to compete for public domestic and some Pacific traffic;
- shifting standard setting to the orbit of the Standards Association of Australia;
- moving toward supplier self-certification instead of Telecom conformance testing.

But there are two important hurdles to be overcome before some of these reforms can occur. First, the rural users cannot be ignored, and for that reason, those who would compete for Telecom's "cream" services will have to pay some kind of social contribution in order to cross-subsidize the remote areas—some single telephone services cost in excess of $30,000 to install.

Second, the needs and concerns of the labor force have to be recognized. Australian unions clamor loudly that privatization in the U.K. has meant reduced service quality and that the U.S. experience meant heavy loss of jobs in the industry. The counterarguments are all too obvious but to make them convincing is not so easy.

So, exciting times are ahead, especially as Australia sees its near neighbor New Zealand throwing off its communications shackles. There is an uncertainty about an unprotected, liberalized industry of the future, but there is also an air of high anticipation that the coming year will see a revitalization and vastly more dynamic industry in Australia.

This is Australia's Bi-Centenary Year. We don't intend to waste it standing still!

Source: Reprinted by permission from *Communique*, volume 41, number 5, February 1988. *Communique* is a publication of the International Communications Association, Dallas, Texas.

Austria

by Georg N. Bruckner
Chairman, Telecommunications
Arbeitsgemeinschaft Für Datenverarbeitung (ADV)

In Austria, the telecommunication scene hasn't changed much over the past two years. It is still marked by a strict monopoly of the government-owned PTT. Because Austria is a small country wth a population of only seven million (equivalent to two thirds of Pennsylvania), it does not take on the role of a front runner either in technological development or in the trend toward liberalization and deregulation. However, the digitalization of the telephone network is progressing quite well, though it will take quite a number of years before every subscriber can enjoy the benefits of the digital network.

Tariffs are still among the highest in Europe, but the trend toward European harmonization is noticeable here too. For example, long distance rates have been going down steadily, and with $1.35 per minute to the U.S. these are at a reasonable level, at least by Austrian standards.

Apart from the telephone network that can be used for data transmission via PTT-approved modems, the PTT also offers a range of specialized public data networks that meet the needs of different applications.

A subscriber can choose between circuit switched and packet switched (X.25) data networks as well as from various types of leased lines. The highest bit rate in the packet switched network is now 64 kbit/second. Leased lines are offered with up to 2 Mbit/second in urban areas.

A sophisticated tariff structure and complicated legal restrictions tend to make it difficult for the customer to select the most economical means of data transmission for a specific telecommunications application.

The introduction of ISDN in Austria has been slowed down considerably. Field trials keep being postponed, and so far it is not foreseeable when ISDN will become commercially available.

As in many other countries, telefax is booming, outpacing more traditional means of communication, such as telex, of which Austria used to enjoy one of the highest subscriber densities in the world. In fact, telefax has become so popular that providers of message handling services have a hard time convincing the users of the added value of services like X.400.

X.400 is the keyword in which Radio Austria, a subsidiary of the PTT, is putting its hopes. Designated as the "Austrian Administration Management Domain" (ADMD), Radio Austria is about to offer the X.400 service as a public VAS, making message handling services worldwide available to Austrian users.

Some ambitious users are already in the process of defining their own EDI applications, according to the international EDI-FACT standard, being ready to go as soon as X.400 becomes available as a public service.

Amidst the monopoly discussion over value added services (VAS), the Austrian government pledges liberalization of telecommunications. It is not yet quite clear how far-reaching the steps toward that goal would be, but it can be expected that Austria will eventually follow its neighbors, such as Germany, where the PTT recently has been restructured along the lines of the "Green Paper" of the EEC.

Despite the technologically advanced state of telecommunications in this country, Austrians will have to continue to live with a public administration that entails red tape in matters such as type approvals, subscriber applications, network maintenance, and so on.

In the past few years, however, the PTT has begun to understand user issues in the segment of computer communications. The continuing dialogue between providers and users will certainly help to foster and improve telecommunications services in Austria.

Source: Reprinted by permission from *Communique*, volume 42, number 6, December 1989. *Communique* is a publication of the International Communications Association, Dallas, Texas.

Belgian Telecommunications

by D. J. Van Steenwegen
President
Telecommunications Managers Association—Belgium

The Telecommunications Managers Association—Belgium is an organization of professional telecommunications managers and other full-time specialists making use of telecommunications services.

Its members are employed to give telecommunications management and operational support to organizations, such as banks, commerce, large corporations, and other agencies, providing that they function as users of the various telecommunications disciplines, such as telephony, telex, telegraph, facsimile, data transmissions, video, radio, etc.

Our association is registered in Belgium as a nonprofit-making body, in accordance with the statutes published in the Annexes of the Moniteur Belge. Our principal objectives are

- Exchanging ideas: to promote a free and fruitful exchange of ideas and experience between members on all telecommunications matters, but with particular regard to evaluating equipment and services;
- RTT representation: to maintain a strong collective working relationship with the RTT in order to encourage a balanced understanding between users and the administration;
- Standards harmony: to provide coherent representation with international bodies, such as the European Commission, on harmonizing standards and practices within Europe and to contribute to international standards via the International Users Group (INTUG) to the CCITT;
- Exhibition support: to represent Belgian Telecommunications Users at national and international exhibitions, seminars and symposiums;
- Tariff data: to inform members of changing national and international rules, regulations, and tariffs.

Working groups examine key aspects of telecommunications in Belgium and report to members through a regularly published newsletter with articles in Dutch, French, and English.

Overall, the association contributes both to individuals in the profession and to their personal careers as well as to the organizations that employ them. It plays an essential role in shaping external issues of vital importance in the rapidly advancing telecommunications technology.

Because of the existence of other local telecommunications users groups in Belgium, mostly operating within the national boundaries, it became essential to regroup these units under one umbrella, called BELTUG. The founder associations setting up the BELTUG organization are TMAB, ASAB-VEBI (Teletran), and very recently ABUT-BVT, which is in the process of being concluded. BELTUG will represent commonly the three different users associations vis á vis the RTT, the respective ministries and cabinets, and be the major interface with the EEC via the ACTUA organization to which BELTUG will subscribe.

On the Belgian national scene, the RTT, which has the telecommunications monopoly, is actively involved in digitizing the nucleus of the telecommunications infrastructure with an 80 percent completion schedule projected for the end of the next decade. On the transatlantic route, TAT-8 will be available for operational use in Belgium by the fourth quarter of 1988 either via England or via France, each of them serving as a backup, if so required. Tariffs for digital circuits are in the process of being reviewed and will result in major reductions in the near future. This also will have a direct impact on international tariffs for telephone calls and leased circuits.

Source: Reprinted by permission from *Communique*, volume 41, number 5, February 1988. *Communique* is a publication of the International Communications Association, Dallas, Texas.

Telecommunications in Canada: A Nation Moving Forward Prudently into the 1990s

by Paul F. Kirvan
ICA Editorial Consultant

Telecommunications in Canada today is often like looking at the U.S. telecommunications market about 10 years ago. Indeed, Canada's strategy of monitoring U.S. actions from its excellent vantage point has helped it provide telecommunications effectively within a nation whose 26 million people occupy the second largest land mass in the world.

Canada has a unique blend of regulation and competition, but in different proportions than the U.S. In many ways, the current environment looks like the U.S. before AT&T's divestiture. The formula has worked successfully for decades. However, various forces in Canada are shaking the telecommunications industry as its very foundation.

Officials in the government and the telephone companies suggest these forces will cause irreparable damage to a fundamentally sound telecommunications environment. From the user side, the Canadian Business Telecommunications Alliance (CBTA), the ICA's peer organization in Canada, believes these forces will indeed prove beneficial to Canada. Still other organizations—the new wave of competitors—are hoping to carve a profitable niche in a traditionally conservative industry. These four forces—government, telephone companies, end users, and competitive suppliers—have transformed Canada's telecommunications industry into an exciting, dynamic, yet confusing place to be.

This article looks at Canadian telecommunications from several perspectives: regulatory, competitive, the impact on Canadian users, and the impact on U.S. firms with Canadian offices or Canadian interests. It will include observations from W. Page Montgomery, Vice President with Economics & Technology, Inc., the ICA's legislative and regulatory consulting firm, and Ian Angus, President of Angus Telemanagement Group, Inc., one of Canada's leading telecommunications consultants, located in Pickering, Ontario. Our No. 1 trading partner to the north is undergoing some significant changes. It is in our best interests as telecommunications professionals to understand these changes and their impact on the U.S. telecom industry.

Years of Progress Dealing with a Unique Environment

Canada is a nation of stark geographic contrasts. Most of its population resides in the lower third of the country. In fact, most of its population lives further south than Minneapolis! The northern territories are characterized by harsh, rugged, yet unblemished country. Providing telecom services to all Canadians has been a series of technological challenges far different from those faced in the U.S. That Canada has successfully developed a wide variety of services reaching all corners of the country is noteworthy in itself. However, it still provides service through a regulated telephone industry that is considered a monopoly despite the growing number of attempts to crack the seemingly impregnable fortress.

Telephone companies are the only providers of long distance and local message telephone service in Canada. CNCP Telecommunications, originally a joint venture of the Canadian National Railway Company and Canadian Pacific Limited (CNCP), and the telephone companies provide voice and data private line service. Cellular phone service is provided by Cantel and Cellnet Canada. Several other smaller carriers also provide private line services. Telesat Canada is the primary satellite transmission carrier in Canada. Teleglobe Canada provides international telecommunications services through several gateways on both coasts. The principal federal regulatory agency in Canada is the Canadian Radio-telephone and Telecommunications Commission (CRTC). Many provinces have their own internal regulatory bodies as well.

The majority of telephone service is provided by Telecom Canada, a consortium of ten provincial telephone organizations, whose offices are in the national capital, Ottawa. The largest member is Bell Canada, which serves Ontario and Quebec, as well as parts of the Northwest Territories with over 8 million access lines. Other members include

- British Columbia Telephone (B.C. Tel)—Canada's second largest telecommunications company, B.C. Tel serves most of British Columbia. It is shareholder-owned and federally regulated by the CRTC. It is a highly diversified company, with over ten subsidiaries in areas ranging from equipment leasing to cellular service.
- Alberta Government Telephones (AGT)—AGT is the designated long distance service to all areas of the province except the capital of Edmonton, which owns and operates its own municipal telephone system. It is regulated by the Alberta Public Utilities Board and has four separate Canadian subsidiaries. It is also part of NovAtel Communications, Ltd., a large manufacturer of cellular phones and systems.
- Saskatchewan Telecommunications (Sask Tel)—SaskTel was the first telecom company in Canada to announce plans for an all-digital switched network by the mid-1990s. SaskTel provides service throughout Saskatchewan. It is regulated by the Saskatchewan Public Utilities Review Commission. SaskTel is a member of Canadian Communications International (CCI), a consortium of Canadian firms with interests in Pacific Rim opportunities, particularly the Peoples Republic of China. The company is regarded as one of the most experienced in fiberoptics.
- Manitoba Telephone System (MTS)—MTS provides telecommunications services throughout the province of Manitoba. Its modernization program targets 1996 as the year when all multiple-party phone lines are eliminated in favor of single-line service. MTS Cellular now offers cellular service to more than 75 percent of the province. MTS is regulated by the Manitoba Public Utilities Board.
- Bell Canada—Leading the way in Canadian telecommunications is Bell Canada, a subsidiary of BCE, Inc. Bell's territory includes most of Ontario and Quebec, reaching into parts of the Northwest Territories. The company is regulated by the CRTC and is a joint owner, along with Northern Telecom,

of Bell Northern Research, a major research and development organization. Bell Canada International is also a subsidiary of BCE, Inc. Bell is the largest supplier of telecommunications services in Canada, with over 8 million access lines in service.

- Maritime Telegraph and Telephone (MT&T)—MT&T provides telecom services throughout Nova Scotia. It's shareholder-owned and regulated by the Board of Commissions of Public Utilities for the Province of Nova Scotia. Like MTS, it has committed over $80 million to make single-party service available throughout the province by 1996.
- New Brunswick Telephone Company (NBTel)—NBTel is a wholly owned subsidiary of Bruncor, Inc., and regulated by the New Brunswick Board of Commissioners of Public Utilities. The company has been installing a fiberoptic network as part of the Telecom Canada fiber system and its commitment to servicing the needs of New Brunswick.
- Island Telephone, P.E.I. (Island Tel)—Island Tell is 51 percent owned by Maritime Tel & Tel, investor owned, and regulated by the Prince Edward Island Public Utilities Commission. The company placed a new Northern Telecom DMS-100 digital central office into service during 1988 and is well into its program of upgrading to single-line service.
- Newfoundland Telephone—The company is a wholly owned subsidiary of NewTel Enterprises, Ltd., and provides service throughout Newfoundland and Labrador. It is regulated by the Board of Commissions of Public Utilities of Newfoundland.
- Telesat Canada—The tenth member of the Telecom family is Telesat Canada, which provides a wide range of telecommunications services via its Anik family of geostationary satellites. Anik C and Anik D satellites support both the C and Ku transmission frequency bands. The next generation of Aniks, E1 and E2, are scheduled for launch during 1990. They will be the largest satellites placed into orbit for domestic communications. It is regulated by the CRTC.

International telecommunications are provided by Teleglobe Canada, a privately owned company, through its network of satellites and undersea cables. It provides voice, data, telex, digital facsimile, and electronic mail services. The company reports directly to the Minister of Communications.

Other regional telephone companies, similar to U.S. independent telcos, include the following

- NorthwesTel—Regulated by the CRTC and owned by the Canadian National Railway Company; the company serves the western region of the Northwest Territories, the Yukon, and northern British Columbia.
- Terra Nova Telecommunications—Regulated by the CRTC and owned by Canadian National Railways, the company serves areas of Newfoundland not covered by Newfoundland Telephone.
- Quebec-Telephone—Regulated by the Regie des services Publics du Quebec, the company is shareholder-owned and an associate member of Telecom Canada. It serves areas in northeastern Quebec.
- Telebec—Like Quebec-Telephone, Telebec is regulated by the Regie des services Publics du Quebec and shareholder-owned. It serves northern Quebec.
- Northern Telephone—Regulated by the Ontario Telephone Services Commission, the company is shareholder-owned and serves areas in northern Ontario.
- Ontario Northland Communications—The company is a crown corporation reporting to the Minister of Northern Development and Mines for the Province of Ontario. It serves areas in northern Ontario.
- Thunder Bay Telephones—Owned by the municipality of Thunder Bay, Ontario, the

company is regulated by the Ontario Telephone Services Commission. It serves the Thunder Bay region.

- The City of Edmonton, Alberta provides its own telecommunications services through a company called Edmonton Telephones. It is regulated by the City Council of Edmonton.

Aside from dealing with a unique geography and a mosaic of telecommunications service providers, Canada has several important regulatory and marketplace attributes that merit analysis. One of the best ways to do that is through the eyes of expert observers of the Canadian scene.

The Issues—A Canadian Perspective

In this article, representives from both sides of the border offer their comments and observations. Leading off is Ian Angus, of Angus Telemanagement, Inc.

Several issues exist in Canadian telecommunications today. The first is deciding just who regulates communications. "Canada has never had the sort of reasonably unitary telecommunications environment found in the U.S.," Angus said. "You have a two-tier structure, in which the FCC (Federal Communications Commission) handles the national issues and the states take care of intrastate activities."

In the U.S., the FCC has functioned as a major focal point for establishing general regulatory structures, whereas the states end up largely setting rates within their boundaries. "Over here, we haven't had that," notes Angus. "In some areas of Canada, provincial regulatory bodies have total responsibility for telecommunications within their boundaries. Still other parts of the country, primarily Ontario, Quebec, and British Columbia, plus several other small areas, are federally regulated by the CRTC."

The result has been a wide variety of regulatory practices from one province to another. "Our environment ranges, on the one hand, all the way from reasonably open competition in terminal equipment, PBXs, even private line services, including resale and sharing, in the federally regulated areas," Angus said, "all the way to provinces where about the only thing you can do is install a second residence phone, and there's even a question about that."

Within the business community, a broad range of rules exists. Businesses throughout Canada have been pushing for some years for a more unified commercial telecommunications environment. This would mean uniform rules throughout the country, if they could be implemented.

"What has shaken that idea up is a recent Supreme Court decision that the federal government has the authority to regulate everywhere," according to Angus. "It implies that issues in telecommunications should be handled by the federal government." The ruling surrounded a very complex case related to an attempt by CNCP Telecommunications to connect customers in several areas of the country.

Now, according to Angus, that situation could be corrected by a one-line amendment to the Railway Act, the legislation comparable to the U.S. Communications Act of 1934. It could mean the development of a formal telecommunications act, something that has been discussed for 25 years but that never gets over the initial hurdles. The Supreme Court decision this past summer was a very positive one for Canada. Most businesses seem to approve of it, according to Angus, but what will eventually happen is still unknown.

Monopoly, Duopoly or Oligopoly?

Telecommunications in Canada is a study in contrasts. In carrier services, the government wants to establish two types of carriers, Type 1 and Type 2. Type 1 are facility-based carriers, which is the present structure. This includes the existing telephone companies plus CNCP Telecommunications and Telesat Canada. Type 1 carriers can have "physical" facilities, which are customer connections.

Type 2 carriers are all other firms that want to be in the business. They would have to obtain physical facilities from Type 1 carriers. (This is very similar to the U.S., in which specialized carriers must obtain local access lines from local telcos.) This scenario is sometimes called the "duopoly" of Canadian service providers. As Angus notes, "In essence, the Type 1/Type 2 carrier issue would create, if not a duopoly, certainly a very limited oligopoly of companies who run things." An example of how this is shaping up is CNCP Telecommunications (which will be renamed) and Rogers Cable (which recently bought approximately 40 percent of CNCP). The combined organizations have probably filed another application to enter the switched long distance business, at present controlled by provincial telephone companies.

"A major concern in Canada is the possibility that a U.S. company could come in and literally take over," according to Angus, "and this could also include MCI or Sprint or anybody like that." One of the most pressing questions in Canada concerns the ultimate survival of the Canadian-owned telephone industry. "The feeling, shared by many in the country, including the government, is that Type 1 carriers should be 80 percent Canadian owned," said Angus.

That poses a potentially significant question: Is the Canadian market large enough to support two carriers? The U.S. market has had its share of difficulties supporting numerous carriers. The Canadian market is smaller than the state of California, yet is spread over 4,000 miles. That suggests an interesting possibility, according to Angus. "If telecommunications was opened wide to competition," he said, "in a couple of years it's just possible we could return to one carrier, another monopoly." Included in this issue is the continuing belief, on the part of the major telephone companies, that long distance subsidizes local service. (Sound familiar?) In Canada, most people, both in the carriers and on the user side, acknowledge that long distance subsidizes local service.

Another hot issue in Canada is the idea of universal service. Telephone networks should be structured so that everybody can afford a phone. This is similar to feelings about socialized medicine in Canada. It is accepted, and most Canadians like it. The same is true with telecommunications: anything that suggests increases in local rates will create a major political tempest.

Competition—A Major Issue

Although many issues in Canadian telecommunications are linked, the major issue is still competition. The two key questions are (1) will Canada endorse competition in all major areas of telecommunications, and (2) if that occurs, what will the rules be? "We already have experience with competition in telephone systems and terminal equipment, equivalent to that in the U.S., since the early 1980s," Angus noted. "In fact, the Canadian market for terminal equipment grew much more rapidly than the U.S. market during the early 1980s, and the penetration of privately owned telephone equipment is probably just as strong as in the U.S."

The Canadian regulatory structure does not require phone companies to sell through separate subsidiaries, as in the U.S. Telephone companies can sell any kind of telecommunications equipment on the market, whether regulated or unregulated, through the same sales representative.

Angus offers an example of how this works in Canada. "I could come out and offer you an SL-1 at the tariffed rate. In that case I would rent it to you. But I can also sell you the same system. If that's the case, all I'm required to do is file a base price list with our federal regulator. The prices are structured so I don't lose money on the sale. It's an interesting situation, and has quite a lot of flexibility for telcos."

Private line service competition has existed in Canada for 10 years. Since 1979, CNCP has offered private line voice and data services. However, many believe competition has been effectively stifled because the CRTC declared that CNCP prices would be no less than 5 percent and no more than 15 percent lower than comparable Bell Canada rates for the same service. In the area of T-1, CNCP has repeatedly offered discounts. Again, a CRTC policy has restrained competition. Specifically, CNCP has tariffs considered "identical" to Telecom Canada, but with a 5 percent discount. In effect, the result is "regulated competition," just not true competition.

Angus notes another factor that has limited CNCP's ability to increase its market share. "Customers say, 'OK, CNCP. You can save us 5 or 10 percent. So what? It's not worth the aggravation to change carriers.'" Angus added that if a specialized carrier like CNCP cannot differentiate itself by offering new services (unavailable elsewhere) for the same price, users will not be able to justify using that carrier.

This arrangement would certainly tend to freeze the long distance market structure the way it is currently. Arguments in favor of that approach are based on the small size of the Canadian economy, relative to the U.S. The logic goes that if the long distance market was suddenly opened up to competition nationwide, it would be difficult for the "native" carriers (such as the provincial telephone companies) to survive. This is especially true when one realizes that the smallest U.S. regional Bell company is bigger than the largest Canadian telephone company.

Canadian Telecommunications—A U.S. Perspective

Offering his observations on Canadian telecommunications is Page Montgomery of ETI. "I think the most significant advantage Canada has with regard to telecommunications is that they have the opportunity to closely observe U.S. activities. In that position, they can serve themselves well by structuring their activities based on lessons learned from U.S. experience."

Canadian regulators generally feel that U.S. policy on competition is ahead of other countries in the world. "In spite of the U.S. approach to competition and regulation, which is often highly spirited," according to Montgomery, "it still makes good sense to Canadians to watch U.S. actions, adopt the policies that are appropriate for Canada, and then avoid making the same mistakes."

Montgomery believes Canadians have come up with better methods of working toward consensus. "This contrasts with the U.S. public policy business, in which we regularly find ourselves working in an environment of conflict where everybody takes fairly extreme positions," notes Montgomery. It is not unusual for the FCC to establish adversary roles among people on tariff and regulatory issues. This is often accomplished by "splitting the difference" between opposing positions. "The result of this is to encourage people to take extreme positions on issues," Montgomery adds.

By contrast, Canadians negotiate differences based on specific principles, which tend to work toward consensus. People who tend to take more extreme positions are less likely to achieve any desired results. This approach is similar to the way consensus is reached in Europe. "The goal is to drive people toward the center," said Montgomery, "whereas in the U.S., and not just in telecommunications, we tend to drive people toward extremes."

The result is that Canada has periodically brought new products and services to market sooner than in the U.S. "In Canada, people say, 'We're going to try this new service out now. We'll probably change it in the future, and we're not going to make up our minds just yet. But we'll try it first,'" said Montgomery. In the U.S., however, introduction of new products and services is often delayed until the regulations, policies, etc. get resolved. Extensive discussion occurs over the rules, which is often not very efficient.

Canada does not have federal/state separation to the same degree as in the U.S. But some of their provinces are regulated by the CRTC, and others are regulated by provincial agencies. In the U.S., for example, much of what occurs in telecommunications, especially from a regulatory perspective, is influenced by the FCC. Some activities may have to be coordinated with state regulators, and in other cases, the states assume full responsibility for certain issues. This is not unlike Canada as well.

"Currently, about half of what is going on in Washington deals with trying to figure out who has jurisdiction over a specific issue," notes Montgomery. "It's a good idea to eventually resolve those turf issues so that the states have jurisdiction over some areas, while the FCC has the final say on others." This is closer to the Canadian system, although jurisdictional disputes still occur.

Montgomery adds, "I think that the Canadians are generally satisfied with their one-step-at-a-time approach to telecom issues. I suspect, however, that there are a growing number of people in the business community who think they should move faster." But the evidence suggests that most of the players, including carriers, suppliers, business users, and regulators at the federal and provincial levels don't really want to move forward too quickly.

Market Opportunities in Canada

Canada is truly Northern Telecom country. In areas like telephone equipment, network hardware, and public switching systems, Northern Telecom owns the market. According to Angus, AT&T has yet to sell a central office in Canada. The only non-Northern switches sold have been smaller units from companies like GTE and Stromberg Carlson. Northern Telecom Centrex is growing in popularity, and the SL-1 is the most popular PBX by a large margin. Virtually the whole telecommunications industry is dominated by one manufacturer.

Further, the natural tendency in Canada is to prefer local Canadian suppliers. It is generally difficult for outside firms to penetrate the Canadian market. According to Angus, "Fujitsu tried and failed. Toshiba is actually doing quite well in small systems because the key system market tends to be more open." AT&T is progressing at a moderate pace in Canada. "They're not doing as well as I thought they would," notes Angus. "They took a while to figure out the Canadian market, deciding to focus more on data products," he said. "In fact, they've done better with their data products here, relatively speaking, than they have in the U.S."

Interestingly, an area virtually untapped in Canada is T-1. One likely reason is that Canadian T-1 service can cost up to seven times the cost for comparable U.S. service. Thus, the T-1 multiplexer market in Canada is rather limited. Only a few systems are currently in use throughout the country.

The Situation for Telecom Professionals

For Canadian telecommunications managers, two factors have impacted the profession. The first, due in large part to the growth of competition, is a dramatic increase in the number of telecom managers and the number of people who are needed in this area. As Angus explains, "The whole profession is growing very rapidly. Just this morning, we were talking to the CBTA. They had more than 1,200 people at this year's conference (held in Toronto, September 25–28, 1989), which was an all-time record." Canada's telecom education programs are just beginning to appear, with community colleges offering diploma programs in telecom management. The nation's first telecommunications degree program is in the development stages.

Telecommunications management in Canada is rapidly becoming more complex. The situation is not unlike what happened in the U.S. back in the early 1980s. The job gets more complicated because the rules keep changing. Many gray areas exist in which the rules have yet to be defined. Organizations like the CBTA have spearheaded efforts to define the new rules and redefine the old ones. More important, until the existing rules are challenged and changed, growth and progress in the telecom profession will be restrained.

Canada and Europe 1992

As the talk about a unified European economic market grows, Canada will indeed have its share of opportunities. The country will need to develop sales skills for the European market, according to Angus. For example, both Northern Telecom and Mitel have been successful in European markets. "I think Canadian manufacturers must build relationships, or partnerships, with European firms, similar to what AT&T did in Italy," said Angus. "Those kinds of partnerships could forge real strength with Europeans." Angus adds that Canada's traditional preferred access to the British market is a thing of the past.

U.S. Telecom Managers with Canadian Offices—The Challenges

As described in this article, the Canadian telecommunications market tends to be controlled by a few key players, i.e., the provincial telephone companies, Telecom Canada and Northern Telecom. Although competition exists, it is much more restrained than in the U.S. Many U.S. companies have offices in Canada. This next section discusses some of the issues for a U.S. firm dealing with Canada.

First, although Canada's telecommunications environment appears to have similarities to the U.S., it is in reality quite different. Telecom managers in the U.S. mistakenly assume that things in Canada are pretty much the same as in the U.S. Angus offers a typical example. "One of the most common things is to ask, 'Why don't you take out most of your local lines and put in WATS?'" On the surface that may make sense, but in Canada, WATS rates are different than in the U.S. "It's a totally different tariff," notes Angus. "In addition, private line rates (for WATS access) here are much higher than in the U.S."

By contrast, packet switched data networks are comparatively less expensive in Canada and very popular. Electronic mail (Envoy) is widely used throughout the country, compared to the U.S. Angus adds, "There's an entire issue relating to data flow management across the borders." That issue is related to the fact that long distance rates in Canada are generally higher than in the U.S. Traditionally, before Canada built its own trans-Canada long distance network, most long distance calls were routed through the U.S. across key border crossings. Add to these the simmering regulatory climate and varying levels of competition. Thus, managing telecommunications cost-effectively in Canada brings its own unique set of challenges for U.S. managers.

Conclusion

The Canadian telecommunications market currently is experiencing growing pains comparable to the U.S. back in the early 1980s. Canadians have had the benefit of watching the trials and tribulations of their No. 1 trading partner. Progress is occurring, but at a typically conservative pace. Canada's place in world markets can be bright or bleak, depending on major decisions to be made in the coming year. For users, telecommunications as a profession is growing rapidly. The challenges are many, and so are the opportunities.

Source: Reprinted by permission from *Communique*, volume 42, number 6, December 1989. *Communique* is a publication of the International Communications Association, Dallas, Texas.

The Caribbean: Technological Change Creates New Services, Applications, and Ventures

by Ruth L. Morrison
Advanced Communications Strategies Group, Inc.

In an area of the world most noted for its warm climate, beautiful beaches, palm trees, and hospitable people just waiting to serve harried tourists seeking a bit of R&R, telecommunications historically has not been one of its selling points. However, the winds of technological change are sweeping across the Caribbean, creating new markets, applications, and services.

Much of the impetus for these initiatives has come from the need to respond to fast-paced developments outside the region. The U.S. government has promoted the expansion of private sector investment in the Caribbean through the Caribbean Basin Economic Recovery Act of 1983 ("the Caribbean Basin Initiative"—"CBI"). The CBI is a broad program to promote economic development through private sector initiatives in the Caribbean Basin. U.S. companies seeking new sources of production are increasingly looking to the Caribbean and Central America, and for a wide range of products manufactured in CBI countries, the CBI provides preferential access to U.S. markets. The ultimate goal, in addition to expanding foreign and domestic investment in nontraditional sectors, is to diversify CBI countries' economies and expand their exports. The focus on economic and social well-being of the Caribbean is not totally altruistic. These countries share the hemisphere with the U.S., and our political and economic stability are interdependent.

To facilitiate corporate investment, many projects to expand and improve regional telecommunications as well as telecommunications projects within individual countries are well underway. These initiatives are creating many opportunities for U.S. companies to market telecommunications products and services and to expand off-shore office operations. Projects in the U.S., in particular, pose the greatest challenge to Caribbean government officials. Caribbean government officials recognizing that an advanced telecommunications infrastructure is essential to the region's economic growth, have established the Caribbean Telecommunications Union (CTU) to address these issues. Caribbean telecommunications ministers agreed on the creation of the CTU within the Caribbean Community Secretariat (CARICOM), which plays a central role in coordinating the development strategies of the 14 Caribbean countries. According to CARICOM Treaty, all instruments for bringing a new organization into being must be approved by heads of government who endorsed the CTU at their annual meeting in 1988.

Moving Toward Global Connectivity

Central to telecommunications infrastructure development in the region is the Transcaribbean fiberoptic cable system (TCS-1) and the Digital Eastern Caribbean Microwave System (DECMS). The FCC in approving the plans for TCS-1, also known as Florico II, approved the first undersea fiberoptic cable system linking Florida, Puerto Rico, the Dominican Republic, Jamaica and Colombia, a project led by AT&T and other U.S. International Service Carriers. This network is expected to meet the primary objectives of facilitating rapid introduction of new technologies and increasing route and media diversity. Additionally, TCS-1 will establish digital links between the western hemisphere and other parts of the world.

Important to meeting the objective of regional and global connectivity is the Digital Eastern Caribbean Microwave System (DECMS). DECMS is owned and operated by Cable and Wireless Plc. The DECMS is a microwave radio relay system interconnecting the islands of the Eastern Caribbean, from the British Virgin Islands in the north to Trinidad and Tobago in the south. The DECMS replaces analog facilities and is the first digital microwave system to cover as many different sovereign states over several bodies of water.

The DECMS will carry traffic into and out of the region on several routes. A digital cable system is expected to connect Tortola to Bermuda in 1990, and the north cable of Bermuda will connect to the Private Trans-Atlantic Telecommunications System, Inc.'s (PTAT) fiberoptic cable carrying traffic to the north Atlantic.

Additionally, DECMS will connect with TCS-1 via Puerto Rico, as there is already microwave connection between Puerto Rico and Tortola via St. Thomas. Satellite earth stations in Trinidad, Barbados, St. Lucia and elsewhere will enable traffic to be carried out of the region via INTELSAT satellites, thus providing alternate transmission channels to ensure reliability and diversity. The ultimate objective is to provide the latest in digital telecommunications technology to every subscriber in the Eastern Caribean with connectivity to all parts of the world.

Teleports and Promising New Services

Teleports in the Caribbean are becoming major players in providing communications applications. Among the many teleports in the region is the Jamaica Digiport International Limited (JDI), a consortium comprised of AT&T (35%), Cable and Wireless (35%) and Telecommunications of Jamaica Limited (30%).

JDI, located in the free trade zone at Montego Bay consists of a large earth station with a 5 ESS switch, a building for data operations and the use of an INTELSAT satellite with digital capability transmitting high-speed data, voice, and video, as well as private line services between Jamaica and points in the U.S. JDI is also expected to expand to include other free trade zones in Jamaica. The consortium was granted a 25 year license by the Jamaican government with a provision that there will be no interconnection with the Jamaican public telephone network.

The Jamaican Digiport, which opened in September 1988, currently lists ten firms, and negotiations with others are underway. Inquiries have also come from Italy, the United Kingdom, and Canada. JDI client firms, primarily representing data processing, airline and hotel reservations, telemarketing, and offshore office services, are companies whose businesses are heavily dependant on information processing.

JDI offers via satellite: T1-45 mbps, 56 kbps dedicated, Reserved T1, Switched 56 digital, international long distance, and international 800 service. JDI also provides support and maintenance for equipment it provides through lease or purchase, equipment such as small PBXs or ACDs. To facilitate a company's move to the Digiport, JDI created the Quick Start Center, a 30 position operation that's fully equipped with state-of-the-art office technology, furniture, etc., especially designed for temporary operations for newly relocated companies.

Regarding JDI's success, there are some critics who are wondering how JDI will be successful with the proliferation of teleports in the region and its own transmission rates. However, according to officials at JDI, tremendous growth is expected from its image scanning operations. Image scanning technology, mass storage devices, a digital transmission system that is economical, state-of-the-art, high quality and with high capacity, and a workforce that is English speaking, highly motivated, literate, and available at a fraction of the cost of the U.S. labor market are important ingredients for the success of JDI, and offer incentives to cut costs that many companies may find hard to pass up.

Many companies are realizing that they must be international in scope in order to compete in a world market. Competition and monetary constraints are the driving issues behind off-shore office operations. Advanced information technology, declining communications costs and low labor costs for data entry operators are the reasons for the proliferation of data entry firms in Jamaica and other Caribbean countries. There was a time when only companies such as American Airlines, who for years has had a data entry facility in Barbados, could take advantage of off-shore data entry because it could ship hard copy at cost. Now JDI believes that image scanning technology will make it feasible for more companies with tremendous document processing needs to seek off-shore data entry services.

For example, an insurance company that processes huge volumes of claim forms is a prime candidate to cut costs using off-shore data entry. Using high-speed scanners, documents can be scanned at 2 seconds per page and loaded in large capacity storage devices. The image is transmitted to Jamaica. Key entry operators in a local area network environment using terminals with split screen capability extract information from the image for indexing and cataloging. The operators transmit the image and extracted information back to the host computer for further processing and storage. This application has significant implications for the realization of the paperless environment and disaster recovery.

Jamaica doesn't have a monopoly on teleports in the region. There are also teleports located in the Dominican Republic and St. Lucia using PanAmSat and INTELSAT. PanAmSat, which got off to a rocky start because of regulatory hurdles, commenced service to a broad range of specialized users of dedicated nonswitched service between the Dominican Republic and the United States in April of this year. Teleports are also on the drawing boards in Costa Rica, Barbados, Belize, and Curacao. A private dedicated teleport using INTELSAT's IBS service is planned for Trinidad and Tobago. An area worth watching is to what extent private satellite systems will impact INTELSAT and how well competition will deliver better services and lower prices.

In-bound telemarketing is another area with room for growth. Declining communications costs, inexpensive labor, international 800 service, and an English speaking labor force that can effectively and efficiently complete an order will also give rise to growth to in-bound telemarketing in the Caribbean.

Cellular

Cellular telephony is gaining fast approval in the region, as some believe it represents the best hope of increasing telephone penetration among the residents in the region. Indira Jhappan of Price Waterhouse points out, "It's too early to point to cellular's success in the region. However, cellular does show great promise." Countries with operating cellular systems include the Bahamas, Costa Rica, and the Dominican Republic. Jamaica has a system on order, and bids were open earlier this year in Guatemala and Trinidad and Tobago.

Opportunities Abound

The winds of technological change also have created opportunities for equipment and service suppliers. However, competition from international equipment suppliers is fierce, according to one U.S. Commerce Department report, as U.S. equipment suppliers' exports represented only 25 percent of the

Caribbean region's telecommunications imports in 1988. According to Ted Johnson, telecommunications specialist for the Caribbean Basin Division at Commerce's International Trade Administration, U.S. companies face aggressive competition from Canada, Japan, Sweden, West Germany, and Brazil. On the service end, investment in the telecommunications infrastructure has opened the door for international consultants, engineers, technicians, and construction experts.

Growing Pains

Modernizing the telecommunications infrastructure is a capital-intensive venture, requiring huge investments in capital and finance. Many Caribbean countries are evaluating options to avoid further national government indebtedness. Privatization, whether it's the sale of telecommunications assets, the provision of licenses, or concessions to private sector companies, is an option that is under discussion and in some cases has been exercised.

Telecommunications companies in Barbados, Belize, Grenada, and Jamaica have been partially privatized. There is much speculation that the government of Mexico will sell its 51 percent interest in Telefonos de Mexico. In Guatemala, the government has announced the intention of providing licenses for data communications services, pay telephone services, in addition to cellular telephony. The thought of privatizing state-owned telephone companies is very difficult for some countries. For many governments, telephone companies are big short-term earners of foreign exchange, and it is this money that often is used in other development areas.

On the other hand, there are countries, such as the Bahamas, where the government-owned and controlled telephone company, The Bahamas Telecommunications Corporation (Batelco), runs efficiently and profitably, and privatization is not a consideration. Batelco employs modern digital technology in its network and provides a range of services from local telephone service to international long distance, packet switching, and cellular mobile telephone service. Barrett A. Russell, Batelco's Deputy General Manager, states that "The economy of the Bahamas depends to a large extent on tourism and international banking. It is not surprising that good telecommunications plays a vital role in keeping the economy of the Bahamas in good shape."

Other challenges facing the Caribbean is the need for skilled technicians to meet the demand for modernization. Although the region does have experienced capable telecommunications managers, the need for skilled technicians to meet the demand for modernization is a challenge facing the region. The region does have telecommunications educational facilities. However, additional money to increase the number of trained instructors, scholarship money for overseas training, and new digital training equipment is critical. Indira Jhappan recently stated that, "An education process is necessary for both lending institutions and some government officials. Telecommunications is more than just infrastructure. It is a vehicle for job creation, foreign exchange earnings, and, in general, it can have an overall positive impact on economic development."

Conclusion

Telecommunications in the Caribbean Basin is not without its problems. However, these are very exciting times as many Caribbean nations have taken important steps to make this sector a priority for the region. Caribbean government officials recognize that continued improvement in regional telecommunications infrastructure enhances the region's ability to capture business from the United States and other areas of the world, including Japan and Taiwan. Both countries have sent several delegations to the region scouting investment opportunities.

Continued private sector investment in the region along with new advanced digital telecommunications networks should have a positive impact on social, economic, and political development. Perhaps, one day in the near future, what may trickle down from all of this activity is an increase in the number of telephones in the local communities and a decrease in the time for installation and repair.

Source: Reprinted by permission from *Communique*, volume 42, number 6, December 1989. *Communique* is a publication of the International Communications Association, Dallas, Texas.

Finland

by Henry Haglund
Managing Director
Central Chamber of Commerce—Finland

The Finnish telecommunication services arena is to a high degree unique—even if the country is following the universal trend of service liberalization.

Finland's 54 regional private telephone companies are developing new strategies for services as the competitiveness of the PTT is increasing. From the beginning of 1990, the PTT will operate as a public owned company responsible for its profitability.

The Telecommunications Act of 1987 relates to liberalization of equipment and services in such a way that new service providers can be licensed in former monopoly areas.

The Association of Telephone Companies in Finland covers around three quarters of telecommunication subscribers in the country, and traditionally the density of telephones in comparison with the population number has been among the highest in the whole world. The PTT is obliged to operate services in sparsely populated and topographically difficult areas. Moreover, the PTT is responsible for international telecommunications services.

According to the promotion of new service provision, there is new business lately established.

- Datatie Oy—a company owned by private corporations and telephone companies providing nationwide data networking services. The license has been granted.
- Yritysverkot Oy—a company owned by the PTT and a number of large companies, providing high capacity connections, which bypass the local networks. The license has been granted.
- Radiolinjat Oy—a company owned by the largest local telephone companies and corporations representing insurance, banking, and commodities distribution, is applying for license to operate a second digital cellular network (in addition to the network managed by the PTT). The license is not granted yet, and the matter may require cabinet level handling.

The licenses for operation are granted by the Ministry of Transport and Communications.

As noted from the foregoing, so far the Telecommunications Act of 1987 has been operative only in the sense that the former national (PTT) and local monopolies have been innovative in the extended competition between each other. There has not yet been much place for true newcomers.

The PTT has been offering 64 kbps "pre-ISDN" network since 1986. ISDN trails using Siemens EWSD exchange began in 1988. Starting in 1989, Nokia's DX200 exchanges conforming to CCITT Blue Book are tested by PTT.

The Helsinki Telephone Company has already placed an order for DX200 switches. The Association of Telephone Companies is active in evaluating potential competitive services, which could benefit from ISDN.

In spite of severe competition, there is, however, much cooperation between the counterparts. It is unanimously seen that common technical standards in the telecommunications infrastructure are a necessity. Options are kept open for competitive services constructed on this standardized base.

During the last few years, the telecommunications user community in Finland has become aware of the need to actively influence the decision-making of new services and the terms of their availability. The Central Chamber of Commerce in Finland has established a working group to take care of the user need evaluation, and the group acts as a full member of the International Telecommunications User Group (INTUG).

Source: Reprinted by permission from *Communique*, volume 42, number 6, December 1989. *Communique* is a publication of the International Communications Association, Dallas, Texas.

France

by J. F. Berry
President
Association Française des Utilîsateurs du Téléphone et des Télécommunications (AFUTT)

Regulation

A wide public debate has been organized by the Minister of Posts, Telecommunications and Space. Mr. Paul Quilès, during the first 6 months of 1989, about the future of all public postal and telecommunications services.

The resulting report recommends transforming the postal and telecommunications administrations into two separate legal entities, to allow more independent management practices and financial policies. It is expected that a law to that effect will be voted by Parliament in the spring of 1990, in despite of strong opposition from labor unions.

For PABXs and terminal equipment, France is one of the most liberal countries in the world. It has always been "interconnect," with freedom to purchase most telecommunications equipment (PTT-certified) directly from authorized private vendors.

Volume

France Télécom has produced revenues of 88.1 billion francs ($14 billion) and a net profit of 2 percent in 1988, with 157,000 employees. Revenue grows 6 percent per year. With over 26 million main telephone lines in service for 55 million inhabitants (an increase of more than 1 million lines in 1988), France has one of the highest ratios (47) of lines per 100 capita. Average traffic per line shows a healthy growth rate of 5 percent per annum.

Mobile Communications

Two competing subscriber link radiotelephone operators France Télécom and SFR (a subsidiary of the powerful Compagnie Générale des Eaux), cover most of metropolitan France. The use of mobile telephony is growing fast although hampered by a shortage of available radio frequencies.

Three nationwide public paging systems are in operation.

- *Alphapage*—started by France Télécom in November 1987, has 60,000 subscribers. It allows them to receive alphanumeric messages displayed on a pocket receiver. The material is entered into the system directly by the sender, with a Minitel videotex terminal, without going through an operator.

- *Operator*—operated by Télédiffusion de France, sends numeric messages.
- *Eurosignal*—sends "beeps" and activates one or several signaling lights.

Videotex

Videotex continues to be a huge success, and its business use is growing. Five million inexpensive Minitel terminals are in use (of which 30 percent is by business), and free distribution by the PTT continues, at the rate of 800,000 units per year. Thirty percent of the population has access to a Minitel at home or at work.

Teletel, as the service is called in France, allows access to 10,000 on-line services, and new ones are announced every day: access to databases, "minitel-order" (instead of mail-order) purchases, opinion polls, questions from the public, and answers from political leaders, etc.

ISDN

With a highly digital public transmission and switching network, France Télécom will offer ISDN commercially in the entire country in 1990. It is connected to Accunet in the U.S. and IDA in Great Britain. Examples of uses are remote surveillance, selection of press photographs made available by an agency, speeding up the processing of orders, and remote access to data stored on digital optic disks.

Data Networks

The large Transpac public packet switched network continues to expand rapidly. It serves some 50,000 subscribers and conveys 1,900 billion characters per month. Videotex traffic accounts for half of that volume.

High-speed, high-performance data networks are available, nationally and internationally, by cable or satellite: Transcom (switched, digital 64 kbps), Transfix (permanent digital circuits, up to 1,920 kbps), Transdyn (digital satellite circuits, up to 2 Mbps).

Rates

National and international rates have been decreasing significantly, and a sizable advantage for business comes from the 18.6 percent value added tax (which most companies can deduct from their tax bill) applied to telecommunications since November, 1987.

Telecommunications Industry

Mergers and alliances have been concluded during the last few years (CGE/ITT, Matra/Ericsson, Jeumont-Schneider/Bosch), and Northern Telecom is building a plant to produce PABXs in France. The French telecommunications industry has a strong position on export markets.

Users

Users are organized, and AFUTT forcefully expresses the needs of both corporate and residential users. It has the good fortune of obtaining excellent response and results from the PTE Ministry and France Télécom.

AFUTT's twentieth anniversary reception in Paris in June 1989 was marked by a very complimentary speech of PTE Minister Paul Quilès and the presence of France Télécom's President, Marcel Roulet.

The association has started groups of corporate users of equipment of specific manufacturers, which are appreciated by members and by vendors. AFUTT publishes each year a directory and buyer's guide of private telecommunications equipment and services.

Source: Reprinted by permission from *Communique*, volume 42, number 6, December 1989. *Communique* is a publication of the International Communications Association, Dallas, Texas.

The Telecommunications Scene in Hong Kong and the Far East

by Richard Sedgwick
Vice Chairman
Hong Kong Telecommunications Users Group

Like other parts of the world, the telecommunications scene in the Asia-Pacific area has been anything but static. A lot has taken place during the past few years, and much more will be occurring in the foreseeable future.

Over the years, Hong Kong has developed as the communications center for a large number of companies and corporations operating in the Far East because of its position, a laissez-faire form of government, and the flexible, cost-effective telecommunications facilities it has to offer.

In 1986, the Hong Kong government decided to call the tenders to provide cable TV for the territory. It soon became apparent to the prospective operators and the major users of telecommunications that, with the latest cable TV technology on the market, there was now an opportunity to provide an alternate telecommunications infrastructure that would see Hong Kong into the 21st century.

As a result of the cable TV issue, the Hong Kong Telecommunications Users Group (HKTUG) was formed in 1987 to be the focal voice of users' needs within Hong Kong and the Asian area. It is now fully established and building itself up into the national association for telecom users within Hong Kong. It is seeking worldwide membership and acceptance as the Hong Kong representative of users at major international telecommunications conferences and relevant government bodies within Hong Kong, such as the Telecommunications Board.

The implications of the cable TV issue eventually led the government to call in consultants, Booz & Hamilton, to undertake an overall study of the telecommunications requirements for the territory and make recommendations. This was completed at the end of the last year, and the government has now asked the United Kingdom's OFTEL to review and comment on the study within the next few months before any formal policy announcements are made. It is believed that some form of deregulation of the existing environment will result, and an opening up of the local infrastructure to competition and benefits will ensue.

At present, all the international telecommunications services are provided by Cable & Wireless (HK) Ltd., with the Hong Kong Telephone Co. providing the domestic services, both of which are subsidiary companies of Cable & Wireless Plc. in London.

In October 1987, Cable & Wireless announced the restructuring of its monopoly in Hong Kong to form a new parent company, called Hong Kong Telecommunications, which would be 74.5 percent owned by Cable & Wireless Plc., 5.5 percent owned by the Hong Kong government, with the remaining 20 percent being offered to the general public on the local Stock Exchange. The announcement declared the move was for the benefit of private investors, enabling them to share in the growth of Hong Kong's telecommunications services. The HKTUG is concerned that Cable & Wireless have improved their monopoly at a time of general deregulation elsewhere. Sixty percent of Cable & Wireless' worldwide profit is derived from the 400-odd square miles of Hong Kong. The users are now eagerly awaiting the outcome of the government's consultancy study.

Looking elsewhere in Asia, the country that has seen the most change is Japan, where the domestic market was opened up just over 2 years ago when three companies were allowed to compete with Nippon Telegraph & Telephone (NTT). This has meant cheaper services, more investment in new technology—especially fiber-optics—and a more vibrant and responsive telecom infrastructure. The Japanese government has not extended this philosophy to the international arena, and since the beginning of 1989, there have been two alternate companies competing with the KDD (Kokusai Denshin Denwa Co., Ltd., the Japanese international carrier). The significant aspect of this is that Cable & Wireless is a shareholder in one of these companies and has been successful in breaking into Japan's telecommunications industry.

In other Asian countries, there has been a tightening up on the interchange between public and private networks, with the emphasis on pushing users into using the public services in preference to private networks, and so forcing up costs. This has generally resulted in users becoming more selective of the types of services they will put over their networks into these countries.

On the technical front, the major thrust has been on ISDN with a number of countries providing trials of various kinds, leaving the users concerned about the impact it will have on the cost of running their networks. The other major technological service that is anxiously awaited is the arrival of fiberoptic cables between the United States and Asia, followed by links within Asia. With these established, the information flood gates will be opened.

The Asia-Pacific area is too large to discuss everything that is going on in the telecommunications arena in this brief report, but if any readers are interested in further details on specific topics or countries, please write to HKTUG, P.O. Box 7649, Hong Kong.

Source: Reprinted by permission from *Communique*, volume 41, number 5, February 1988. *Communique* is a publication of the International Communications Association, Dallas, Texas.

The Netherlands

by J. Scheltus and Jan A. Tiems
Nederlandse Vereniging van Bedrijfs Telecommunicatie Grootgebruikers (NVBTG)

Legislation

Of the Western European countries, the Netherlands was second in liberalizing its legislation of telecommunication. After a long period of preparation, which started in 1983, the Dutch Parliament accepted a new legislation that became effective January 1, 1989.

This new legislation reflects in general the Green Book issued by the European Committee, from which the essential (main) points are given here:

- A separation has been made, also in organizational point of view, between PTT as supplier of basic telecom infrastructure and services and PTT as supplier of Value Added Services and terminal equipment.
- Because of its basic infrastructure and services, the PTT has been granted an exclusive concession from the Dutch Government.
- Value Added Services and terminal equipment have become free and can also be

offered by other parties at the market under the provision, however, that the equipment has been type-approved.

Open Market

At this moment, a reasonable offer of equipment is already available, and besides the traditional suppliers, there are more sellers on the market. However, within the realm of bigger PABXs, the offer is still limited. This may be partly due to an extended period of planning, which often precedes the purchase of equipment. Possibly, careful movements of clients together with fear of leaving well-trodden paths may play a role here as well.

Nevertheless, a number of installations have been delivered by companies that did so without involving the PTT.

The Dutch PTT pursues an active policy in making the Netherlands attractive for business purposes. Recently, for example, the international traffic tariffs decreased considerably (in the order of 30 percent), and in Rotterdam a first pilot project with ISDN was started in October 1989.

Dutch Business User Association—NVBTG

The Dutch Users Association actively concerns itself serving the users' interests. Within the 4 years the association has been active, the number of members increased to almost 80, particularly multinational companies and institutions. The result of this active policy can be seen in the fact that users groups for Philips and Ericsson PABXs also have been established, and Northern Telecom will soon follow.

There also has been established a consultation group, which focuses on international telecommunications. Some of these groups together investigate such items as the PABX-market and network management.

There is frequent consultation with PTT. Attention is paid to matters that are done by PTT as concessionaire, particularly the national infrastructure.

An important part of the activities of NVBTG during the next few months will be centered around how to achieve a solid constructive climate suitable for consultation with the PTT's national infrastructure operations.

International

As a member of the International Telecommunications Users Group (INTUG), NVBTG plays an active role in the European role of getting across the business users' needs for open telecom markets, infrastructures, and services.

Source: Reprinted by permission from *Communique*, volume 42, number 6, December 1989. *Communique* is a publication of the International Communications Association, Dallas, Texas.

Norway

by Jan Henrik Nyheim
Secretary General
Norsk Telebrukerforening (NORTEB)

Deregulation reached Norwegian telecommunications in earnest in January 1988, when in-house installations and choice of equipment became the sole responsibility of the owner of the premises. The reform was proposed by a government commission that looked into how Norwegian telecommunications should adjust to rapid technological change and new—and more articulate—user requirements.

As 1988 began, Norway had already seen the introduction of a new government agency, The Norwegian Telecommunications Regulatory Authority (Statens Teleforvaltning, STF). This, too, flows from the same 1983–1984 study, chaired by professor of telecommunications Gunnar Stette (University of Trondheim). Although a Conservative government introduced the review, it was left to a Socialist government to introduce the actual legislation—proof of how generally accepted the need for reform was.

In late 1988, the winds of change led to a study into the telecommunications requirements of the business community by the Confederation of Norwegian Business and Industry in close cooperation with the Norwegian Telecommunications User Group (Norsk Telebrukerforening, NORTEB). The report will highlight the demands of the business community on today's and tomorrow's telecommunications, as well as address the task of informing that community of the importance of applying modern telecommunications to its endeavors.

In early summer of 1989, the national assembly decided to let the private sector on Value Added Networks (VAN) be on equal terms with the Norwegian Telecommunications Administration (NTA). Although regarded with mixed emotions by those who would rather not include the NTA in VAN, the decision still reflects the prevailing mood of change in Norwegian telecommunications policies.

One dynamic element in these changes is the unending technological process. Another is the need for Norwegian trade and industry to adjust to energetic telecommunications deregulation within the European Common Market—the destination of some 70 percent of Norweigan exports.

Source: Reprinted by permission from *Communique,* volume 42, number 6, December 1989. *Communique* is a publication of the International Communications Association, Dallas, Texas.

Spain

by Fernando de Elzaburu
President
Asociación Española de Usuarios de Telecomunicaciones

Until not long ago, talking about telecommunications was equivalent to talking about the telephone and telex services exclusively. However, today, the technological development of the so-called Informations Technologies offers such a variety of possible services that it seems difficult to deal with this concept in an overall manner and define its situation in a certain country. Although in a concise way, we must deal with each service in a different way.

In Spain, the situation is more complex because we are slowly going from a strong monopoly situation, with a dominant position for Telefónica, a private company functionally controlled by the State with a very special contract since 1924, to another, more in line with the liberalizing spirit of the European Community. The State, thanks to the Telecommunications Ordinance Law of 1987, is regaining its lost sovereignty, although some resolutions are still to be completed (such as the Regulation, which is why in practice the law has still not become effective and the users are still disoriented), as are agreements, such as the new contract that will rule the relationship between the Ministry and Telefónica, so that the new judicial and regulatory framework is defined.

The traditional telephone service in Spain, with a little over 39 million inhabitants at the end of 1988, had almost 11 million local lines in service, which makes the Spanish telephone network the ninth largest worldwide, but only in numbers. The local lines are geographically not well distributed, hindering those who do not live in the large cities.

The situation is worsened by the still weak level of digitalization and by network infrastructure not growing in the rhythm required by the development of the country in the past years. To get a telephone installed means waiting many months—even though there are growing investment rhythms—and the waiting list does not go below 600,000 applications.

In data transmission, Spain was the first country in the 1970s, together with Belgium, to have a packet switched network, with its own protocol—RSAN—before the X.25 existed. By the end of 1988, the present Iberpac network had around 45,000 connections. As AUTEL's president, I must point out that this network has the inconvenience of being slowed down by the fees, a consequence of the government's interacting policy that, for incorrect social reasons, has reduced the fees in the local telephone service and relatively increased the high prices in the business communication services as well as on international traffic.

The integrated voice-data services, or narrow band ISDN, are at a trial stage, and until 1991 it seems there will be no commercial offer available. There is the Ibercom network offered by Telefónica, but the majority of users are satisfied neither with its philosophy, against open access nor with the fact that the standard is mainly national.

The highly important mobile communications are not well introduced in Spain, with a very reduced cellular telephone network, to such an extent that there are already saturation problems in cities such as Madrid and Barcelona because they are waiting to optimize the radioelectric spectrum, which is also suffering a transitory period in relation to the existing concessions.

If we refer to the data banks or the speeds at which our telephone lines support the data transmission, with the present techniques of understanding information, the prices also rise in an unjustified manner for Spanish users. This worries us because in the year 1993, in which the single European market law will become valid, the competitiveness for Spanish companies will obviously be greater and because all these technologies require time to become familiarized with them.

The same could be said for the reduced number of data banks or the absence of electronic mail if we do not go abroad. This is the reason for being a member of the Spanish Association of Telecommunications Users (AUTEL). AUTEL is made up of thousands of professionals who try to make the government sensitive and convince it of the possibilities for the progress of society, which they are closing, including the world of integrated communications and the aim of having wideband transmission for all services and hopefully for all villages.

I think that at the end of this century communications will have stopped being merely an instrument and will become an inherent factor or value for the development of people, which should be part of what we know as human rights. Information and communications are two aspects of the same inherent factor. The validity of one implies the validity of the other. So that human beings may fulfill themselves in accordance with human nature, information and communications should be placed above the boundaries of ideology, doctrine, or ways of thinking.

Source: Reprinted by permission from *Communique*, volume 42, number 6, December 1989. *Communique* is a publication of the International Communications Association, Dallas, Texas.

Sweden

by Roland Linderoth
Swedish Telecommunications User's Group (NTK)

Various steps and initiatives are being taken by the Swedish government as a reaction to the new situation in the telecommunications field.

- The responsibility for technical regulations concerning terminal equipment is being changed, and the current monopoly for Televerket (Swedish Telecommunications Administration) to connect large PABXs (private automatic branch exchanges) to the public network is to be abolished during 1989.
- Decisions in readjustments of tariffs have been made.
- Initiatives have been taken to fully admit third-party traffic on its leased lines and to change the international accounting rates.
- The Postal and Telecommunication Commission reviews the social and regional policy responsibility of the government with these areas.

Equipment

The Swedish Parliament (Riksdagen) established a new government agency called the Government Telecommunication Council (abbreviated STN in Swedish) as per July 1, 1989. The STN shall be assigned the following duties:

- issuing directives governing connections to Swedish Telecom facilities;
- registering equipment that can be connected to the public telecommunication network;
- monitoring technical developments as well as international and national standardization work and proposing appropriate measures to the government;
- supporting the government agencies responsible for maintaining a truly competitive market by providing them with technical expertise.

Services

There is no special legislation regulating the conduct of telecommunication operations. Accordingly, there is no statutory monopoly or concession system as regards the installation of facilities or the provision of services. Consequently, there are no parliamentary or government regulations aimed at drawing a distinction between monopoly and competitive telecommunications services.

Although the Swedish Telecommunications Administration has no legal monopoly to build or supply telecommunications network services, it has a very dominant position. But there are other operators.

For the providers of value added services—including information and data-processing services as well as electronic mail—who use the public switched data and text services (PSTN, CSDN, PSDN, videotex, and telex), there are no limitations on the use of telecommunications services.

For those who use private networks based on flat-rate leased lines from The Swedish Telecommunications Administration (Televerket), the straight resale of capacity to a third party (no value added) is in principle not permitted. There is, however, a general exemption to this rule for data service bureau operations. Those who want to carry third party traffic on leased lines may do so after authorization in each case by Televerket. Special rates are then negotiated. This is in accordance with CCITT recommendations. These recommendations are, however, becoming increasingly difficult to maintain in the new telecommunications environment. Televerket has consequently recently presented a proposal to the government to allow third party traffic on certain conditions. One of them is that the tariffs for leased lines would be adjusted to the costs of providing them. The Postal and Telecommunication Commission is reviewing the proposal.

Standardization

SIS-ITS, the Information Technology Standardization body, is an independent legal entity affiliated with the Swedish Standards Institution.

Members of SIS-ITS represent a broad spectrum of organizations, including governmental agencies, The Swedish Telecommunications Administration, manufacturers, and users. The basic philosophy guiding SIS-ITS participation in international and regional standardization is that international standards should be taken over as Swedish standards, without changes, as far as possible.

EFTA Activities

EFTA will create two working parties in the telecommunications field. One group will deal with telecommunications and the other group with new technologies and services.

Source: Reprinted by permission from *Communique,* volume 42, number 6, December 1989. *Communique* is a publication of the International Communications Association, Dallas, Texas.

Switzerland

by Max Kunz
Swiss Telecom User Association (ASUT)

Switzerland, this beautiful little country in the heart of Europe, is elbowing its way into the Information Age. Despite its central location, it is not part of the European Economic Community (EEC) and has to find its own way to create an Information Technology playground most suited to its environment.

It is beyond this paper to discuss why Switzerland is not part of the EEC, but very briefly and simply, it could be compared with two incompatible education (school) systems, where in some domains it is far in advance and in others it is lagging behind. Switzerland's strategy to cope with this issue is to catch up where it feels it is behind, but the other way around is a much more difficult process. That this approach is being taken very seriously can be illustrated by the fact that, to my knowledge, Switzerland has more agreements with the EEC than any other non-EEC country in the world.

Switzerland used to have a reputation for good quality and widespread telecommunications services. In such a situation, I am tempted to say that it must be expected that a monopoly PTT administration would not take measures that impinge on their exclusive right to provide telecom services with the inherent risk of quality degradation for services beyond their control—as in baseball, one does not change a winning team!

Only the rules have changed. We no longer talk about POTS only. Information Technology (IT) has become a key success factor for almost every business, and no one single organization (including PTTs) can meet efficiently all the diverse telecom services needs of all the businesses. Today, there is no question that every country where business is competing in the world market must adopt telecom legislation that allows IT services to be provided efficiently to support the various businesses. This requires a large variety of different services that can best be provided competitively.

Switzerland is lagging behind in Telecom Regulation! This criticism often is heard in international forums. I would agree that it started late. This process was finally triggered by industry and USER pressures. This late start, however, is also mainly due to the situation described previously. Now it is well advanced, with almost a risk of being finished too fast.

For Switzerland, it is important that the Telecom bill does not put Swiss business at a competitive disadvantage with its major business partners (Europe, USA), and hence some coordination is needed especially with the still evolving regulations of the EEC.

I think it is important to understand the process for a new Telecom bill in Switzerland to better see the progress being made: As a European Free Trade Association (EFTA) country, Switzerland participates in the EFTA - EEC telecom coordination effort and participates in many EEC-sponsored high tech programs. Of course, it is also a member of European Telecommunications Standards Institute (ETSI), which is open to all interested parties, including users, and Swiss Telecom User Association (ASUT) has elected to participate. As such, it can influence national voting where it applies. Remember, the goal of ETSI is to base its work on International Standards (mainly CCITT) and not to create regional standards barriers.

With respect to the evolution of the Telecom Regulation, ASUT already had a chance to influence the Telecom draft bill, which was first published the end of 1987. During a public feedback period of several months, ASUT made an extensive contribution and reserved its annual seminar to this topic. Among the 1,000 participants, there were more than a dozen parliamentarians present, and it was no surprise that many aspects of the ASUT feedback got incorporated into the Telecom bill presented to parliament. One of the chambers just finished the preparatory work in a special commission, and it is expected that the official debate will take place in one of the next sessions. It is further expected that the other chamber will deal with it in early 1990.

In the meantime, and this is most important, more liberal measures are being implemented progressively in view of this new Telecom bill. Users would like to see a bit of faster progression, but the important thing to understand is that, when the new Telecom bill finally gets approved, it will already be largely implemented. It then will provide a new telecom framework that should be sufficiently flexible to continuously adapt to technological and environmental evolution well into the next century, with the following advantages to business users.

Terminal Liberalization

By the time the Swiss Telecom bill is approved, it can be expected that all terminals can be obtained from the competitive market and can be connected to the network under the condition that it will not harm the network. Type approval will still be necessary, but it will be much simplified and comply with EEC procedures (which are still under development).

With ETSI defining harmonized interfaces for Europe, this not only will increase the choice for the users but also will make terminals transportable. Hence, the user will be able to choose the most appropriate equipment and will be able to order the same equipment for all of Europe, resulting in meeting the required performance at much lower prices to satisfy the business requirement. The simplified approval should also result in earlier availability of equipment not only in one country, but throughout Europe. There will be additional benefits, e.g., flexible maintenance possibilities adapted to the user's business requirement.

There are still some potential issues that are not cleared yet, but most of them are not related to Switzerland specifically, rather to the EEC process. The only major open question is the exact timetable for complete equipment liberalization (PABXs), which, however, might coincide with the approval of the Telecom bill.

Services Liberalization (VANs)

Using the underlying infrastructure, there will be no restrictions to provide IT services, and no licensing to do so is foreseen. Since infrastructure tariffs evolve to be more cost based, business users will be able to choose the most appropriate IT services or operate their own network, whatever best meets their requirement and is most cost-effective to them. It is expected that both public and private IT services will compete fairly in this market.

This evolution will most likely generate many more new business solutions being offered via networked IT offerings, which in turn will make these respective businesses more attractive, all of which will contribute to increased wealth for everyone.

If this IT evolution progresses fairly, Switzerland is planning to create a special consultative committee to support the government in which users should be properly represented.

Conclusion

Not everything is settled, and there are still many hurdles to be taken. Obviously, business users would like to see a faster liberalization process. However, in the end, I believe the new Telecom bill in Switzerland will be a typical Swiss compromise—pleasing no one entirely, but a solution with which everyone can live. I think this is called democracy.

Source: Reprinted by permission from *Communique*, volume 42, number 6, December 1989. *Communique* is a publication of the International Communications Association, Dallas, Texas.

United Kingdom

by Ron Bell
Chairman
Telecommunications Managers Association (TMA)

Most national regulations are now in place, but with the move to create a single market environment across Europe in 1992, attempts are being made to produce regulations that will apply in all community countries. This process has commenced with type approval standards. Indications to date show that rather than simplifying type approval across Europe, the nationalistic attitudes of various telecommunications authorities are rendering standards more complex.

Within the U.K. the first Telepoint networks—that allow cordless telephones to be used in conjunction with remote base stations located in public places, such as gas and railroad stations, bus depots, etc., for originating calls—are opening up.

Later this year, licenses for three Personal Communications Networks (PCNs) will be granted. PCNs are small diameter cellular networks allowing the use of very small, hand-held personal telephones. They may supersede Telepoint but will be subsidiary to the two vehicle cellular networks that already cover all of the U.K. and have today 600,000 users, growing by 500,000 per month.

A review of the competitive position of the two long distance carriers—British Telecom (BT) and Mercury Communications Limited (MCL)—is scheduled for November 1990.

Interested parties, including TMA, are already positioning themselves.

As competition in the "Local Loop," i.e., the last link serving users' premises, is limited, this will be given special attention. In this regard two-way earth satellite links (VSAT etc.) could be an option. Currently, this is not permitted, although one-way broadcast is, by certain licensees.

ISDN is starting to gain some interest but is not receiving the same attention as in the U.S. The problem is twofold:

- lack of applications ideas by both users and vendors;
- little widespread offerings by BT and MCL.

Building wiring systems and automated wiring design is currently the subject of a marketing hype. This is probably as a result of intensive building development work in most central city areas, particularly London.

Videoconferencing is also getting a lot of publicity. It is viewed by most top executives as an essential tool in the process of greater geographic dispersion. Financial justification through travel and hotel accommodation savings is being given less emphasis over strategic advantages.

Electronic Document Interchange (EDI) also is starting to take off. Many large companies are refusing to trade with their suppliers unless EDI is used.

A rationalization of main telecommunications vendors is underway. A consortium of GEC and Siemens recently acquired Plessey. AT&T has acquired ISTEL, a U.K. system house with its own extensive value added network, which works in conjunction with EDS' network. This is obviously another attempt by AT&T to break into the European market, something they have not had much success in achieving to date.

Source: Reprinted by permission from *Communique,* volume 42, number 6, December 1989. *Communique* is a publication of the International Communications Association, Dallas, Texas.

Appendix B

Telecommunications Trade-offs

Throughout this book there have been many references to the trade-offs made in planning, designing, or operating a communications system. Trade-offs occur because there are alternative solutions to many communication problems and different ways to design products and provide services. In some cases, the trade-off requires that an absolute *either/or* decision be made. For example, either vendor A or vendor B will be selected to provide the front-end processor. There is no way to buy half of the FEP from one vendor and half from another. In most situations, however, an absolute choice does not have to be made between the alternatives. For example, a company might well decide to have some leased circuits and some switched circuits; it doesn't have to select only one type or the other.

The important thing is to be aware of the options that exist so that you can make a proper selection to solve a problem optimally. Knowledge is the key, for if you don't know that options exist and there are trade-offs to be made, you can't choose between the alternatives.

Many trade-offs are interrelated. For example, the choice of an asynchronous versus a synchronous protocol may be directly related to the choice of terminals to be used for the application. The use of a vendor's proprietary protocol can occur only if that vendor has been selected to provide communications products. Some trade-off decisions are primarily made by the company that is using the communication products. Others are determined by the company in combination with the vendor. Still others are primarily made by the vendors, typically when they are designing their products or services.

The following list itemizes many communications trade-offs and alternatives. Depending on the level of detail, such a list could go on for pages. The trade-offs included here are the most important ones faced by network managers and designers.

CUSTOMER TRADE-OFFS

Circuits

- Private network versus public network
- Private versus leased versus switched circuits
- Two-wire versus four-wire circuits
- Circuit speed versus cost
- Point-to-point circuits versus multipoint circuits
- Conditioned versus unconditioned circuits
- Compressing versus not compressing data transmissions
- Multiplexing versus concentrating versus inverse concentrating

Terminals

- Intelligent versus smart versus dumb terminals
- General-purpose terminal versus application-oriented terminal
- VDT versus printing terminal

Applications

- Voice messaging versus text messaging versus both
- Teleconferencing versus travel
- Encrypting versus not encrypting data
- The level of reliability, availability, and responsiveness that is required versus the cost of providing it

Other Trade-offs

Custom software versus off-the-shelf programs

Leasing versus buying hardware

Vendor A versus vendor B

Performing certain telecommunications services inside the company versus contracting them to outsiders

CUSTOMER–VENDOR TRADE-OFFS

Architected versus nonarchitected communications approach

Terrestrial versus satellite circuits

Analog versus digital transmission

Asynchronous versus synchronous transmission

VENDOR TRADE-OFFS

Front-end processor versus direct connection of circuits to a computer

Hardware versus software implementation of certain functions

Entering versus not entering a certain geographic market

Providing versus not providing a certain telecommunications product or service

Telecommunications Periodicals and Newsletters

The rapid growth of the telecommunications industry in recent years has brought with it an explosion in the number of trade journals and specialized newsletters aimed at providing information to telecommunications professionals. This list contains information about the most widely read telecommunications publications.

Business Communications Review. Published monthly by BCR Enterprises, Inc., 950 York Road, Hinsdale, IL 60521, 1971–.

Communications News. Published monthly by Harcourt Brace Jovanovich Publications, Inc., 124 S. First Street, Geneva, IL 60134, 1964–.

Communications Week. Published weekly by CMP Publications, Inc., 600 Community Drive, Manhasset, NY 11030.

"Datacomm Advisor: IDC's Newsletter Covering Network Management—Products, Services, Applications." Published monthly by International Data Corporation, 5 Speen Street, Framingham, MA 01701, 1977–.

Data Communications. Published monthly by McGraw-Hill Publications Co., 1221 Avenue of the Americas, New York, NY 10020, 1972–.

Datapro Reports on Data Communications. Published monthly by Datapro Research Corp., 600 Delran Parkway, Delran, NJ 08075.

"The LOCALNetter Newsletter." Published monthly by Architecture Technology Corporation, P.O. Box 24344, Minneapolis, MN 55424.

"The MAPNetter Newsletter." Published monthly by Architecture Technology Corporation, P.O. Box 24344, Minneapolis, MN 55424.

Network Strategy Report. Published monthly by Forrester Research, Inc., 185 Alewife Brook Parkway, Cambridge, MA 02230.

Network World. Published weekly by CW Communications Inc., 375 Cochituate Road, Box 9171, Framingham, MA 01701–9171.

Networking Management. Published monthly by PennWell Publishing Company, 1421 South Sheridan, Tulsa, OK 74112.

"The OSINetter Newsletter." Published monthly by Architecture Technology Corporation, P.O. Box 24344, Minneapolis, MN 55424.

"The PCNetter Newsletter." Published monthly by Architecture Technology Corporation, P.O. Box 24344, Minneapolis, MN 55424.

Satellite Communications. Published monthly by Cardiff Publishing Co., 3900 S. Wadsworth, Suite 560, Denver, CO 90235, 1977–.

Telecommunications. Published monthly by Horizon House, 685 Canton Street, Norwood, MA 02062, 1968–.

"Telecommunications Alert." Published monthly by United Communications Group, 4550 Montgomery Road, Suite 700 North, Bethesda, MD 20814.

Teleconnect: A Monthly Telecommunications Magazine. Published monthly by Telecom Library, Inc., and G. A. Friesen, Inc., 12 West 21st Street, New York, NY 10010.

Telephone News. Published biweekly by Phillips Publishing, Inc., 7811 Montrose Road, Potomac, MD 20854, 1980–.

Telephony: Journal of the Telephone Industry. Published weekly by Telephony Publishing Corp., 53 E. Jackson Boulevard, Chicago, IL 60604, 1901–.

"Token Perspectives Newsletter." Published monthly by Architecture Technology Corporation, P.O. Box 24344, Minneapolis, MN 55424.

Telecommunications Professional and User Associations

This list contains the names, addresses, and telephone numbers of the major national telecommunications associations. Information about membership in these organizations can be obtained by contacting the association at the address listed. In addition to the groups listed here, there are numerous regional, state, and local telecommunications organizations throughout the country.

Association of College and University Telecommunications Administrators (ACUTA)
Lexington Financial Center
250 West Main Street
Lexington, KY 40587
(606) 252-2882

Canadian Business Telecommunications Alliance (CBTA)
67 Yonge Street
Suite 1102
Toronto, Ontario
Canada M5E 1J8
(416) 865–9993

Communications Managers' Association (CMA)
40 Morristown Road
Bernardsville, NJ 07924
(201) 766–3824

Computer and Communications Industry Association
1500 Wilson Boulevard
Suite 512
Arlington, VA 22209
(703) 524–1360

Energy Telecommunications and Electrical Association (ENTELEC)
P.O. Box 795038
Dallas, TX 75379–5038
(214) 578–1900

Institute of Electrical and Electronics Engineers (IEEE) Council on Communications
345 E. 47 Street
New York, NY 10017
(212) 705–7900

International Communications Association (ICA)
12750 Merit Drive
Suite 710
LB–89
Dallas, TX 75251
(800) 422–4636

International Telecommunications Users' Group (INTUG)
INTUG Secretary
18 Westminster Palace Gardens
Artillery Row
London SW1P 1RR
England
44–1–799–2446

National Association of Radio and Telecommunications Engineers, Inc.
P.O. Box 15029
Salem, OR 97309
(503) 581–3336

North American Telephone Association (NATA)
2000 M Street N.W.
Washington, DC 20036
(202) 296–9800

Organization for the Protection and Advancement of Small Telephone Companies (OPASTCO)
2000 K Street
Suite 205
Washington, DC 20056
(202) 659-5990

Society of Telecommunications Consultants (STC)
1841 Broadway
Suite 1203
New York, NY 10023
(212) 582–3909

Tele-Communications Association (TCA)
1515 West Cameron
Suite B-140
West Covina, CA 91790

Telecommunications Industry Association (TIA)
150 North Michigan Avenue
Suite 600
Chicago, IL 60601
(312) 782-8597

United States Telephone Association (USTA)
1133 15th Street N.W.
Suite 1200
Washington, DC 20005
(202) 862-3419

Organizations That Conduct Telecommunications Seminars

Many public and private organizations conduct telecommunications education courses and seminars. In addition to those listed here, many colleges and universities have regularly scheduled classes and occasional seminars covering the gamut of telecommunications topics.

American Institute for Professional Education
437 Madison Avenue, 23rd Floor
New York, NY 10032
(212) 883-1770

AT&T
Customer Education Center
15 West 6th Street
Cincinnati, OH 45202
(513) 352–7419

Architecture Technology Corporation
P.O. Box 24344
Minneapolis, MN 55424
(612) 935–2035

BCR Enterprises, Inc.
950 York Road
Hinsdale, IL 60521
(800) 227–1234
In Illinois (312) 986–1432

Datapro Seminars
Datapro Research Corporation
600 Delran Parkway
Delran, NJ 08075
(609) 746–0100

Data-Tech Institute
Lakeview Plaza
P.O. Box 2429
Clifton, NJ 07015
(201) 478–5400

Friesen's School of Generic Telephony
Gerry Friesen
Building B, 1300 Chinquapin Road
Churchville, PA 18966
(215) 355–2886

Institute for Advanced Technology
Control Data Corporation
6003 Executive Boulevard
Rockwood, MD 20852
(301) 468–8576

International Communications Association (ICA)
12750 Merit Drive
Suite 710, LB-89
Dallas, TX 75251
(214) 233–3889

Management Development Foundation
2920 N. Academy
Colorado Springs, CO 80907
(719) 597–7350

Systems Technology Forum
10201 Lee Highway
Suite 150
Fairfax, VA 22030
(703) 591–3666

Technology Transfer Institute
741 Tenth Street
Santa Monica, CA 90402–2899
(213) 394–8305

Telco Research Corporation
1207 17th Avenue South
Nashville, TN 37212
(615) 329–0031

Telecom Library, Inc.
12 West 21st Street
New York, NY 10010
(212) 691–8215

Telenet Communications Corporation
12490 Sunrise Valley Drive
Reston, VA 22096
(800) 835-3638

Tele-Strategies
1355 Beverly Road
Suite 110
McLean, VA 22101
(703) 734–7050

Tellabs, Inc.
4951 Indiana Avenue
Lisle, IL 60532
(312) 969–8800, ext 241

United States Telephone Association (USTA)
1133 15th Street N.W.
Suite 1200
Washington, DC 20005
(202) 862–2419

Acronyms in the Text

A/D	analog to digital
ACD	automatic call distributor
ACK	acknowledge
ACP	Airline Control Program (IBM)
ADPCM	adaptive differential pulse code modulation
AHS	American Hospital Supply
AM	amplitude modulation
ANSI	American National Standards Institute
APA	all points addressable
ARPA	Advanced Research Projects Agency
ARQ	automatic repeat request
ASCII	American Standard Code for Information Interchange
ASR	automatic send-receive
AT&T	American Telephone and Telegraph Company
ATM	automatic teller machine
ATTCOM	AT&T Communications
ATTIS	AT&T Information Systems
BCC	block check character
BCD	binary coded decimal
BDLC	Burroughs data link control (Burroughs/Unisys)
BELLCORE	Bell Communications Research
BISYNC	binary synchronous communications
BNA	Burroughs Network Architecture (Burroughs/Unisys)
BOC	Bell Operating Company
BSC	binary synchronous communications
CAD	computer aided design or computer assisted drafting
CAM	communications access method
CASE	common application service elements
CBX	computerized branch exchange
CCIS	common channel interoffice signaling
CCITT	Consultative Committee on International Telegraphy and Telephony
CCS	centa call seconds
CCU	cluster control unit
CDR	call detail recording
CI–I	Computer Inquiry I
CI–II	Computer Inquiry II
CI–III	Computer Inquiry III
CICS	Customer Information Control System (IBM)
CIO	chief information officer
CNA	Communication Network Architecture (NCR)
COE	Council of Europe
COP	central order processing (Dow Corning)
CRC	cyclic redundancy checking
CRT	cathode ray tube
CRTC	Canadian Radio-television and Telecommunications Commission
CSMA/CA	carrier sense multiple access with collision avoidance
CSU	channel service unit
D/A	digital to analog
DC	direct current
DCA	Distributed Communication Architecture (Sperry/Unisys)
DCE	data circuit-terminating equipment
DDCMP	Digital Data Communications Message Protocol (DEC)
DDD	direct distance dialing
DDP	distributed data processing
DEC	Digital Equipment Corporation
DES	data encryption standard

DID	direct inward dialing
DLE	data link escape
DNA	Digital Network Architecture (DEC)
DNHR	dynamic nonhierarchical routing (AT&T)
DOC	Department of Communications (Australia)
DPSK	differential phase shift keying
DSE	Distributed Systems Environment (Honeywell)
DSP	digital signal processor
DSU	data service unit
DTE	data terminal equipment
DTMF	dual-tone-multifrequency
E-Mail	electronic mail
EBCDIC	Extended Binary Coded Decimal Interchange Code
EDI	electronic document interchange
EIA	Electrical Industries Association
EMA	Enterprise Network Management Architecture
EOT	end of transmission
EPSCS	enhanced private switched communication service
ETB	end of text block
ETN	electronic tandem network
ETX	end of text
EUC	End User Computing (Dow Corning)
FAX	facsimile
FCC	Federal Communications Commission
FDDI	fiber distributed data interface
FDM	frequency division multiplexing
FDX	full-duplex
FEC	forward error correction
FEP	front-end processor
FIFO	first-in-first-out
FM	frequency modulation
FSK	frequency shift keying
FTAM	file transfer, access, and management
FX	foreign exchange
GE	General Electric Company
GEIS	General Electric Information Services
GM	General Motors Corporation
HBO	Home Box Office
HDLC	high-level data link control
HDX	half-duplex
Hz	hertz
IBM	IBM Corporation
ICC	Interstate Commerce Commission
IDDD	international direct distance dialing
IEEE	Institute of Electrical and Electronic Engineers
IIN	IBM Information Network
IMS	Information Management System (IBM)
IRC	international record carrier
IRM	information resource management
IS&CG	Information Systems & Communications Group (IBM)
ISDN	Integrated Service Digital Network
ISO	International Standards Organization
ITT	International Telephone and Telegraph Company
IVDT	integrated video display terminal
IXC	interexchange carrier
JTM	job transfer and manipulation
KSR	keyboard send-receive
LAN	local area network
LATA	local access and transport area
LEC	local exchange carrier (or company)
LED	light emitting diode
LLC	logical link control
LPC	linear predictive coding
LRC	longitudinal redundancy checking
LU	logical unit (IBM)
MAC	media access control
MAN	metropolitan area network
MAP	Manufacturing Automation Protocol
MCI	MCI Communications Corporation
MFJ	modified final judgment
MOTIS	message oriented interchange systems
MPT	Ministry of Posts and Telecommunications (Japan)
MSNF	Multi System Networking Facility (IBM)
MTBF	mean time between failures
MTTR	mean time to repair
NAK	negative acknowledge
NANP	North American numbering plan
NAU	network addressable unit (IBM)
NCC	network control center
NCP	network control program
NEC	Nippon Electric Company
NetBIOS	Network Basic Input Output System
NRZ	nonreturn to zero
NSA	National Security Agency
OCC	other common carrier
OCR	optical character recognition
ODA	office document architecture

ODIF	office document interchange facility
OECD	Organization for Economic Cooperation and Development
OFTEL	Office of Telecommunications (United Kingdom)
ONA	open network architecture
OSI	Open Systems Interconnection
PABX	private automatic branch exchange
PACTEL	Pacific Telesis Corporation
PAM	phase amplitude modulation
PBX	private branch exchange
PCM	pulse code modulation
PDN	packet data network
PDU	protocol data unit
PEL	picture element
PERT	program evaluation and review technique
Pixel	picture element
PM	phase modulation
POP	point of presence
POS	point of sale
POTS	plain old telephone service
PROFS	Professional Office System (IBM)
PSC	Public Service Commission
PSK	phase shift keying
PSTN	public switched telephone network
PTT	Post, Telephones and Telegraph
PU	physical unit (IBM)
PUC	public utility commission
QAM	quadrature amplitude modulation
RBOC	Regional Bell Operating Company
RFP	request for proposal
RFQ	request for quotation
RJE	remote job entry
RO	receive only
RTT	Regie des Telegraphes et des Telephones (Belgium)
RVI	reverse interrupt
SBS	Satellite Business Systems
SBT	six-bit transcode
SCC	specialized common carrier
SDLC	synchronous data link control (IBM)
SDN	software defined network
SEI	Special Secretariat for Informatics (Brazil)
SIM	Systems and Information Management Department (Dow Corning)
SMDR	station message detail recording
SNA	Systems Network Architecture (IBM)
SOH	start of header
SSCP	system service control point (IBM)
STDM	statistical time division multiplexing
STX	start of text
SYN	synchronization
TASI	time assignment speech interpolation
TCAM	telecommunications access method
TCM	telecommunications monitor
TCM	trellis coding modulation
TCP/IP	transmission control protocol/internet protocol
TDM	time division multiplexing
TNDF	transnational data flow
TOP	Technical Office Protocol
TPF	Transaction Processing Facility (IBM)
TPNS	teleprocessing network simulator (IBM)
TTD	temporary text delay
TWX	teletypewriter exchange system
UDLC	Universal Data Link Control (Sperry/Unisys)
UNMA	Unified Network Management Architecture (AT&T)
VAN	value-added network
VDT	video display terminal
VRC	vertical redundancy checking
VTAM	Virtual Telecommunications Access Method (IBM)
VTS	virtual terminal services
WAN	wide area network
WATS	wide area telecommunication service
bps	bits per second
dB	decibel
gHz	gigahertz
kHz	kilohertz
kbps	thousand of bits per second
MHz	megahertz
Mbps	millions of bits per second

GLOSSARY

access line. A telecommunications line that continuously connects a remote station to a switching exchange. A telephone number is associated with the access line.
access method. Computer software that moves data between main storage and input/output devices.
acknowledgment. The transmission, by a receiver, of acknowledgment characters as an affirmative response to a sender.
acknowledgment character (ACK). A transmission control character transmitted by a receiver as an affirmative response to a sender.
acoustic coupler. Telecommunications equipment that permits use of a telephone handset to connect a terminal to a telephone network.
acoustic coupling. A method of coupling data terminal equipment (DTE) or a similar device to a telephone line by means of transducers that use sound waves to or from a telephone handset or equivalent.
activation. In a network, the process by which a component of a node is made ready to perform the functions for which it was designed.
active line. A telecommunications line that is currently available for transmission of data.
adaptive differential pulse code modulation (ADPCM). A variation of pulse code modulation in which only the difference in signal samples is coded.
address. (1) A character or group of characters that identifies a data source or destination. (2) To refer to a device or an item of data by its address. (3) The part of the selection signals that indicates the destination of a call.
addressing. The means by which the originator or control station selects the unit to which it is going to send a message.
addressing characters. Identifying characters sent by the computer over a line that cause a particular station (or component) to accept a message sent by the computer.
aerial cable. A telecommunications cable connected to poles or similar overhead structures.
airline reservation system. An online application in which a computing system is used to keep track of seat inventories, flight schedules, passenger records, and other information. The reservation system is designed to maintain up-to-date data files and to respond, within seconds, to inquiries from ticket agents at locations remote from the computing system.
algorithm. A set of mathematical rules.
all-points-addressable (APA). An attribute of a VDT or printer that allows each individual dot on the screen or spot on a page to be individually addressed for output or input.
alphanumeric. Pertaining to a character set that contains letters, digits, and usually other characters, such as punctuation marks.
American National Standards Institute (ANSI). An organization formed for the purpose of establishing voluntary industry standards.
amplifier. A device that, by enabling a received wave to control a local source of power, is capable of delivering an enlarged reproduction of the wave.
amplitude. The size or magnitude of a voltage or current analog waveform.
amplitude modulation (AM). (1) Modulation in which the amplitude of an alternating current is the characteristic varied. (2) The variation of a carrier signal's strength (amplitude) as a function of an information signal.
analog. Pertaining to data in the form of continuously variable physical quantities.
analog channel. A data communications channel on which the information transmitted can take any value between the limits defined by the channel. Voice-grade channels are analog channels.
analog signal. A signal that varies in a continuous manner. Examples are voice and music. *Contrast with* digital signal.
analog-to-digital (A/D) converter. A device that senses an analog signal and converts it to a proportional representation in digital form.
analysis. The methodical investigation of a problem and the separation of the problem into smaller related units for further detailed study.
analyst. A person who defines problems and develops algorithms and procedures for their solution.
answerback. The response of a terminal to remote control signals.
architecture. A plan or direction oriented toward the needs of a user. An architecture describes "what"; it does not tell "how".
area code. A three-digit number identifying a geographic area of the USA and Canada to permit direct distance dialing on the telephone system.
ASCII. American Standard Code for Information Interchange. The standard code, using a coded character set consisting of 7-bit coded characters (8 bits, including the parity check), used for information interchange among data processing systems, data communications systems, and associated equipment. The ASCII character set consists of control characters and graphic characters.
asynchronous. Without a regular time relationship.
asynchronous transmission. (1) Transmission in which the time of occurrence of the start of each character or block of characters is arbitrary. (2) Transmission in which each information character is individually synchronized (usually by the use of start elements and stop elements).
attention interruption. An I/O interruption caused by a terminal user pressing an attention key or its equivalent.
attenuation. A decrease in magnitude of current, voltage, or power

of a signal in transmission between points. It is normally expressed in decibels.

attenuation distortion. The deformation of an analog signal that occurs when the signal does not attenuate evenly across its frequency range.

audio frequencies. Frequencies that can be heard by the human ear (approximately 15 hertz to 20,000 hertz).

audio response unit. An output device that provides a spoken response to digital inquiries from a telephone or other device. The response is composed from a prerecorded vocabulary of words and can be transmitted over telecommunications lines to the location from which the inquiry originated.

audiotex. A voice messaging system that can access a database on a computer.

audit. To review and examine the activities of a system, mainly to test the adequacy and effectiveness of control procedures.

audit trail. A manual or computerized means for tracing the transactions affecting the contents of a record.

authorization code. A code, typically made up of the user's identification and password, used to protect against unauthorized access to data and system facilities.

auto-answer. *See* automatic answering.

auto-dial. *See* automatic dialing.

auto-poll. A machine feature of a transmission control unit or front-end processor that permits it to handle negative responses to polling without interrupting the processing unit.

automatic answering. (1) Answering in which the called data terminal equipment (DTE) automatically responds to the calling signal; the call may be established whether or not the called DTE is attended. (2) A machine feature that allows a transmission control unit or a station to respond automatically to a call that it receives over a switched line.

automatic call distribution unit (ACD). A device attached to a telephone system that routes the next incoming call to the next available agent.

automatic dialing. A capability that allows a computer program or an operator using a keyboard to send commands to a modem, causing it to dial a telephone number.

automatic repeat request (ARQ). An error correction technique. When the receiving DTE detects an error, it signals the transmitting DTE to resend the data.

automatic send-receive (ASR). A teletypewriter unit with keyboard, printer, paper tape reader/transmitter, and paper tape punch. This combination of units may be used online or offline and, in some cases, online and offline concurrently.

automatic teller machine (ATM). A specialized computer terminal that enables consumers to conduct banking transactions without the assistance of a bank teller.

availability. Having a system or service operational when a user wants to use it.

background noise. Phenomena in all electrical circuitry resulting from the movement of electrons. *Also known as* white noise *or* Gaussian noise.

bandwidth. The difference, expressed in hertz, between the two limiting frequencies of a band.

bar code reader. A device that reads codes printed in the form of bars on merchandise or tags.

baseband. A form of modulation in which signals are pulsed directly on the transmission medium. In local area networks, baseband also implies the digital transmission of data.

baseband transmission. Transmission using baseband techniques. The signal is transmitted in digital form using the entire bandwidth of a circuit or cable. Typically used in local area networks.

basic business transactions. Fundamental operational units of business activity.

basic services. Services performed by the common carriers to provide the transportation of information. Basic services are regulated. *Contrast with* enhanced services.

batch. A set of data accumulated over a period of time.

batch processing. (1) Processing data or performing jobs accumulated in advance so that each accumulation is processed or accomplished in the same run. (2) Processing data accumulated over a period of time.

batched communication. A large body of data sent from one station to another station in a network, without intervening responses from the receiving unit. *Contrast with* inquiry/response communication.

baud. A unit of signaling speed equal to the number of discrete conditions or signal events per second. If the duration of a signal event is 20 milliseconds, the modulation rate is 1 second ÷ 20 milliseconds = 50 baud.

Baudot code. A code for the transmission of data in which five equal-length bits represent one character. This code is used in some teletypewriter machines, where one start element and one stop element are added.

Bell Operating Companies (BOCs). The 22 telephone companies that were members of the Bell system before divestiture.

Bell System. The collection of companies headed by AT&T and consisting of the 22 Bell Operating Companies and the Western Electric Corporation. The Bell System was dismantled by divestiture on January 1, 1984.

bid. In the contention form of invitation or selection, an attempt by the computer or by a station to gain control of the line so that it can transmit data. A bid may be successful or unsuccessful in seizing a circuit in that group. *Contrast with* seize.

binary. (1) Pertaining to a selection, choice, or condition that has two possible values or states. (2) Pertaining to the base two numbering system.

binary code. A code that makes use of exactly two distinct characters, usually 0 and 1.

binary-coded decimal (BCD) code. A binary-coded notation in which each of the decimal digits is represented by a binary numeral, for example, in binary-coded decimal notation that uses the weights 8-4-2-1, the number 23 is represented by 0010 0011 Compare this to its representation in the pure binary numeration system, which is 10111.

binary digit. (1) In binary notation, either the character 0 or 1. (2) *Synonym for* bit.

binary synchronous communications (BSC, BISYNC). (1) Communications using binary synchronous protocol. (2) A uniform procedure, using a standardized set of control characters and con-

trol character sequences, for synchronous transmission of binary-coded data between stations.

bipolar, nonreturn to zero (NRZ) signals. Signals that have the 1 bits represented by a positive voltage and the 0 bits represented by a negative voltage. *Contrast with* bipolar, return to zero.

bipolar, return to zero signals. Signals that have the 1 bits represented by a positive voltage and the 0 bits represented by a negative voltage. Between pulses, the voltage always returns to zero. *Contrast with* bipolar, nonreturn to zero.

bit. Synonym for binary digit.

bit-oriented protocol. A communications protocol that uses only one special character, called the *flag character,* to mark the beginning and end of a message. All other combinations of bits are treated as valid data characters.

bit rate. The speed at which bits are transmitted, usually expressed in bits per second.

bit stream. A binary signal without regard to grouping by character.

bit stuffing. (1) The occasional insertion of a dummy bit in a bit stream. (2) In SDLC, a 0 bit inserted after all strings of five consecutive 1 bits in the header and data portion of the message. At the receiving end, the extra 0 bit is removed by the hardware.

bit synchronization. A method of ensuring that a communications circuit is sampled at the appropriate time to determine the presence or absence of a bit.

bits per second (bps). The basic unit of speed on a data communications circuit.

blank character. A graphic representation of the space character.

blink. Varying the intensity of one or more characters displayed on a VDT several times per second to catch the operator's attention.

block. (1) A string of records, a string of words, or a character string formed for technical or logic reasons to be treated as an entity. (2) A set of things, such as words, characters, or digits, handled as a unit. (3) A group of bits, or characters, transmitted as a unit. An encoding procedure generally is applied to the group of bits or characters for error-control purposes. (4) That portion of a message terminated by an EOB or ETB line-control character or, if it is the last block in the message, by an EOT or ETX line-control character.

block check. That part of the error control procedure used for determining that a data block is structured according to given rules.

block check character (BCC). In longitudinal redundancy checking and cyclic redundancy checking, a character that is transmitted by the sender after each message block and is compared with a block check character computed by the receiver to determine if the transmission was successful.

block error rate. The ratio of the number of blocks incorrectly received to the total number of blocks sent.

block length. (1) The number of records, words, or characters in a block. (2) A measure of the size of a block, usually specified in units, such as records, words, computer words, or characters.

blocking. (1) The process of combining incoming messages into a single message. (2) In a telephone switching system, the inability to make a connection or obtain a service because the devices needed for the connection are in use.

bridge. A device that allows data to be sent from one network to another so terminals on both networks can communicate as though a single network existed.

broadband. (1) A communications channel having a bandwidth greater than a voice-grade channel, and therefore capable of higher-speed data transmission. (2) In local area networks, an analog transmission with frequency division multiplexing.

broadband transmission. A transmission technique of a local area network in which the signal is transmitted in analog form with frequency division multiplexing.

broadcast. The simultaneous transmission to a number of stations.

buffering. The storage of bits or characters until they are specifically released. For example, a buffered terminal is one in which the keyed characters are stored in an internal storage area or buffer until a special key, such as the CARRIAGE RETURN or ENTER key, is pressed. Then all of the characters stored in the buffer are transmitted to the host computer in one operation.

bus network. A network topology in which multiple nodes are attached to a single circuit of limited length. A bus network is typically a local area network that transmits data at high speed.

business machine. Customer-provided data terminal equipment (DTE) that connects to a communications common carrier's telecommunications equipment for the purpose of data movement.

business machine clocking. An oscillator supplied by the business machine for regulating the bit rate of transmission. *Contrast with* data set clocking.

busy hour. The hour of the day when the traffic carried on a network is the highest.

bypass. Installing private telecommunications circuits to avoid using those of a carrier.

byte. An 8-bit binary character operated on as a unit.

byte-count-oriented protocol. A protocol that uses a special character to mark the beginning of the header, followed by a count field that indicates how many characters are in the data portion of the message.

call back. A security technique used with dial-up lines. After a user calls and identifies himself or herself, the computer breaks the connection and calls the user back at a predetermined telephone number. In some systems, the number at which the user wishes to be called back can be specified when the initial connection is made and before the computer disconnects.

call control procedure. The implementation of a set of protocols necessary to establish, maintain, and release a call.

call detail recording (CDR). *See* station message detail recording (SMDR).

call progress signal. A call control signal transmitted from the data circuit-terminating equipment (DCE) to the calling data terminal equipment (DTE) to indicate the progress of the establishment of a call, the reason why the connection could not be established, or any other network condition.

call setup time. The time taken to connect a switched telephone call. The time between the end of dialing and answering by the received party.

camp-on. A method of holding a call for a line that is in use and of signaling when it becomes free.

carriage-return character (CR). A format effector that causes the print or display position to move to the first position on the same line. *Contrast with* line feed character.

carrier. (1) A company that provide the telecommunications networks. *See* communications common carrier. (2) A communications signal. *See* carrier wave.

carrier sense multiple access with collision avoidance (CSMA/CA). A communications protocol used on local area networks in which a station listens to the circuit before transmitting in an attempt to avoid collisions.

carrier sense multiple access with collision detection (CSMA/CD). A communications protocol frequently used on local area networks in which stations, on detecting a collision of data caused by multiple simultaneous transmissions, wait a random period of time before retransmitting.

carrier system. A means of obtaining a number of channels over a single circuit by modulating each channel on a different carrier frequency and demodulating at the receiving point to restore the signals to their original form.

carrier wave. An analog signal that in itself contains no information.

cathode ray tube (CRT) terminal. A particular type of video display terminal that uses a vacuum tube display in which a beam of electrons can be controlled to form alphanumeric characters or symbols on a luminescent screen, for example, by use of a dot matrix.

CCITT. *See* Consultative Committee on International Telegraphy and Telephony (CCITT).

cellular telephone service. A system for handling telephone calls to and from moving automobiles. Cities are divided into small geographic areas called *cells*. Telephone calls are transmitted to and from low-power radio transmitters in each cell. Calls are passed from one transmitter to another as the automobile leaves one cell and enters another.

centa. One hundred.

centa call seconds. A measure of equipment or circuit utilization. One centa call second is 100 seconds of utilization.

centi. One hundredth.

central office. In the United States, the place where communications common carriers terminate customer lines and locate the equipment that interconnects those lines.

central office switch. The equipment in a telephone company central office that allows any circuit to be connected to any other.

centralized network. *Synonym for* star network.

centrex. Central office telephone equipment serving subscribers at one location on a private automatic branch exchange basis. The system allows such services as direct inward dialing, direct distance dialing, and console switchboards.

change coordination meeting. A meeting held to ensure that changes to a system are properly approved and communicated to all interested parties.

change management. The application of management principles to ensure that changes in a system are controlled to minimize the impact on system users.

channel. (1) A one-way communications path. (2) In information theory, that part of a communications system that connects the message source with the message sink.

channel service unit (CSU). *See* data service unit.

character. A member of a set of elements upon which agreement has been reached and that is used for the organization, control or representation of data. Characters may be letters, digits, punctuation marks, or other symbols, often represented in the form of a spatial arrangement of adjacent or connected strokes or in the form of other physical conditions in data media.

character assignments. Unique groups of bits assigned to represent the various characters in a code.

character-oriented protocol. A communications protocol that uses special characters to indicate the beginning and end of messages. BISYNC is a character-oriented protocol.

character set. A set of unique representations called *characters,* for example, the 26 letters of the English alphabet, 0 and 1 of the Boolean alphabet, the set of signals in the Morse code alphabet, and the 128 ASCII characters.

character synchronization. A technique for ensuring that the proper sets of bits on a communications line are grouped to form characters.

check bit. (1) A binary check digit, for example, a parity bit. (2) A bit associated with a character or block for the purpose of checking for the absence of error within the character or block.

checkpoint restart. The process of resuming a job or process at a checkpoint within the job or process that terminated abnormally. The restart may be automatic or deferred.

chief information officer (CIO). A title sometimes given to the highest ranking executive in charge of a company's information resources.

chip. (1) A minute piece of semiconductive material used in the manufacture of electronic components. (2) An integrated circuit on a piece of semiconductive material.

circuit. The path over which two-way communications take place.

circuit grade. The information-carrying capability of a circuit, in speed or type of signal. The grades of circuits are broadband, voice, subvoice, and telegraph. For data use, these grades are identified with certain speed ranges.

circuit noise level. The ratio of the circuit noise to some arbitrary amount chosen as the reference. This ratio normally is indicated in decibels above the reference noise.

circuit speed. The number of bits that a circuit can carry per unit of time, typically 1 second. Circuit speed is measured in bits per second.

circuit-switched data transmission service. A service using circuit switching to establish and maintain a connection before data can be transferred between data terminal equipments (DTEs).

circuit switching. The temporary establishment of a connection between two pieces of equipment that permits the exclusive use until the connection is released. The connection is set up on demand and discontinued when the transmission is complete. An example is a dial-up telephone connection.

cladding. The glass that surrounds the core of an optical fiber and acts as a mirror to the core.

Class 5 central office. A local central office designed to serve consumers or businesses.

clock. (1) A device that measures and indicates time. (2) A device that generates periodic signals used for synchronization. (3)

Equipment that provides a time base used in a transmission system to control the timing of certain functions, such as sampling, and to control the duration of signal elements.
clock pulse. A synchronization signal provided by a clock.
clocking. The use of clock pulses to control synchronization of data and control characters.
cluster. A station that consists of a control unit (cluster controller) and the terminals attached to it.
cluster control unit (CCU). A device that can control the input/output operations of more than one device connected to it. A cluster control unit may be controlled by a program stored and executed in the unit, or it may be controlled entirely by hardware.
coaxial cable. A cable consisting of one conductor, usually a small copper tube or wire, within and insulated from another conductor of larger diameter, usually copper tubing or copper braid.
code. (1) A set of unambiguous rules specifying the manner in which data may be represented in a discrete form. (2) A predetermined set of symbols that have specific meanings.
code conversion. A process for changing the bit grouping for a character in one code into the corresponding bit grouping for a character in a second code.
code converter. A device that changes the representation of data, using one code in the place of another or one coded character set in the place of another.
code efficiency. Using the least number of bits to convey the meaning of a character with accuracy.
code-independent data communications. A mode of data communications that uses a character-oriented link protocol that does not depend on the character set or code used by the data source.
code points. The number of possible combinations in a coding system.
code transparent data communication. A mode of data communications that uses a bit-oriented link protocol that does not depend on the bit sequence structure used by the data source.
coded character set. A set of unambiguous rules that establish a character set and the one-to-one relationships between the characters of the set and their coded representations. *Synonymous with* code.
coding scheme. *See* code (1).
collision. Two (or more) terminals trying to transmit a message at the same time, thereby causing both messages to be garbled and unintelligible at the receiving end.
common carrier. *See* communications common carrier.
common channel interoffice signaling (CCIS). A system for sending signals between central offices in a telephone network.
communications. (1) A process that allows information to pass between a sender and one or more receivers. (2) The transfer of meaningful information from one location to a second location. (3) The art of expressing ideas, especially in speech and writing. (4) The science of transmitting information, especially in symbols.
communications access method (CAM). Computer software that reads and writes data from and to communications lines. *Synonym for* telecommunications access method (TCAM).
communications adapter. An optional hardware feature, available on certain processors, that permits telecommunications lines to be attached to the processors.
communications common carrier. In the USA and Canada, a public data transmission service that provides the general public with transmission service facilities, for example, a telephone or telegraph company.
communications controller. A type of communications control unit whose operations are controlled by one or more programs stored and executed in the unit. The communications controller manages the details of line control and data routing through a network. *See also* front-end processor.
communications directory. An online or hardcopy document that lists the names and telephone numbers of a company's employees and departments, as well as other information, such as terminal names or user IDs, which is pertinent to communications.
communications facility. *See* telecommunications facility.
communications line. Deprecated term for telecommunications line or transmission line.
communications network. A collection of communications circuits managed as a single entity.
communications standards. Standards established to ensure compatibility among several communications services or several types of communications equipment.
communications theory. The mathematical discipline dealing with the probabilistic features of data transmission in the presence of noise.
compaction. The process of eliminating redundant characters from a data stream before it is stored or transmitted.
compandor (compressor-expandor). Equipment that compresses the outgoing speech volume range and expands the incoming speech volume range on a long distance telephone circuit. Such equipment can make more efficient use of voice telecommunications channels.
compression. *See* compaction.
computer aided design (CAD). The use of a computer with special terminals and software for engineering design and drafting. *Synonym for* computer assisted drafting (CAD).
computer assisted drafting (CAD). *See* computer aided design.
computer branch exchange (CBX). *See* private automatic branch exchange (PABX) *and* private branch exchange (PBX).
Computer Inquiry I (CI–I). A study conducted by the FCC, concluded in 1971, that examined the relationship between the telecommunications and data processing industries to determine which aspects of both industries should be regulated for the long term.
Computer Inquiry II (CI–II). A study conducted by the FCC, concluded in 1981, that accelerated the deregulation of the telecommunications industry.
Computer Inquiry III (CI–III). A study conducted by the FCC to determine to what extent the AT&T and the BOCs are allowed to provide enhanced (data processing) services in the network.
computer network. A complex consisting of two or more interconnected computing units.
concentration. The process of combining multiple messages into a single message for transmission. *Contrast with* deconcentration.
concentrator. (1) In data transmission, a functional unit that per-

mits a common transmission medium to serve more data sources than there are channels currently available within the transmission medium. (2) Any device that combines incoming messages into a single message (concentration) or extracts individual messages from the data sent in a single transmission sequence (deconcentration).

conditioned line. A communications line on which the specifications for amplitude and distortion have been tightened. Signals traveling on a conditioned circuit are less likely to encounter errors than on an unconditioned circuit.

conditioning. The addition of equipment to a leased voice-grade circuit to provide minimum values of line characteristics required for data transmission.

configuration control. The maintenance of records that identify and keep track of all of the equipment in a system.

Consultative Committee on International Telegraphy and Telephony (CCITT). An international standards organization that is part of the International Telecommunications Union, which is an arm of the United Nations.

continuous ARQ. An error correction technique in which data blocks are continuously sent over the forward channel, while ACKs and NAKs are sent over the reverse channel. *Contrast with* stop and wait ARQ.

control character. A character whose occurrence in a particular context initiates, modifies, or stops a control operation. A control character may be recorded for use in a subsequent action, and it may have a graphic representation in some circumstances.

control station. A station on a network that assumes control of the network's operation. A typical control station exerts its control by polling and addressing. *Contrast with* slave station.

control terminal. Any active terminal on a network at which a user is authorized to enter commands affecting system operation.

control unit. A device that controls input/output operations at one or more devices. *See also* controller *and* cluster control unit (CCU).

controller. A device that directs the transmission of data over the data links of a network. Its operations may be controlled by a program executed in a processor to which the controller is connected, or they may be controlled by a program executed within the device. *See* cluster control unit (CCU) *and* communications controller.

conversational mode. A mode of operation of a data processing system in which a sequence of alternating entries and responses between a user and the system takes place in a manner similar to a dialog between two persons. *Synonym for* interactive mode.

core. The glass or plastic center conductor of an optical fiber that provides the transmission carrying capability.

cost center. An accounting term used to designate a department or other entity where costs are accumulated. Departments that are cost centers do not make a profit by selling or recharging their services. *Contrast with* profit center.

creeping commitment of resources. A concept of project management that suggests that the resources dedicated to a project should only be increased as the scope of project becomes better defined. The intention is to minimize the amount of resources spent in case the project is determined to be infeasible or is otherwise canceled.

crossbar switch. A relay-operated device that makes a connection between one line in each of two sets of lines. The two sets are physically arranged along adjacent sides of a matrix of contacts or switch points.

crosstalk. The unwanted energy transferred from one circuit, called the *disturbing circuit,* to another circuit, called the *disturbed circuit.*

current beam position. On a CRT display device, the coordinates on the display surface at which the electron beam is aimed.

current loop. An interface between a terminal and a circuit that indicates 1 and 0 bits by the presence or absence of an electrical current.

cursor. (1) In computer graphics, a movable marker that indicates a position on a display space. (2) A displayed symbol that acts as a marker to help the user locate a point in text, in a system command, or in storage. (3) A movable spot of light on the screen of a display device, usually indicating where the next character is to be entered, replaced, or deleted.

cursor control keys. The keys that control the movement of the cursor.

cycle stealing. Interrupting a computer to store each character coming from a telecommunications line in the computer's memory.

cyclic redundancy check (CRC). (1) An error checking technique in which the check key is generated by a cyclic algorithm. (2) A system of error checking performed at both the sending and receiving stations after a block check character has been accumulated.

data. (1) A representation of facts, concepts, or instructions in a formalized manner suitable for communication, interpretation, or processing by human or automatic means. (2) Any representations, such as characters or analog quantities, to which meaning is, or might be, assigned.

data access arrangement (DAA). Equipment that permits attachment of privately owned data terminal equipment and telecommunications equipment to the public telephone network.

data circuit. Associated transmit and receive channels that provide a means of two-way data communications.

data circuit-terminating equipment (DCE). The equipment installed at the user's premises that provides all the functions required to establish, maintain, and terminate a connection and the signal conversion and coding between the data terminal equipment (DTE) and the line.

data communications. (1) The transmission and reception of data. (2) The transmission, reception, and validation of data. (3) Data transfer between data source and data sink via one or more data links according to appropriate protocols.

data communications channel. A means of one-way transmission.

data encryption standard (DES). A cryptographic algorithm designed to encipher and decipher data using a 64-bit cryptographic key, as specified in the Federal Information Processing Standard Publication 46, January 15, 1977.

data integrity. The quality of data that exists as long as accidental or malicious destruction, alteration, or loss of data is prevented.

data link. (1) The physical means of connecting one location to another to transmit and receive data. (2) The interconnecting data circuit between two or more pieces of equipment operating in accordance with a link protocol. It does not include the data source and the data sink.
data link control character. A control character intended to control or facilitate transmission of data over a network.
data link escape (DLE) character. A transmission control character that changes the meaning of a limited number of contiguous following characters or coded representations and that is used exclusively to provide supplementary transmission control characters.
data network. The assembly of functional units that establishes data circuits between pieces of data terminal equipment (DTE).
data PBX. A switch especially designed for switching data calls. Data PBXs do not handle voice calls.
Data-Phone. Both a service mark and a trademark of AT&T and the Bell System. As a service mark, it indicates the transmission of data over the telephone network. As a trademark, it identifies the telecommunications equipment furnished by the Bell System for transmission services.
data processing (DP). The systematic performance of operations upon data, for example, handling, merging, sorting, and computing. *Synonym for* information processing.
data processing system. A system, including computer systems and associated personnel, that performs input, processing, storage, output, and control functions to accomplish a sequence of operations on data.
data security. The protection of data against unauthorized disclosure, transfer, modifications, or destruction, whether accidental or intentional.
data service unit/channel service unit (DSU/CSU). An interface device that ensures that the digital signal entering a communications line is properly shaped into square pulses and precisely timed.
data set clocking. A time-based oscillator supplied by the modem for regulating the bit rate of transmission.
data sink. (1) A functional unit that accepts data after transmission. It may originate error control signals. (2) That part of a data terminal equipment (DTE) that receives data from a data link.
data terminal equipment (DTE). That part of a data station that serves as a data source, data sink, or both and provides for the data communications control function according to protocols.
data transfer rate. The average number of bits, characters, or blocks per unit of time transferred from a data source to a data sink. The rate is usually expressed as bits, characters, or blocks per second, minute, or hour.
datagram. In packet switching, a self-contained packet that is independent of other packets, that does not require acknowledgment, and that carries information sufficient for routing from the originating data terminal equipment (DTE) to the destination DTE without relying on earlier exchanges between the DTEs and the network.
dB meter. A meter having a scale calibrated to read directly in decibel values at a reference level that must be specified (usually 1 milliwatt equals zero dB). Used in audio-frequency amplifier circuits of broadcast stations, public address systems, and receiver output circuits to indicate volume level.
dBm. Decibel based on 1 milliwatt.
deactivation. In a network, the process of taking any element out of service, rendering it inoperable, or placing it in a state in which it cannot perform the functions for which it was designed.
decibel (dB). (1) A unit that expresses the ratio of two power levels on a logarithmic scale. (2) A unit for measuring relative power. The number of decibels is 10 times the logarithm (base 10) of the ratio of the measured power levels. If the measured levels are voltages (across the same or equal resistance), the number of decibels is 20 times the log of the ratio.
DECNET. A family of hardware and software that implement Digital Network Architecture (DNA).
deconcentration. The process of extracting individual messages from data sent in a single transmission sequence. *Contrast with* concentrator.
decryption. Converting encrypted data into clear data. *Contrast with* encryption.
delta modulation. A technique of digitizing an analog signal by comparing the values of two successive samples and assigning a 1 bit if the second sample has a greater value and a 0 bit if the second sample has a lesser value.
demarcation point. The physical and electrical boundary between the telephone company "responsibility and the customer" responsibility.
demodulation. The process of retrieving intelligence (data) from a modulated carrier wave. *Reverse of* modulation.
demodulator. A device that performs demodulation.
destination code. A code in a message header containing the name of a terminal or application program to which the message is directed.
destructive cursor. On a VDT device, a cursor that erases any character through which it passes as it is advanced, backspaced, or otherwise moved. *Contrast with* nondestructive cursor.
device control character. A control character used for the control of ancillary devices associated with a data processing system or data communications system, for example, for switching such devices on or off.
dial. To use a dial or pushbutton telephone to initiate a telephone call. In telecommunications, this action is taken to attempt to establish a connection between a terminal and a telecommunications device over a switched line.
dial backup. A technique for bypassing the failure of a private circuit. When a failure occurs, a switched connection is made so that communications can be reinstated.
dial line. *Synonym for* switched connection.
dial pulse. An interruption in the dc loop of a calling telephone. The interruption is produced by breaking and making the dial pulse contacts of a calling telephone when a digit is dialed. The loop current is interrupted once for each unit of value of the digit.
dial tone. An audible signal indicating that a device is ready to be dialed.

dial-up. The use of a dial or pushbutton telephone to initiate a station-to-station telephone call.
dial-up line. A line on which the connection is made by dialing. *See also* switched line.
dial-up terminal. A terminal on a switched line.
dialing. Deprecated term for calling.
dialog. In an interactive system, a series of interrelated inquiries and responses analogous to a conversation between two people.
dibit. A group of two bits. In four-phase modulation, each possible dibit is encoded as one of four unique carrier phase shifts. The four possible states for a dibit are 00, 01, 10, and 11. *Contrast with* quadbit *and* tribit
differential phase shift keying (DPSK). A modulation technique in which the relative changes of the carrier signal phase are coded according to the data to be transmitted.
digital circuit. A circuit expressly designed to carry the pulses of digital signals.
Digital Data Communications Message Protocol (DDCMP). A byte-count-oriented protocol developed by the Digital Equipment Corporation.
Digital Network Architecture (DNA). A communications architecture developed by the Digital Equipment Corporation as a framework for all of the company's communications products.
digital signal. A discrete or discontinuous signal; one whose various states are pulses that are discrete intervals apart. *Contrast with* analog signal.
digital signal processor (DSP). A microprocessor especially designed to analyze, enhance, or otherwise manipulate sounds, images, or other signals.
digital switching. A process in which connections are established by operations on digital signals without converting them to analog signals.
digital-to-analog (D/A) converter. A device that converts a digital value to a proportional analog signal.
digitize. To express or represent in a digital form data that are not discrete data, for example, to obtain a digital representation of the magnitude of a physical quantity from an analog representation of that magnitude.
digitizing distortion. *See* quantizing noise.
direct current (DC) signaling. Signaling caused by opening and closing a direct current electrical circuit.
direct distance dialing (DDD). A telephone exchange service that enables the telephone user to call subscribers outside of the user's local service area without operator assistance.
direct inward dialing (DID). A facility that allows a telephone call to pass through a telephone system directly to an extension without operator intervention.
direct outward dialing (DOD). A facility that allows an internal caller at an extension to dial an external number without operator assistance.
disconnect. To disengage the apparatus used in a connection and to restore it to its ready condition.
disconnect signal. A signal transmitted from one end of a subscriber line or trunk to indicate at the other end that the established connection is to be disconnected.
dispersion. The difference in the arrival time between signals that travel straight through the core of a fiberoptic cable and those that reflect off the cladding and, therefore, travel a slightly longer path.
display device. (1) An output unit that gives a visual representation of data. (2) In computer graphics, a device capable of presenting display elements on a display surface, for example, a terminal screen, plotter, microfilm viewer, or printer.
distinctive ringing. A ringing cadence that indicates whether a call is internal or external.
distortion. The unwanted change in wave form that occurs between two points in a transmission system. The six major forms of distortion are (1) *bias:* a type of telegraph distortion resulting when the significant intervals of the modulation do not all have their exact theoretical durations, (2) *characteristic:* distortion caused by transients that, as a result of the modulation, are present in the transmission channel and depend on its transmission qualities, (3) *delay:* distortion that occurs when the envelope delay of a circuit or system is not constant over the frequency range required for transmission, (4) *end:* distortion of start-stop teletypewriter signals. The shifting of the end of all marking pulses from their proper positions in relation to the beginning of the start pulse, (5) *fortuitous ("jitter"):* a type of distortion that results in the intermittent shortening or lengthening of the signals. This distortion is entirely random in nature and can be caused by such things as battery fluctuations, hits on the line, and power induction, (6) *harmonic:* the resultant presence of harmonic frequencies (due to nonlinear characteristics of a transmission line) in the response when a sine wave is applied.
distributed data processing (DDP). Data processing in which some or all of the processing, storage, and control functions, in addition to input/output functions, are situated in different places and connected by communications facilities.
distributed data processing network. A network in which some or all of the processing, storage, and control functions, in addition to input/output functions, are dispersed among its nodes.
distribution cable. A subgrouping of individual telephone lines as they approach a central office.
distribution frame. A structure for terminating permanent wires of a telephone central office, private branch exchange, or private exchange and for permitting the easy change of connections between them by means of cross-connecting wires.
domain. In an SNA network, the resources that are under the control of one or more associated host processors.
downlink. The rebroadcast of a microwave radio signal from a satellite back to earth.
downloading. The transmission of a file of data from a mainframe computer to a personal computer.
driver. *See* workload generator.
drop out. In data communication, a momentary loss in signal, usually due to the effect of noise or system malfunction.
drop wire. The wire running from a residence or business to a telephone pole or its underground equivalent.
dual-tone-multifrequency (DTMF). A method of signaling a desired telephone number by sending tones on the telephone line.
dumb terminal. A terminal that has little or no memory and is

not programmable. A dumb terminal is totally dependent on the host computer for all processing capability.

duplex. *See* full-duplex (FDX).

duplex transmission. Data transmission in both directions at the same time.

E&M signaling. A type of signaling between a switch or PBX and a trunk in which the signaling information is transferred via two-state voltage conditions on two wires.

EBCDIC. Extended Binary Coded Decimal Interchange Code. A coded character set consisting of 8-bit coded characters.

echo. The reversal of a signal, bouncing it back to the sender, caused by an electrical wave bouncing back from an intermediate point or the distant end of a circuit.

echo check. A check to determine the correctness of the transmission of data in which the received data are returned to the source for comparison with the originally transmitted data.

echo suppressor. A device that permits transmission in only one direction at a time, thus eliminating the problems caused by the echo.

effective data transfer rate. The average number of bits, characters, or blocks per unit time transferred from a data source to a data sink and accepted as valid.

800 service (in-WATS). *See* in-WATS (800 service).

electronic document interchange (EDI). The use of telecommunications to transmit documents electronically.

electronic mail (E-Mail). The use of telecommunications for sending textual messages from one person to another. The capability to store the messages in an electronic mailbox is normally a part of the electronic mail system.

electronic mailbox. Space on the disk of a computer to store electronic mail messages.

electronic switching system (ESS). Electronic switching computer for central office functions.

encryption. Transformation of data from the meaningful code that is normally transmitted, called *clear text*, to a meaningless sequence of digits and letters that must be decrypted before it becomes meaningful again. *Contrast with* decryption.

end of address (EOA). One or more transmission control characters transmitted on a line to indicate the end of nontext characters (for example, addressing characters).

end of block (EOB). A transmission control character that marks the end of a block of data.

end of message code (EOM). The specific character or sequence of characters that indicates the end of a message or record.

end of text (ETX). A transmission control character sent to mark the end of the text of the message.

end of text block (ETB). A transmission control character sent to mark the end of a portion of the text of a message.

end of text character (ETX). A control character that marks the end of a message's text.

end of transmission block character (ETB). A transmission control character used to indicate the end of a transmission block of data when data is divided into such blocks for transmission purposes.

end of transmission character (EOT). A transmission control character used to indicate the conclusion of a transmission that may have included one or more messages.

end office. A class 5 office of a local telephone exchange where a subscriber's loop terminates.

enhanced services. Communications services in which some processing of the information being transmitted takes place. *Contrast with* basic services.

enquiry character (ENQ). A transmission control character used as a request for a response from the station with which the connection has been set up; the response may include station identification, the type of equipment in service, and the status of the remote station.

enter. To place a message on a circuit to be transmitted from a terminal to the computer.

envelope delay distortion. Distortion caused by the electrical phenomenon that not all frequencies propagate down a telecommunications circuit at exactly the same speed.

equal access. A part of the modified final judgment that specified local telephone companies must provide all of the long distance companies access equal in type, quality, and price to that provided AT&T.

equalization. Compensation for differences in attenuation (reduction or loss of signal) at different frequencies.

equalizer. Any combination of devices, such as coils, capacitors, or resistors, inserted in a transmission line or amplifier circuit to improve its frequency response.

equivalent four-wire system. A transmission system using frequency division to obtain full-duplex operation over only one pair of wires.

ergonomics. The study of the problems of people in adjusting to their environment, especially the science that seeks to adapt work or working conditions to suit the worker.

Erlang. A measure of communications equipment or circuit usage. One Erlang is 1 hour of equipment usage or 36 CCS.

Erlang B capacity table. A table for determining the number of circuits required to carry a certain level of telephone traffic. The Erlang B table assumes that the sources of traffic are infinite and that all unsuccessful call attempts are abandoned and not retried.

error. A discrepancy between a computed, observed, or measured value or condition and the true, specified, or theoretically correct value or condition.

error correcting code. A code in which each telegraph or data signal conforms to specific rules of construction so that departures from this construction in the receive signals can be automatically detected, permitting the automatic correction, at the receiving terminal, of some or all of the errors. Such codes require more signal elements than are necessary to convey the basic information.

error correction system. A system employing an error detecting code and so arranged that some or all of the signals detected as being in error are automatically corrected at the receiving terminal before delivery to the data sink or to the telegraph receiver. *Note:* In a packet switched data service, the error correcting system might result in the retransmission of at least one or more complete packets should an error be detected.

error detection. The techniques employed to ensure that transmission and other errors are identified.
error detecting code. A code in which each element that is represented conforms to specific rules of construction so that if certain errors occur, the resulting representation will not conform to the rules, thereby indicating the presence of errors. Such codes require more signal elements than are necessary to convey the fundamental information.
error message. An indication that an error has been detected.
error rate. A measure of the quality of a circuit or system; the number of erroneous bits or characters in a sample, frequently taken per 100,000 characters.
error ratio. The ratio of the number of data units in error to the total number of data units.
error recovery. The process of correcting or bypassing a fault to restore a computer system to a prescribed condition.
escape character (ESC). A code extension character used to indicate that the following character or group of characters is to be interpreted in a nonstandard way.
escape mechanism. A method of assigning an alternate meaning to characters in a coding system. *See* escape character (ESC).
Ethernet. A local area network that uses CSMA/CD protocol on a baseband bus.
exchange. A room or building equipped so that telecommunications lines terminated there may be interconnected as required. The equipment may include manual or automatic switching equipment.
exchange code. The first three digits of a seven-digit telephone number. The exchange code designates the exchange that serves the customer.
extended binary coded decimal interchange code (EBCDIC). A coding system consisting of 256 characters, each represented by eight bits.
external modem. A modem that exists in its own box or cabinet.

facsimile machine (FAX). A machine that scans a sheet of paper and converts the light and dark areas to electrical signals that can be transmitted over telephone lines.
Federal Communications Commission (FCC). A board of commissioners appointed by the president under the Communications Act of 1934. The commissioners regulate all interstate and foreign electrical telecommunications systems originating in the United States.
feeder cable. A grouping of several distribution cables as they approach a central office.
fiber distributed data interface (FDDI). An emerging standard for transmitting data on an optical fiber.
figures shift (FIGS). A physical shift in a teletypewriter that enables the printing of images, such as numbers, symbols, and uppercase characters.
first-in-first-out (FIFO) queuing. Queuing in the order that calls or transactions arrive. Calls that arrive first get serviced first.
five-level code. A telegraph code that uses five impulses for describing a character. Start and stop elements may be added for asynchronous transmission. A common five-level code is the Baudot code.
flag. (1) Any of various types of indicators used for identification. (2) A bit sequence that signals the occurrence of some condition, such as the end of a word. (3) In high-level data link control (HDLC), the initial and final octets of a frame with the specific bit configuration of 01111110. A single flag may be used to denote the end of one frame and the start of another.
flat panel display. A technology for VDTs yielding a display that is much flatter and takes up less space on a desk than a CRT.
flat rate service. A method of charging for local calls that gives the user an unlimited number of calls for a flat monthly fee.
foreign exchange (FX) lines. A service that connects a customer's telephone system to a telephone company central office normally not serving the customer's location.
format effector character. A character that controls the positioning of information on a terminal screen or paper.
formatted mode. A method of displaying output on a VDT in which the entire screen can be arranged in any desired configuration and transmitted to the terminal at one time. *Contrast with* line-by-line mode.
forward channel. The primary transmission channel in a data circuit. *Contrast with* reverse channel.
forward error correction (FEC). A technique of transmitting extra bits or characters with a block of data so that transmission errors can be corrected at the receiving end.
four-wire circuit. A path in which four wires (two for each direction of transmission) are presented to the station equipment. Leased circuits are four-wire circuits.
four-wire terminating set. An arrangement by which four-wire circuits are terminated on a two-wire basis for interconnection with two-wire circuits.
frame. (1) In SDLC, the vehicle for every command, every response, and all information that is transmitted using SDLC procedures. Each frame begins and ends with a flag. (2) In high-level data link control (HDLC), the sequence of contiguous bits bracketed by and including opening and closing flag (01111110) sequences.
fractional T-1. The subdivision or multiplexing of T-1 circuits to provide circuit speeds that are a fraction of the T-1's capacity.
framing bits. Noninformation-carrying bits used to make possible the separation of characters in a bit stream.
freeze-frame television. A television system in which the picture is only updated as needed, typically every 30 to 90 seconds.
frequency. An attribute of analog signals that describes the rate at which the current alternates. Frequency is measured in hertz.
frequency division multiplexing (FDM). A technique of putting several analog signals on a circuit by shifting the frequencies of the signals to different ranges so that they do not interfere with one another.
frequency modulation (FM). Modulation in which the frequency of an alternating current is the characteristic varied.
frequency shift keying (FSK). Frequency modulation of a carrier by a signal that varies between a fixed number of discrete values.

front-end processor (FEP). A processor that can relieve a host computer of certain processing tasks, such as line control, message handling, code conversion, error control, and application functions. *See also* communications controller.
full-duplex (FDX). A mode of operation of a data link in which data may be transmitted simultaneously in both directions over two channels.
full-duplex transmission. Data transmission in both directions simultaneously on a circuit.
full motion television. Television pictures in which 30 pictures are sent every second.
function key. On a terminal, a key, such as an ATTENTION or an ENTER key, that when pressed transmits a signal not associated with a printable or displayable character. Detection of the signal usually causes the system to perform some predefined function.

Gantt chart. A project management tool that shows projects, activities or tasks (normally listed chronologically) on the left and dates across the top. Each activity is indicated by a bar on the chart that shows its starting and ending dates.
gateway. The connection between two networks that use different protocols. The gateway translates the protocols in order to allow terminals on the two networks to communicate.
Gaussian noise. *See* background noise.
general poll. A technique in which special invitation characters are sent to solicit transmission of data from all attached remote devices that are ready to send.
geosynchronous orbit. A satellite orbit that exactly matches the rotation speed of the earth. Thus, from the earth, the satellite appears to be stationary.
giga (G). One billion. For example, 1 gigahertz equals 1,000,000,000 hertz. One gigahertz also equals 1,000 megahertz and 1,000,000 kilohertz.
grade of service. A measure of the traffic handling capability of a network from the point of view of sufficiency of equipment and trunking throughout a multiplicity of nodes.
graphic character. A character that can be displayed on a terminal screen or printed on paper.
Gray code. A binary code in which sequential numbers are represented by binary expressions, each of which differs from the preceding expression in one place only.
group. Twelve 4 kilohertz voice signals multiplexed together into a 48 kHz signal.
guard band. *See* guard channel.
guard channel. The space between the primary signal and the edge of an analog channel.

hacker. A term originally denoting a technically inclined individual who enjoyed pushing computers to their limits and making them perform tasks no one thought possible. Recently, a term describing a person with mischievous, malevolent intention to access computers in order to change or destroy data or perform other unauthorized operations.
half-duplex (HDX). A mode of operation of a data link in which data may be transmitted in both directions but only in one direction at a time.
half-duplex transmission. Data transmission in either direction, one direction at a time.
hamming code. A data code that is capable of being corrected automatically.
handset. A telephone mouthpiece and receiver in a single unit that can be held in one hand.
handshake. A security technique, used on dial-up circuits, that requires that terminal hardware identify itself to the computer by automatically sending a predetermined identification code. The handshake technique is not controlled by the terminal operator.
handshaking. Exchange of predetermined signals when a connection is established between two dataset devices.
harmonic. The resultant presence of harmonic frequencies (due to nonlinear characteristics of a transmission line) in the response when a sine wave is applied.
header. The part of a data message containing information about the message, such as its destination, a sequence number, and perhaps a date or time.
help desk. The single point of contact for users when problems occur.
hertz (Hz). A unit of frequency equal to one cycle per second.
hierarchical network. A network in which processing and control functions are performed at several levels by computers specially suited for the functions performed, for example, in factory or laboratory automation.
high-level data link control (HDLC). A bit-oriented data link protocol that exercises control of data links by the use of a specified series of bits rather than by the control characters. HDLC is the protocol standardized by ISO.
high-speed circuit. A circuit designed to carry data at speeds greater than voice-grade circuits. *Synonym for* wideband circuit.
hit. A transient disturbance to a data communications medium that could mutilate characters being transmitted.
holding time. The duration of a switched call. Most often applied in traffic studies to the duration of a telephone call.
host computer. In a network, a computer that primarily provides services, such as computation, data base access, or special programs or programming languages.
hot key. A key or combination of keys that allows the user to switch from one computer or session to view information for another computer or session.
hot standby. A standby computer or telecommunications line in place ready to automatically take over in case of failure.
hub polling. A type of polling in which each station polls the next station in succession on the communications circuit. The last station polls the first station on the circuit.
hybrid network. A network made up of a combination of various network topologies.

identification (ID) characters. Characters sent by a station to identify itself. TWX, BSC, and SDLC stations use ID characters.
idle character. (1) A control character that is sent when there is

no information to be transmitted. (2) A character transmitted on a telecommunications line that does not print or punch at the output component of the accepting terminal.

idle line. *Synonym for* inactive line.

image. A faithful likeness of the subject matter of the original.

impulse noise. A sudden spike on the communications circuit when the received amplitude goes beyond a certain level, caused by transient electrical impulses, such as lightning, switching equipment, or a motor starting.

IMS/VS. (Information Management System/Virtual Storage). A database-data–communications product developed and marketed by IBM. It allows users to access a computer-maintained database through remote terminals.

in-band signaling. Signaling that occurs within the frequency range allowed for a voice signal. *Contrast with* out-of-band signaling.

in-WATS (800 service). A telephone service in which the called party pays the telephone charges. *See* wide area telecommunications service (WATS).

inactive line. A telecommunications line that is not currently available for transmitting data. *Contrast with* active line.

inactive node. In a network, a node that is neither connected to nor available for connection to another node.

information bearer channel. A channel provided for data transmission that is capable of carrying all the necessary information to permit communications, including such information as users' data synchronizing sequences and control signals. It may, therefore, operate at a greater signaling rate than that required solely for the users' data.

information bits. In data communications, those bits that are generated by the data source and that are not used for error control by the data transmission system.

information interchange. The process of sending and receiving data in such a manner that the information content or meaning assigned to the data is not altered during the transmission.

information processing. *Synonym for* data processing.

information resource management (IRM). An organization of the information-related resources of a company usually incorporating data processing, data communications, voice communications, office automation, and sometimes the company's libraries.

information security. The protection of information against unauthorized disclosure, transfer, modifications, or destruction, whether accidental or intentional.

inquiry. A request for information from storage, for example, a request for the number of available airline seats or a machine statement to initiate a search of library documents.

inquiry and transaction processing. A type of application in which inquiries and records of transactions received from a number of terminals are used to interrogate or update one or more master files.

inquiry/response communication. In a network, the process of exchanging messages and responses, with one exchange usually involving a request for information (an inquiry) and a response that provides the information. *Contrast with* batched communication.

integrated circuit. A combination of interconnected circuit elements inseparably associated on or within a continuous substrate.

Integrated Service Digital Network (ISDN). An evolving set of standards for a digital, public telephone network.

integrated video display terminal (IVDT). A terminal that combines a VDT with a telephone.

intelligent terminal. A terminal that can be programmed.

intensifying. A method for highlighting characters on the screen of a VDT for easy identification by the user. A character, or any collection of dots on an all points addressable (APA) screen, is made brighter than the other characters around it.

inter-LATA calls. Long distance calls between LATAs. Inter-LATA calls are handled by an interexchange carrier.

interactive. Pertaining to an application in which each entry calls forth a response from a system or program, as in an inquiry system or an airline reservation system. An interactive system may also be conversational, implying a continuous dialogue between the user and the system.

interconnect industry. A segment of the communications industry that makes equipment for attachment to the telephone network that provides customers with alternatives such as decorative telephones and private telephone systems for business.

interexchange carrier (IXC). Long distance carriers. *Contrast with* local exchange carrier (LEC).

interface. A shared boundary. An interface might be a hardware component to link two devices, or it might be a portion of storage or registers accessed by two or more computer programs.

intermediate text block (ITB). A character used to terminate an intermediate block of characters. The block check character is sent immediately following ITB, but no line turnaround occurs. The response following ETB or ETX also applies to all of the ITB checks immediately preceding the block terminated by ETB or ETX.

intermessage delay. The elapsed time between the receipt at a terminal of a system response and the time that a new transaction is entered. *Synonym for* think time.

internal modem. A modem contained on a single circuit card that can be inserted into a personal computer or other device.

international carrier. *See* international record carrier (IRC).

international direct distance dialing (IDDD). A telephone exchange service that enables the telephone user to call subscribers in other countries without operator assistance.

international record carriers (IRC). U.S. companies that provide communications services to other countries.

International Standards Organization (ISO). An organization established to promote the development of standards to facilitate the international exchange of goods and services and to develop mutual cooperation in areas of intellectual, scientific, technological, and economic activity.

International Telecommunications Union (ITU). The specialized telecommunications agency of the United Nations, established to provide standardized communications procedures and practices, including frequency allocation and radio regulations, on a worldwide basis.

internet. An interconnected set of networks.

interoffice trunk. A direct trunk between local central offices in the same exchange.
intertoll trunk. A trunk between toll offices in different telephone exchanges.
inverse concentrator. Equipment that takes a high-speed data stream, for example, from a computer, and breaks it apart for transmission over multiple slower-speed circuits.
invitation. The process in which a processor contacts a station in order to allow the station to transmit a message if it has one ready. *See also* polling.
invitation list. A series of sets of polling characters or identification sequences associated with the stations on a line. The order in which sets of polling characters are specified determines the order in which polled stations are invited to enter messages on the line.
isochronous transmission. A data transmission process in which there is always an integral number of unit intervals between any two significant instants.

jack. A connecting device to which a wire or wires of a circuit may be attached and that is arranged for the insertion of a plug.
jitter. Small, rapid, unwanted amplitude of phase changes in an analog signal. Small variations of the pulses of a digital signal from their ideal positions in time.
job. A set of data that completely defines a unit of work for a computer. A job usually includes all necessary computer programs, linkages, files, and instructions to the operating system.
journaling. Recording transactions against a dataset so that the dataset can be reconstructed by applying transactions in the journal against a previous version of the dataset.
joy stick. In computer graphics, a lever that can pivot in all directions and that is used as a locator device.

Kanji. A character set of symbols used in Japanese ideographic alphabets.
key. (1) On a keyboard, a control or switch by means of which a specified function is performed. (2) To enter characters or data from a keyboard.
key-encrypting key. A key used in sessions with cryptography to encipher and decipher other keys.
key system. A small, private telephone system.
keyboard. (1) On a typewriter or terminal, an arrangement of typing and function keys laid out in a specified manner. (2) A systematic arrangement of keys by which a machine is operated or by which data is entered. (3) A device for the encoding of data by key depression, which causes the generation of the selected code element. (4) A group of numeric keys, alphabetic keys, and function keys used for entering information into a terminal and into the system.
keyboard send/receive (KSR). A combination teletypewriter transmitter and receiver with transmission capability from keyboard only.

kilo. One thousand. For example, 1 kilohertz equals 1,000 hertz.

laser. A device that transmits an extremely narrow beam of energy in the visible light spectrum.
layer. (1) In the open systems interconnection (OSI) architecture, a collection of related functions that comprise one level of a hierarchy of functions. Each layer specifies its own functions and assumes that lower-level functions are provided. (2) In SNA, a grouping of related functions that are logically separate from the functions in other layers. The implementation of the functions in one layer can be changed without affecting functions in other layers.
leased circuit. A circuit that is owned by a common carrier but leased from them by another organization for full-time, exclusive use. *See also* private line.
leased circuit data transmission service. A service whereby a circuit, or circuits, of the public data network are made available to a user or group of users for their exclusive use.
leased line. *See* leased circuit.
least cost routing. Routing a telephone call so that the cost of the call is minimized.
letters shift (LTRS). A physical shift in a teletypewriter that enables the printing of lowercase characters.
level. The amplitude of a signal.
levels of support. A concept related to a support organization that suggests that the minimum skills necessary to solve a problem should be used. *See also* problem escalation.
light emitting diode (LED). A semiconductor chip that gives off visible or infrared light when activated.
lightpen. A specialized input device that is attached to a VDT by cable. It is held by the operator and pressed against the screen of the terminal to mark a spot or indicate a selection from several choices.
line. (1) On a terminal, one or more characters entered before a return to the first printing or display position. (2) A string of characters accepted by the system as a single block of input from a terminal, for example, all characters entered before a carriage return or all characters entered before the terminal user presses the ATTENTION key. (3) *See* circuit.
line-by-line mode. A mode of operation for terminals in which one line at a time is sent to or received from the computer.
line control. *Synonym for* protocol.
line feed. On a teletypewriter, one line space increment.
line feed character (LF). A format effector that causes the print or display position to move to the corresponding position on the next line. *See also* carriage-return character.
line group. One or more telecommunications lines of the same type that can be activated and deactivated as a unit.
line level. The signal level in decibels at a particular position on a telecommunications line.
line load. Usually a percentage of maximum circuit capability to reflect actual use during a span of time; for example, peak hour line load.
line noise. Noise originating in a telecommunications line.

line switching. *Synonym for* circuit switching.
line trace. In the network control program, an optional function that logs online diagnostic information.
line turnaround. A process for half-duplex transmission in which one modem stops transmitting and becomes the receiver, and the receiving modem becomes the transmitter.
linear predictive coding (LPC). A technique of digitizing an analog signal by predicting which direction the analog signal will take. LPC samples the analog signal less often than other digitizing techniques, allowing the transmitted bit rate to be lower.
link. A segment of a circuit between two points.
local access and transport area (LATA). The local calling areas that were defined originally within the United States when divestiture occurred.
local area network (LAN). A limited distance network, usually existing within a building or several buildings in close proximity to one another. Transmission on a LAN normally occurs at speeds of 1 Mbps and up.
local calls. Calls within a local service area.
local central office. A central office arranged for terminating subscriber lines and provided with trunks of establishing connections to and from other central offices.
local exchange carrier (LEC). The BOCs and the independent telephone companies.
local loop. A channel connecting the subscriber's equipment to the line-terminating equipment in the central office exchange.
local service area. Telephones served by a particular central office and (usually) several surrounding central offices.
lockout. In a telephone circuit controlled by an echo suppressor, the inability of one or both subscribers to get through because of either excessive local circuit noise or continuous speech from one subscriber.
logical circuit. In packet mode operation, a means of duplex transmission across a data link, comprising associated send and receive channels. A number of logical circuits may be derived from a data link by packet interleaving. Several logical circuits may exist on the same data link.
logical link control (LLC). The data link control protocol defined by the IEEE for use on LANs.
logical unit (LU). SNA's view of users of a communications user.
logoff. The procedure by which a user ends a terminal session.
logon. The procedure by which a user begins a terminal session.
long distance calls. Calls outside of the local service area.
longitudinal parity check. (1) A parity check performed on a group of binary digits in a longitudinal direction for each track. (2) A system of error checking performed at the receiving station after a block check character has been accumulated.
longitudinal redundancy check (LRC). *Synonym for* longitudinal parity check.
loop back test. A procedure in which signals are looped from a test instrument through a modem or loopback switch and back to the test instrument for measurement.
loop network. A network configuration in which there is a single path between all nodes, and the path is a closed circuit.
low speed. Usually, a data transmission speed of 600 bps or less.
low-speed circuit. A circuit that is designed for telegraph and teletypewriter usage at speeds of from 45 to 200 bps and that cannot handle a voice transmission. Used by the public telex network. *Synonym for* subvoice-grade circuits.

main distribution frame. A frame that has one part on which the permanent outside lines entering the central office building terminate and another part on which cabling, such as the subscriber line cabling or trunk cabling, terminates. In a PBX, the main distribution frame is for similar purposes.
manufacturing automation protocol (MAP). A communications protocol, based on the OSI reference model, specifically oriented toward use in an automated manufacturing environment.
mark. The normal no-traffic line condition by which a steady signal is transmitted.
master group. Ten supergroups, each of which contains 60 voice channels. The bandwidth of a master group is 2.4 MHz.
master station. *See* control station.
mean time between failures (MTBF). For a stated period in the life of a function unit, the mean value of the lengths of time between consecutive failures under stated conditions.
mean time to repair (MTTR). The average time required for corrective maintenance.
measured rate service. A method of charging for local calls based on the number of calls, their duration, and the distance.
medium. *See* transmission medium.
medium speed. Usually, a data transmission rate between 600 bps and the limit of a voice-grade facility.
mega. One million. For example, 1 megahertz equals 1,000,000 hertz. Also 1 megahertz equals 1,000 kilohertz.
mesh network. A network configuration in which there are two or more paths between any two nodes.
message. (1) An arbitrary amount of information whose beginning and end are defined or implied. (2) A group of characters and control bit sequences transferred as an entity.
message center. A location where messages are received from a communications network and either forwarded to another location or delivered to the intended recipient.
message queue. A line of messages that are awaiting processing or waiting to be sent to a terminal.
message routing. The process of selecting the correct circuit path for a message.
message switching. (1) In a data network, the process of routing messages by receiving, storing, and forwarding complete messages. (2) The technique of receiving a complete message, storing it, and then forwarding it to its destination unaltered.
message text. The part of a message that is of concern to the party ultimately receiving the message, that is, the message exclusive of the header or control information.
metropolitan area network (MAN). A network of limited geographic scope, generally defined as within a 50-mile radius. Standards for MANs are being defined by the IEEE.
micro. One millionth.
microcomputer. A computer system whose processing unit is a microprocessor. A basic microcomputer includes a microproces-

sor, storage, and an input/output facility, which may or may not be on one chip.

microprocessor. An integrated circuit that accepts coded instructions for execution. The instructions may be entered, integrated, or stored internally.

microwave radio. Radio transmissions in the 4 to 28 gHz range. Microwave radio transmissions require that the transmitting and receiving antennas be within sight of each other.

milli. One thousandth.

modem (modulator–demodulator). A device that modulates and demodulates signals transmitted over data communications lines. One of the functions of a modem is to enable digital data to be transmitted over analog transmission facilities.

modified final judgment (MFJ). The stipulation that on January 1, 1984, AT&T would divest itself of all 22 of its associated operating companies in the Bell System.

modulation. The process by which some characteristic of one wave is varied in accordance with another wave or signal. This technique is used in modems to make DTE signals compatible with communications facilities.

modulation rate. The reciprocal of the measure of the shortest nominal time interval between successive significant instants of the modulated signal. If this measure is expressed in seconds, the modulation rate is given in bauds.

modulator. A functional unit that converts a signal into a modulated signal suitable for transmission. *Contrast with* demodulator.

monitor. Software or hardware that observes, supervises, controls, or verifies the operations of a system.

mouse. In computer graphics, a locator device operated by moving it on a surface.

multidrop circuit. *See* multipoint circuit.

multimode. A type of fiber used in fiberoptics with a core approximately 50 microns (.050 millimeters) in diameter.

multiplexer (MUX). A device capable of interleaving the events of two or more activities or capable of distributing the events of an interleaved sequence to the respective activities.

multiplexing. (1) In data transmission, a function that permits two or more data sources to share a common transmission medium such that each data source has its own channel. (2) The division of a transmission facility into two or more channels either by splitting the frequency band transmitted by the channel into narrower bands, each of which is used to constitute a distinct channel (frequency division multiplexing), or by allotting this common channel to several different information channels, one at a time (time division multiplexing).

multipoint circuit. A circuit with several nodes connected to it.

multiport. *See* split steam operation.

negative acknowledge character (NAK). A transmission control character transmitted by a station as a negative response to the station with which the connection has been set up.

negative polling limit. For a start-stop or BSC terminal, the maximum number of consecutive negative responses to polling that the communications controller accepts before suspending polling operations.

network. (1) An interconnected group of nodes. (2) The assembly of equipment through which connections are made between data stations.

network addressable unit (NAU). In SNA, a logical unit, a physical unit, or a system services control point. It is the origin or the destination of information transmitted by the path control network.

network application. The use to which a network is put, such as data collection or inquiry/update.

network architecture. A set of design principles, including the organization of functions and the description of data formats and procedures, used as the basis for design and implementation of a telecommunications application network.

network congestion. A network condition when traffic is greater than the network can carry, for any reason.

network control center (NCC). A place from which a communications network is operated and monitored.

network control mode. The functions of a network control program that enable it to direct a communications controller to perform activities, such as polling, device addressing, dialing, and answering.

network control program (NCP). A program that controls the operation of a front-end processor or communications controller.

network node. *Synonym for* node.

network operator. A person or program responsible for controlling the operation of all or part of a network.

network operator console. A system console or terminal in the network from which an operator controls the network.

network simulator. *See* workload generator.

network topology. The schematic arrangement of the links and nodes of a network.

node. In a network, a point at which one or more functional units interconnect transmission lines. The term *node* derives from graph theory, in which a node is a junction point of links, areas, or edges.

noise. (1) Random variations of one or more characteristics of any entity, such as voltage, current, or data. (2) A random signal of known statistical properties of amplitude, distribution, and spectral density. (3) Loosely, any disturbance tending to interfere with the normal operation of a device or system.

nondestructive cursor. On a VDT device, a cursor that does not erase characters through which it passes as it is advanced, backspaced, or otherwise moved. *Contrast with* destructive cursor.

noninformation bits. In data communications, those bits that are used for error control by the data transmission system.

nonswitched connection. A connection that does not have to be established by dialing. *Contrast with* switched connection.

nontransparent mode. A mode of transmission in which all control characters are treated as control characters (that is, not treated as text). *Contrast with* transparent text mode.

null character (NUL). A control character that is used to accomplish media-fill or time-fill and that may be inserted into or removed from a sequence of characters without affecting the meaning of the sequence. However, the control of equipment or the format may be affected by this character.

numbering plan. A uniform numbering system in which each telephone central office has a unique designation similar in form to that of all other offices connected to the nationwide dialing network. In the numbering plan, the first 3 of 10 dialed digits are the area code, the next 3, the office code; and the remaining 4, the station number.

numeric keypad. Extra keys on a keyboard that function like a 10-key calculator.

off-hook. Activated (in regard to a telephone set). By extension, a data set automatically answering on a public switched system is said to go off-hook. *Contrast with* on-hook.

offline. Pertaining to the operation of a functional unit without the continual control of a computer.

offline system. A system in which human operations are required between the original recording functions and the ultimate data processing function. This includes conversion operations as well as the necessary loading and unloading operations incident to the use of point-to-point or data-gathering systems. *Contrast with* online system.

on-hook. Deactivated (in reference to a telephone set). A telephone not in use is on-hook. *Contrast with* off-hook.

one-way communication. Communications in which information is always transferred in one preassigned direction.

one-way trunk. A trunk between central exchanges where traffic can originate on only one end.

online. (1) The state of being connected, usually to a computer. (2) Pertaining to the operation of a functional unit that is under the continual control of a computer.

online system. A system in which the input data enters the computer directly from the point of origin or in which output data is transmitted directly to where it is used.

open network architecture (ONA). A set of provisions imposed by the FCC on the BOCs and AT&T to ensure the competitive availability of and access to unregulated, enhanced network services.

Open System Interconnection (OSI) reference model. A telecommunications architecture proposed by the International Standards Organization (ISO).

open wire. (1) A conductor separately supported above the surface of the ground; that is, on insulators. (2) A broken wire.

open-wire line. A pole line in which the conductors are principally in the form of bare, uninsulated wire. Ceramic, glass, or plastic insulators are used to physically attach the bare wire to the telephone poles. Short circuits between the individual conductors are avoided by appropriate spacing.

operating system. The central control program that governs a computer hardware's operation.

optical character recognition (OCR). The process of scanning a document with a beam of light and detecting individual characters.

optical fiber. A communications medium made of very thin glass or plastic fiber that conducts light waves.

optical recognition. A device that can detect individual data items or characters and convert them into ASCII or another code for transmission to a computer.

oscilloscope. An instrument for displaying the changes in a varying current or voltage.

other common carrier (OCC). *See* specialized common carrier (SCC).

out-of-band signaling. Signaling outside of the frequency range allowed for a voice signal. *Contrast with* in-band signaling.

out-pulsing. The pulses caused by a rotary dial opening and closing an electrical circuit when the dial is turned and released.

out-WATS. A WATS capability that provides for outgoing calls only. *See also* wide area telecommunications service (WATS).

overrun. Loss of data because a receiving device is unable to accept data at the rate it is transmitted.

pacing. A technique by which a receiving station controls the rate of transmission of a sending station to prevent overrun.

packet. A sequence of binary digits (including data and call control signals) that is switched as a composite whole. The data, call control signals, and possibly error control information are arranged in a specific format.

packet data network (PDN). A network that uses packet switching techniques for transmitting data.

packet sequencing. A process of ensuring that packets are delivered to the receiving data terminal equipment (DTE) in the same sequence as they were transmitted by the sending DTE.

packet switched data transmission service. A user service involving the transmission and, if necessary, the assembly and disassembly of data in the form of packets.

page mode. *See* formatted mode.

parallel transmission. (1) In data communications, the simultaneous transmission of a certain number of signal elements constituting the same telegraph or data signal. (2) The simultaneous transmission of the bits constituting an entity of data over a data circuit. *Contrast with* serial transmission.

parity bit. The binary digit appended to a group of binary digits to make the sum of all the digits either always odd (odd parity) or always even (even parity).

parity check. A redundancy check that uses a parity bit.

patch. (1) A temporary electrical connection. (2) To make an improvised modification.

patch cord. A cable with plugs at both ends that is used to connect two devices together.

patch panel. Equipment that allows a piece of equipment to be temporarily connected to other equipment or a circuit using patch cords.

path. In a network, a route between any two nodes.

peer-to-peer. The ability of two computers to communicate directly without passing through or using the capability of a mainframe computer.

pel. *See* picture element (pel, pixel).

performance management. The application of management principles to ensure that the performance of a system meets the required parameters.

phase. An attribute of an analog signal that describes its relative position measured in degrees.

phase jitter. An unwanted change in the phase of the signal.

phase modulation. Modulation in which the phase angle of a carrier is the characteristic varied.

phase shift. The offset of an analog signal from its previous location. Phase shifts are measured in degrees.
phase shift keying (PSK). A modulation technique in which the phase of an analog signal is varied.
physical unit (PU). SNA's view of communications hardware.
picture element (pel, pixel). (1) The part of the area of the original document that coincides with the scanning spot at a given instant and that is of one intensity only, with no distinction of the details that may be included. (2) In computer graphics, the smallest element of a display space that can be independently assigned color and intensity. (3) The area of the finest detail that can be effectively reproduced on the recording medium.
pixel. *See* picture element.
plain old telephone service (POTS). Basic telephone service with no special features.
point of presence (POP). The location within a LATA at which customers are connected to an IXC.
point of sale (POS) terminal. A specialized terminal designed to be used by a clerk in a store and used to enter sales transactions into a computer.
point-to-point circuit. A circuit connecting two nodes. *Contrast with* multipoint circuit.
Poisson capacity table. A table for determining the number of circuits required to carry a certain level of telephone traffic. The Poisson table assumes that the sources of traffic are infinite and that all blocked calls will be retried within a short period of time.
polling. (1) Interrogation of devices for purposes such as to avoid contention, to determine operational status, or to determine readiness to send or receive data. (2) The process in which stations are invited, one at a time, to transmit.
polling characters. A set of characters peculiar to a terminal and the polling operation. Response to these characters indicates to the computer whether the terminal has a message to enter.
polling delay. A user-specified delay between polling passes through an invitation list for either a line or a line group.
polling ID. The unique character or characters associated with a particular station.
polling list. A list that specifies the sequence in which stations are to be polled.
polynomial error checking. An error checking technique in which the bits of a block of data are processed by a mathematical algorithm using a polynomial function to calculate the block check character.
port. An access point for a circuit.
Post, Telephone and Telegraph Administration (PTT). A generic term for the government-operated common carriers in countries other than the USA and Canada.
preventive software maintenance. A philosophy of software maintenance that advocates the application of software corrections, whether the problem being corrected has been seen or not. The intention is to correct software problems before they are seen by users.
private automatic branch exchange (PABX). A private automatic telephone exchange that provides for the transmission of calls to and from the public telephone network. *See also* private branch exchange (PBX).
private branch exchange (PBX). A private telephone exchange connected to the public telephone network on the user's premises.
private line. (1) A communications circuit that is leased from a common carrier. (2) A communications circuit that is owned by a company other than a common carrier.
private network. A network built by a company for its exclusive use using circuits available from a variety of sources.
problem escalation. The process of bringing a problem to the attention of higher levels of management and/or bringing more highly trained technical resources to work on the problem.
problem management. The application of management principles to ensure that problems are resolved as quickly as possible with the minimum resources.
problem tracking meeting. A periodic meeting to discuss the status of all open problems that have not yet been resolved.
profit center. A department or other entity that generates more revenue than it spends. Profit centers typically sell their services or recharge them to other departments in the company. *Contrast with* cost center.
program evaluation review technique (PERT). A project management technique that shows on a chart the interrelationships of activities.
program function keys (PFKs). Special keys on a terminal keyboard that direct the computer to perform specific actions determined by the computer program.
project team. A group of people organized for the purpose of completing a project.
propagation delay. The time necessary for a signal to travel from one point to another.
protocol. (1) A specification for the format and relative timing of information exchanged between communicating parties. (2) The set of rules governing the operation of functional units of a communications system that must be followed if communications are to be achieved.
protocol analyzer. Test equipment that examines the bits on a communications circuit to determine whether the rules of a particular protocol are being followed.
protocol converter. Hardware or software that converts a data transmission from one protocol to another.
protocol data unit (PDU). A frame of logical link control protocol.
public data transmission service. A data transmission service established and operated by an administration and provided by means of a public data network. Circuit-switched, packet-switched, and leased circuit data transmission services are feasible.
public network. A network established and operated by communications common carriers or telecommunications administrations for the specific purpose of providing circuit-switched, packet-switched, and leased circuit services to the public.
public service commission (PSC). *See* public utility commission (PUC).
public switched network (PSN). A network that provides circuits switched to many customers. In the United States, there are four: telex, TWX, telephone, and broadband exchange.
public switched telephone network (PSTN). *See* public switched network (PSN).
public telephone network. *See* public switched network (PSN).

public utility commission (PUC). An arm of state government that has jurisdiction over intrastate rates and services. *Also known as* public service commission (PSC).
pulse. A variation in the value of a quantity, short in relation to the time schedule of interest, the final value being the same as the initial value.
pulse code modulation (PCM). A process in which a signal is sampled, and the magnitude of each sample with respect to a fixed reference is quantized and converted by coding to a digital signal.
punchdown block. A connector for telephone wiring on which the connection is made by pushing (punching) the wire between two prongs with a special tool.
punched paper tape. A medium used on older teletypewriters. Characters were coded in the tape by punched holes.
pushbutton dialing. The use of keys or pushbuttons instead of a rotary dial to generate a sequence of digits to establish a circuit connection. The signal form is usually tones. *Contrast with* rotary dial.
pushbutton dialing pad. A 12-key device used to originate tone keying signals. It usually is attached to rotary dial telephones for use in originating data signals.

quadbit. A group of four bits. In 16-phase modulation, each possible quadbit is encoded as one of 16 unique carrier phase shifts. *Contrast with* dibit *and* tribit.
quadrature amplitude modulation (QAM). A combination of phase and amplitude modulation used to achieve high data rates while maintaining relatively low signaling rates.
quantization. The subdivision of the range of values of a variable into a finite number of nonoverlapping, and not necessarily equal, subranges or intervals, each of which is represented by an assigned value within the subrange. For example, a person's age is quantized for most purposes with a quantum of 1 year.
quantizing noise. The error introduced when an analog signal is digitized.
queue. A line or list formed by items in a system waiting for service, for example, tasks to be performed or messages to be transmitted in a message routing system.
queuing. The process of placing items that cannot be handled into a queue to await service.

RS-232-C. A specification for the physical, mechanical, and electrical interface between data terminal equipment (DTE) and circuit-terminating equipment (DCE). *See also* V.24.
RD-232-D. A 1987 revision and update to the RS-232-C specification that is exactly compatible with the V.24 standard.
RS-336. A specification for the interface between a modem (DCE) and a terminal or computer (DTE). This interface, unlike the RS-232-C, has a provision for the automatic dialing of calls under modem control.
RS-449. A specification for the interface between a modem (DCE) and a terminal or computer (DTE). This specification was designed to overcome some of the problems with the RS-232-C interface specification.
radio paging. The broadcast of a special radio signal that activates a small portable receiver carried by the person being paged.
rate center. A specified geographic location used by telephone companies to determine mileage measurements for the application of interexchange mileage rates.
real enough time. Response time that is fast enough to meet the requirements of a particular application.
realtime. Pertaining to an application in which response to input is fast enough to affect subsequent input, such as a process control system or a computer assisted instruction system.
receive-only (RO) device. A teletypewriter that has no keyboard. It is used where no input to the computer is desired or necessary.
reed relay. A switch that has contacts that open or close when an electrical current is applied.
regenerative repeater. *See* repeater.
Regional Bell Operating Company (RBOC). One of the seven corporations formed when divestiture occurred and that comprise the 22 Bell Operating Companies.
regional center. A control center (class 1 office) connecting sectional centers of the telephone system together. Every pair of regional centers in the United States has a direct circuit group running from one center to the other.
relative transmission level. The ratio of the test-tone power at one point to the test-tone power at some other point in the system chosen as a reference point. The ratio is expressed in decibels. The transmission level at the transmitting switchboard is frequently taken as zero level reference point.
relay center. A central point at which message switching takes place; a message switching center.
reliability. Trouble-free operation.
remote batch. Data collected in a batch and then transmitted to the computer as a unit.
remote job entry (RJE). The process of submitting a job to a computer for processing using telecommunications lines. Normally the output of the processing is returned on the lines to the terminal from which the job was submitted.
remote terminal. A terminal attached to a computer via a telecommunications line.
repeater. A device that performs digital signal regeneration together with ancillary functions. Its function is to retime and retransmit the received signal impulses restored to their original shape and strength.
request for proposal (RFP). A letter or document sent to vendors asking them to show how a (communications) problem or situation can be addressed. Normally, the vendor's response to an RFP proposes a solution and quotes estimated prices.
request for quotation (RFQ). *See* request for proposal.
response. An answer to an inquiry.
response time. The elapsed time between the end of an inquiry or demand on a data processing system and the beginning of the response, for example, the length of time between an indication of the end of an inquiry and the display of the first character of the response at a user terminal.
reverse channel. A means of simultaneous communications from the receiver to the transmitter over half-duplex data transmission systems. The reverse channel is generally used only for the transmission of control information and operates at a much slower speed than the primary channel.

reverse video. A technique used with VDTs that reverses the character and background colors for highlighting purposes.
ring. (1) The signal made by a telephone to indicate an incoming call. (2) A part of a plug used to make circuit connections in a manual switchboard or patch panel. The ring is the connector attached to the negative side of the common battery that powers the station equipment. By extension, it is the negative battery side of a telecommunications line. *Contrast with* tip.
ring network. A network in which each node is connected to two adjacent nodes.
ringback tone. An audible signal indicating that the called party is being rung.
roll call polling. The most common implementation of a polling system, in which one station on a line is designated as the master and the others are slaves.
rotary dial. In a switched system, the conventional dialing method that creates a series of pulses to identify the called station. *Contrast with* pushbutton dialing *and* dual-tone-multifrequency.
router. A piece of hardware or software that directs messages toward their destination, often from one network to another.

satellite carrier. A company that offers communications services using satellites.
scrambler. A voice encryption device that makes the voice unintelligible to anyone without a descrambler, effectively rendering wiretapping useless.
screen. An illuminated display surface; for example, the display surface of a VDT or plasma panel.
seize. To gain control of a line in order to transmit data. *Contrast with* bid.
serial. (1) Pertaining to the sequential performance of two or more activities in a single device. In English, the modifiers serial and parallel usually refer to devices, as opposed to sequential and consecutive, which refer to processes. (2) Pertaining to the sequential processing of the individual parts of a whole, such as the bits of a character or the characters of a word, using the same facilities for successive parts.
serial system. A system made up of a number of components connected in series. In telecommunications, a terminal, modem, and computer may be connected in a series forming a serial system.
serial transmission. (1) In data communications, transmission at successive intervals of signal elements constituting the same telegraph or data signal. The sequential elements may be transmitted with or without interruption, provided that they are not transmitted simultaneously. (2) The sequential transmission of the bits constituting an entity of data over a data circuit. *Contrast with* parallel transmission.
server. On a local area network, a computer with software that provides service to other devices on the LAN. Typical servers are file servers, print servers, and communications servers.
service level agreement. Performance objectives reached by consensus between the user and the provider of a service.
serving central office. A telephone subscriber's local, Class 5 central office.
session. (1) A connection between two stations that allows them to communicate. (2) The period of time during which a user of a terminal can communicate with an interactive system; usually, the elapsed time between logon and logoff.
shielded twisted pair. Twisted pair wires surrounded by a metallic shield.
shielding. A metallic sheath that surrounds the center conductor of a cable. Coaxial cable has shielding around the center conductor.
sidetone. The small amount of signal fed back from the mouthpiece to the receiver of a telephone handset.
signal. A variation of a physical quantity, used to convey data.
signal-to-noise ratio. The ratio of signal strength to noise strength.
signal transformation. The action of modifying one or more characteristics of a signal, such as its maximum value, shape, or timing.
signaling rate. The number of times per second that a signal changes. Signaling rate is measured in baud.
simplex circuit. A circuit that carries communication in one direction only.
simplex communication. *Synonym for* one-way communication.
simultaneous transmission. Transmission of control characters or data in one direction while information is being received in the other direction.
sine wave. The waveform of a single-frequency analog signal of constant amplitude and phase.
single-address message. A message that is to be delivered to only one destination.
single mode. A type of optical fiber that has a glass or plastic core approximately 5 microns (.005 millimeters) in diameter.
sink. In telecommunications, the receiver.
slave station. A data station that operates under the control of a master or control station. *Contrast with* control station.
smart terminal. A terminal that is not programmable but that has memory capable of loading with information.
source. The transmitting station in a telecommunications system.
specialized common carrier (SCC). Companies that offer alternative communications service to that of AT&T and the BOCs. *Also known as* other common carrier (OCC).
specific polling. A polling technique that sends invitation characters to a device to find out whether the device is ready to enter data.
split stream operation. A modem feature that allows several slower-speed data streams to be combined into one higher-speed data stream for transmission, and split apart at the receiving end. The total data rate of the slower-speed data streams must not exceed the capacity of the circuit.
spooling. A technique of queuing input or output between slow-speed and high-speed computer hardware. Print spooling is most common. Output from several computers is queued (spooled) to a disk until a printer is free.
standard test-tone power. One milliwatt (0 dBm) at 1,000 cycles per second.
star network. A network configuration in which there is only one path between a central or controlling node and each endpoint node.
start bit. *See* start signal.
start-of-heading character (SOH). A transmission control character used as the first character of a message heading.
start-of-text character (STX). A transmission control character

that precedes text and that may be used to terminate the message heading.

start signal. (1) A signal to a receiving mechanism to get ready to receive data or perform a function. (2) In a start-stop system, a signal preceding a character or block that prepares the receiving device for the reception of the code elements. A start signal is limited to one signal element generally having the duration of a unit interval. *Synonym for* start bit.

start-stop transmission. (1) Asynchronous transmission such that a group of signals representing a character is preceded by a start element and is followed by a stop element. (2) Asynchronous transmission in which a group of bits is preceded by a start bit that prepares the receiving mechanism for the reception and registration of a character and is followed by at least one stop bit that enables the receiving mechanism to come to an idle condition pending the reception of the next character.

station. One of the input or output points of a system that uses telecommunications facilities; for example, the telephone set in the telephone system or the point at which the business machine interfaces with the channel on a leased private line.

station features. Features of a telephone system that are activated by the user of the system.

station message detail recording (SMDR). A feature of a telephone system that records detailed information about telephone calls placed through the system.

station selection code. A Western Union term for an identifying call that is transmitted to an outlying telegraph receiver and automatically turns its printer on.

statistical time division multiplexing (STDM). A device that combines signals from several terminals. An STDM does not reserve specific time slots for each device but assigns time only when a device has data to send.

step-by-step switch. A switch that moves in synch with a pulse device, such as a rotary telephone dial. Each digit dialed moves successive selector switches to carry the connection forward until the desired line is reached.

stop and wait ARQ. An error checking technique in which each block of data must be acknowledged before the next block can be sent. *Contrast with* continuous ARQ.

stop bit. In start-stop transmission, the bit that indicates the end of a character. *See also* start-stop transmission.

stop signal. (1) A signal to a receiving mechanism to wait for the next signal. (2) In a start-stop system, a signal following a character or block that prepares the receiving device for the reception of a subsequent character or block.

store-and-forward system. An application in which input is transmitted, usually to a computer, stored, and then later delivered to the recipient.

stress testing. Placing a heavy load on a system to see if it performs properly.

Strowger switch. A step-by-step switch named after its inventor, Almon B. Strowger. *See also* step-by-step switch.

subnetwork. (1) In the OSI reference model, layers 1, 2, and 3 together constitute the subnetwork. (2) A portion of a network.

subordinate station. *See* slave station.

subscriber's loop. *See* local loop.

subvoice-grade circuit. A circuit of bandwidth narrower than that of voice-grade circuits. Such circuits are usually subchannels of a voice-grade line.

supergroup. Five 48 kHz groups of 12 voice channels each. The total bandwidth of a supergroup is 240 kHz.

superhighway effect. A concept that suggests that the capacity of new facilities is often exceeded faster than anticipated because users, finding the capability better than expected, use it to a greater extent or for purposes beyond those for which it was originally designed.

switched connection. (1) A mode of operating a data link in which a circuit or channel is established to switching facilities, as, for example, in a public switched network. (2) A connection that is established by dialing. *Contrast with* nonswitched connection.

switched line. A telecommunications line in which the connection is established by dialing. *See also* dial-up line *and* leased circuit.

switched telecommunications network. A switched network furnished by communication common carriers or telecommunications administrations.

switchhook. A switch on a telephone set, associated with the structure supporting the receiver or handset. It is operated by the removal or replacement of the receiver or handset on the support.

switching center. A location that terminates multiple circuits. It is capable of interconnecting circuits or transferring traffic between circuits.

synchronization character (SYN). In a data message, a character that is inserted in the data stream from time to time by the transmitting station to ensure that the receiver is maintaining character synchronization and properly grouping the bits into characters. The synchronization characters are removed by the receiver and do not remain in the received message.

synchronous. (1) Pertaining to two or more processes that depend on the occurrences of specific events, such as common timing signals. (2) Occurring with a regular or predictable time relationship.

synchronous data link control (SDLC). A bit-oriented data link protocol developed by IBM. SDLC is a proper subset of HDLC.

synchronous line control. A scheme of operating procedures and control signals by which telecommunications lines are controlled.

synchronous transmission. (1) Data transmission in which the time of occurrence of each signal representing a bit is related to a fixed time frame. (2) Data transmission in which the sending and receiving instruments are operating continuously at substantially the same frequency and are maintained, by means of correction, in a desired phase relationship.

system features. Features of a telephone system that are available to all users and that may be automatically activated on behalf of the user.

Systems Network Architecture (SNA). A seven-layer communications architecture developed by IBM to serve as a basis for future telecommunications products.

T-carrier system. A family of high-speed, digital transmission systems, designated according to their transmission capacity.

tariff. The published rate for a specific unit of equipment, facility, or type of service provided by a telecommunications carrier. Also, the vehicle by which the regulating agencies approve or disapprove such facilities or services. Thus, the tariff becomes a contact between the customer and the telecommunications facility.

technical office protocol (TOP). A communications architecture, based on the OSI reference model, specifically oriented toward office automation.

telecommunications. (1) Any transmission, emission, or reception of signs, signals, writing, images, and sounds or intelligence of any nature by wire, radio, optical, or other electromagnetic systems. (2) Communication, as by telegraph or telephone.

telecommunications access method (TCAM). Communications software that controls communications lines. *See also* communications access method (CAM).

telecommunications control unit. *See* front-end processor (FEP).

telecommunications facility. Transmission capabilities or the means for providing such capabilities.

telecommunications monitor (TCM). Computer software that governs the overall operation of a network and may provide transaction processing or other services.

telecommuting. Using telecommunications to work from home or other locations instead of on the business's premises.

telediagnosis. Using telecommunications to diagnose a problem from a remote location.

telegraph. A system employing the interruption or change in polarity of direct current for the transmission of signals.

telegraph-grade circuit. A circuit suitable for transmission by teletypewriter equipment. Normally, the circuit is considered to employ DC signaling at a maximum speed of 75 baud.

telephone company. Any common carrier providing public telephone system service.

telephony. Transmission of speech or other sounds.

teleprinter. Equipment used in a printing telegraph system.

teleprocessing. Remote access data processing.

teletex. A standardized communications messaging technology allowing automatic, error-free transmissions between terminals at speeds 48 times greater than telex. Teletex is the logical successor to telex and TWX.

Teletype. Trademark of AT&T, usually referring to a series of teleprinter equipment, such as tape punches, reperforators, and page printers, that is used for telecommunications.

teletypewriter. A slow-speed terminal with a keyboard for input and paper for receiving printed output.

teletypewriter exchange service (TWX). Teletypewriter service provided by Western Union in which suitably arranged teletypewriter stations are provided with lines to a central office for access to other such stations throughout the United States and Canada. Both Baudot and ASCII coded machines are used. Business machines may also be used, with certain restrictions.

telex network. An international public messaging service using slow-speed teletypewriter equipment and Baudot code to exchange messages between subscribers. In the United States, telex service is provided by Western Union.

terminal. (1) A device, usually equipped with a keyboard and a display device, capable of sending and receiving information over a link. (2) A point in a system or network at which data can either enter or leave.

terminal component. A separately addressable part of a terminal that performs an input or output function, such as the display component of a keyboard-display device or a printer component of a keyboard-printer device.

terminal emulation program. A program that makes a personal computer act like a terminal that is recognized by the host.

terminal session. *See* session.

test tone. A tone used in identifying circuits for trouble location or for circuit adjustment.

text. The part of a data message containing the subject matter of interest to the user.

think time. The time between the receipt of a message at a terminal until the next message is entered by the user. *Synonym for* intermessage delay *and for* user response time.

tie line. *See* tie trunk.

tie trunk. A telephone line or channel directly connecting two branch exchanges.

time assignment speech interpolation (TASI). A technique of multiplexing telephone calls by taking advantage of the pauses in normal speech and assigning the channel to another call during the pause.

time division multiplexing (TDM). A technique that divides a circuit's capacity into time slots, each of which is used by a different voice or data signal.

timesharing. (1) Pertaining to the interleaved use of time on a computer system that enables two or more users to execute computer programs concurrently. (2) A mode of operation of a data processing system that provides for the interleaving in time of two or more processes in one processor. (3) A method of using a computing system that allows a number of users to execute programs concurrently and to interact with the programs during execution.

tip. The end of the plug used to make circuit connections in a manual switchboard or patch panel. The tip is the connector attached to the positive side of the common battery that powers the station equipment. By extension, it is the positive battery side of a telecommunications line. *Contrast with* ring.

token. A particular character in a token-oriented protocol. The terminal that has the token has the right to use the communications circuit.

toll. In public switched systems, a charge for a connection beyond an exchange boundary, based on time and distance.

toll calls. Calls outside a local service area.

toll office. A central office at which channels and toll circuits terminate. Whereas this is usually one particular central office in a city, larger cities may have several central offices where toll message circuits terminate. A class 4 office.

toll trunk. A communications circuit between telephone company toll offices.

tone dialing. *See* dual-tone-multifrequency (DTMF).

tone signaling. Signaling performed by sending tones on a circuit.

topology. The way in which a network's circuits are configured.

torn tape message system. Deprecated name for a teletypewriter-based message switching system in which paper tape was torn off one teletypewriter and read into another.

torn tape switching center. A location where operators tear off the incoming printed and punched paper tape and transfer it manually to the proper outgoing circuit.

touch-sensitive screen. A VDT screen that can detect the location of the user's finger using either a photosensitive or resistive technique.

Touchtone. AT&T's tradename for dual-tone-multifrequency dialing.

trackball. In computer graphics, a ball, movable about its center, that is used as a locator device.

traffic. Transmitted and received messages.

trailer. The part of a data message following the text.

transaction. An exchange between a terminal and another device, such as a computer, that accomplishes a particular action or result; for example, the entry of a customer's deposit and the updating of the customer's balance.

transaction processing systems. Systems in which the users run prewritten programs to perform business transactions, generally of a somewhat repetitive nature.

transceiver. A terminal that can transmit and receive traffic.

transient error. An error that occurs once or at unpredictable intervals.

transition. The switching from one state (for example, positive voltage) to another (negative voltage).

transmission. (1) The process of sending data from one place for reception elsewhere. (2) In data communications, a series of characters including headings and texts. (3) The process of dispatching of a signal, message, or other form of intelligence by wire, radio, telegraphy, telephony, facsimile, or other means. (4) One or more blocks or messages. Note: Transmission implies only the sending of data; the data may or may not be received.

transmission code. A code for sending information over telecommunications lines.

transmission control character. (1) Any control character used to control or facilitate transmission of data between data terminal equipment (DTEs). (2) Characters transmitted over a line that are not message data but that cause certain control operations to be performed when encountered. Among such operations are addressing, polling, message delimiting and blocking, transmission error checking, and carriage return.

Transmission Control Protocol/Internet Protocol (TCP/IP). A set of transmission rules for interconnecting communications networks. TCP/IP is heavily supported by the United States government.

transmission control unit. *See* front-end processor (FEP).

transmission efficiency. The ratio of information bits to total bits transmitted.

transmission medium. Any material substance that can be, or is, used for the propagation of signals, usually in the form of modulated radio, light, or acoustic waves, from one point to another, such as an optical fiber, cable, bundle, wire, dielectric slab, water, or air. *Note:* Free space can also be considered as a transmission medium for electromagnetic waves.

transmit. (1) To send data from one place for reception elsewhere. (2) To move an entity from one place to another; for example, to broadcast radio waves, to dispatch data via a transmission medium, or to transfer data from one data station to another via a line.

transnational data flow (TNDF). The transmission of data across national borders.

transparent data. Data that is not recognized as containing transmission control characters. Transparent data is sometimes preceded by a control byte and a count of the amount of data following.

transparent text mode. A mode of binary synchronous text transmission in which data, including normally restricted data link control characters, are transmitted only as specific bit patterns. Control characters that are intended to be effective are preceded by a DLE character. *Contrast with* nontransparent mode.

trellis code modulation (TCM). A specialized form of quadrature amplitude modulation that codes the data so that many bit combinations are invalid. TCM is used for high-speed data communications.

tribit. A group of three bits. In eight-phase modulation, each possible tribit is encoded as 1 of 8 unique carrier phase shifts. *Contrast with* dibit *and* quadbit.

trouble ticket. An online record or paper form that is filled out to document the symptoms or a problem and the action being taken to correct it.

trunk. A telephone channel between two central offices or switching devices that is used in providing a telephone connection between subscribers.

trunk group. Those trunks between two switching centers, individual message distribution points, or both, that use the same multiplex terminal equipment.

tuning. The process of making adjustments to a system to improve its performance.

turnaround time. The actual time required to reverse the direction of transmission from send to receive or vice versa when a half-duplex circuit is used. For most telecommunications facilities, there will be time required by line propagation and line effects, modem timing, and machine reaction. A typical time is 200 milliseconds on a half-duplex telephone connection.

twisted pair wires. A pair of wires insulated with a plastic coating and twisted together, used as a medium for telecommunications circuits.

two-wire circuit. A metallic circuit formed by two conductors insulated from each other. It is possible to use the two conductors as a one-way transmission path, a half-duplex, or a duplex path.

unbuffered. A terminal in which a character is transmitted to the computer as soon as a key on the keyboard is pressed.

unipolar. A digital signaling technique in which a 1 bit is represented by a positive voltage pulse and a 0 bit by no voltage.

universal service. The attribute of the telephone system that allows any station to connect to any other.

uplink. The microwave radio signal beamed up to a satellite.

uploading. The transmission of a file of data from a personal computer to a mainframe.

user. (1) The ultimate source or destination of information flowing

through a system. (2) A person, process, program, device, or system that employs a user application network for data processing and information exchange.
user friendly. A terminal or system that is easy to learn and easy to use.
user response time. The time it takes the user to see what the computer displayed, interpret it, type the next transaction, and press the ENTER key. *See also* think time.

V.24. A CCITT specification for the interface between a modem (DCE) and a terminal or computer (DTE). The interface is identical to the RS-232-D. *See also* RS-232-D *and* RS-232-C.
V.32. A CCITT standard for medium speed data transmission.
value-added carrier. Carriers that provide enhanced communications services. Normally, some type of computation is provided in addition to the basic communications service.
value-added network (VAN). A public data network that contains intelligence that provides enhanced communications services.
vertical redundancy check (VRC). A parity check performed on each character of data as the block is received.
video conferencing. Meetings conducted in rooms equipped with television cameras and receivers for remote users' participation.
video display terminal (VDT). A computer terminal with a screen on which characters or graphics are displayed, and (normally) a keyboard that is used to enter data.
video signal compression. The process of reducing the number of bits required to carry a digitized video signal while maintaining adequate quality.
videotex. An application in which the computer is able to store text and images in digital form and transmit them to remote terminals for display or interaction.
virtual circuit. In packet switching, those facilities provided by a network that give the appearance to the user of an actual connection.
virtual telecommunications access method (VTAM). IBM's primary telecommunications access method.
virtual terminal. A concept that allows an application program to send or receive data to or from a generic terminal. Other software transforms the input and output to correspond to the actual characteristics of the real terminal being used.
voice-band. The 300 Hz to 3,300 Hz band used on telephone equipment for the transmission of voice and data.
voice compression. The process of reducing the number of bits required to carry a digitized voice signal while maintaining the essential characteristics of speech.
voice-grade circuit. A circuit suitable for transmission of speech, digital or analog data, or facsimile, generally with a frequency range of about 300 Hz to 3,300 Hz. Voice-grade circuits can transmit data at speeds up to 19,200 bps.
voice messaging. A messaging service that people use to leave voice messages for others. The system provides a voice mailbox on a computer in which the voice messages are digitized and stored. The voice equivalent of an electronic mail system for textual messages.

WATS. *See* wide area telecommunications service (WATS).
white noise. *See* background noise.
wide area network (WAN). A network that covers a large geographic area, requiring the crossing of public right-of-ways and the use of circuits provided by a common carrier.
wide area telecommunications service (WATS). A service that provides a special line on which a subscriber may make calls to certain zones using direct distance dialing. Charges for the service are a combination of a flat monthly charge and a usage charge.
wideband circuit. A circuit designed to carry data at speeds greater than voice-grade circuits. *Synonym for* high-speed circuit.
workload generator. Computer software designed to generate transactions or other work for a computer or network for testing purposes.
workstation. The place where a terminal operator sits or stands to do work. It contains the working surface, terminal, chair, and other equipment or supplies needed by the person to do his or her job.

X.21. A specification for an interface between data terminal equipment (DTE) and a digital public telephone network.
X.21 BIS. A specification for an interface between data terminal equipment (DTE) and an analog telephone network. X.21 BIS is electrically virtually identical to the RS-232-C and V.24 interface specifications.
X.25. The first three layers of the OSI reference model. A standard for data transmission using a packet switching network.
X.400. A standard for the transmission of electronic mail.
Xmodem protocol. An asynchronous protocol developed for use between microcomputers, especially for transfers of data files between them.

INDEX